BIRKHÄUSER

Science Networks . Historical Studies
Founded by Erwin Hiebert and Hans Wußing
Volume 35

Edited by Eberhard Knobloch, Erhard Scholz and Helge Kragh

João Caramalho Domingues

Lacroix
and the Calculus

Birkhäuser
Basel · Boston · Berlin

João Caramalho Domingues
Centro de Matemática da Universidade do Minho
Campus de Gualtar
4710-057 Braga
Portugal
e-mail: jcd@math.uminho.pt

2000 Mathematical Subject Classification: Primary: 01A50; Secondary: 01A55

Library of Congress Control Number: 2008920493

Bibliographic information published by Die Deutsche Bibliothek
Die Deutsche Bibliothek lists this publication in the Deutsche Nationalbibliografie;
detailed bibliographic data is available in the Internet at http://dnb.ddb.de

ISBN 978-3-7643-8637-5 Birkhäuser Verlag AG, Basel - Boston - Berlin

1005451568

© 2008 Birkhäuser Verlag AG
Basel · Boston · Berlin
P.O. Box 133, CH-4010 Basel, Switzerland
Part of Springer Science+Business Media
Printed on acid-free paper produced from chlorine-free pulp. TCF∞
Cover illustration: Figure 30 in volume II of Lacroix's *Traité du Calcul différentiel
et du Calcul intégral*: construction of a first-order partial differential equation in
three variables (see pages 247-248).
Printed in Germany

ISBN 978-3-7643-8637-5   e-ISBN 978-3-7643-8638-2

9 8 7 6 5 4 3 2 1   www.birkhauser.ch

Abridged contents

Contents

Acknowledgments

First of all, I must thank Professor Ivor Grattan-Guinness for the supervision of the PhD thesis that resulted in this book: for all the pieces of information, suggestions, criticisms, and corrections; for many offprints and even some books; and for having proposed to me the topic of Lacroix's *Traité* in the first place. Being retired, Professor Grattan-Guinness could not be director of studies; this position was assumed by Dr Tony Crilly, for which I thank him.

I also thank Carlos Correia de Sá, who introduced me to history of mathematics when I was an undergraduate student, who supervised my master thesis, and who encouraged me to do my doctorate abroad.

For three years I was fortunate to be able to use the British Library on a daily basis; not only is its collection remarkable, but it is also a wonderful working place. Thanks to the Sconul (formerly M25) scheme, I was able to use other libraries in London, two of which stand out for their excellent collections in the history of science: the Library of University College, London (particularly the Graves collection), and the Science Museum Library. In my two trips to France, access to the Archives of the *Académie des Sciences*, to the library of the *Institut de France*, and to the Archives of the *École Polytechnique*, was invaluable. Back in Portugal, I was surprised and grateful for the 18th- and 19th-century science collections at the *Museu de Ciência da Universidade de Lisboa*, *Faculdade de Ciências da Universidade do Porto*, and *Biblioteca Municipal do Porto*. I am thankful to the staff of all these libraries and archives.

Even with all these physical libraries, their online counterparts offer many obvious advantages. *Gallica* <http://gallica.bnf.fr>, the digital library of the *Bibliothèque nationale de France*, proved invaluable. Also extremely helpful were the *Euler Archive* <http://www.eulerarchive.org> and the collection of publications of the Berlin Academy <http://bibliothek.bbaw.de/bibliothek-digital/digitalquellen/schriften>. Although I did not use it much as a library, the *Perseus Digital Library* <http://www.perseus.tufts.edu> was a great aid in translating from Latin, through its "morphological analysis" tool. I am very grateful for these online resources.

I also thank Abhilasha Aggarwal, Christian Gilain, Pierre Lamandé, and Iolanda Nagliati for sending me copies of their works; and Marco Panza for sending me a copy of Arbogast's 1789 memoir.

I am grateful to the British Society for the History of Mathematics, Mathematisches Forschungsinstitut Oberwolfach, Centro de História da Ciência da Universidade de Lisboa, Sociedade Portuguesa de Matemática, and both the Portuguese Seminário Nacional de História da Matemática and the Seminário Brasileiro de História da Matemática (jointly), for opportunities to present my research.

I thank my colleagues at the *Departamento de Matemática* for their support; I would like to mention in particular Filipe, who has to put up with me in the office we share, and Cláudia, who helped me solve many bureaucratic problems while I was in London.

I thank all those who made my life in London more enjoyable. Among others, Marta, Tiago, Alex, Gaby, and Florence.

I thank my mother for all the administrative and postal support while I was in London, and for other logistic and motherly support while in Portugal. I thank Sofia for too many and too important reasons to be mentioned here.

I am thankful to the *Centro de Matemática* and *Departamento de Matemática* of the *Universidade do Minho* (Braga, Portugal), where I work, for several facilities granted; in particular, I thank the department for allowing me to accumulate my teaching duties for the year 2006/2007 in the second semester, thus releasing me for five fundamental months.

Finally, I thank *Fundação para a Ciência e a Tecnologia* for the doctoral scholarship SFRH/BD/8518/2002, funded by the programme POCI2010 and the European Social Fund.

Chapter 0

Introduction

SILVESTRE FRANÇOIS LACROIX (Paris, 1765 - *ibid.*, 1843) was not a prominent mathematician, in the sense of someone who creates (or discovers) new mathematics, but he was certainly a most influential mathematical book author. The revolutionary times he lived in, changing political and social structures, changed also the social role of mathematicians and mathematics, through a great expansion of education. Lacroix dedicated his career to the teaching of mathematics, both in person (he taught at numerous institutions, from the *École des Gardes de La Marine* to the *École Polytechnique* and the *Collège de France*) and as a prolific (and much read) textbook writer. He also showed much concern for the history of mathematics, namely writing biographies of several mathematicians for Michaud's *Biographie Universelle*.

One of the most successful of his textbooks was the *Traité élémentaire du calcul différentiel et du calcul intégral* [*1802a*]. It had several editions throughout the 19th century, being widely used in teaching even after Cauchy's radical transformation of the subject: its first edition was in 1802; the last edition during Lacroix's lifetime (the 5th) was in 1837; in 1861-1862 a 6th one was published with notes added by Charles Hermite and Joseph Alfred Serret; in 1881 the 9th edition was reached. Translations were published in Portuguese (in 1812, in Rio de Janeiro), English (in 1816, as part of an effort to introduce Continental analysis into Britain), German (twice, in 1817 and 1830-1831), Polish (in 1824, in Vilnius), and Italian (in 1829).

Prior to [*1802a*], Lacroix had published a monumental *Traité du calcul différentiel et du calcul intégral* (three large volumes, 1797-1800; a second edition appeared in 1810-1819) [Lacroix *Traité*]. This is not a textbook: in the preface to the first volume of the second edition Lacroix, comparing it to elementary books, says that "such a voluminous treatise as this one, can hardly be consulted but by people to whom the subject is not entirely new, or that have an unwavering taste for this kind of study" [Lacroix *Traité*, 2nd ed, I, xx]. It is rather a reference work – an encyclopedia of 18th-century calculus. In an *encyclopédiste* style, Lacroix aimed at presenting a comprehensive account of the differential and integral calculus, but not as a simple compilation of methods: he thought it necessary to choose between different but equivalent methods or to show how they relate to one another, as well as give all of them a "uniform hue" that would not allow tracing the respective authors. It was a major appraisal of the calculus just before this subject was radically transformed by Cauchy in the 1820s.

Throughout this work I will often refer to [Lacroix *1802a*] as "Lacroix's *Traité élémentaire*", and to [Lacroix *Traité*] as "Lacroix's *Traité*" or (to better distinguish from the former) as his "large *Traité*".

0.1   Lacroix and his *Traité* in the literature

In spite of the great influence of Lacroix in early 19th-century mathematics, "no major study has been written of [him]" [Grattan-Guinness *1990*, I, 112]. Grattan-Guinness adds that "the most useful studies" are [Taton *1953a*], [Taton *1953b*] and [Taton *1959*]; but even these are mainly biographical, focusing on Lacroix's career but not studying his works (and yet, they do not constitute a complete biography of Lacroix, which is still lacking).

Meanwhile, Lacroix's textbooks have received some more attention. In [*1987*] Schubring presented Lacroix as a very good example of a textbook author to be analysed, due to the extension and influence of his textbook *œuvre* (however, he hardly touched on the mathematical content of any of Lacroix's books). Pierre Lamandé has written several papers where he addresses Lacroix's textbooks; the most important for us here are [*1988*; *1998*], where he compares the *Traité élémentaire* with the older texts on the calculus by l'Hôpital and Bézout.

But the large *Traité* has not been studied thoroughly, which is a serious omission, both in itself and as a necessary step before a global study of Lacroix can be achieved.

True, a considerable number of references to [Lacroix *Traité*] can be found in the historical literature: when studying the history of some aspect of the calculus in a period of time that includes the turn of the 18th to the 19th century, it is not uncommon to briefly address Lacroix's account of the aspect under study, taking it as typical of the period. For instance, [Gilain *1981*] uses the second edition of Lacroix's *Traité* to highlight the novelties in Cauchy's treatment of differential equations. But each of those references concerns only one or another particular aspect, and most of them are extremely small: [Grabiner *1981*] gives several examples of influence on Cauchy (in details and terminology), but Lacroix's *Traité* is still quite secondary; [Boyer *1956*] attributes to [Lacroix *Traité*] a very important place in the history of analytic geometry, but this is only a very particular aspect of the *Traité*.

0.2 This study

The main purpose of this book is to study Lacroix's large *Traité*, focusing on the first edition and on the process of its composition (much more than on its aftermath). There is also a chapter on the second edition (chapter 9), but it plays a secondary role here. The chapter on the *Traité élémentaire* is more important – but that is mainly because it offers the opportunity to compare the large *Traité* with another text, by the same author, on the same subject, actually based on the large *Traité*, but with a very different intended audience.

0.2.1 Comparisons

If every historical research must be set in a context, the more so when the object of research is a scientific text that did not intend to be original but rather an appraisal of an existing subject. Thus, a great part of this study consists of comparisons:

1. The obvious model for Lacroix's larger *Traité* was Euler's six-volume set of treatises on the calculus [*Introductio*; *Differentialis*; *Integralis*], published between 1748 and 1770. Lacroix himself admitted having taken passages from there [*Traité*, I, xxiv]. But he was following different foundations and he wanted to incorporate recent developments as well as alternative methods. How did this affect the structure of the *Traité*? We will see in chapter 2 that the difference in foundations hardly affected it: Lacroix kept much of the structure of Euler's set of works. He did not include an early section on finite difference calculus (vital for Euler's foundation, but not for Lacroix's), but other than that he departed mostly in his systematic inclusion of geometrical applications, and in the inclusion of a final volume on "differences and series"; these departures are related to the incorporation of recent developments (namely Monge's differential geometry, finite difference equations, and several studies involving definite integrals, mostly by Euler himself and by Laplace).

2. One of the choices Lacroix actually made between methods was that the foundational approach would be the one suggested by Lagrange in [*1772a*], based on the power-series expansion of arbitrary functions. Lagrange used this method in his lectures at the *École Polytechnique* from 1794, but only published it in detail in [*Fonctions*], in 1797 – the same year in which the first volume of Lacroix' *Traité* appeared. Lacroix attended Lagrange's lectures at least in 1795, but he was working on the *Traité* since 1787, and therefore he probably had already written its first chapters. The question of Lacroix's relatively independent development of details for the Lagrangian foundations of the calculus (a comparison with [Lagrange *Fonctions*]) is addressed in chapter 3.

3. In the 1790s two other books were published in France with similar titles: Cousin's *Traité de Calcul Différentiel et de Calcul Intégral* [*1796*] and Bossut's *Traités de Calcul Différentiel et de Calcul Intégral* [*1798*]. The latter

was more a textbook, but Cousin's was truly a treatise. Both (and, up to a certain point, also the section on the calculus in [Bézout *1796*, IV]) offer points of comparison with [Lacroix *Traité*], representing more traditional and/or less advanced accounts.

4. A more general comparison is that between Lacroix's text and his sources, which is facilitated by the inclusion of a wonderful bibliography. We will see that in most cases Lacroix simply summarizes those sources, adapting terminology and notation so as to give the *Traité* the "uniform hue" required. But there are also several instances of originality – in some cases in content (for instance, total differential equations in three variables that do not satisfy the conditions of integrability), in other cases in systematization (for instance, analytic geometry).

Besides these, there are two more comparisons that must be made, and that have already been mentioned:

5. We will see in chapter 8 how Lacroix reduced and adapted his large *Traité* for teaching, and how this resulted in the *Traité élémentaire*.

6. In chapter 9 we will take a brief look at the second edition. There were no major differences, but Lacroix improved the organization of the material, and included many new developments by Lagrange, Poisson, and others.

0.2.2   Structure

This book starts with a short biography of Lacroix (chapter 1), followed by an overview of the first edition of the *Traité* (chapter 2). Next come five chapters where particular aspects are analysed in detail. Then we examine the *Traité élémentaire* (chapter 8), we take a look at the second edition of the large *Traité* (chapter 9), and conclude the study with some final remarks (chapter 10). The book closes with some appendices, mainly containing relevant documentation.

The five chapters 3-7 constitute the bulk of the study. Their subjects were chosen taking several issues in consideration. First of all, Lacroix's possible originalities would have to be addressed, but this could not be reduced to a study of possible originalities. The topics chosen should allow one to form a prospect of the whole *Traité*: they should cover both Lagrangean and Mongean topics (Lagrange being an acknowledged influence, and Monge being a mentor of Lacroix), and the three volumes should be present (even if not with the same weights). There was an attempt at having topics dealing more with concepts than with methods; of course, in several situations methods have to be addressed, because they have interesting conceptual consequences (as in the case of Euler's approximate integration) or underpinnings (as in the case of the several methods for calculating tangents to curves). But this is the main reason for the lesser weight of volume III in these chapters – that volume consisting almost exclusively in a collection of methods; the other reason, actually related to the former, is that volume III does not offer much opportunity to study possible originalities by Lacroix – its great originality residing in its existence and structure (for which see section 2.5).

Chapter 3 analyses the foundations of differential calculus – the most *classical* topic here; it has already been mentioned (item 2 above). After this Lagrange-related topic, chapter 4 deals with analytic and differential geometry – the most direct influences from Monge; analytic geometry is included because of the important role of Lacroix's *Traité* in its history.

Chapter 5 addresses two subjects that seem to be more closely related in Lacroix's *Traité* than before: approximate integration, and conceptions of the integral – Lacroix used Euler's method of approximation (the one Cauchy would later use to define the definite integral) to explore "the nature of integrals". Chapter 6 combines several issues on what types of objects can be solutions of differential equations – the distinctions between complete, general, particular, and singular integrals/solutions, the geometrical interpretations of all these, what types of arbitrary functions (and how many) may occur in integrals of partial differential equations, the special case of total differential equations in three variables that do not satisfy the conditions of integrability, and finally Fontaine's conception of formation of (ordinary) differential equations by elimination of arbitrary constants (with different adaptations to partial differential equations). Lacroix regarded Fontaine's conception as the basis of the theory of differential equations, and used it to build his own analytical theory for total differential equations in three variables that do not satisfy the conditions of integrability.

Chapter 7 explores three aspects of "differences and series". The first is the subscript index notation, whose introduction has been misattributed to Lacroix. The other two are partly a follow-up of chapter 6: studies of the solutions of (finite) difference equations and of mixed difference equations.

0.2.3 Notations

An effort has been made to be as faithful as possible to original notations. Thus, sometimes we have expressions such as

$$fx, \qquad X_1k + X_2k^2 + X_3k^3 + \text{etc.}, \qquad \text{or} \qquad 1 + \frac{1}{2} + \frac{1}{3} + \frac{1}{4} \ldots + \frac{1}{x},$$

while we would nowadays write them as

$$f(x), \qquad X_1k + X_2k^2 + X_3k^3 + \cdots, \qquad \text{and} \qquad 1 + \frac{1}{2} + \frac{1}{3} + \frac{1}{4} + \cdots + \frac{1}{x}.$$

There is only one notable exception: it was common in late 18th-century to print the *d* of *differential* as ∂ (particularly in publications of the Paris Academy of Sciences, for instance the ordinary differential equation $\partial y = p\,\partial x$ in [Laplace *1772a*, 343]); since this would now be easily and systematically confused with notation for partial differentiation, I have substituted *d* or d for ∂.

Chapter 1

A short biography of Silvestre-François Lacroix

A detailed biography of Lacroix is still lacking, despite the articles by René Taton [*1953a*; *1953b*; *1959*]. In this chapter the main focus is on his education (in a broad sense) and his career until the publication of the large *Traité*.

1.1 Youth and early career (1765-1793)

Silvestre François de Lacroix[1] was born on April 28th, 1765, in Paris. His parents were Jean François De la Croix (a "bourgeois", that is, a burgher – an urban member of the third estate) and his wife Marie Jeanne Antoinette Tarlay. They lived in the rue de la Lune, parish of Notre Dame de Bonne Nouvelle, nowadays

---

[1]In his *Procés-verbal d'individualité* for the *Légion d'Honneur* (probably the most official document one may hope for), dating of 1837, Lacroix's surname appears as "Lacroix (de)", and the christian names as "Silvestre François". In a transcript of his baptism certificate the christian names are written "Silvestre françois", and the family name is "De la Croix" [Lacroix *LH*] According to his own statement, Lacroix stopped using the particle "de" when addressing a petition to a court in Besançon in 1793 (a time when any hint at *aristocracy* would not be favourable); having published several works afterwards without the particle, he never retook it [Lacroix *IF*, ms 2399]. Variations in capitalization and word splitting in names like Lacroix/La Croix/la Croix (or Lagrange/La Grange/la Grange) were common in the 18th and early 19th centuries. As for whether his first name was "Silvestre" or "Sylvestre" (most modern authors refer to him as "Sylvestre"): late 18th/early 19th century Frenchmen had the annoying habit of almost never using their christian names in public, at least not in full – nearly all of Lacroix's books appeared under the name "S. F. Lacroix"; in manuscript sources there are some (not many) occurrences of his christian names in full, and both "Sylvestre" and "Silvestre" occur (even within his Légion d'Honneur file [Lacroix *LH*]), but the more official documents tend to have "Silvestre"; this is also how the name appears in its *two* contemporary printed occurrences that I know of – [Anonymous *1818*] and the title page of the first edition of [Lacroix *1795*] (see fig. 1.1). I have decided to stick with "Silvestre". Of course, this is not a very important issue – but one must acknowledge it in an era of computerized searches.

in the 2nd *arrondissement*. There is no mention of Lacroix's father later than the
baptism certificate (while his mother is mentioned in a letter by Monge from 1783
[Lacroix *IF*, ms 2396]); it is likely that he died when Lacroix was still young. We
know that Lacroix was protected by a nobleman, the *chevalier de* Champigny
(1712-1787).[2] In a letter written in 1783, Lacroix goes as far as addressing him
as "mon cher papa" [Lacroix *IF*, ms 2397][3]; and Lacroix's close friend (and also
Champigny's *protégé*) Jean-Henri Hassenfratz (1755-1827), in a letter written to
Lacroix in 1785 speaks of the "bon papa Mr le Ch. de Champigni" [Lacroix *IF*,
ms 2396; Grison *1996*, 51-52]. But we do not know when this protection started.

The only information available on Lacroix's early studies comes from a spe-
ech read by Guglielmo/Guillaume Libri (1803-1869) at Lacroix's funeral – a source
open to anecdotes and exaggerations (and which does not mention Champigny),
but with the advantage of the author having known Lacroix personally (he was his
substitute at the *Collège de France* for several years). According to Libri [*1843*,
5-6], Lacroix often recalled the humble conditions in which he spent his childhood,
living with his poor mother. But "cet enfant, qui avait à peine de quoi se nourrir,
était dominé par le besoin de lire et d'apprendre"[4]. Having read *Robinson Crusoe*,
he wished to sail the seas. So, he tried to read a treatise on navigation. But in
order to understand it he needed to know geometry, and so he started attending
Mauduit's course at the *Collège Royal de France*[5]. Antoine-René Mauduit (1731-
1815) occupied two chairs there. From 1775 to 1779 he taught, in the chair of
mathematics, on conic sections (1775), integral calculus (1776), nature and construction of equations and elements of differential calculus (1778) and spherical
trigonometry (1779); and in the Ramus Chair[6] he taught on "elements of the art
of analysis" (1775-1778) and "elements of curves" (1779) [Torlais *1964*, 283, 285].
Lacroix may also have attended lectures by other professors at the *Collège Royal*,
like Lalande (astronomy), Le Monnier and Cousin (both professors of "universal
physics") [Torlais *1964*, 283]; we know that Le Monnier transmitted astronomical
observations to Lacroix not later than 1779 (see below).

Thanks to a letter from the *abbé* Joseph-François Marie (1738-1801), kept in
[Lacroix *IF*, ms 2396], we know that Lacroix also followed lectures by him. Marie
was professor of mathematics at the *Collège Mazarin* of the University of Paris.
He had published a much revised and enlarged edition of a one-volume course of
pure mathematics by his predecessor La Caille, which went from arithmetic to the
elements of differential and integral calculus [La Caille & Marie *1772*]. But we do
not know what he taught Lacroix.

Mauduit, Le Monnier, and Marie notwithstanding, Lacroix's great educatio-

---

[2]On Champigny, see [Grison *1996*, 24].

[3]All or nearly all the letters kept in [Lacroix *IF*, ms 2397] are in fact drafts of letters. It will
be assumed that there were not significant changes in the versions posted.

[4]"this child, who had barely anything to eat, was dominated by a need to read and learn"

[5]The courses of the *Collège Royal* were open to anyone, and had traditionally been free. It
appears that fees were introduced precisely around this time [Torlais *1964*, 267]; but presumably
these newly introduced fees were not very high.

[6]Named after the 16th century mathematician Petrus Ramus (Pierre de La Ramée).

nal influence was Gaspard Monge. Since the late 1760's Monge had been teaching at the *École Royale du Génie* (*Royal Engineering School*) at Mézières, where he developed descriptive geometry. But in January 1780 Monge was elected *adjoint* to the Geometry section of the *Académie des Sciences* of Paris; this meant that he had to live in the capital for at least five months per year, and Bossut, who had been Monge's predecessor at Mézières and who was in charge of a chair of hydrodynamics at the Louvre, arranged for Monge an assistant post there. During his half-year stays in Paris Monge did much more than attend Academy meetings and help teaching hydrodynamics. In particular, in 1780 he gave some sort of extraordinary lectures in mathematics to a group of students that included Lacroix.[7] Lacroix became a disciple and lifelong admirer of Monge, who was an excellent teacher. As for the contents of Monge's lectures, a letter written by Lacroix to Monge in 1789 [Lacroix *IF*, ms 2397; Belhoste *1992*, 565] indicates that they covered geometry in space – certainly analytic and differential geometry. One of the indications of their high level is given by Lacroix [*Traité*, II, 487], recollecting that Monge had integrated a partial differential equation using an early version of what was to be his method of characteristics. Descriptive geometry was excluded, since Monge was not authorized by the *École du Génie* (a military school) to divulge it [Taton *1951*, 14-15]; he could only allude to the fact that he was able to solve graphically the problems that he was solving analytically [Belhoste *1992*, 565].

Lacroix's first attempts at research predate his acquaintance with Monge. Pierre-Charles Le Monnier (1715-1799), astronomer and professor of "universal physics" at the *Collège Royal*, had given him a notebook with lunar observations that led Lacroix to conduct long calculations during the years of 1779 to 1781. Lacroix would later tell Le Monnier that he was diverted from this labour because of his application to pure mathematics [Lacroix *IF*, ms 2397; Taton *1959*, 129].

In a letter to Marie dated 4 August 1781 [*IF*, ms 2397], Lacroix still declared that

> "je me destine entièrement a l'astronomie étant a present très difficile de devenir geometre. Je veux pourtant apprendre autant de geometrie que je pourrai car les ouvrages de Mr Euler [et] Clairaut m'ont bien persuadé de ce qu'on peut faire en astronomie lorsqu'on posse[de] bien la geometrie."[8]

This letter accompanied a work by Lacroix on ballistics (now lost), where (if I understand correctly his summary) he used approximation techniques inspired by Clairaut's treatment of the three-body problem. It is worth quoting his own contextualization, as it shows some of his strong early influences:

---

[7]Taton [*1951*, 24] indicates the years 1781-1782, but Lacroix [*Traité*, II, 487] spoke clearly of 1780. Since Monge spent the winters in Paris and summers in Mézières, the autumn-winter of 1780-1781 is the most likely. But they certainly contacted again in 1781-1782.

[8]"I fully intend to pursue astronomy, as it is very difficult nowadays to become a geometer. Nevertheless, I wish to learn as much geometry as I can, since the works of Mr Euler and Clairaut have convinced me of how much one can do in astronomy if one really dominates geometry."

"J'etais plein des methodes de Mr Monge et sur-tout de sa geometrie dans l'espace. Je venais d'étudier la théorie de la Lune de Mr Clairaut que j'avais assez bien entendu. Je voulais simple[ment] m exercer sur cette matiere et faire usage des principes que j'avais tirez de cet excellent ouvrage. Je m'avisai de transporter tout d'un coup la question dans l'espace et de soumettre mes [?] a des coordonnées rectangulaires comme étant plus faciles et plus simetriques que les angles."[9]

Remember: Lacroix was only 16 years old.

Reaching what in the 18th century was adulthood, and not being rich, Lacroix needed to obtain a source of income. On the 1st December 1782, under recommendation of Monge and/or Champigny,[10] Lacroix was appointed for his first job: teaching mathematics at the *École des Gardes de la Marine* in Rochefort.[11]

Lacroix stayed in Rochefort until the end of 1785. During these three years he maintained continued correspondence with Monge[12] – who also became his superior in October 1783, being appointed examiner of the navy students. This correspondence dealt mainly with scientific issues, but occasionally it included also more personal advice from Monge. Lacroix was not happy in Rochefort. Later he would recall the lack of authority that the teachers had over their pupils (due to social differences – the pupils were young noblemen [Hahn *1964*, 547], while the teachers, like Lacroix, were not), and the poor methods of teaching, based on memory alone [Lacroix *1805*, 128, 217-220]. The only positive comment on his location is in one of his earlier letters, dated 28 April 1783, where he says that "l'analogie que ma situation a avec la votre de Mezieres m'encourage"[13].

Throughout 1783 Lacroix studied nonlinear partial differential equations, following Monge's methods – including viewing them as resulting from the eli-

---

[9]"I was engrossed with Mr Monge's methods, especially with his space geometry. I had just studied Mr Clairaut's Lunar theory and had understood it quite well. I simply wished to train myself on that matter and to use the principles that I had acquired from that excellent work. I dared to transfer all at once the question to space, and to bring my [?] into rectangular coordinates, as they are easier and more symmetrical than angles."

[10]According to Libri [*1843*, 6] it was Monge who recommended him for the post. Grison [*1996*, 27] has attributed that recommendation to Champigny, citing a letter from the minister of navy to Champigny, dated 8 October 1782 [Lacroix *IF*, ms 2398]. This letter shows that Lacroix's protector was interceding in his favour, although with a different place in view, and unsuccessfully: Champigny tried to secure Lacroix a place as "aspirant élève ingénieur constructeur" – something like cadet student of (ship-)building engineering. The minister was sympathetic, but there were no vacancies at the moment.

[11]Rochefort is a port town on the river Charente, only a few kilometers inland from the Atlantic ocean, in southwestern France.

[12]From this period, four letters from Monge to Lacroix survive, dated: 27 January 1783 [Lacroix *IF*, ms 2396], c. 12 January 1784 [Éc. Pol. *Arch*, IX GM 1.19], end of August 1785 [Éc. Pol. *Arch*, IX GM 1.20-21] (partly transcribed in [Taton *1959*, 130], wrongly cited as being kept in the *Institut*), and end of 1785 [Lacroix *IF*, ms 2396; Taton *1959*, 138-142]; while five drafts of letters from Lacroix to Monge are kept in [Lacroix *IF*, ms 2397], dated: 10 March 1783, 28 April 1783, 5 August 1783, 11 July 1785, and 9 October 1785 (extract). Their content makes it clear that there were more.

[13]"the analogy between my situation and yours in Mézières encourages me"

mination of arbitrary functions, and interpreting them geometrically (see sections 6.1.3.4 and 6.1.4.2). By the end of the year Lacroix asked Monge whether his results would make a memoir worthy of being submitted to the *Académie des Sciences*. Monge (in his letter of January 1784) was not too encouraging: "ces matières ne sont pas très accueillies aujourd'hui, à cause de leur peu d'utilité prochain"[14]. Instead, Monge suggested, now that Lacroix knew enough geometry, he should study mechanics.

Lacroix did not follow this advice. Instead, he returned temporarily to astronomy. During 1784 he constructed solar tables, using observations by Le Monnier and La Caille [Taton *1959*; Wilson *1994*, 280]. By the end of the year he sent them to the *Académie des Sciences*;[15] they were presented in the meeting of 15 January 1785.[16] This was a good move: that same day an election was held for a place of "adjoint astronôme" and Lacroix ran fifth – that he was considered at all was excellent. Now the members of the *Académie* had heard of him.

A very prominent member of the *Académie* – its perpetual secretary, the marquis de Condorcet – took an interest in Lacroix. Monge spoke well of Lacroix's talents, and Condorcet asked for works by Lacroix. In July Lacroix sent to Monge a new version of the research on partial differential equations that he had conducted in 1783 (much revised, according to his letter of 11 July 1785). This memoir (transcribed in appendix A.1) was presented to the *Académie* in December. In February 1786 Monge and Condorcet reported favourably on it, recommending that it be published in the *Savans Étrangers* series. However, this did not happen, because the publication of the *Savans Étrangers* stopped.

But the other goal of this memoir was accomplished: Condorcet (who must have seen the memoir before its presentation to the *Académie*) was convinced of Lacroix's capabilities. This had very good consequences for Lacroix. The first was that Condorcet employed Lacroix as his substitute at the newly founded *Lycée*. This *Lycée* is not to be confused with the later secondary education institutions; it was a private school for gentlemen who wished to acquire a general culture; it had renowned professors who in fact nominated their substitutes to give all lectures under their general direction; Condorcet was in charge of mathematics.

Thus, Lacroix returned to Paris in January 1786 to teach at the *Lycée*. He stayed in Paris until August 1788. During this time Condorcet became another great influence for him. Scientifically, this influence resulted mainly in Lacroix gaining an interest for probability (which does not concern us much here); the influence of Condorcet's work on integral calculus is ambiguous – Lacroix used a few details from Condorcet in his *Traité*, namely on considerations about the number of arbitrary functions in integrals of higher-order partial differential equations, but he expressly omitted Condorcet's "general method of integration" (see sections 6.1.4.1 and 6.2.2.3). But the most important aspect of this influence is

---

[14]"those matters are not very well received, nowadays, because of their little immediate utility"

[15]Following a complicated path: Lacroix - Champigny - Hassenfratz - Monge - Le Monnier [Grison *1996*, 52].

[16]Not 15 July, as Wilson [*1994*, 280] has it.

probably philosophical: Lacroix always admired in Condorcet the *encyclopédiste* and the educationalist, probably more than the mathematician.

The course of mathematics at the *Lycée* was far from successful, because of the natural difficulty of teaching mathematics to an audience who wished only to acquire a "general culture"; it was cancelled at the end of the second year of the *Lycée* [Taton *1959*, 143-153]. As a supporting text, Condorcet and Lacroix prepared a new edition of Euler's popularization book *Lettres à une princesse d'Allemagne* – cutting out most of Euler's theological considerations [Taton *1959*, 153-155].

In February 1787 Lacroix added to his post at the *Lycée* another at the *École Royale Militaire de Paris* (to which he was appointed also by recommendation of Condorcet). This proved fortunate when the course at the *Lycée* was cancelled in August.

One of the main topics in the course of mathematics at the *Lycée* was the calculus of probabilities. There is no indication that Lacroix had ever taken an interest in this. But he taught it according to Condorcet's instructions, in the following years kept a correspondence with Condorcet on the subject, and even later (1815) published an influential textbook. Still in 1786, he submitted an entry for a prize competition on the theory of marine insurance proposed by the *Académie des Sciences*, and he received the best classification.[17] Taton [*1959*, 245] suggests that it was Condorcet who pressed Lacroix to write his entry (as well as probably giving some guidance).

During this period in Paris, besides Monge and Condorcet, Lacroix met other mathematicians and astronomers: Laplace, Legendre, Cassini and Lalande [Taton *1959*, 248].[18] It was an active period. According to his later statements, it was in 1787 that he started collecting material for writing his *Traité* (see section 2.1). In the same year, he submitted to the *Académie des Sciences* a memoir containing corrections to his solar tables [Wilson *1994*, 280].

Besides all this working activity, Lacroix married in 1787,[19] to Marie Nicole Sophie Arcambal, one year older than him. She outlived her husband, dying in 1846. There is no indication of any children.

In 1788 the *École Militaire* of Paris was closed. This time, it was under Laplace's recommendation that Lacroix obtained a new appointment, teaching mathematics, physics and chemistry at the *École Royale d'Artillerie* in Besançon[20] (Laplace was examiner of the artillery students, and thus became Lacroix's superior). Lacroix was forced to go once again into *exile*. He stayed in Besançon until 1793.

In Besançon Lacroix felt isolated from the scientific community. He com-

---

[17]There was no absolute winner. Only half of the prize was conferred – Lacroix received 30%, and another contestant received 20%.

[18]We have seen above that Lacroix may have known Lalande from his period at the *Collège Royal*.

[19]The marriage contract was signed on 5 June 1787 [Lacroix *LH*].

[20]Besançon is a city in eastern France, close to Switzerland and to Alsace.

plained in letters to Laplace and Monge about the lack of good libraries and the difficulty in having access to recent books when away from Paris [Lacroix *IF*, ms 2397; Taton *1953b*, 352-353]. In 1792 he told Laplace that he had not been able to advance much on his *Traité*, because of the difficulty in accessing the sources he needed.

But he kept postal contact with Monge, Condorcet, Cassini, Lalande, Legendre and Laplace (often asking them for books or off-prints of memoirs). The correspondence with Condorcet (namely on statistics of the population of Besançon) earned him in August 1789 the official title of correspondent of Condorcet by the *Académie des Sciences* – this gave him access to the meetings when he might be in Paris (namely during Summer holidays), and was of course a nice encouragement.

However, Lacroix's next submission to the *Académie des Sciences* was not Condorcetian, but rather Mongean: a memoir on developable surfaces and total differential equations in three variables (transcribed in appendix A.2), which he read himself at the meeting of 1 September 1790. Lagrange, Condorcet and Monge were charged with reporting on it, but apparently they never did. Lacroix himself may be to blame: some months later he wrote to Monge telling him that he had not yet done a fair copy of the memoir ("mis au net le memoire"), because he wanted to redo the second part; he had found out that he could use the theory of particular (i.e., singular) integrals to study the total differential equations that do not satisfy the conditions of integrability [Lacroix *IF*, ms 2397].[21] He probably took too much time to complete this, and in August 1793 the *Académie des Sciences* was dissolved (together with the other academies). But we will see that he carried on with this idea.

Through other letters, we know that in Besançon Lacroix occupied himself also with descriptive geometry: he already knew the basic principles, and which problems Monge solved with it; now he tried to reconstitute the solutions [Lacroix *IF*, ms 2396-2397; Belhoste *1992*]. He had some help from Monge (who was not allowed to say much about it), as well as from two of Monge's former pupils at Mézières, Girod-Chantrans and Charles Tinseau, who were stationed near Besançon. Finally, he studied the "new chemistry" of Lavoisier (a favorite subject of Monge also), with the help of his friend Hassenfratz (who had worked in Lavoisier's laboratory).[22]

From November 1792 to early 1793 Lacroix was in Paris to acquire books and scientific equipment for the *École* of Besançon; during that stay (22 December) he was elected a corresponding member of the *Société Philomathique de Paris* [Taton *1959*, 258; *1990*]. This was a scientific society that was about to become quite important, because of the closure of the *Académie des Sciences*. The only work we know to have been submitted from Besançon is a chemical analysis of confervae

---

[21] This letter draft does not have a precise date, but it carries the indication "90-91", and Lacroix speaks of the memoir that he had read "last summer".

[22] Later, Hassenfratz taught "general physics" at the *École Polytechnique*, and mineralogy and metallurgy at the *École des Mines*.

(a kind of algae) – a joint work with Chantrans [Soc. Phil. *Rapp*, II, 58-59].

1.2   The most productive years (1793-1806)

Lacroix returned definitively to Paris in October 1793. This was the period of Terror – the most radical in the French Revolution, dominated by Robespierre and the Jacobins. Laplace did not feel safe and withdrew from Paris to the countryside at an uncertain date in 1793, until mid 1794 (after the fall of Robespierre in July) [Gillispie *1997*, 154-155]. On the 1st October 1793 Lacroix was chosen to replace Laplace as examiner of the artillery students [Lacroix *IF*, ms 2398]. According to Libri [*1843*, 4], Lacroix took the noble and dangerous attitude of refusing the place and making an effort for its restitution to Laplace. There is no evidence supporting this story. It is possible that Lacroix offered this post back to Laplace after the latter had returned to Paris and the political situation had changed (Laplace was reinstated in July 1795 [Lamandé *2004*, 51]); but in October 1793 he took the chance to move back to Paris, and in January 1794 he was fulfilling his duties, examining artillery students and candidates in Chalons-sur-Marne [Lacroix *IF*, ms 2399].

This does not mean that Lacroix was a Jacobin. Quite the contrary: he held moderate, progressive opinions, in line with the tradition of 18th-century enlightenment. In June 1791 (when the king fled from Paris and was then arrested) he had expressed to Monge his uneasiness about the unrolling of the revolution [Taton *1948*]. His philosophical mentor, Condorcet, was persecuted during the Terror, and committed suicide while imprisoned, in March 1794. But his other mentor, Monge, was a Jacobin, as well as his friend Hassenfratz. It was probably due to these two friendships, as well as to his moderation, that Lacroix traversed safely through the Terror. But he was certainly much more at home with the moderate republican regimes of the Thermidorian Convention (July 1794 - October 1795) and of the Executive Directory (October 1795 - November 1799).

In these final years of the 18th century, and in the beginning of the 19th, Lacroix held several posts related to education (all in Paris), often accumulating.[23] On 18 Vendemiaire of year 3 of the French Republic (9 October 1794), he was appointed *chef de bureau* at the Executive Commission for Public Instruction, where he stayed until 1799; there he played an important role in the educational reforms, namely on the establishment of the *École Normale* (of year 3), and of programmes for the *Écoles Centrales* (secondary schools) [Taton *1953a*, 589; Belhoste *1992*, 564]. In the *École Normale* that functioned in year 3 (1794-1795), he assisted Monge in the teaching of descriptive geometry, together with Hachette. On 6 Prairial year 3 (25 May 1795) he was appointed a teacher at the *Écoles Centrales*; this was confirmed the next year, when these schools were regulated, and he taught mathematics at the *École Centrale des Quatre-Nations*; when the

---

[23]Two lists of his public posts (omitting private jobs, namely at the *Lycée*), are kept at [Lacroix *LH*; *UF*].

ESSAIS
DE GÉOMÉTRIE,
SUR LES PLANS
ET

LES SURFACES COURBES;
(Ou *Élémens de Géométrie descriptive*) :

Par Silvestre-François LACROIX.

A PARIS,

Chez { Fuchs, Libraire, quai des Augustins, N°. 28 ;

Régent et Bernard, Libraires, même quai, N°. 37.

L'An IIIe de la République.
M. DCC. XCV.

Figure 1.1: Title page of Lacroix's first textbook.

Écoles Centrales were replaced by the *Lycées* (a move towards a more classical education that he disapproved of), Lacroix was appointed teacher of transcendental mathematics at the *Lycée Bonaparte* (3 Vendemiaire year 13 = 25 September 1804). He was an admission examiner for the *École Polytechnique* in the years 3 to 6 (1794-1795 to 1797-1798). Finally, on 24 Brumaire year 8 (15 October 1799) he was appointed professor of analysis at the *École Polytechnique*.

    A consequence of these pedagogical activities was the writing of a series of remarkably successful textbooks (besides what follows, see section 8.1). The first of these appeared in 1795, and resulted from his teaching at the *École Normale*: it was the *Essais de Géométrie sur les plans et les surfaces courbes*, also called *Élémens de Géométrie descriptive* (fig. 1.1); this was the first textbook on descriptive geometry to be published, and the only one directed to secondary schools until the 1820's [Belhoste *1992*, 568]; from the second edition (1802) onwards it was included in

Lacroix's *Cours de Mathématiques*, with a new subtitle – *Complément des Élémens de Géométrie*. Most of Lacroix's other texbooks resulted from the need of good textbooks to be used in the *École Centrale des Quatre-Nations*, and appeared between 1797 and 1800: textbooks on arithmetic, algebra (both an *elementary* textbook and a volume of *complements*), geometry, and trigonometry and analytic geometry [Schubring *1987*; Lamandé *2004*]. When the last of those directed to the *École Centrale des Quatre-Nations* (the *Complément des Élémens d'Algèbre*) appeared in 1800, some of the others already had two editions, and they all ran to many more.[24] Another textbook was published in 1802, mainly directed to the students of the *École Polytechnique*: the *Traité élémentaire de Calcul différentiel et de Calcul intégral*, which will be the subject of chapter 8. This textbook activity culminated with [Lacroix *1805*], a complementary book addressed not to students, but to teachers, containing pedagogical reflections and an analysis of his textbook series.

Besides all those textbooks, being in Paris allowed Lacroix to finally complete his great project: the *Traité du Calcul différentiel et du Calcul intégral*. Printing started in 1795, although the first volume only appeared in 1797; the second appeared in 1798 and the third in 1800.

During this period Lacroix still carried out some mathematical research, but not much – and all of it in the context of the *Société Philomathique*. Not later than 1797 he communicated some "elucidations about a passage in Lagrange's *méchanique analytique*, related to rotation of bodies", and "observations on the number of arbitrary functions in the integrals of partial differential equations" [Soc. Phil. *Rapp*, II, 25]; I do not know of any trace of the elucidations about Lagrange's passage, but the observations on integrals of partial differential equations were certainly those included in the second volume of the *Traité* (see section 6.2.2.3). In 1798 he submitted a memoir on total differential equations resulting from the idea that he had communicated to Monge in 1790 or 1791 [Soc. Phil. *Rapp*, III, 9-10]; a slightly abridged version was published in the *Bulletin* of the *Société Philomathique* [Lacroix *1798a*], and a fuller version in the second volume of the *Traité* (see section 6.2.4). In 1799 he read two memoirs: one on geographical maps,[25] and another about curves traced on developable surfaces [Soc. Phil. *Rapp*, IV, 13]; the latter was certainly the first part of the one he had read to the *Académie des Sciences* in 1790, or a new version of it; in 1810 he published a second or third version as the final section in the first volume of the second edition of the *Traité*. Although classified as "physics" in the *Bulletin*, we may also mention a "note on fluid resistance" [Lacroix *1802b*].

These seem to have been his last attempts at original research. As Taton

---

[24]The most impressive figures are those of the *Arithmétique*, which reached the 20th edition in 1846, the *Éléments d'algèbre*, which reached the 23rd in 1871, and the *Éléments de géométrie*, which reached the 22nd in 1884.

[25]In 1804 Lacroix published an introduction to mathematical and physical geography, as a first volume of a larger geographical work directed by J. Pinkerton and C. Walkenaer [Lamandé *2004*, 105].

said in [*1953a*, 590], writing his large *Traité* and his textbooks, Lacroix realized
"que son érudition si étendue et son talent si remarquable de mise au point et de
présentation lui permettrait de faire là une oeuvre plus utile que celle qu'il aurait
réalisée en se confinant dans des recherches de détail"[26]. Yet, we should stress that
he did some research, and that he included in the *Traité* most of that that was
related to analysis.

In spite of his reduced research career, Lacroix gained the respect of the
mathematical community. His *Traité* was certainly a major factor in this. On
16 Germinal year 7 (5 April 1799) he was elected a member of the first class
("mathematical and physical sciences") of the *Institut National* (founded in 1795
in replacement of the *Académie des Sciences*). As we have seen, he did not present
any mathematical research to the *Institut*; but he was an active member – mostly
participating in commissions for reporting on works submitted by non-members;
in addition, he was *secrétaire* of the mathematical section between 1 Germinal year
10 (22 March 1802) and 11 Pluviose year 11 (31 January 1803). It was probably in
that capacity that he wrote a "Compte rendu à la section de Géométrie de l'Institut
national, des progrès que les mathématiques ont faits depuis 1789 jusqu'au 1.er
Vendemiaire an 10" (that is, a "report to the Geometry section of the *Institut
National*, on the progress made in mathematics from 1789 to *Vendemiaire* 1st,
year 10 [= September 23rd, 1801]"); most of it was eventually incorporated in
[Delambre *1810*] (see appendix B).

Speaking of this "Compte rendu..." is a good cue to mention Lacroix's historical activities. The reading programme that he must have carried out to write
his *Traité*, and the impressive bibliography that he included in it, indicate that he
acquired a very good knowledge of the history of the calculus in the process of its
composition. And this should have been obvious for everyone at the time. When
Lalande set to complete the second, enlarged edition of Montucla's *Histoire des
Mathématiques*, after Montucla's death in 1799, he asked Lacroix to revise the article on partial differential equations [Montucla & Lalande *1802*, 342-352], as this
was "un des plus difficiles de tout l'ouvrage"[27] [Montucla & Lalande *1802*, 342];
Lacroix added a couple of footnotes with his name (one of which is quite substantial and interesting [Montucla & Lalande *1802*, 344]), and he may also have
changed a few details in the main text – the article uses Lacroix's terminology,
speaking of "differential coefficients" and "partial differential equations" (rather
than "partial difference equations" as was usual at the time).[28]

But Lacroix's historical output was not restricted to the calculus. We have
already mentioned his "Compte rendu..." on the recent progress of mathematics,
which covered all branches of pure mathematics. Apart from this, he read to

---

[26]"that his so wide erudition and his so remarkable talent for clarification and presentation
would allow him to make there a work more useful than that he would have achieved had he
confined himself to researches on details"

[27]"one of the most difficult in the whole work"

[28]Grattan-Guinness [*1990*, I, 143] suggests that Lacroix's participation in the third volume of
Montucla's *Histoire* was more extensive. However, I have not seen any other traces of it.

the *Société Philomathique* a "historical summary" ("précis historique") of physical astronomy (that is, celestial mechanics) that was published in the *Décade Philosophique* [Lacroix *1797*] – a short piece where he tried to explain to laymen (hence without any formulas) the development of the methods used to approximate planetary movements (especially those by Lagrange and Laplace). He also wrote a historical eulogy of the applied mathematician Jean-Charles Borda (1733-1799), whose vacancy in the *Institut* he had occupied; this succession was the obvious motivation for the eulogy but, oddly, it was again published by the *Société Philomathique* [Soc. Phil. *Rapp*, IV, 92-135], rather than the *Institut*.[29]

1.3   Second editions and prestige (1806-1820)

After 1805 Lacroix's productivity clearly dropped. Most of his publications until 1820 were second (or third, or fourth,...) editions of his books. And in most cases the changes were not very significant; for instance, the relevant changes in his algebra textbook had all been introduced in the second (1800) and third (1802) editions [Lamandé *2004*, 68]. The second and third editions of the *Traité élémentaire de calcul...* (1806, 1820) and the second edition of the large *Traité* (1810-1819) clearly stand out (those of the former demarcate this period). But even the long period between the publication of the first and the third volumes of the latter suggests a decrease in productivity.

The only new book published by Lacroix in this period was his textbook on probability: the *Traité élémentaire du Calcul des Probabilités* (1816).

On the other hand, in this period Lacroix participated in a huge historical enterprise: the 52-volume biographical dictionary published by Louis-Gabriel Michaud [Michaud *Biographie*]. Actually, Lacroix's participation was limited to volumes 1 to 13 (published between 1811 and 1815); he authored the entries for d'Alembert, Apollonios, Arbogast, Archimedes, Barrow, de Beaune, the Bernoullis, Bézout, Bombelli, Cardano, Cavalieri, Clairaut, John Craig, Diophantos, Euclid, Euler, and Eutocios of Ascalon.[30] The reason for the interruption of his participation must have been the rejection of his entry on Condorcet: it was too favourable to the philosopher, and risked causing problems with the censorship; it was replaced by an anonymous and much more neutral text [Taton *1959*, 259-261].[31] Lacroix published his own text elsewhere [Lacroix *1813*].

In contrast to the decrease in productivity, we notice an increase in prestige of Lacroix's appointments. In 1809 he exchanged the position as professor at the *École Polytechnique* for that of permanent examiner – which was more prestigious

---

[29]Taton [*1953a*, 593] mentioned an *Essai sur l'Histoire des Mathématiques* written by Lacroix, unpublished and whose manuscript had apparently vanished. Itard [*1973*, 550] repeated this. I do not know Taton's source, but I find it likely that this *Essai* was simply the *Compte rendu...* (see page 397).

[30]I cannot guarantee that this list is exhaustive.

[31]Itard [*1973*, 550] wrongly gives Borda and Condorcet as examples of Lacroix's contributions to [Michaud *Biographie*]. Borda's entry is in fact by Biot and De Rossel.

and meant an increase in salary [Grattan-Guinness *1990*, I, 97]; he kept this post until 1815. Also in 1809 he was appointed professor of differential and integral calculus at the newly-founded *Faculté des Sciences de Paris* – with this appointment came an automatic degree of doctor[32]. The *Faculté des Sciences* was actually less prestigious than the *École Polytechnique*, but Lacroix was also made its first dean. Finally, in 1812 he replaced Poisson as a substitute for his old teacher Mauduit at the by then *Collège Impérial de France*; and when Mauduit died in March 1815 he was appointed for the vacant chair of mathematics (which was confirmed a few months later for the again *Collège Royal de France*).

1.4 Declining years (1820-1843)

After 1820 Lacroix's activities decreased even more. In 1815 he had already quit his posts of teacher at the *Lycée Bonaparte* and of permanent examiner at the *École Polytechnique*. In 1821 he quit the post of dean of the *Faculté des Sciences*, and from 1825 onwards, invoking health reasons, he was substituted as professor there by Lefébure de Fourcy. He may have kept his teaching at the *Collège de France*, but he was substituted there by Francoeur in 1828 [Lamandé *2004*, 54] and from 1836 by Libri [Lützen *1990*, 84].

In 1826 Abel, then visiting Paris, wrote to his former teacher Holmboe describing the mathematical scene in the French capital. Lacroix was only 61 years old, but appeared "terribly bald and extremely old" [Grattan-Guinness *1990*, II, 1275]. Itard [*1973*, 550] interprets this as indicating that "his astonishing activity since adolescence had affected his health".

He still published in 1826 a book on surveying and in 1828 an introduction to the "knowledge of the sphere" [Lamandé *2004*, 54]. None of these are among Lacroix's most famous books.

In addition, of course, he kept publishing new editions of his older textbooks. Those of the *Traité élémentaire de calcul différentiel et de calcul intégral* still brought a few changes, particularly through the inclusion of new endnotes on some special topics.

As for historical work, in 1831 he published a new edition of Montucla's history of the squaring of the circle, with several additions of his own; according to Sarton [*1936*, 533], "Lacroix's edition superseded completely the original one".

Lacroix died on the 24th May 1843, at his home in Paris.

---

[32]The diploma is kept at [Lacroix *IF*, ms 2398]

Figure 1.2: A medallion by David d'Angers, the only known portrait of Lacroix, made two years prior to his death. [Académie des Sciences de l'Institut de France]

Chapter 2

An overview of Lacroix's *Traité*

2.1 The project of the *Traité*

According to his own statement, Lacroix started collecting material for his *Traité* in 1787, while employed at the *École Royale Militaire* in Paris [*Traité*, I, xxiv]. This is confirmed by his correspondence: during his stay in Besançon (1788-1793) he wrote to mathematicians in Paris asking them to send him material or information on how to find it. In October 1789 Lacroix thanked Legendre for information on a work by Landen, and explained that he wished to use the tables of integrals included there for a project "dans lequel j'ai pour objet de rassembler dans un corps d'ouvrage les materiaux sur le calcul integral qui se trouvent dans les memoires des societes savantes"[1] [Lacroix *IF*, ms 2397]; in 1792 he communicated the same intent to Laplace [Taton *1953b*, 353].[2]

In both these letters, as well as in the Preface to the first edition of the *Traité*, Lacroix indicated as the trigger for this project his reading of Lagrange's "Sur une nouvelle espèce de calcul relatif à la differentiation et à la intégration des quantités variables" [*1772a*] – the memoir where Lagrange first suggested a power-series foundation for the calculus. Thus, he intended to write a complete treatise under this unifying principle.

However, it is clear that the purpose was not simply to apply Lagrange's suggestion. The reason for assembling the material dispersed in the volumes of memoirs of learned societies was that this had not been done, at least not recently. In the 1789 letter to Legendre, Lacroix declared: "les livres elementaires les plus complets, le Calcul Integral d'Euler, celui de M. Cousin ont besoin d'adition"[3]. In the Preface to the second edition, he stressed this motivation [*Traité*, 2nd

---

[1]"in which my goal is to assemble in a single work the materials on integral calculus that are found in the memoirs of learned societies"

[2]Presumably "calcul integral" is to be read here as short for "differential and integral calculus".

[3]"the most complete elementary books, the integral calculus of Euler, that of M. Cousin, need to be supplemented"

ed, I, xviii-xix]: in the 1780's there was an enormous gap between elementary books and research memoirs on "analysis and transcendental geometry", and this made their (advanced) study very difficult. This was especially true for those not living in Paris, because those research memoirs were available only in academic collections and books with low print runs; in his 1792 letter to Laplace, Lacroix had complained about the scientific indigence of Besançon – the only public library did not have even the memoirs of the Paris Academy of Sciences. One might suspect that this was the main motivation only *a posteriori* (and invoked especially in the second edition, when Lacroix's enthusiasm for the power-series foundation had cooled off); but it is easy to imagine how his bad experiences far from Paris would have led to this plan.

The "livres elementaires les plus complets" mentioned by Lacroix were [Euler *Introductio*; *Differentialis*; *Integralis*] and [Cousin *1777*].[4] Euler's set, six volumes in total, published between 1748 and 1770, was hard to reproach. But in 1792 Laplace would agree with Lacroix that it was beginning to grow old [Taton *1953b*, 355]. Moreover, there were topics that Euler had never included there, such as differential geometry, or finite difference equations. As for Cousin's *Leçons de Calcul Différentiel et de Calcul Intégral* [*1777*], it was probably the most comprehensive survey of the calculus (apart from Euler's), but it still lacked some topics, and the order of subjects is confusing, making it difficult to use as a reference work. Lacroix was fair when assessing it thus: "L'ouvrage, remarquable d'abord par le grand nombre de choses que l'auteur avoit réunies dans un petit espace, laissoit à désirer un ordre plus sévère et quelques développemens indispensables à la clarté de l'exposition"[5] [Delambre *1810*, 95][6]. He was more critical in a letter to Prony dated 1791 [Lacroix *IF*, ms 2396], accusing Cousin of slavishly copying everything in his "compilations" (to the point of employing a particular notation only once, just because it was used in the article he was copying). A second, enlarged edition appeared under the title *Traité de Calcul Différentiel et de Calcul Intégral* [Cousin *1796*], but these shortcomings persisted.

Lacroix's plan was different from Cousin's: not only to compile all the major methods, but also to choose between different but equivalent ones or to show how they relate to one another, as well as to give all of them a uniform hue that would not allow tracing of the respective authors [*Traité*, I, iii-iv].

His model was clearly Euler's six-volume set, except that it should include

---

[4]"Elementary" here must be understood in the sense that they start with the first notions, the "elements" of the calculus, rather than assuming them and addressing original research straight away. After the educational reforms of the 1790's and 1800's, "elementary" would mean simple, or introductory – see for example [Lacroix *Traité*, 2nd ed, I, xx], where the *Traité* is specifically opposed to "elementary books"; see also section 8.1.

[5]"This work, remarkable above all for the great number of topics assembled in a small space, wanted a stricter order and some developments essential for the clarity of the exposition"

[6]This sentence can be found in fl. 19v of Lacroix's "Compte rendu [...] des progrès que les mathématiques ont faits depuis 1789 [...]". See appendix B for the relation between the "Compte rendu" and [Delambre *1810*].

geometrical applications[7]. Physical applications, on the other hand, were entirely omitted.

An important point, made in the Preface of the second edition but likely to be applicable also to the first edition, is that this *Traité* was not intended to be a first introduction to the calculus: "un Traité aussi volumineux que celui-ci, ne peut guère être consulté que par des personnes auxquelles le sujet n'est pas tout-à-fait étranger, ou qui ont un goût décidé pour ce genre d'étude"[8] [*Traité*, 2nd ed, I, xx]. In fact, the three volumes of the first edition add up to around 1800 quarto pages.

A remarkable feature is the subject index included at the end of the third volume. It is not completely unprecedented: La Caille's book on astronomy [*1764*] also has one. But this was certainly uncommon. Moreover, it is a substantial index: 34 pages long [Lacroix *Traité*, III, 545-578]! In the Preface of the second edition Lacroix explained that with this index he hoped to make the whole book "a sort of dictionary of analysis and transcendental geometry" [*Traité*, 2nd ed, I, xlviii] – we would call it an encyclopedia.

Speaking of encyclopedia: the title pages of the three volumes bear the motto "Tantùm series juncturaque pollet. HORAT." (see figure 2.1). This is a quotation from Horatio's *De Arte Poetica*, and translates as "Such power has a just arrangement and connection of the parts". This is an interesting clue on Lacroix's views. But it becomes even more interesting when we notice that the motto of Diderot and d'Alembert's [*Encyclopédie*] was "Tantùm series juncturaque pollet, Tantùm de medio sumptis accedit honoris! HORAT." – "Such power has a just arrangement and connection of the parts: such grace may be added to subjects merely common"[9]. In 1797-1800 probably any reader would understand the allusion.

The result of this grand plan was a monumental reference work: an encyclopedic appraisal of the calculus at the turn of the century.

2.2 The bibliography

Another remarkable feature in Lacroix's *Traité*, one that does seem to be unprecedented in mathematical books, is the bibliography attached to the table of contents: for each chapter and section, Lacroix gives a list of the main works related to its subject.

All the major 18th-century works on the calculus are included there, as well as many minor and even some obscure ones. Typically, in the list for a given chapter/section one will find the corresponding chapters in one of Euler's three

---

[7][Euler *Introductio*] does include geometrical applications (analytic geometry); but they are missing from [Euler *Differentialis*] and [Euler *Integralis*].

[8]"such a voluminous treatise as this one can hardly be consulted but by persons to whom the subject is not entirely new, or that have an unwavering taste for this kind of study"

[9]This translation, and of course the previous one, were taken from *Perseus* <http://www.perseus.tufts.edu/cgi-bin/ptext?lookup=Hor.+Ars+220> (accessed 21 February 2007).

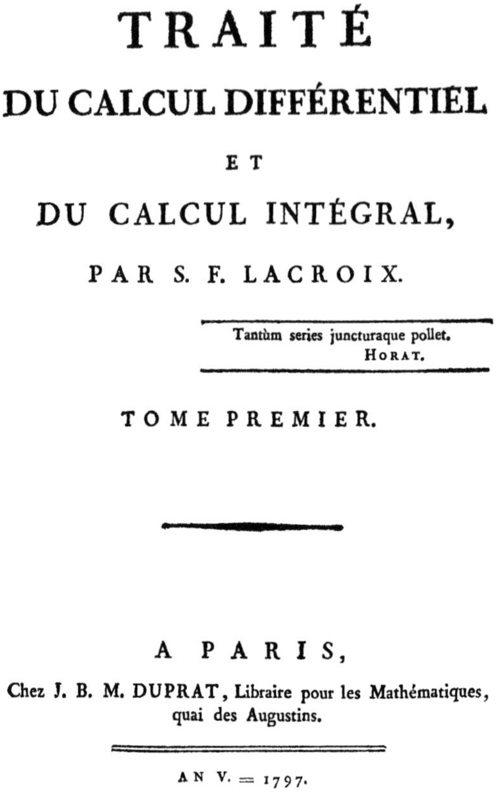

TRAITÉ

DU CALCUL DIFFÉRENTIEL

E T

DU CALCUL INTÉGRAL,

PAR S. F. LACROIX.

Tantùm series juncturaque pollet.
HORAT.

TOME PREMIER.

———

A PARIS,

Chez J. B. M. DUPRAT, Libraire pour les Mathématiques,
quai des Augustins.

===

AN V. = 1797.

Figure 2.1: Title page of volume I

books, some other relevant books (say, Lagrange's *Théorie des Fonctions Analytiques*, Jacob Bernoulli's *Opera* or Stirling's *3rd-order lines*) and memoirs drawn
from the volumes published by the *Académie des Sciences de Paris*, by the Berlin
Academy, by the St. Petersburg Academy, by the Turin Academy, and so on. The
most cited authors are those that one would expect: Euler, Lagrange, Laplace,
d'Alembert, Monge; but it is also possible to find references to such authors as
Fagnano [Lacroix *Traité*, II, v] or even Oechlitius [Lacroix *Traité*, III, viii].

An interesting issue is that of the languages of the works included. Me-
moirs are cited only as, say, "Nouv. Mém de Petersbourg, T. XV et XVI. (Lexell)"
[Lacroix *Traité*, I, xxx] – thus not indicating in which language they were written.
Therefore, it would be impracticable to give precise quantitative data. But it is

safe to say that French is the most common language, followed by Latin. Of course, this only reflects the weight of these languages in the scientific community at the time (the memoirs of the Berlin Academy for instance, were usually in French). At a long distance come English and Italian, languages that Lacroix clearly could read.[10] No other languages appear. In particular, no work in German – the few works of the German Combinatorial School included are in Latin [*Traité*, III, vi].

This bibliography shows how incredibly well-read Lacroix was. But note that not all of the works appearing there are used in the main text. As an extreme example, take the section on "application of the calculus of differences to summation of sequences", in chapter 1 of the third volume: it is 29 pages long, and has about 40 bibliographical entries! As Lacroix explains, the titles indicated are of the works used in writing the text *or* of works somehow related to it [*Traité*, I, xxix]. Some works appear to be included in the bibliography solely for their "classic" nature: for example, l'Hôpital's *Analyse des Infiniment Petits* [*1696*] for chapter 1 of the first volume; being the first textbook ever written on the differential calculus it had to be included, but by the late 18th century it was utterly out-dated; Lacroix does not include it for chapter 2, which is where "l'Hôpital's rule" is given (however, it had aged much better as a reference for differential geometry of plane curves, and it appears again in the bibliography for chapter 4).

We should also note that the bibliography is restricted to printed works. There are a few cases in which Lacroix made use of manuscripts (for instance, Biot's memoirs on difference and mixed difference equations, that were still unpublished – see sections 7.2.2 and 7.3.2); but, although he acknowledges them in the main text, they do not appear in the bibliography.

2.3 Volume I: differential calculus (1797)

Tables 2.1 and 2.2 show the contents of the first volume, dedicated to differential calculus. It must be noticed that in the text we will usually follow the division of chapters into sections, but that these are not shown exactly in the tables; the horizontal lines often correspond to them, but sometimes to "subsections" (inspired by the rather better divided sections in the second edition).

The first volume starts with a general Preface to the whole *Traité*. This includes an explanation of the aims of the work and the plan for the three volumes, but is mostly taken up with a long account of the history of the calculus [*Traité*, I, iv-xxiii]. Having a historical introduction is consistent with Lacroix's *encyclopédisme*, but it is hardly original: both Cousin [*1777*, xiv-xxx; *1796*, I, x-xvi] and Bossut [*1798*, I, iii-lxxvii] do the same.

After the table of contents (with bibliography) comes an Introduction. Its purpose is to give series expansions of algebraic, exponential, logarithmic and

---

[10] As an aside, it is curious to know that in 1818-1819 Lacroix took a course in Chinese by Rémusat (the first professor of Chinese at the *Collège de France*) [Lacroix *IF*, ms 2402, fls 380-465].

Volume I		
topics	chapter	pages
History of the calculus; overview of the *Traité*	Preface	iii-xxix
Table of contents and bibliography	Table	xxx-xxxii
General notions on functions, series and limits	Introduction	1-19
Series expansion of algebraic functions		19-32
Series expansion of exponential and logarithmic functions		33-52
Series expansion of trigonometric functions		52-61 78-80
Relationships between trigonometric and logarithmic functions		61-75
Reversion of series		75-78
Changes on a function of x when x becomes $x + k$	Chapter 1: Principles of differential calculus	82-87
Recursion between the coefficients of $f(x+k)$ (derivation)		87-94
Differentials and differential coefficients; differentiation of algebraic functions		94-107
Differentiation of logarithmic, exponential and trigonometric functions		107-114
Differentiation of explicit functions of two variables		114-131
Differentiation of explicit functions of any number of variables		131-134
Differentiation of equations; change of independent variable; elimination of constants, irrational exponents, and functions		134-178
Condition equations for a formula to be an exact differential		178-189
Method of limits; infinitesimals		189-194
Expansion of functions of one variable in series	Chapter 2: Main analytical uses of the differential calculus	195-232
Particular cases in the expansion of $f(x + k)$ (infinite values of the differential coefficients)		232-240
Indeterminacies ($\frac{0}{0}, 0 \times \infty$, etc.)		241-255
Expansion of functions of two variables in series		255-264
Maxima and minima of functions of one or several variables		264-276
Symmetric functions of the roots of an equation	Chapter 3: Digression on equations	277-286
"Imaginary expressions" (i.e., complex numbers); inc. the fundamental theorem of algebra and Cotes's theorem		286-326

Table 2.1: Volume I of Lacroix's *Traité* (continued in table 2.2)

trigonometric functions. The idea was to make the *Traité* accessible to readers who knew algebra only as it was treated in the textbooks of Bézout and Bossut [Lacroix *Traité*, I, xxiv], that is, elementary algebra – mainly equation solving. Thus the Introduction plays a role broadly equivalent to the first volume of Euler's *Introductio in Analysin Infinitorum* [Euler *Introductio*, I]. But with an important difference: Euler had used infinite and infinitesimal quantities extensively, while

Lacroix wished to avoid them.

This Introduction starts with a section about "general notions on functions and series" [*Traité*, I, 1-18], which includes definitions for function, implicit and explicit functions, and also, apropos of series, a fairly extensive treatment of limits. But this does not mean that limits are to be used as the foundational concept for what follows: Lacroix believes that if the expansion of a function results in a nonconvergent series, this series can still be used to represent that function – just not its "value" [*Traité*, I, 7] (see section 3.2.6). The section on series expansion of algebraic functions [*Traité*, I, 19-32] is dedicated to the binomial theorem, for the case of rational exponent (the case of irrational "or even imaginary" n appears later as an application of the expansion of the logarithm[11]). The section on series expansion of exponential and logarithmic functions [*Traité*, I, 33-52] is more interesting, because it was more challenging: Lacroix expands a^x using the functional equation $a^x \times a^u = a^{x+u}$ and the method of indeterminate coefficients; he was quite proud of how he had avoided the notions of infinite and of limits in this expansion (see section 7.1.2). Similar procedures are used for the logarithm, and for the sine and cosine in the section on expansion of "circular" functions [*Traité*, I, 52-80]. This latter section also addresses several trigonometric formulas (including $\sin nx = \frac{e^{nx\sqrt{-1}} - e^{-nx\sqrt{-1}}}{2\sqrt{-1}}$ and similar ones), and the important method of *reversion of series*.

Chapter 1 is entitled "analytical exposition of the principles of differential calculus". In the Preface Lacroix announces that he will give this "purely analytical exposition", "complete" and "d'un seul jet"[12] [*Traité*, I, xxiv]. He likens this comprehensiveness to what Euler had done (obviously in [*Differentialis*]). The alternative would be to include some applications in between – that is what Lacroix would later do in [*1802a*], where both analytical and geometrical applications of differential calculus of functions of one variable precede the analytical exposition of the differential calculus of functions of two variables. The separation between theory and applications is one of the characteristics that marks this as a treatise, rather than a textbook.

As for the exposition being "purely analytical", it may partly be an allusion to the separation from geometrical applications. But it is most likely a reference to the foundation followed, which does not appeal to geometrical or mechanical notions. In fact, Lacroix builds the differential calculus on the basis suggested by Lagrange in [*1772a*] – power series. This will be treated in section 3.2: let us only summarize the chapter here. First comes the expansion

$$f(x + k) - f(x) = X_1 k + X_2 k^2 + X_3 k^3 + \text{etc.} \tag{2.1}$$

---

[11]In [Domingues *2005*, 281] I said that "a 'weak' version of the binomial theorem, stating $(1 + x)^n = 1 + nx^{n-1} + \text{etc.}$ is proven (for 'any n'; the full expansion is given for integer n)". Apart from the fact that one should read "rational" instead of "integer", this is misleading because Lacroix shows the recursive relation between the coefficients independently of n being integer or not [*Traité*, I, 19-22]. My mistake resulted from the physical separation between this and the general proof that the first two terms in the expansion of $(1 + x)^n$ are $1 + nx$ [*Traité*, I, 49].

[12]"at one stroke"

Then, after establishing the iterative relation between the coefficients and thus renaming them to

$$f(x+k) - f(x) = f'(x)k + \frac{f''(x)}{2}k^2 + \frac{f'''(x)}{1\cdot 2\cdot 3}k^3 + \text{etc.}$$

the first term $f'(x)k$ is christened *differential* "because it is only a portion of the difference" and is given the symbol $df(x)$. "For uniformity of symbols [...] dx will be written instead of k", so that

$$f'(x) = \frac{df(x)}{dx}$$

is an immediate *conclusion*. Occasionally $f'(x)$, $f''(x)$, etc. are called "derived functions" (as in [Lagrange *1772a*, § 1-4; *Fonctions*]), because of the recursive process of derivation; but in page 98 Lacroix introduces the name *differential coefficients* for them, and uses it throughout the three volumes. The differential notation is also much more frequent than the use of accents. Overall this foundation for the calculus is Lagrangian, but much closer to [Lagrange *1772a*] than to [Lagrange *Fonctions*], where differentials have no place. The results obtained in the Introduction allow easy deductions of the differentials of algebraic, logarithmic, exponential and trigonometric functions of one variable: it is only necessary to expand $f(x+dx)$ and extract the term with the first power of dx. Differentiation of functions of two variables is also inspired by [Lagrange *1772a*], but without resorting to the cumbersome notation that Lagrange had employed ($u'^{,''}$ for our $\frac{\partial u^3}{\partial x \partial y^2}$). $f(x+h, y+k)$ is expanded in two steps and in two ways (via $f(x+h, y)$ and via $f(x, y+k)$), whence the conclusion that $\frac{d^2 u}{dx dy} = \frac{d^2 u}{dy dx}$. The definition of differential as the first-order term in the series expansion of the incremented function is extended to $u = f(x, y)$ giving

$$df(x, y) = du = \frac{du}{dx}dx + \frac{du}{dy}dy$$

(the ∂ notation is still absent). The largest section in this chapter is dedicated to "differentiation of equations" [*Traité*, I, 134-178]. It covers several topics, namely: differentiation of implicit functions; change of independent variable – Lacroix was proud of the way he had treated this without infinitesimals (see section 3.2.4); and use of differentiation to eliminate constants, irrational exponents, transcendental functions, and unknown functions. Elimination of constants and unknown (i.e., arbitrary) functions will play a relatively important part in the second volume, as they furnish a theory for the formation of differential equations (see sections 6.2.1.1 and 6.2.2.1). The next section, on condition equations for a formula to be an exact differential, proceeds in the direction of preparing the way for the treatment of differential equations in volume II. Chapter 1 ends with a section about alternative foundations for the calculus. Both d'Alembert's limit approach and Leibniz's infinitesimals are treated. This is typical of Lacroix's *encyclopédiste* approach: to expound all relevant alternative methods or theories. It is also an

essential instance of that approach because in future chapters Lacroix will sometimes need to resort to one or another of those alternative foundations in order to
explain some particular method.

Chapter 2 is dedicated to some analytic applications of the differential calculus. First, its use in expanding functions in series, for which of course Taylor's theorem (or rather Maclaurin's) is central. But this section has a lot more
to offer, including Lagrange's formula for expanding $\psi(y)$ in powers of x, where
$\alpha - y + x\varphi(y) = 0$. Oddly, the section finishes with a non-differential, approximation method by Lagrange [*1776*] for expanding implicit functions in continued
fractions, adapted to give also power-series expansions. After this comes an examination of certain cases in which the differential coefficient "becomes infinite" (as
with $f(x) = \sqrt{x-a}$ for $x = a$) and why the expansion (2.1), "although true in general", is not valid in such cases. The explanation for this rests on the irrationality
of the function involved disappearing for certain values of the variable, dragging a
collapse of multiple values of the function. Lacroix attributes this to Lagrange and
in fact it appears in his *Théorie des fonctions analytiques*: it may be one of the few
remarks drawn from Lagrange's lectures at the *École Polytechnique* that Lacroix
was able to include in the first volume (see section 3.2.5). This is followed by a
section on indeterminacies ($\frac{0}{0}, 0 \times \infty, ...$) and how to raise them. After this we have
a section on series expansion of functions of two variables (much shorter than the
one for functions of one variable). And the chapter finishes with the investigation
of maxima and minima of functions of one or several variables.

After analytical applications, we would expect to see geometrical applications. And they eventually appear. But chapter 3 is a "digression on algebraic
equations" – an interlude in the natural sequence of topics. Lacroix justifies this
chapter by the "imperfection" of the available textbooks on algebra, and by the
want for these methods in integral calculus [*Traité*, I, xxv]. But why not include
them in the Introduction? There are a couple of uses of differential calculus, but
they could have been avoided (if this were a chapter on applications of differential
calculus to algebraic equations, it could have been merged into chapter 2). In the
Preface to the second edition Lacroix explains the arrangement in the first as being
due to his fear that the Introduction might become too long and retard too much
the entry of the main subject – differential calculus [*Traité*, 2nd ed, I, xx] (this
changed in the second edition: Lacroix omitted several of these topics, because
meanwhile he had included them in his *Complément des élémens d'algèbre* [*1800*];
while the rest was moved precisely to the Introduction). This explanation is quite
unsatisfactory; Lacroix should not be too worried with the length of the Introduction in this kind of treatise. One must consider the possibility of chapter 3 not
being in the original plans, and having been included only after the Introduction
was printed.

Chapter 3 has two sections. The first, on "similar functions of the roots of
equations" (i.e., all the roots appear in a similar form) is about symmetric functions
(incidentally, Lacroix appears to introduce the expression "symmetric functions"
[*Traité*, 277]). Here Lacroix gives a proof, which he claims to be original, of New-

ton's theorem on the sums of powers of the roots of an equation;[13] Lacroix's proof
does not use differential calculus or infinite series, and he thought it worthy of
mention in his *Compte rendu [...] des progrès que les mathématiques ont faits de-*
puis 1789 (see appendix B, under "algèbre", or [Delambre *1810*, 90]). In the second
section, on "imaginary expressions" (i.e., complex numbers), Lacroix gives, among
other things, a proof by Laplace of the fundamental theorem of algebra, Cotes'
theorem, Descartes's sign rule, and Euler's solution to the problem of logarithms
of negative numbers.

The two final chapters are devoted to analytic and differential geometry:
chapter 4 on the plane; chapter 5 in the space. They will be treated at length in
chapter 4 below (sections 4.1.2, 4.2.1.2, 4.2.2.2, and 4.2.2.3). The determination in
including geometrical applications (which also serve as illustrations of the analytical theory), and at the same time in keeping them separate (trying not to derive
any analytical result from geometry), are important characteristics of Lacroix's
Traité.

Volume I		
topics	chapter	pages
Analytic geometry: coordinates and fundamental formulas for points and straight lines	Chapter 4: Theory of curved lines (plane curves)	327-332
Analytic geometry: curves		332-341
Analytic geometry: change of coordinates		341-362
Applications of series expansion to the theory of curves		362-369
Use of differential calculus to find tangents		369-377
Use of differential calculus to find singular points		377-388
Contact and osculation		388-394
Properties of the osculating circle; evolutes		394-401
Transcendental curves (logarithmic, cycloid, spirals); polar coordinates; diff. of arc-length and of the area under a curve		401-419
Method of limits applied to curves		419-422
Curves as polygons; roulettes		422-434
Analytic geometry: coordinates and fundamental formulas for points, planes and straight lines	Chapter 5: Curved surfaces and curves of double curvature	435-448
Analytic geometry: "curved surfaces of second order" (quadrics); change of coordinates		448-465
Application of differential calculus to the theory of contact of surfaces		465-471
Theory of curvature of surfaces		471-482
Generation of surfaces (envelopes; developable surfaces; etc.)		482-504
Curves of double curvature		504-519

Table 2.2: Volume I of Lacroix's *Traité* (continued from table 2.1)

---

[13]Nowadays often called Newton-Girard formulas (not by Lacroix, who ignores Girard).

Here the influence from Monge is most marked. What was still generally known as "'application of algebra to geometry" was then being transformed into *analytic geometry*. Monge was the main architect of this change (with an important suggestion by Lagrange in a 1773 memoir on tetrahedra), but Lacroix played an important role in its systematization, precisely in the *Traité* [Taton *1951*, ch. 3]. As he explains in the Preface, he tried to keep apart all geometric constructions and synthetic reasonings, and to deduce all geometry by purely analytic methods [*Traité*, I, xxv]. That is why chapter 4 starts with an extensive study of fundamental formulas for points, straight lines and distances, to be used in what follows, instead of "geometric constructions". These elementary subjects were usually regarded as belonging to the realm of synthetic geometry. After these preliminaries, Lacroix develops the analytic geometry of plane curves, including plotting, classification of singular points and changes of coordinates. Changes of coordinates have several applications, including finding tangents and multiple points.

Before differential geometry properly speaking, comes the application of series expansions (which because of their approximative nature supply a way of finding tangents and asymptotes). But the central part of chapter 4 is the application of differential calculus (that is, the use of differential coefficients) to find properties of the curves: their tangents, normals, singular points, the differentials of their arclength and of the area under them; and to develop a theory of osculation, and hence of curvature via the osculating circle. The chapter concludes in a manner very typical of Lacroix: presenting alternative points of view, namely an application of the method of limits to find tangents and osculating curves and the Leibnizian consideration of curves as polygons. It is significant that in total this chapter has five approaches to the determination of tangents. In this last section is included a study of envelopes of one-parameter families of curves, the language alternating between limit-oriented and infinitesimal. A very important special case is that of the evolute of a given curve, formed by the consecutive intersections of its normals.

The matter of chapter 5, a theory of surfaces and space curves, is mostly due to Monge, according to Lacroix [*Traité*, I, 435]. In fact, in spite of some isolated studies by Euler and others, it was Monge who set spatial differential geometry going, and made it a discipline [Struik *1933*, 105-113; Taton *1951*, ch. 4]; and for this he needed to develop also three-dimensional analytic geometry.

The fundamental formulas for planes and points, straight lines and distances in space are followed by more traditional subjects: second-order surfaces (that is, quadrics), and changes of coordinates.

There is some discussion of contact of surfaces using their series expansions, but as the chapter proceeds power series lose ground to limits and infinitesimals. Alternatively to comparison of coefficients in series expansions, the tangent plane through a point with coordinates x', y', z' is determined by the tangents to the sections parallel to the vertical coordinate planes (these tangents have slopes $\frac{dz'}{dx'}, \frac{dz'}{dy'}$, so that

$$z - z' = \frac{dz'}{dx'}(x - x') + \frac{dz'}{dy'}(y - y')$$

is the equation of the plane). Not surprisingly, curvature of a surface on a point is
studied through the radii of curvature of plane sections through that point: these
have a maximum and a minimum, which allow us to calculate the curvature of
any other plane section. There is no discussion yet of kinds of curvature or of the
possibilities of the centres of curvature being on the same or on different sides
of the surface. Envelopes of one-parameter families of surfaces are studied as the
"limits" of their consecutive intersections (these intersections are called, following
Monge, "characteristics"). A special case is that in which the generating surfaces
are planes: the envelope is then called a "developable surface".

Three approaches are given to study curves in space ("curves of double curvature"). But two of them only briefly (through their projections on the coordinate
planes; and through the series expansions of two coordinates as functions of the
third). The bulk of the section follows Monge in regarding space curves as polygons where three consecutive sides are not coplanar. This allows Lacroix not only
to study tangents, osculating planes, and differentials of arc-length, but also the
developable surface generated by a curve's normal planes, and evolutes.

2.4   Volume II: integral calculus (1798)

Although the second volume of Lacroix's *Traité* is the largest of the three, it is
the one that receives the least attention in the general Preface at the beginning of
volume I.[14] The integral calculus, being just the inverse of the differential calculus,
did not offer much occasion for reflection: it consisted only of a "collection de
procédés analytiques, qu'il suffit d'ordonner de manière à en faire appercevoir les
rapports"[15] [Lacroix *Traité*, I, xxvii]. Lacroix proposes then to follow the ordering
of [Euler *Integralis*], adding new developments and replacing some methods by
more recent and general ones. In the second edition Lacroix would be a little
more explicit in the characterization of Euler's order: the methods are classified
according to the form of the functions to which they apply [Lacroix *Traité*, 2nd
ed, I, xxxix].

There are however two significant differences in structure from Euler's integral calculus. One is the inclusion of a chapter on calculation of areas, lengths, and
volumes (chapter 2); [Euler *Integralis*] does not include geometrical applications.

The other difference lies in the way the material is divided, in particular the
structural relevance of integration of explicit functions versus integration of differential equations. [Euler *Integralis*] is divided into two "books", the first (volumes 1
and 2) on problems involving functions of one variable and the second (volume 3)

---

[14]This would change in the second edition, where the coverage of the second volume increases
from one small paragraph [Lacroix *Traité*, I, xxvii] to about six pages [Lacroix *Traité*, 2nd ed,
I, xxxviii-xliv]. This is more than the three pages for the third volume (one page in the first
edition), but still much less than the nineteen pages for the first volume (about three pages in
the first edition).

[15]"collection of analytical procedures, which is enough to order so as to make perceive their
connections"

Volume II		
topics	chapter	pages
Table of contents and bibliography	Table	iii-viii
Integration of polynomial functions		2-5
Integration of rational functions		5-29
Integration of irrational functions		29-33
Integration of binomial differentials		33-48
Irrational polynomial differentials (inc. elliptic integrals)	Chapter 1:	48-66
Integration by series	Integration	66-88
Integration of logarithmic and exponential functions	of functions	89-100
Integration of trigonometric functions	of one	100-118
Expansion of $(a + b\cos z)^m$	variable	118-135
General method for approximating integrals; integrals as limits of sums; definite and indefinite integrals; Bernoulli series		135-156
Integration of higher-order differentials		156-160
Quadrature of curves (calculation of areas under curves)	Chapter 2:	161-176
Rectification of curves (arc lengths)	Quadratures,	176-188
Volumes of solids and areas of surfaces; rectification of curves of double curvature; double and triple integration	cubatures and	189-206
Functions with algebraic integrals – squarable curves, etc.	rectifications	206-220
Separation of variables		221-230
Integrating factors for 1st-order differential equations		230-251
1st-order eqs. with differentials raised to powers above 1		251-262
Particular solutions of 1st-order differential equations		262-284
Approximate solutions of 1st-order differential equations		284-296
Geometrical construction of 1st-order diff. equations		296-307
Integration of 2nd-order diff. eqs. through transformations (the simplest differential equations of order higher than 1)	Chapter 3: Integration	307-332 364-365
Integrating factors for 2nd-order differential equations	of differ.	332-349
Approximate solutions of 2nd-order differential equations	equations	349-364
1st-degree differential equations of any order	in two variables	365-378 389-394
Systems of first-degree differential equations		378-389
Use of 1st-degree diff. eqs. for approximate integration		394-407
Particular solutions of diff. eqs. of order higher than 1		408-418
Diff. eqs. that are easier to integrate after being differentiated		418-423
On logarithmic and trigon. functions (from their diff. eqs.)		423-427
On elliptic transcendents		427-452

Table 2.3: Volume II of Lacroix's *Traité* (continued in table 2.4)

on problems involving functions of two or more variables; the first "book" is then divided into two parts (corresponding to volumes 1 and 2), the first on first-order problems and the second on higher-order problems; thus, integration of explicit

functions does not have – at least in the table of contents – the prominence that
a modern reader might expect, being the subject only of the first section of the
first part of the first book and of both chapters 1 of the first and second sections
of the second part of the first book. In [Lacroix *Traité*, II], on the other hand, integration of explicit functions is awarded the entire first chapter out of 5, ranking
at the same level as integration of ordinary differential equations (chapter 3) and
integration of partial differential equations (chapter 4).

Apart from the ordering, Lacroix also admitted taking his examples from
Euler – in an explicit reference to chapters 2 and 3 (which should rather be to
chapters 1 and 3) of the second volume [Lacroix *Traité*, 2nd ed, I, xli].

Most of Chapter 1 is dedicated to finding antiderivatives of functions of one
variable: algebraic, rational, irrational, and transcendental (exponential, logarithmic and trigonometric). On the formalistic character of these procedures, see sections 5.1.1 and 5.2.3. It is in the section on integration of irrational functions that
the elliptic integrals

$$
\int \frac{dx}{(x^2 + a)\sqrt{\alpha + \beta x + \gamma x^4}}, \quad \int \frac{dx}{\sqrt{\alpha + \beta x + \gamma x^4}}, \quad \int \frac{x^2\, dx}{\sqrt{\alpha + \beta x + \gamma x^4}}
$$

first appear (with no particular name here; in chapter 2 they gain the name *elliptic
transcendents*, after Legendre); Lacroix remarks that they are new transcendental
functions that must be introduced in the calculus [Lacroix *Traité*, II, 59]. The
subject of elliptic integrals is resumed several times later, most importantly in
chapter 3.

There is also a section on "integration by series" (see section 5.2.1); and
another, on a "general method" by Euler for approximating integrals, which includes some very interesting remarks on the "nature of integrals" and the definitions
of *definite* and *indefinite integrals* (see sections 5.2.2 and 5.2.3).

Chapter 2 is dedicated to calculation of areas under curves, arc-lengths, and
volumes and areas of surfaces. Since the methods of integration had been studied
in the previous chapter, and the differentials of the area under a curve and of the
arc-length had already been found in the first volume, a large part of this chapter
consists of examples. But it still remained to derive the differentials of the volume
of a surface of revolution, of the volume under a surface, and of the area of a
surface.

It is in this context that double integration is introduced, as repeated integration [Lacroix *Traité*, II, 192-193].[16] Geometrical meaning is lost when Lacroix
analogously introduces also triple integration (because of its frequent occurrence in
mechanics) [Lacroix *Traité*, II, 204-205]. Change of variables is discussed for both
double and triple integrals, arriving at the expressions nowadays called *jacobians*
[Lacroix *Traité*, II, 203-206].

---

[16]Multiple integration of functions of only one variable had already appeared at the end of
chapter 1, but that is a very special case.

This chapter ends with a small section on squarable curves (that is, functions with algebraic integrals), rectifiable curves (algebraic arc-length), and spatial counterparts.

Chapter 3, dedicated to integration of differential equations in two variables, is the largest in volume II. This is not surprising, as it corresponds to about half of [Euler *Integralis*] (second and third sections of volume 1 and the whole volume 2). Like Euler's work, this chapter is broadly organized by the order of the differential equations: first order first; then higher, mostly second; and finally methods unrelated to order (but mostly related to degree, namely "first degree"). Still, the presence, location and relative weight of the latter methods are noteworthy departures from the more strictly order-based Eulerian organization. Naturally, it is in connection to these methods that we notice the most significant novelties relative to Euler's work.

A certain peculiarity in terminology must be mentioned at once: Lacroix [*Traité*, II, 225] rejects the application of the adjective "linear" to differential equations, since that word refers to straight lines (as in algebraic "linear equations"), and of course linear differential equations usually belong to transcendental curves. Instead, he uses the expression "first-degree differential equations". This may be particularly confusing to the modern reader, because Lacroix [*Traité*, II, 365-366] even restricts this expression to equations that are of first degree in regard to the dependent variable and all its differentials (and thus, in modern terms, strictly "linear", as opposed to "quasi-linear" or "first-degree", which need only be linear in regard to the highest-order derivative). However, it is a quite fitting stand for someone so concerned as Lacroix with geometrical interpretations of analytical concepts.

Naturally this chapter starts with the most classic methods: separation of variables and integrating factors, applied to first-order and first-degree equations. But even in regard to these simpler cases, Lacroix complains about the imperfection of analysis, which does not provide a better algorithm than groping for an integrating factor [*Traité*, II, 251]. He alludes to general methods proposed by Fontaine and Condorcet,[17] but justifies not saying anything about them with their unpracticality; still, their references appear in the table of contents [Lacroix *Traité*, II, vi].

After some considerations on "first-order equations where the differentials are raised to powers higher than one" (either solving them algebraically for $\frac{dy}{dx}$ first, or using "analytical artifices", particularly for homogeneous equations), come three sections on special topics of first-order equations: singular solutions are examined following mainly [Lagrange *1774*], but using Laplace's name "particular solutions", instead of Lagrange's "particular integrals" (see section 6.2.1.2); a section on approximate integration includes the use of Taylor series, Euler's "general method" (which also serves to show that all first-order equations "are possible"), and a

---

[17]Very briefly, these methods relied on obtaining all possible forms for the solutions (or integrating factors) of differential equations, and then trying to adequate one of those to the equation to be solved (using the method of indeterminate coefficients) [Gilain *1988*, 91-97].

method of expansion in continued fractions (see section 5.2.4); a section on "geometrical constructions" includes some historical remarks, trajectory problems, and the geometrical interpretation of "particular solutions" as envelopes of the families of curves given by the "complete integrals" (see section 6.2.3).

As for second-order equations, Lacroix starts by addressing several particular cases that are easier to treat (for instance, by considering a new variable $p = \frac{dy}{dx}$). This is followed by integrating factors. To finish come approximation methods (mostly by expansion in series, but also including a brief mention to Euler's "general method", and hence a "general construction" of second-order equations, that shows their possibility and that they represent an infinity of curves – see section 5.2.4).

A section on "integration of differential equations of order higher than 2" [Lacroix *Traité*, II, 364-394] is in fact almost entirely dedicated to "first-degree" equations of any order – both isolated and systems of such equations (including what Gilain [*2004; to appear*] calls "d'Alembert's theory"[18]).

The next section is still on "first-degree equations", more precisely their use for approximate integration. This refers to a method much used in astronomy. Unfortunately, several mistakes occur here (see section 5.2.4, pages 175 ff.).

The final section in chapter 3 ("general reflections on differential equations and on transcendents") is a medley. First, particular (i. e., singular) solutions of differential equations of order higher than 1 (section 6.2.1.3), followed by certain equations that are easier to integrate after being differentiated. To finish, Lacroix studies some transcendental functions from differential equations that characterize them (particularly elliptic integrals). For motivation, he expresses the opinion that the most useful result in integral calculus would be the exact classification of the distinct transcendental functions [Lacroix *Traité*, II, 423].

The second largest chapter in the second volume, chapter 4, is mostly dedicated to differential equations in more than two variables (both partial and total). It is named "integration of functions of two or more variables", probably because of about two pages in the beginning, addressing the case in which the (first-order) differential coefficients of the function are given explicitly – that is, the integration of exact differentials like $p \, dx + q \, dy$ or $n \, du + p \, dx + q \, dy$. But it turns out to be a misnomer, because of its last section, on "total differential equations that do not satisfy the conditions of integrability" – in the case of three variables (the most common) these correspond to two functions of *one* independent variable.

Just after explicit functions, Lacroix addresses at some length the conditions of integrability for total differential equations and the integration of those that

---

[18]Consisting essentially in a method to solve systems of 1st-order linear equations using multipliers, and in the reduction of systems of higher-order equations to first order, considering new variables $p = \frac{dy}{dx}, q = \frac{dp}{dx}$, etc. Gilain stresses Lacroix's role in the transmission of d'Alembert's theory, which was not particularly well known by his contemporaries (still, it appears in [Cousin *1796*, I, 234-238]). Gilain focuses especially on the transmission through [Lacroix *1802a*], and especially to Lacroix's student Cauchy, who would give it in [*1981*] an importance much greater than the marginal place it occupies in [Lacroix *1802a*] (and, it may be added, in [Lacroix *Traité*]).

Volume II		
topics	chapter	pages
Integration of explicit differential functions of several variables		453-456
Integration of total differential equations in three variables		456-466
Total differential equations in more than three variables	Chapter 4:	466-471
Total differential equations of higher orders	Integration	471-476
1st-order partial diff. eqs. (1st degree rel. to diff. coeffs.)	of functions	476-496
1st-order partial diff. eqs. (with raised diff. coefficients)	of two or	496-520
Integration of higher-order partial differential equations	more	520-608
Geometrical construction of partial diff. eqs.; determination of the arbitrary functions contained in their integrals	variables	608-624
Total diff. eqs. not satisfying the conditions of integrability		624-643
Geometrical remarks on the previous section		643-654
Principles of the calculus of variations	Chapter 5:	655-689
Application to problems of maxima and minima	Method of	689-718
Distinguishing maxima from minima	variations	718-724
Additions (on total and partial differential equations)	Additions	725-727
Corrections to volumes I and II	Errata	728-732

Table 2.4: Volume II of Lacroix's *Traité* (continued from table 2.3)

satisfy them (that is, those in which one variable may be taken as a function of the others). Another issue of terminology: Lacroix never explains nor introduces the expression "total differential equations", and he does not even use it at this point, although in the index he refers to these articles as being about "total differential equations" [*Traité*, III, 555-556]; and he uses it without further ado in page 492 and in the title of the last section of the chapter. In spite of such a familiar use, this may be the first appearance of the adjective "total" in this context – at least a contemporary author, the Belgian Nieuport [*Mélanges*, II, xiii], attributed it to Lacroix. It certainly was not at all common at the time – for instance Monge [*1784c*] spoke of "équations aux différences ordinaires à trois variables"[19]. Perhaps Lacroix was just using "total" as the natural opposite of "partial".

But of course most of the chapter is dedicated to partial differential equations. There are three sections on these: first order, higher orders, and a much smaller one on geometrical constructions and determination of the arbitrary functions that appear in integrals. For the most simple first-order equations, Lacroix uses Euler and d'Alembert's early method of reducing to a total differential equation, to which is then applied an integrating factor [Demidov *1982*, 329][20]. This works for all linear ("first-degree") equations, but not for all quasi-linear ones, and naturally Lacroix [*Traité*, II, 482-484] expounds Lagrange's method for quasi-linear first-order partial differential equations (reducing them to a system of total differential

---

[19]"equations of ordinary differences in three variables"
[20]For an example see equation (6.27), page 236 below.

equations), minding to remark that Monge had also independently obtained it [Lacroix *Traité*, II, 487].

As for nonlinear equations, we find one of the most directly influential passages of Lacroix's *Traité*. In [*1772b*] Lagrange had reduced the integration of a general first-order partial differential equation to that of a quasi-linear first-order partial differential equation; but strangely, he did not combine this with the method mentioned above. This was done by the young mathematician Paul Charpit in a memoir presented to the *Académie des Sciences* of Paris in 1784. Unfortunately, Charpit died soon after, and his memoir was never published. His name might have been entirely forgotten, if Lacroix had not reported his work, citing his name, in [*Traité*, II, 496-520 (esp. 496-497, 513-516)]; instead, this combination became known as the "Lagrange-Charpit method" [Demidov *1982*, 332; Grattan-Guinness & Engelsman *1982*][21].

Thus, Lacroix was fortunate enough to have at hand a theory of first-order partial differential equations. Higher-order equations were a different matter altogether, but in the long section (88 pages) dedicated to them Lacroix still tries to have as much of a structure as possible, focusing on what we call linear and quasi-linear second-order equations. What is perhaps most striking is the neglect of physical motivations.

After considering a few cases in which the order may be lowered, Lacroix addresses second-order equations in three variables, of first degree in regard to the second-order differential coefficients (in modern terms, quasi-linear) [*Traité*, II, 524-535]. For these, he uses Monge's method [*1784b*, 126-155], which is analogous to Lagrange's (and Monge's) method for first-order quasi-linear equations, and which gives (when it works) one or two first-order integrals.[22] But Lacroix [*Traité*, II, 526] admits that this second-order version is less general than the first-order one (it fails when a certain auxiliary differential equation in three variables does not satisfy the integrability condition). This method is also extended to third-order equations in three variables and to second-order ones in four variables [Lacroix *Traité*, II, 535-546].

The failures of this method motivate a discussion about why sometimes there are no first-order integrals of second-order differential equations (or fewer integrals than expected), even if there are finite integrals. The way this is discussed leads to the distinction between "complete" and "general" integrals, and to the consideration of "particular" (i.e., singular) solutions (see sections 6.2.2.3 and 6.2.2.4).

After this theoretical interlude, Lacroix turns his attention to "first-degree" second-order equations. He had already applied Monge's method to them [*Traité*,

---

[21]Kline [*1972*, II, 535] also tells this story but, ignoring the existence of two manuscript copies of Charpit's memoir [Grattan-Guinness & Engelsman *1982*], he still relies exclusively on Lacroix's information (carefully adding not to "know whether Lacroix's statement is correct").

[22]Lacroix's basic version [*Traité*, II, 524-526] is as usual much clearer and/or easier to follow than Monge's. Kline's account [*1972*, II, 538-539], who claims to follow [Monge *Feuilles*] rather than [Monge *1784b*], in fact seems to draw on Lacroix. I also do not understand why Kline calls "nonlinear" these equations which are "linear only in the second derivatives", while a few pages earlier he had used "linear" for first-order equations which are linear only in the derivatives.

II, 531-535]; but now [*Traité*, II, 565-590] he reports at length Laplace's cascade method [*1773c*] (with a few complements by Legendre [*1787*]), based on a reduction to a simpler form $\frac{d^2 z}{du\,dv} + P\frac{dz}{du} + Q\frac{dz}{dv} + Nz = M$ via an appropriate change of variables, which facilitates the use of indeterminate coefficients to find a solution in the form of a finite series $z = A + B\varphi(u) + C\varphi'(u) + D\varphi''(u) + \text{etc.} + B_1\psi(v) + C_1\psi'(v) + D_1\psi''(v) + \text{etc.}$

The situation is more complicated for "first-degree" third-order equations, but Lacroix still presents attempts at analogous finite series solutions [*Traité*, II, 590-594], and wider uses for Laplace's change of variables [*Traité*, II, 595-596]. The section finishes with miscellaneous integrations of particular equations, especially of degree above 1 [*Traité*, II, 596-608].

After this comes a small section with the long title "on the geometrical construction of partial differential equations, and on the determination of the arbitrary functions that appear in their integrals". This deals mostly with Monge's constructions of surfaces corresponding to partial differential equations, forcing them to pass through given curves. An offshoot is the argument that these curves, and the arbitrary functions appearing in the integrals, need not be "continuous". (See section 6.2.3.3.)

The final section in chapter 4 is on "total differential equations that do not satisfy the conditions of integrability". Once again, this is based on Monge's work: in total differential equations in three variables that do not satisfy those conditions, it is not possible to consider one of the variables as a function of the other two (or, in Mongean fashion, these equations do not represent surfaces); but Monge had shown that they represent families of curves in space. Lacroix gives his own analytical theory of these equations (of which he was rather proud), followed by the geometrical interpretations. (See section 6.2.4.)

Chapter 5, the last in the second volume, is dedicated to the "method of variations", an obligatory subject in any treatise of integral calculus at this time. It is divided into two sections, the first [*Traité*, II, 656-688] on calculating variations (interchangeability of d and δ, formulas for $\delta \int V\,dx$, Euler-Lagrange equations), and the second [*Traité*, II, 689-724] on applications to problems of maxima and minima. It must be remarked that (in this first edition) Lacroix makes no attempt to suit the calculus of variations to the Lagrangian power-series foundation of the calculus. Accordingly, he presents Lagrange's δ algorithm (which Lagrange was abandoning by then [Fraser *1985*]), in Leibnizian shape: $\delta dy = d\delta y$ is justified using infinitesimal considerations; the rules of δ-differentiation come from those of d-differentiation by plain analogy. Todhunter [*1861*, 11-27] examined at length the version of this chapter in the second edition, concluding that "on the whole the calculus of variations does not seem to have been very successfully expounded by Lacroix, and this is perhaps one of the least satisfactory parts of his great work"; he also seemed to agree with another author, Richard Abbatt, who had called Lacroix's treatment of this subject "prolix and inelegant". These negative opinions may have been somewhat influenced by the fact that in the second edition Lacroix

added a section to conform with Lagrange's new foundation, but also maintained
the old treatment; but this is not a full justification – it does seem to be one of
the less clear parts of Lacroix's *Traité*.

2.5   Volume III: differences and series (1800)

The third volume of Lacroix's *Traité* bears, in the first edition, a separate title –
"Traité des Différences et des Séries"[23], followed by the indication "faisant suite
au Traité du Calcul différentiel et du Calcul intégral"[24]. This has given rise to
bibliographical descriptions in which it appears as a separate work. For example:
Taton [*1953a*, 589] mentions the *Traité du calcul différentiel et du calcul intégral*,
composed of two volumes, 1797-1798, the *Traité des Différences et des Séries*, one
volume, 1800, and then a "nouvelle édition de l'ensemble"[25], three volumes, 1800-
1814-1819; somewhat more radically, Jean Itard, in his list of works by Lacroix,
has "*Traité du calcul différentiel et du calcul intégral*, 2 vols. (Paris, 1797-1798);
2nd ed., 3 vols. (Paris, 1810-1819); *Traité des différences et des séries* (Paris,
1800)" [*1973*, 551] – the relationship between the *Traité des différences et des
séries* and the *Traité du calcul...* is only explained in the main text [*1973*, 550].
Although these bibliographical separations make sense, they are misleading. It is
clear enough that Lacroix viewed the *Traité des différences et des séries* as part
of the *Traité du calcul...*: its summary is included in the general Preface in the
first volume (calling it an "Appendix") [*Traité*, I, xxvii-xxviii]; the numbering of its
articles follows directly that of the second volume; the subject index at its end is
for the entire set of three volumes; in the "corrections and additions" it is referred
to as "tome III" [*Traité*, III, 581]. Thus, it is called throughout this work simply
as the third volume of Lacroix's *Traité*, or [Lacroix *Traité*, III].

The reason for the particular title of the third volume is probably that Lacroix wished to call attention to its greatest originality, namely its very subject
– a complete treatise on series (studied for themselves, rather than regarded as
expansions of functions) and finite differences. He remarked in the general Preface that no one had assembled the whole "theory of sequences" in a single "corps
de doctrine" after Jacob Bernoulli and James Stirling (an obvious reference to
[Jac. Bernoulli *Series*] and [Stirling *1730*]), in spite of the "prodigious" growth of
the area through later work by Euler, Lagrange, Laplace, and more recently Prony
[Lacroix *Traité*, I, xxvii]; Lacroix repeated this claim for originality in his *Compte
rendu [...] des progrès que les mathématiques ont faits depuis 1789* (see appendix
B, page 400, or [Delambre *1810*, 109]).

In fact, finite differences were a topic sometimes found in books on differential calculus, but not as an autonomous subject with one dedicated section. The
most typical appearances happened in early chapters, preparing the way for diffe-

---

[23]"Treatise on Differences and Series"
[24]"being a continuation of the Treatise on differential and integral calculus"
[25]"new edition of whole set"

rentials, which might be introduced as infinitely small differences or as the terms in the limit $\frac{dy}{dx}$ of a ratio of decreasing finite differences $\frac{\Delta y}{\Delta x}$ (see sections 3.1.1 and 3.1.2). In advanced works we may find some other, scattered, occurrences: in [Euler *Differentialis*], chapters 1 and 2 of the first part address finite differences (in that typical introductory manner), while several chapters of the second part address applications of the differential calculus to finite differences or to closely related topics (such as interpolation, or summation of series), interspersed with applications to unrelated issues (such as maxima and minima, or indeterminacies); in [Cousin *1777*; *1796*] we find an introductory chapter on the "calculus of differences in general" [*1777*, ch. 1; *1796*, I, Intr., ch. 3], a section on finite difference equations in the chapter on "integral calculus in general" [*1777*, 313-321; *1796*, I, 271-277], and finally, near the end, a chapter wholly dedicated to these equations [*1777*, ch. 11; *1796*, II, ch. 7]. Lacroix, on the other hand, thought it was "convenient" to separate the calculus of differences from the first principles of the differential calculus, and not to cut up ("morceler") the former (see again appendix B, page 400, or [Delambre *1810*, 109]).

[Prony *1795a*] is a different case, and quite unique. It is almost entirely dedicated to the calculus of finite differences; but, perhaps because it was intended as an introductory course in analysis[26], there are several subjects absent – such as "second-order powers" (i.e., factorials), Bernoulli numbers, generating functions, mixed difference equations – so that Lacroix apparently did not count it as containing "the whole theory of sequences".

Before entering in the contents of [Lacroix *Traité*, III], we must address an issue of terminology: Lacroix keeps the 18th-century tradition of not distinguishing between the words "series" and "sequence", using both interchangeably (here I will try to make a modern distinction, except when referring to the whole subject, usually the "theory of series", and of course in quotations). More confusingly still, both words were applied not only to infinite series or sequences, but also to finite sums or progressions. Thus, the "theory of series" was a theory of summations, both finite and infinite – and closely linked to the inverse calculus of differences.

The main chapter in [Lacroix *Traité*, III] is by very far chapter one, "on the calculus of differences". It occupies more than half of the volume, and contains a full account of the calculus of differences. In the second edition it was divided into three chapters, and even in the first edition we can see clearly the three parts corresponding to those future chapters: direct calculus of differences; inverse calculus of differences of explicit functions; and difference equations. This organization, of course, reflects the perspective of the difference calculus as a discrete analogue of the differential and integral calculus.

The first section [Lacroix *Traité*, III, 2-26] is dedicated to the *pure* direct calculus of differences: the definition of differences of first and higher orders, and several formulas for calculating them, and relations between the differential and

---

[26]The differential calculus is introduced at the end as the infinitesimal case [Prony *1795a*, IV, 543-551].

Volume III		
topics	chapter	pages
Table of contents and bibliography	Table	iii-viii
Direct calculus of differences – basic notions and analogy between differences and powers		1-26
Interpolation of sequences of one variable		26-60
Differences and interpolation of functs. of several variables		60-64
Integration of rational functions		65-74 83-84
Digression on *2nd-order powers* or *factorials*		74-82
Integration of transcendental functions		84-92
Expansions of Σ integrals by differences and differentials	Chapter 1:	92-122
Applic. of difference calculus to summation of "sequences"	Calculus of	122-151
Application of summation of series to interpolation	differences	151-175
Digression on elimination in algebraic equations		175-183
Integration of 1st-degree difference eqs. in two variables		184-210
Equations where the difference of the independent variable is not constant		210-215
Determination of the arbitrary functions in integrals of partial differential equations		215-225
Systems of first-degree equations		225-229
Integrating factors for first-degree difference equations		229-231
On the nature of the arbitrary quantities introduced by the integration of difference eqs., and on their construction		231-237
The different types of integrals of difference equations		237-247
Integration of difference eqs. in three or more variables		247-288
Condition eqs. for integrability of functions of differences		289-300
Functions of one variable	Chapter 2: Theory of	301-326
Transformation of series	sequences fr.	326-333
Expansions of differences, differentials, and integrals	generating	333-338
Functions of two variables	functions	338-355

Table 2.5: Volume III of Lacroix's *Traité* (continued in table 2.6)

difference calculi (namely a new deduction of Taylor series). These relations lead to formal expressions such as

$$\Delta^n u = \left(e^{\frac{du}{dx}h} - 1 \right)^n,$$

where, after expanding the right-hand binomial, the powers du^k of du must be replaced by higher differentials $d^k u$. This formula, and this kind of analogy between powers and differences, had been introduced by Lagrange [*1772a*]; Lacroix acknowledges this, but gives also a demonstration by Laplace [*1773b*, 534-540]. The next, longer section [Lacroix *Traité*, III, 26-64] addresses the main application of

the direct calculus of differences – that is, its application to interpolation of sequences. We can find here the most familiar formulas – the Gregory-Newton formula
(without any specific name) [*Traité*, III, 28], Newton's and Lagrange's interpolation polynomials (with these attributions) [*Traité*, III, 32, 34], the Newton-Stirling
formula (attributed to Stirling) [*Traité*, III, 39] – as well as less familiar work –
such as an account of Mouton's method, with developments by Prony [*Traité*, III,
55-60].

Next comes the inverse calculus of differences, for differences given explicitly.
Again, Lacroix starts by a section dedicated to the pure calculus [*Traité*, III,
65-122], followed by sections on applications. There are two operators here: the
"integral" Σ is the inverse of the difference operator Δ, i.e. an analogue of the
indefinite integral – if $\Delta u = f(x, h)$ (where $h = \Delta x$) then $u = \Sigma f(x, h) + const.$;[27]
the "summatory term" S is closer to the definite integral – $S f(x, h)$ is the sum
$\Delta u + \Delta u_1 + \ldots + \Delta u_n$,[28] where again the generic difference Δu is given by f(x, h);[29]
they are related by the equality $S f(x, h) = \Sigma f(x, h) + f(x, h) - const.$[30] Naturally,
in the section on the pure inverse calculus, the integral receives almost exclusive
attention. Integration of polynomials leads to a detailed study of "second-order
powers", that is, generalized factorials – products of equally spaced factors $x(x +
\Delta x) \ldots (x + n\Delta x)$; Lacroix focuses mostly on the falling factorial

$$p(p-1)(p-2)\ldots(p-n+1),$$

using Vandermonde's notation $[p]^{n}$ – which is quite convenient for enhancing analogies between falling factorials in difference calculus and (common) powers in
differential calculus.[31] After reporting the integration of the trigonometric functions and integration by parts (giving formulas by Taylor and Condorcet), Lacroix
addresses ways to express Σu through the differences and the differentials of u –
including Lagrange's

$$\Sigma^m u = \frac{1}{\left(e^{\frac{du}{dx}h} - 1\right)^m},$$

with similar provisions as above, for changing positive powers $\frac{du^p}{dx^p}$ into $\frac{d^p u}{dx^p}$ and
negative powers $\frac{du^{-p}}{dx^{-p}}$ into $\int^p u \, dx^p$. The search for the coefficients in the series
expansion of Σu leads, through the particular case of Σx^m, to the Bernoulli numbers.

In the section on the application of difference calculus to summation of series
[*Traité*, III, 122-151], the S operator comes to the foreground. This application

---

[27] Jordan [*1947*, 100-101] calls this the "indefinite sum".

[28] That is, $\Delta u_0 + \Delta u_1 + \ldots + \Delta u_n$.

[29] But in $S f(x, h)$, x is presumably at its *last* value, that is such that $f(x, h) = \Delta u_n$.

[30] Thus, we do not find here the *true* analogue of the definite integral, namely the modern
definite sum $S_a^b f(x) = f(a) + f(a + 1) + \ldots + f(b - 1)$ [Jordan *1947*, 116; Goldstine *1977*, 99].

[31] But he also gives the notation $[x, \Delta]^{n}$ (his own?) for $x(x + \Delta x) \ldots (x + (n - 1)\Delta x)$.

consists essentially in substituting the expressions obtained in the previous section for $\Sigma\,\mathrm{f}(x,h)$ in the equation $S\,\mathrm{f}(x,h) = \Sigma\,\mathrm{f}(x,h) + \mathrm{f}(x,h) - const$ (one of the most important results is the Euler-Maclaurin summation formula [*Traité*, III, 125][32]). It must be kept in mind that the "series" (or "sequences") to be summed are usually finite. Occasionally x is made infinite, so that the number of terms in the sum $S\,\mathrm{f}(x,h)$ is infinite; but infinite series occur mainly because the integration process introduces them, that is, because the expression for $\Sigma\,\mathrm{f}(x,h)$ is an infinite series. Thus, the finite sum $S\frac{1}{x} = 1 + \frac{1}{2} + \frac{1}{3} + \frac{1}{4} \ldots + \frac{1}{x}$ is obtained as the infinite series $1x + \frac{1}{2x} - \frac{B_1}{2x^2} + \frac{B_3}{4x^4} - \frac{B_5}{6x^6} + $ etc. $+ A$ (A being what is nowadays called the Euler, or Euler-Mascheroni, constant).[33] As in volume 1, convergence of series is a practical matter: convergent series are preferable because they provide approximate values.

$S\frac{1}{x}$ is an example of what Euler had called "inexplicable functions": not possessing a determinate expression or equation; in practice they corresponded to sums and products of a variable number of terms not expressible algebraically [Euler *Differentialis*, II, § 367; Ferraro *1998*, 311]. In a section called "application of summation of series to interpolation" [*Traité*, III, 151-175], Lacroix reports some of Euler's work on those sums such that the general term, or its differences of some order, tend to a constant, and on their interpolation. The last section before difference equations, a "digression on elimination in algebraic equations" [*Traité*, III, 175-183], may seem out of place, at first; but it is still an application of the calculus of differences, making ample use of "second-order powers" – it gives a short account of Bézout's elimination method, and a proof of Bézout's theorem, both of which had been announced in [Lacroix *Traité*, I, 324] but needed preliminary notions of difference calculus.[34]

As has already been mentioned, the third, and larger, part of this chapter is dedicated to difference equations [Lacroix *Traité*, III, 184-300]. In the treatises of Euler there is nothing on difference equations, which is not so surprising, as the subject was inaugurated not much prior to the publication of [Euler *Integralis*]: it was Lagrange, in [*1759b*], who started applying to difference equations (namely linear equations) methods originally intended for differential equations [Cousin *1796*, I, 272].[35] Through the rest of the 18th century, most of the work done on difference equations consisted in transferring methods and concepts of differential equations [Wallner *1908*, 1052].

This does not mean that Lacroix follows the same order as for differential equations – there is a significant difference, caused by the much greater

---

[32]With a typo, not mentioned in the errata: the coefficient of $\frac{du}{dx}$ is written $B_1[1]^{\overset{1}{}}$, that is $\frac{1}{6}$, instead of the correct $B_1[1]^{-1} = \frac{1}{12}$.

[33]In modern notation, this series is written $\log x + \frac{1}{2x} - \frac{B_2}{2x^2} - \frac{B_4}{4x^4} - \frac{B_6}{6x^6} - \ldots + \gamma$.

[34]Notice that the Introduction of [Bézout *1779*] is a short account of the direct and inverse calculus of differences.

[35]Much earlier, Moivre had determined the general term of recurrent sequences, which is equivalent to solving linear finite difference equations with constant coefficients. But apparently it was Lagrange who first made the connection, and treated them as difference equations [Laplace *1773a*, 38].

importance of linearity (or "first degree"[36]). The section entitled "on the integration of difference equations in two variables" [Lacroix *Traité*, III, 184-231] is almost entirely devoted precisely to "first-degree" difference equations. It starts with a few preliminaries, and then Lagrange's integration of $\Delta y + Py = Q$ (the historical beginning of the subject) and his later treatment of the general first-degree equation $y_{x+n} + P_x y_{x+n-1} + Q_x y_{x+n-2} \ldots + U_x y_x = V_x$, reduced to $z_{x+n} + P_x z_{x+n-1} + Q_x z_{x+n-2} \ldots + U_x z_x = 0$, and especially of the equation with constant coefficients $z_{x+n} + P z_{x+n-1} + Q z_{x+n-2} \ldots + U z_x = 0$ (the one most effectively treated by Lagrange) [Grattan-Guinness *1990*, I, 172-175]. Special attention is then given to Laplace's research on equations with variable coefficients [*1773a*], as it had been him who had gone farther in that direction [Lacroix *Traité*, III, 195]. Equations where the increment of the independent variable is not constant are reduced to equations where it is constant, again using a procedure by Laplace. The main situation in which nonconstant increments of the independent variable occur is also one of the most important analytical applications of difference equations: the determination of the arbitrary functions in integrals of partial differential equations; naturally, Lacroix reports Monge's work on this. Systems of first-degree difference equations are also treated using procedures analogous to those for differential equations (including d'Alembert's method [*Traité*, III, 227-229]). The section ends with a short account of a method by Paoli, using a sort of integrating factor.

The next two sections (quite short) address special topics where the analogies with differential equations are weaker or less straightforward. One is "on the nature of the arbitraries introduced by the integration of difference equations, and on the construction of those quantities" [*Traité*, III, 231-237]: Euler had remarked that difference equations are not "completed" by arbitrary constants, but rather by arbitrary periodic functions $\varphi(\sin \frac{\pi x}{h}, \cos \frac{\pi x}{h})$, in the case of constant $\Delta x = h$ (and rather more complicated expressions in the case of nonconstant Δx); the determination of these functions requires data about an interval of length Δx; likewise, the construction of a difference equation uses not just an arbitrary first point, but rather an arbitrary first curve (whose projection onto the x axis has length Δx). The other section is "on the multiplicity of integrals of which difference equations are capable" [Lacroix *Traité*, III, 237-247]: Jacques Charles had discovered the existence of new complete integrals of difference equations whose formation was analogous to that of singular integrals of differential equations; but he had taken the analogies too far and had fallen into paradoxes; Lacroix's protégé Jean-Baptiste Biot clarified them, and Lacroix reported his work (before its publication in full) – see section 7.2.

The section "on integration of difference equations in three or more variables" [*Traité*, III, 247-288] addresses extensions of methods already exposed for equations in two variables. Firstly, Lacroix reports the extension of Lagrange's integration of first-degree difference equations with constant coefficients. Then,

---

[36]Naturally, Lacroix had not changed his mind about the use of the word "linear".

the extension of Laplace's method for equations with variable coefficients. Lacroix remarks that although Laplace's method is more complicated, it is not only
more general, as it "offers a real procedure of integration", while the success of Lagrange's rests on a particular substitution [*Traité*, III, 279]. The rest of the section
is dedicated to a method by Paoli which comprises Lagrange's.

Chapter 1 finally finishes with a section "on condition equations relative to
the integration of functions of differences" [*Traité*, III, 289-300]. These equations
are the work of Condorcet – for whom integrability conditions was a favorite
topic. Lacroix explains having left them to last because they are "more curious
than useful". But the connection between equations of integrability and those
for maxima and minima of integrals [Fraser *1985*, 177-180] justifies that most of
this short section is in fact on the calculus of variations applied to integrals of
differences. It is a proper ending – volume II had ended with the common calculus
of variations.

The much shorter chapter 2 – "Theory of sequences, derived from the consideration of their generating functions" [*Traité*, III, 301-355] – is yet another example
of the encyclopedic character of Lacroix's *Traité*: it consists in readdressing matter
from chapter 1, this time following an approach by Laplace [*1779*], namely using
generating functions [Goldstine *1977*, 185-209]: u is the generating function of y_x
if

$$u = y_0 + y_1 t + y_2 t^2 + \dots + y_x t^x + y_{x+1} t^{x+1} + \text{etc.}$$

The connection with differences and series comes easily: if u is the generating
function of y_x, then $u \left(\frac{1}{t} - 1 \right)^p$ is the generating function of $\Delta^p y_x$ and $u \left(\frac{1}{t} - 1 \right)^{-p}$
is the generating function of $\Sigma^p y_x$ [Lacroix *Traité*, III, 302-305]. In the preface to
the second edition, Lacroix explained that the "state of science" did not recommend
making a choice between generating functions and the calculus of differences: one
did not know which of these approaches would permit overcoming the difficulties
posed to science; that is why he exposed both, the second chapter being "for a
great part an abridgment of the first" [Lacroix *Traité*, 2nd ed, I, xlvi].

Chapter 3 [Lacroix *Traité*, III, 356-529] is an odd piece. It mixes the "theory of
series" with the integral calculus, in several ways, but often with little connection to
series or differences, making its title, "application of integral calculus to the theory
of sequences", too restrictive and not quite correct. Lacroix explained later that he
had included here "quelques méthodes pour ainsi dire *anomales*, qu'on ne pouvait
rapporter que difficilement aux procédés d'intégration déduits du renversement
de la différentiation"[37] (see appendix B, page 400, or [Delambre *1810*, 109]) –
an allusion to the large role played by definite integrals in this chapter. In the
preface to the second edition, he confirmed that the inclusion of these "anomalous
methods" would not only make a treatise on integral calculus (i.e., his second
volume) too large, as it would cause "une espèce de désordre, par le mélange

---

[37]"some *anomalous* methods, so to speak, which could only hardly be reported to the procedures of integration derived from the reversal of differentiation"

Volume III		
topics	chapter	pages
Summation of series		356-385
Interpolation of series		385-392
Investigation of values of definite integrals	Chapter 3:	392-418
Digression on infinite products for sines and cosines	Application	418-445
Continuation of the investig. of values of definite integrals	of integral	445-461
Series for evaluat. integrals that are functs. of large numbers	calculus to	461-475
Examination of the transcendent $\int \frac{e^x\,dx}{x}$	the theory	475-483
Use of definite integrals to express functions given by differential equations	of sequences	483-519
Application of the formulas $\int e^{-ux}v\,du$, $\int u^x v\,du$, etc. to integrate difference and differential equations		519-529
Analytical theory of mixed difference equations	Chapter 4:	530-534
Application of mixed difference equations to geometrical questions	Mixed difference	535-543
Partial and mixed difference equations and conclusion	equations	543-544
Subject index for the three-volume set	Subject table	545-578
Corrections and additions to vols. II and III	Corr. & addit.	579-582

Table 2.6: Volume III of Lacroix's *Traité* (continued from table 2.5)

continuel de procédés trop différens de ceux de l'intégration proprement dite"[38]
[*Traité*, 2nd ed, I, xlvi]. The best way to try to understand the structure and
contents of this chapter is to divide it into three parts, corresponding to the three
chapters into which Lacroix split it in the second edition.

The first of these parts kept the title "application of integral calculus to the
theory of sequences"; it consists of the two sections that best fit under that name.
The first of these sections [*Traité*, III, 356-385] is "on summation of series" – with
the aid of integral calculus, of course. Lacroix reports some methods by Euler,
consisting in manipulations of sums and series so as to transform them into others
known to be expansions of certain integrals. He also gives here Parseval's formula
(in its pre-Fourier sense, of course) [Grattan-Guinness *1990*, I, 204, 206], an "analogous but less general" formula by Euler, and the remainder of the Taylor series,
in both "integral" and "Lagrange" forms (not using these names, of course). The
second section [Lacroix *Traité*, III, 356-385], even more Eulerian, is "on interpolation of series" – using definite integrals that represent those series; we find here for
example the integral $\int dx(1\frac{1}{x})^p$ (to be taken between 0 and 1) for the "second-order
power" $[p]^{p}$, which provides the Euler Gamma function. We also find here Euler's
interpolation of differentials, often misattributed to Lacroix (see section 10.1.2).

The second part of this chapter [*Traité*, III, 392-483] corresponds to the
chapter "investigation on the values of definite integrals" of the second edition.

---

[38]"a kind of disorder, by the continued mixture of procedures too different from those of
integration in the strict sense"

Its first section has that same title, and the third is a "continuation". These two sections give an abridged account of a favorite subject of Euler: the evaluation of certain definite integrals of functions whose indefinite integrals cannot be obtained in finite form. The last example studied is Euler's gamma function (without this name) [*Traité*, III, 453-460]. The intermediate section is a "digression on the expressions of sines and cosines as indefinite products"; it deals with various applications of the expressions for the functions sine and cosine as *infinite* products. It is still Eulerian but, interestingly, Lacroix substitutes some of l'Huilier's limit considerations [*1795*] for Euler's uses of infinity. The fourth section is "on series appropriate to evaluate integrals that are functions of large numbers": this is a method by Laplace for approximating functions given by definite integrals where some terms are raised to very high powers, making exact calculations impracticable [Gillispie *1997*, 81, 89-91]. The final section in this part is an "examination of the transcendent $\int \frac{e^x}{dx}$". This examination is done through several determinations of limits of integration (the allocation of a separate section for this may be due to the fact that it reports work by Mascheroni rather than Euler).

The third part of the chapter [*Traité*, III, 483-529] corresponds to the chapter "on definite integrals applied to solving differential and difference equations" of the second edition. It contains two sections. The first is on the "use of definite integrals to express functions given by differential equations"; Lacroix reports a method by Laplace [*1779*] for finding solutions to second-order linear (and some quasi-linear) partial differential equations as definite integrals, antecedents by Euler, and some developments by Parseval. The second section (the last in this chapter) is on the "application of the formulas $\int e^{-ux} v \, du$, $\int u^x v \, du$, etc. to the integration of difference and differential equations" – once again Laplace's work [*1782*], namely the ancestors of the Laplace transform [Grattan-Guinness *1997*, 261-262].

It is interesting to remark that although so much of chapter 3 is dedicated to definite integrals, only in two articles [*Traité*, III, 446-447, 475] (both in what was called here the second part) does Lacroix use Euler's notation

$$\int \frac{x^{m-1} dx}{1+x^n} \left[\begin{array}{l} x = 0 \\ x = \inf \end{array} \right]$$

(that is, the integral taken from 0 to $+\infty$). Elsewhere, the limits of integration – and the plain fact that there are limits of integration – is only indicated in the main text.

Chapter 4, the last one, is also the shortest [*Traité*, III, 530-544]. It is "on mixed difference equations", that is, equations involving both differentials and differences: an analytical theory followed by some geometrical applications. Lacroix acknowledges that most of the chapter is taken from a memoir by Jean-Baptiste Biot that had not yet been published (see section 7.3.2) – a very similar situation to the one above on multiple integrals of difference equations.

2.6   (Partial) translations of the *Traité*

Several of Lacroix's textbooks were translated into other languages. We will see
in section 8.10 that his *Traité élémentaire de Calcul...* was translated into six
languages. But translating his large *Traité* would have been quite a different task,
given the difference in size. Moreover, not being a textbook, the public for such
a translation would be small. It is not a wonder that no complete translation is
known. Still, there were attempts, in Germany and Greece.

2.6.1   One or two German partial translations

2.6.1.1   J. P. Grüson's translation of volume I

A German translation of the first volume of Lacroix's *Traité* was published in
Berlin with remarkable rapidity: 1799-1800.

The translator was Johann Philipp Grüson (Neustadt-Magdeburg, 1768 –
Berlin, 1857). Grüson moved to Berlin in 1794 to teach mathematics, first at the
Cadet School, from 1799 at the *Bauakademie* (Architecture/Construction Academy), later at the University (1816) and at the French Gymnasium (1817). In
1798 he became a member of the Berlin Academy of Sciences. He was a prolific
mathematician, but not a very good one: Moritz Cantor said in [*1879*] that his
original writings had justly fallen into oblivion. Neither was he very honest: in
1813 he plagiarized two papers by Parseval [Grattan-Guinness *1990*, I, 208].

Apart from his original (and pseudo-original) works, Grüson published several translations from the French. Among them are a translation of [Lagrange
Fonctions] in two volumes (1798 and 1799), and that of [Lacroix *Traité*, I]. This
translation, under the title *Lehrbegriff des Differential- und Integralcalculs* was published in Berlin by F. T. Lagarde, also in two volumes [Lacroix *1799-1800*]. The
first volume (1799) goes up to chapter 2 of [Lacroix *Traité*, I], while the second
volume (1800) contains chapters 3, 4 and 5.[39] Their format is octavo – half of the
original edition's quarto.

Grüson made an explicit connection between the translations of Lagrange's
and Lacroix's books: the latter was to function as an introduction and elucidation
("Erläuterung") of the former [Lacroix *1799-1800*, I, xlviii].

It is clear that Grüson planned to publish the translation of the whole *Traité*,
or at least of the second volume also (not in the least because of the title used). I
do not know why he did not accomplish it (possibly, as I have suggested above, it
was not very successful commercially; or he may have lost courage when the third
volume appeared in 1800). He also promised a translation of Lacroix's textbook
on descriptive geometry [Lacroix *1799-1800*, II, 256-257], but I have not found
any trace of it.

---

[39]Both I [Domingues *2005*, 277] and Grattan-Guinness [*1990*, I, 140] have been tricked, by
the fact that the translation has two volumes, into thinking that it was a translation of the first
and second volumes of Lacroix's *Traité*.

The title pages of both volumes promise some additions and notes ("mit eini-
gen Zusätzen und Anmerkungen"). But in the second volume the only addition or
note that I have found is the promise mentioned in the previous paragraph. In the
first there are some, not many, notes by Grüson – always signed "G". In the table
of contents he indicates some German translations of books cited by Lacroix. An
interesting short note appears at the end of chapter 1. Lacroix finishes that chapter by explaining that he will not speak of Newton's theory of fluxions because of
its use of movement, a concept alien to analysis and geometry. Grüson disagrees:
movement without consideration of forces belongs in geometry – as in the formation of the circle, sphere, cone, Archimedes' spirals and Dinostratos' quadratrix;
but he does not proceed to explain Newton's fluxions [Lacroix 1799-1800, I, 329].

2.6.1.2   A possible partial translation by F. Funck

Both the German national bibliographical catalogue [GV, LXXXIII, 198] and a
collective online catalogue *Gemeinsamer Verbundkatalog*[40] mention an *Einleitung
in die Differential- und Integralrechnung* (i.e., *Introduction to differential and integral calculus*) by Lacroix, translated into German by Franz Funck, and published
in Berlin by Reimer in 1833. I have not seen this book, so I can only make some
conjectures, based on the information given in these catalogues.

The word *Einleitung* in the title suggests that this might be a translation of
Lacroix's *Traité élémentaire du calcul...* [Lacroix *1802a*], rather than of the large
Traité. But there are several details that do not fit well with that possibility. First
of all, both catalogues also indicate that this translation was made from the *second*
edition (of whatever the original book was), and that the same publisher Reimer
had published a translation of [Lacroix *1802a*] in 1830-1831, made from the *fourth*
edition (see section 8.10.3). In addition, the *Gemeinsamer Verbundkatalog* informs
that the book has iv+167 pages and one folding plate; this is far too small to be
a translation of [Lacroix *1802a*] (whose second edition has xii+606 pages and five
folding plates). But it fits very well with the possibility of being a translation of
the Introduction in [Lacroix *Traité*, 2nd ed, I] – which has 138 pages, and three
figures in the first folding plate.[41]

Franz Funck (1803-1886) had studied at the University of Bonn from 1821
to 1823, and was a teacher of mathematics in the towns of Recklinghausen and
Kulm [Schubring *2005*, 518].

2.6.2   The Greek partial and unpublished translation

Volume I and part of volume II of Lacroix's *Traité* were translated by Ioannis
Carandinos, "l'initiateur des mathématiques modernes en Grèce"[42], who coined

---

[40]<http://gso.gbv.de> (accessed on 22 January 2007).
[41]Chapters 1, 2 and 3 have no figures, which excludes the possibility of this being a translation
of chapter 1, or chapters 1 and 2, for instance.
[42]"the initiator of modern mathematics in Greece"

the Greek words in use for such concepts as function and series [Phili *1996*, 305] – this section is based on this paper.

Ioannis Carandinos (Ἰωάννης Καραντινός)[43] was born in the Ionian island of Cephalonia in 1784. From 1807 to 1814 the Ionian islands were occupied by the French, who instituted in the chief island of Corfu an *Ionian Academy*. Teaching at this academy was Charles Dupin (1784-1873), a graduate of the *École Polytechnique* and admirer of Monge. Carandinos had started his studies of mathematics in Corfu before the French period, but under Dupin he acquired contemporary mathematics. In the 1810's Carandinos taught at a public school in Corfu, following Lacroix, Laplace, and other French authors. In 1815 the British replaced the French as occupiers of the islands. The new governor, Lord Guilford, instituted a new Ionian Academy, and he appointed Carandinos as rector and professor of mathematics. The academy started functioning in 1823; but before that Guilford sponsored periods of study abroad for the future professors. In spite of being British, the place where he sent Carandinos was Paris. In 1820 Carandinos was at the *École Polytechnique*. Returning to Corfu he taught higher mathematics at the Academy from 1824 to 1832. In 1833 he suffered some mental problem, and was sent to a psychiatric hospital in Naples, where he died in 1834.

In the 1820's Carandinos published a few original works (namely, on the "nature" of differential calculus, on combinations, on polygonometry, and on equations of degree higher than 4), and translations of textbooks: Bourdon's arithmetic, Legendre's geometry and trigonometry, and John Leslie's geometrical analysis. Phili [*1996*, 314-316] has noted Carandinos general preference for Lacroix's textbooks, but also his dislike of Lacroix's *Essais sur l'enseignement...* [*1805*], and his choice of the authors above for several reasons.

Still, starting in 1824 he translated several of Lacroix's textbooks, as well as the first volume of the *Traité*, and started translating the second volume [Phili *1996*, 318]. Unfortunately, this remained unpublished, along with his translations of [Lagrange *Fonctions*], Poisson's mechanics, and others. The manuscripts appear to have been destroyed during the German bombardment of Corfu in World War II.

---

[43] Phili [*1996*, 305] also gives the alternative spelling Καρανδίνος. The online library *Hellinomnimon* <http://www.lib.uoa.gr/hellinomnimon/main.htm> (accessed on 23 January 2007) uses Καραντηνός. The title pages of his books available there seem to alternate between Καρανδίνος, Καραντινός, and Καρανδηνός.

Chapter 3

The principles of the calculus

3.1 The principles of the calculus in the late 18th century

In the late 18th century there were various competing foundational approaches for the differential calculus. In this section I will try to present them, drawing mainly upon works that were published (not necessarily for the first time) while Lacroix was preparing the first edition of his *Traité*, or that were then still widely used.

As for the integral calculus, it will not be mentioned here, since there were no fundamental differences in opinion about it – integration was generally viewed simply as the opposite operation of differentiation (or derivation) and no discussions arose about this. The few relevant issues on the conception of the integral will be discussed in chapter 5.

3.1.1 Infinitesimals

The approach that was most widely followed, at least at the educational level, was still that of the Leibnizian infinitesimals.[1] It was well represented by Bézout's hugely successful *Cours de Mathématiques* [*1796*], on the section covering the calculus (opening the fourth volume). Bézout's *Cours* was a multi-volume textbook (4 to 6 volumes, depending on the edition), which had multiple editions[2] in the second half of the 18th century and even in the 19th. The section on the calculus was translated into English in the United States as late as 1824 [Bézout *1824*].

---

[1]"Leibnizian" here does not refer necessarily to adherence to Leibniz's personal views, but rather to the "Leibnizian tradition", which had other authors, among whom Jacob (I) and Johann (I) Bernoulli. Leibniz's personal views on infinitesimals are a quite complicated subject [Bos *1974*, 52-66].

[2]With variants: there was one version to be used by the *Gardes du Pavillon et de la Marine*, another by the Artillery, and there were separate editions and translations of some volumes or sections.

The main tool for Bézout is the consideration of *infinitely great* or *infinitely small* quantities:

> "Nous disons qu'une quantité est infinie ou infiniment petite à l'égard d'une autre, lorsqu'il n'est pas possible d'assigner aucune quantité assez grande ou assez petite pour exprimer le rapport de ces deux-là, c'est-à-dire, le nombre de fois que l'une contient l'autre."[3] [Bézout *1796*, IV, 3]

Of course, if x is infinitely great with regard to a, then $\frac{x^2}{a}$ is infinitely great with regard to x, since $a : x :: x : \frac{x^2}{a}$, and $\frac{a^2}{x}$ is infinitely small with regard to a, since $x : a :: a : \frac{a^2}{x}$. This entails the consideration of infinitely great or infinitely small quantities of *different orders*. In order to express these relations it is necessary to neglect, in algebraic expressions, the infinite quantities of the inferior orders, that is, if a is infinitely small with regard to x, then x should be taken for $x + a$. Bézout tries to convince the reader that this neglect is in fact necessary to reflect the supposition of infinitely smallness, but he does not seem to have any doubts about the validity of the supposition itself.

Bézout then considers "a variable quantity as increasing by infinitely small degrees", and, wishing to know its increments, he simply calculates its values for any one instant and the "instant immediately following"; their difference is the increment or decrement of the quantity and it is called its *differential* [Bézout *1796*, IV, 11-12; *1824*, 13]. For example, the differential of xy, $d(xy)$, is $x\,dy + y\,dx$, because the difference between two successive states of xy is $(x + dx)(y + dy) - xy = x\,dy + y\,dx + dy\,dx$, and $dy\,dx$ is infinitely small with regard to both $x\,dy$ and $y\,dx$.

When applying the calculus to calculate tangents, Bézout conceives a "curve to be a polygon of an infinite number of infinitely small sides". A tangent is a prolongation (to finite size) of one of these sides [Bézout *1796*, IV, 34; *1824*, 28].

The differential of a variable, being itself a variable, can be differentiated: the differential of dx is ddx, that of ddx is $dddx$, or d^3x, and so on; ddx is infinitely small with regard to dx, so that ddx, dx^2 (which means $(dx)^2$), and $dxdx$ are all infinitely small of the second order [Bézout *1796*, IV, 20-21; *1824*, 18-19]. When several variables are involved, it is customary to suppose that one of the first differentials — say, dx — is constant, so that $ddx = d^3x = \ldots = 0$. This is possible because "on peut toujours prendre une des différences premieres, pour terme fixe de comparaison des autres différences premieres"[4] [Bézout *1796*, IV, 22]. What this means is that one can assume that the *successive* values of one of the variables are equally spaced, or in other words, that that variable varies *uniformly*; this can be done because *a priori* the progression of any variable (the spacing between its *successive* values) is arbitrary.

---

[3]"We say that a quantity is infinitely great or infinitely small with regard to another, when it is not possible to assign any quantity sufficiently large or sufficiently small to express the ratio of the two, that is, the number of times that one contains the other" [Bézout *1824*, 8].

[4]"we may always take one of the first differentials as a fixed term of comparison for the other first differentials"[Bézout *1824*, 20]

Of course this entails a fundamental indeterminacy, since different results occur according to the choice made about the progression of the variables. Bézout [*1796*, IV, 22-23; *1824*, 20] gives an example: the differential of $\frac{dx}{dy}$ is $-\frac{dx\,ddy}{dy^2}$ if dx is taken as constant; but it is $\frac{ddx}{dy}$ if dy is taken as constant. There is a more serious aspect of this indeterminacy that Bézout does not mention: when faced with an expression like $\frac{ddx}{dy}$, in order to know its meaning, one needs to know whether it is dy that is taken as constant, or some other differential (certainly not dx, because ddx occurs in the formula; but it could well be $ds = \sqrt{dx^2 + dy^2}$, a common case when studying curves; or it could be that no differential is taken as constant). Of course usually one will know by the context which choice has been made about the progression of the variables.

Bézout's version of the differential calculus is essentially the same that had been published in the first textbook on this subject: [l'Hôpital *1696*].

A variant on this approach is presented in [Euler *Differentialis*]. For Euler, those quantities usually called infinitely small were in fact equal to zero; however, this did not mean that one could not reckon with them, since what really mattered in the calculus was not the values of differentials, but rather those of their ratios. For example, if $dy = 2dx$, although both dy and dx are null, $dy : dx = 2 : 1$. From the fact that they are zeros comes the neglect of infinitesimals of higher orders: the ratio of $dx + dx^2$ to dx is $\frac{dx+dx^2}{dx} = 1 + dx = 1$ [Euler *Differentialis*, I, § 88], therefore dx may be taken for $dx + dx^2$. In fact Euler only used these arguments involving zeros in order to justify the validity of the rules for reckoning with infinitely great and infinitely small quantities. His differential calculus is presented as a particular case of the method of (usually finite) differences, the case in which these are infinitely small.[5]

The most important aspect of his discussion is his assumption of the prominent role of ratios of differentials, as opposed to differentials themselves. There is a subtle distinction to be made here between ratios of differentials and quotients of differentials. In spite of the $\frac{dx+dx^2}{dx}$ example above, Euler's ratios are usually not the result of division between differentials; his point is that there is always a finite P such that $dy : dx = P : 1$ [Euler *Differentialis*, I, § 120]; and this P is usually introduced as the finite quantity such that $dy = Pdx$.[6]

These differential ratios were especially useful for dealing with higher-order differentiation; or perhaps we should say for dispensing with higher-order diffe-

---

[5]In the preface to [Euler *Differentialis*], Euler referred also to limits to explain the differential calculus: the ratio of $2xdx + dx^2$ to dx is exactly $2x + dx$, but the smaller dx becomes the more this ratio approaches $2x$, and when dx finally vanishes the ratio effectively arrives at the value $2x$. However, not only is this very vague and a very naïve version of limits, but also Euler does not use limits at all in the development of the calculus, so that his adherence to them seems to be entirely rhetorical.

[6]Euler did not use any particular name for the differential ratios. In [Bos *1974*] they are called *differential coefficients* (opposed to *differential quotients*). But it seems that it was Lacroix who introduced the expression *differential coefficients* (see page 73 below). Therefore, here I will use the expression *differential ratios* when referring to Euler.

rentials. Euler faced the fact that the meaning of a formula involving higher-order differentials depends on the underlying choice made about the progression of the variables, and concluded that because of this, higher-order differentials were undesirable in analysis. He did not exclude them completely — and in fact their consideration was indispensable for some problems, such as changing the independent variable (see page 77 below) — but he gave a method for removing them and tried to avoid them as much as possible. This method used the differential ratios: if p is a finite quantity such that $dy = p\,dx$, then it can be differentiated giving something as $dp = q\,dx$, where q is once again finite and can be differentiated giving something as $dq = r\,dx$, and so on; if x is taken as the independent variable, so that $ddx = 0$, then $ddy = dp\,dx = q\,dx^2$, $d^3y = dq\,dx^2 = r\,dx^3$, and so on... [Euler *Differentialis*, § 126-133, 264]. In this way the differential calculus can be seen as being not so much about infinitesimal differentials as about the finite quantities $p, q, r, ...$, which are functions of x. This was a major step in the evolution of the calculus towards a subject about *functions*, rather than *variable quantities*, and a first step in setting as its main concept what would later be known as the derivative [Bos *1974*].

Lacroix was quite aware of this, as is clear from the preface to [Lacroix *Traité*] where he claims that it was Euler "qui le premier sépara ce Calcul de son application aux courbes, et qui, en exprimant par des lettres les rapports des différentielles, avoit délivré des quantités infiniment petites, les équations que en contenoient"[7] [Lacroix *Traité*, I, xxiii].

Because of what was explained above, it is natural to identify *independent variable* and *variable with constant differential*. This identification helps modern readers in making sense of many calculations in Leibnizian calculus, and it is quite straightforward in one-variable calculus. It is trickier in multivariate calculus, and in that situation it was actually rejected by Euler [*Differentialis*, I, § 246]; but it was adopted by Lagrange [*1759a*, 4-5] and later mathematicians [Domingues *2004b*].

Euler's version of the infinitesimal approach (reckoning with zeros) was not often followed by other authors, but one of those that did follow him was Charles Bossut (1730-1814), in a treatise published almost at the same time as Lacroix's [Bossut *1798*]. Like Euler, Bossut starts by expounding the calculus of finite differences, supposing later that those differences become infinitely small, and then "peuvent être regardées ou traitées comme de véritables zéros, qui ont entr'eux des rapports déterminables par l'état d'une question"[8] [*1798*, I, 94]. However, the insistence on the finite quantities $p, q, r, ...$ as the true object of the calculus is entirely absent, perhaps due to Bossut's less theoretical exposition, based essentially on examples.

---

[7]"the first who separated this calculus from its application to curves, and who, using letters to denote the ratios of differentials, delivered the equations containing them from infinitely small quantities"

[8]"can be viewed or treated as true zeros, which have between them ratios determinable by the state of a question".

3.1.2 Limits

For most of the 18th century the most serious competitor to infinitesimals was the method of limits. These had been propounded in 1754 by d'Alembert as the basis for the true metaphysics of the differential calculus, in the article "Différentiel" of the [*Encyclopédie*]. D'Alembert retraced this metaphysics to Newton, "quoiqu'il se soit contenté de la faire entre-voir"[9], referring to the theory of "ultimate ratios" of "vanishing quantities" in *Quadratura curvarum* and *Principia Mathematica* [Boyer *1939*, 195-201]. D'Alembert may have given a larger glimpse than Newton of this metaphysics, but still only a glimpse: he proved the uniqueness of the limit and gave an example of how limits could be used to calculate the tangent to a parabola, but gave only an intuitive argument for the limit of $\frac{a}{2y+z}$ being $\frac{a}{2y}$, and was satisfied to conclude, from that single example, that the differential calculus (with infinitesimals) reached the same results as the method of limits.

D'Alembert's suggestion was taken up by a few mathematicians, among whom was Cousin, in both [*1777*] and [*1796*] – the sections on the metaphysics of the calculus are essentially the same.

The first chapter in [Cousin *1777*][10] is, just like that of [Euler *Differentialis*], dedicated to the calculus of differences "in general". The second is then devoted to the method of limits. It starts by a definition of limit that is essentially the same that the Abbé de la Chapelle had given in the article "Limite" of the [*Encyclopédie*]:

> "On dit d'une grandeur qu'elle a pour *limite* une autre grandeur, quand on conçoit qu'elle peut en approcher jusqu'à n'en différer que d'une quantité aussi petite qu'on voudra, sans pouvoir jamais coïncider avec elle."[11] [Cousin *1777*, 17; *1796*, I, 84]

Cousin concludes very quickly that the limit of a given magnitude is unique and that if two magnitudes have a constant ratio, then their limits have the same ratio. In spite of these being "the two propositions on which the whole method of limits is founded", for the first only a slim argumentation is given and for the second not even that: it is plainly evident. He proceeds to give geometrical examples, in which the handling of limits is extremely naïve: to calculate the limit of a given formula, he simply replaces magnitudes occurring in that formula with their limits. A cone with base $ABDE$ is simply stated, without any argumentation, to be the limit of pyramids with the same vertex and having as bases polygons inscribed in $ABDE$ [Cousin *1777*, 19; *1796*, I, 85].

Much of the chapter on limits is heavily based on geometrical considerations. Moving towards the "transcendental geometry of the Moderns", Cousin proposes to

---

[9]"although he was satisfied to give only a glimpse of it".

[10]Third in [Cousin *1796*], after two introductory chapters on analytic geometry and the method of undetermined coefficients.

[11]"It is said of a magnitude that it has another as *limit*, when it is regarded as being able to approach the latter until they differ by a quantity as little as wished, without ever being able to coincide with it.

find the subtangent of a curve, and is led to consider the limit of the ratio between the ordinate and the abscissa, $\frac{\Delta y}{\Delta x}$. He takes $\frac{dy}{dx}$ as a special symbol ("signe") to represent the limit of the ratio between the differences of the variables x and y [Cousin *1777*, 32][12]. "The terms dy, dx of the limit $\frac{dy}{dx}$" [Cousin *1777*, 73; *1796*, I, 151] are then called differentials and are used throughout the rest of the book, in spite of not having more than this vague definition (if it can be called a definition at all).

This kind of naïve consideration of limits did not usually lead to mistakes, because the examples were very simple. But in section 7.2 we will see serious mistakes being committed by a somewhat obscure member of the Academy of Sciences of Paris, Jacques Charles. Of course, his examples were much less simple – he dealt with the finite-difference equivalent of singular solutions of differential equations, and tried to take their limits.

A quite different limit-based approach, and less naïve, was that of the Swiss mathematician Simon l'Huilier (1750-1840), in [l'Huilier *1786*]. The Mathematics Section of the Academy of Berlin, of which Lagrange was the director, had proposed a competition for 1786 on the subject of establishing a "clear and precise theory of what is called Infinite in Mathematics", namely an explanation for the strange fact that so many correct theorems had been deduced from the contradictory supposition of the existence of infinite magnitudes. L'Huilier won this competition[13] and his entry, *Exposition élémentaire des principes des calculs supérieurs*, was published as [l'Huilier *1786*]. An expanded Latin translation was later published as [l'Huilier *1795*].

L'Huilier proposed to establish the "higher calculi" on the basis of the Greek method of exhaustion developing the ideas that d'Alembert had only sketched [l'Huilier *1786*, 6, 167; *1795*, ii, 7]. L'Huilier is much more careful than Cousin, and his work is thus much more rigorous. However, his views on rigour and on the method of limits are too much based on the ancient Greeks and on the method of exhaustion. L'Huilier insists on a distinction between quantities and ratios of quantities (focusing his attention mainly on the latter). Instead of a single definition of limit, he has two, for *limit of a variable quantity* and for *limit of a variable ratio*, which in fact turn into four, since each is split into two cases: *limit in greatness* and *limit in smallness*.[14] To give an example:

> "Soit un rapport variable toujours plus petit qu'un rapport donné, mais qui puisse être rendu plus grand qu'aucun rapport assigné plus petit que ce dernier: le *rapport* donné est appelé la *limite en grandeur* du rapport variable."[15] [l'Huilier *1786*, 7]

---

[12]In [*1796*] Cousin uses $\frac{p}{q}$ in the chapter on the method of limits, and changes to $\frac{dy}{dx}$ later on, when explicitly addressing the differential calculus.

[13]Although the judges spoke in their report of his text not as the best, but as the least unsatisfactory of the entries to the prize [Acad. Berlin *1786*].

[14]L'Huilier took these definitions from a small tract by Robert Simson (*De Limitibus Quantitatum et Rationum Fragmentum*), published posthumously in [Simson *1776*].

[15]"Let a variable ratio be always smaller than a given ratio, but capable of being rendered

In the Latin versions of these definitions [*1795*, 1] it is even more obvious that L'Huilier was assuming that the approaching quantity or ratio was monotonic: apparently he viewed any limiting process as similar to those of either inscribed or circumscribed polygons. He was certainly not the only one at the time, as is suggested by the assumption of la Chapelle in the article "Limite" in the [*Encyclopédie*], that the approaching magnitude can never surpass its limit. But it was in fact l'Huilier who, apparently for the first time, remarked that the approaching ratio or variable need not be monotonic. He did so precisely in the Latin edition, where he supplied a separate definition for the limit of an alternating ratio[16], remarking that a similar definition could be given for the limit of an alternating quantity [l'Huilier *1795*, 16-18].

L'Huilier introduced, very casually, the abbreviation 'lim.' (or 'Lim.') for 'limit' [*1786*, 24], which would later be turned into the standard symbol for limit (namely after its use by Cauchy in the 1820's).

Contrary to what was common practice at the time, l'Huilier did use his definitions of limits to prove theorems about them. That is, to prove that lim. $A : X = A : B$ ($A : X$ increasing, say) he would propose an arbitrary ratio $A : Y < A : B$ and prove that it was possible to take X such that $A : X > A : B$. The problem is that these demonstrations needed to be split into several different cases and were too fastidious for any supporter of the modern mathematics.

Like Cousin, l'Huilier defined $\frac{dy}{dx}$ as the limit of $\frac{\Delta y}{\Delta x}$ but, unlike Cousin, he saw $\frac{dy}{dx}$ as a "single and non decomposable" symbol [l'Huilier *1786*, 31-32; *1795*, 36], avoiding the use of dy and dx. He did call $\frac{dy}{dx}$ a *differential ratio*, but that was probably motivated by concerns on homogeneity: the limit of a ratio could not be anything else; and a ratio could be treated as a single entity.

3.1.3 Carnot on the compensation of errors

Lazare-Nicolas-Marguerite Carnot (1753-1823), a French mathematician, engineer, and politician, was another competitor for the Berlin Academy prize of 1786. His entry, defeated, would stay forgotten in the Academy's archives; but in 1797, while Carnot was a member of the Executive Directory (then the governing body of the French Republic), it was published in a revised version as *Réflexions sur la Métaphysique du Calcul Infinitésimal* [Carnot *1797*]. The original version was published in facsimile in [Gillispie *1971*, 171-262].

Carnot adhered to the idea that the differential calculus worked by *compensation of errors*: in the traditional process of infinitesimal calculus, we start by regarding a curve as a polygonal line; here an error is being committed; afterwards, during the calculations, the neglect of infinitesimals introduces a second error that

---

greater than any assigned ratio that is smaller than the latter: the given *ratio* is called the *limit in greatness* of the variable ratio."

[16]This was prompted from the study of the ratio of two decreasing quantities AX, CY, with limits AB, CD, respectively; $AX : CY$ may be made as close as wished to $AB : CD$, but it is not necessarily always greater or always smaller [l'Huilier *1795*, 16-17].

cancels the first. This justification had been proposed by the idealist philosopher George Berkeley (1685-1753), Anglican bishop of Cloyne, Ireland, in *The Analyst* (London, 1734), a sharp critique on the logical inconsistencies of the method of fluxions or differential calculus. Around 1760 Lagrange agreed that compensation of errors was the true "metaphysics of the calculus with infinitely small [quantities]" [Lagrange *1760-61b*, 598]. But Carnot decided to *prove* that it worked.

Carnot's argumentation ran around what he called *imperfect equations*. The members of one of these were in fact not equal, but had the same limit, which means that they had to involve variables, or as Carnot said, "auxiliary quantities"; imperfect equations were operated upon by replacing quantities with other, infinitely close, quantities; once all the auxiliary quantities had disappeared, an exact equation would remain. Apparently Carnot did not truly convince his readers, judging from the fact that he had no followers. Moreover, in 1797 (and still in 1813, when Carnot's work was widely known) Lagrange reasserted his opinion that the compensation of errors explained the infinitesimal calculus, but adding that "it would perhaps be difficult to give a general demonstration of that" [Lagrange *Fonctions*, 1st ed, 3; 2nd ed, 17] – implying that Carnot had not given one.

Nevertheless, Carnot's book was quite successful, judging from the facts that it had a second and enlarged edition in 1813 that was reprinted a few times until 1921, and that it was translated into Portuguese, German, English, Italian and Russian [Youschkevitch *1971*, 149]. It was also praised by Lacroix, who had read a manuscript version (possibly the 1786 prize entry) and urged it to be published[17] [Lacroix *Traité*, I, xxi-xxii]. But what Lacroix probably liked most in Carnot's work (and possibly what made it popular) was not so much the "compensation of errors", as his discussion and comparison of the several points of view then available for the calculus.

3.1.4 Power series

Joseph-Louis Lagrange had a special interest in the principles of the calculus, and, being the most important mathematician at this time (or, at least, one of the two most important, with Laplace), he was very influential in making the issue fashionable, as it were, in the late 18th century.

As we have seen above, around 1760 Lagrange thought of compensation of errors as the true metaphysics (that is, the reason why it works) of the Leibnizian infinitesimal calculus; while the Newtonian method (that of ultimate ratios) was perfectly rigorous, but entailed long and complicated demonstrations, which was a reason to use infinitesimals instead [Lagrange *1760-61b*].

Later, Lagrange showed himself dissatisfied with these explanations. Compensation of errors did not seem capable of demonstration [*Fonctions*, 3] and, for

---

[17]Carnot's book appeared in print that same year of 1797 as [Lacroix *Traité*, I] and [Lagrange *Fonctions*].

the method of limits, it was not clear enough what happened to $\frac{a}{b}$ when both a and b became null [*Fonctions*, 3-4].

In 1772 Lagrange published in the *Nouveaux Mémoires de l'Académie de Berlin* a memoir that would be central to this story. Its title was "Sur une nouvelle espèce de calcul relatif à la différentiation et à l'intégration des quantités variables". Its subject was not the principles, or metaphysics, of the calculus, but rather results taken from analogies between power-raising and differentiation (and between root-extracting and integration). However, Lagrange thought best to start by establishing "quelques notions générales et préliminaires sur la nature des fonctions d'une ou de plusiers variables, lesquelles pourraient servir d'introduction à une théorie générale des fonctions"[18] [Lagrange *1772a*, 442].

This was the first appearance of his power-series version of the differential calculus. Lagrange *knew* from the theory of series that if u is a function of x and we substitute $x + \xi$ for x, it will become

$$u + p\xi + p'\xi^2 + p''\xi^3 + p'''\xi^4 + \dots \tag{3.1}$$

"où $p, p', p'', \dots$ seront de nouvelles fonctions de x, dérivées d'une certaine manière de la fonction u"[19] [Lagrange *1772a*, §1]. He then characterized the differential calculus as concerned with finding the functions $p, p', p'', \dots$ derived from u. He saw this as the clearest and simplest conception of the calculus ever given, being "indépendante de toute métaphysique et de toute théorie des quantités infiniment petites ou évanouissantes"[20] [Lagrange *1772a*, §3].

Lagrange then proceeded to simultaneously explain how come this was a definition of the calculus and arrive at Taylor's formula: substituting $x + \xi + \omega$ for x in the function u and expanding the result in two different ways – namely substituting $x + \omega$ for x and substituting $\xi + \omega$ for ξ in (3.1) – and equating the resulting power-series, comes

$$p' = \frac{\varpi}{2}, \quad p'' = \frac{\varpi'}{3}, \quad p''' = \frac{\varpi''}{4}, \dots;$$

$\varpi, \varpi', \varpi'', \dots$ had appeared in the expansions: ϖ was derived from p, ϖ' from p', ϖ'' from p'', and so on, in the same manner that p was derived from u. This prompted a change in notation that would be remarkably enduring: u' instead of p, the accent signifying this one-step derivation (and u'' signifying $(u')'$), so that the p' of (3.1) became $\frac{u''}{2}$, p'' became $\frac{u'''}{2\cdot 3}$, and so on, giving

$$u + u'\xi + \frac{u''\xi^2}{2} + \frac{u'''\xi^3}{2\cdot 3} + \frac{u^{IV}\xi^4}{2\cdot 3\cdot 4} + \dots \tag{3.2}$$

---

[18]"some general preliminary notions on the nature of functions of one or more variables, which might serve as an introduction to a general theory of functions".

[19]"where $p, p', p'', \dots$ will be new functions of x, derived in a certain way from the function u"

[20]"independent of all metaphysics and of any theory of infinitely small or vanishing quantities".

for the result of substituting $x + \xi$ for x in the function u.[21] Now, taking ξ to be infinitesimal and neglecting its powers $\xi^2, \xi^3, \ldots$, (3.2) gives only $u'\xi$ for the increment of u; using the traditional notations of du, dx, we get

$$du = u'\,dx \quad \text{and} \quad u' = \frac{du}{dx};$$

"ainsi, pour avoir la fonction u', il n'y aura qu'à chercher la différentielle du par les règles du calcul des infiniment petits, et la diviser ensuite par la différentielle dx"[22] [Lagrange *1772a*, §6]. Notice how $u' = \frac{du}{dx}$ had to be *proved*, and how Lagrange resorts to the infinitesimal calculus, including a differential *quotient*.

At this point it is clear enough that

$$u'' = \frac{d\frac{du}{dx}}{dx} = \frac{d^2 u}{dx^2}, \quad u''' = \frac{d\frac{d^2 u}{dx^2}}{dx} = \frac{d^3 u}{dx^3}, \ldots$$

so that (3.2) becomes

$$u + \frac{du}{dx}\xi + \frac{d^2 u}{dx^2}\frac{\xi^2}{2} + \frac{d^3 u}{dx^3}\frac{\xi^3}{2 \cdot 3} + \ldots$$

Lagrange remarks that this seemed to him one of the simplest demonstrations of Taylor's theorem.

All of the above have multivariate equivalents, with the notation $u'^{,''}$ for modern $\frac{\partial u^3}{\partial x\,\partial y^2}$. This allows a *proof* of $\frac{d^2 u}{dx\,dy} = \frac{d^2 u}{dy\,dx}$ that relies heavily on the ambiguity of $u'^{,'}$.

From then onwards the memoir proceeds on its true subject, ignoring these foundational digressions and using only occasionally the notation u'.

It must be noted that the assumption that the increment of any function may be expanded into a power series, or the use of such power series in the development of the principles of differential calculus, are *not* exclusive of works following a power-series foundation. We can see that assumption and uses of it for fundamental results, for instance in [Euler *Differentialis*, I], and in [Cousin *1777; 1796*]. The distinction between a *technical* use of power series and a *foundation* of the calculus based on power series may sometimes be subtle; we will see borderline examples shortly (Condorcet) and in sections 8.2 (Fourier and Garnier) and 8.5 (Lacroix). The cases of Lagrange and Arbogast, treated below, are more clear-cut.

---

[21] *Change* of notation *within* this memoir. The accent notation had already been used by Lagrange in 1770 and possibly 1759 [Cajori *1928-1929*, II, 208]. And also, very clearly, by Euler [*Integralis*, III, §138]: "in designandis functionibus hac lege utemur, ut sit $d.f\!:\!v = dv\,f':v$, sicque porro $d.f':v = dv\,f'':v$ et $d.f'':v = dv\,f''':v$ etc." ("we will use this rule in designating functions, so that $d.f\!:\!v = dv\,f':v$, and so forth $d.f':v = dv\,f'':v$ et $d.f'':v = dv\,f''':v$ etc."). But most often Euler used p, q, etc.; and of course it was [Lagrange *Fonctions*] that made the accent notation popular.

[22] "therefore, to find the function u', it is enough to find the differential du using the rules of the infinitesimal calculus, and then divide it by the differential dx"

A few years after publishing [*1772a*], Lagrange took a major part in proposing the 1786 competition of the Berlin Academy on a "clear and precise theory of what is called Infinite in Mathematics" and in judging the entries. It has been suggested that this indicates that Lagrange was not entirely satisfied with his own suggestion of basing the calculus on power series.[23] It is possible that this interpretation is correct, but it should be taken into account that, for Lagrange, a "theory of the Infinite in Mathematics" and a sound foundation of the calculus were not the same thing: his power-series version of the principles of the calculus was, in his own words, "reduced to the algebraic analysis of *finite* quantities"[24] (my emphasis) and, as we have seen above, even after he had published it, he still thought the *infinitesimal* calculus worked because of compensation of errors. His power-series approach could not be the basis for an entry for the competition, because, as he saw it, it had nothing to do with the *infinite*.

After giving his foundational suggestion in [*1772a*], Lagrange did not develop it until 1795 (as will be seen below). Meanwhile, one or two other mathematicians took it up, in works that unfortunately have remained unpublished. In the late 1770's and early 1780's the marquis de Condorcet wrote the first two parts (five parts were planned) of a very large *Traité du Calcul intégral*; the first few sections (containing, in spite of the title, the principles of the differential calculus) were printed [Condorcet *Traité*], apparently around 1786. In 1810 Lacroix attributed to this work by Condorcet the priority in a purely analytical exposition of the principles of the differential calculus [*Traité*, 2nd ed, I, xxii-xiii]. Youschkevitch [*1976*, 76] confirms that in this treatise "Condorcet attempts to derive a Taylor series formally for an arbitrary function, almost in the way Lagrange had done". But Gilain [*1988*, 135], while acknowledging Condorcet's use of series expansions for differentiation, thinks that there was not a foundational concern involved (so that it would be an example of technical, rather than foundational, use of power series). I have not been able to study [Condorcet *Traité*] thoroughly, but from what I could gather I would say that Condorcet does give a power-series foundation for the differential calculus, even though that is far from being his main concern – after all, this was just an introduction to a treatise on the integral calculus. He expands the finite difference of a function F of x as

$$\Delta F = A\Delta x + B\Delta x^2 + C\Delta x^3 \ldots;$$

then he considers another increment for x, dx (presumably also finite) and observes that expansions for $F' = F:(x + \Delta x + dx)$ may be obtained either by substituting $x + dx$ for x, or $\Delta x + dx$ for Δx (just as Lagrange had done, with a different notation); calling $\frac{dF}{dx}$ the coefficient of dx in the expansion of $F:(x + dx)$, he concludes that $A = \frac{dF}{dx}$, $B = \frac{dA}{2dx}$, and so on; and writing $\frac{d^2F}{dx^2}$ for $\frac{d(\frac{dF}{dx})}{dx} = \frac{dA}{dx}$ (and

---

[23]For instance, in [Grabiner *1966*, 40-46] or in [Grattan-Guinness *1980*, 101].

[24]From the full title of [Lagrange *Fonctions*]: *Théorie des Fontions Analytiques, contenant les Principes du Calcul Différentiel, dégagés de toute considération d'Infiniment Petits ou d'Évanouissans, de Limites ou de Fluxions, et réduits à l'Analyse Algébrique des Quantités Finies.*

so on) he arrives at Taylor's theorem [Condorcet *Traité*, 26-28]. He then proceeds using power series to obtain such results as what we would call the equality of mixed derivatives, and the derivatives of the common functions. However, he fails (as far as I could see) to do something as simple as giving a *name* to $\frac{dF}{dx}$; or to argue for the possibility of expanding ΔF in a series of powers of Δx – which may be one of the reasons why Gilain does not recognize it a foundational character.

A different issue is whether Lacroix read the printed pages of Condorcet's treatise before the publication of the first edition of his *Traité*. Given his close association to Condorcet, this is very much possible. But the fact that he does not mention it in the first edition casts serious doubts on this possibility. Moreover I have not noticed any obvious influence from [Condorcet *Traité*] in [Lacroix *Traité*].[25]

The first person to devote a piece of work to the development of Lagrange's suggestion was L. F. A. Arbogast; he did this in a memoir entitled "Essai sur de nouveaux principes de Calcul différentiel et intégral, indépendans de la théorie des infiniment-petits et de celle des limites", submitted to the *Académie des Sciences* of Paris in 1789. This memoir was never published, although a book by Arbogast, *Du Calcul des Dérivations*, in which he expanded and generalized his thoughts on the subject, appeared in 1800.[26] In his 1789 memoir, Arbogast tried to effectively improve on [Lagrange *1772a*]: he tried, for instance, to prove that (3.1) was valid, whatever the function u, something that in [Lagrange *1772a*] was simply assumed. However, Arbogast's attempt at a proof rested on a general validity of the binomial formula and on the assumption that any function y of x could be written as

$$y = Ax^{\alpha} + Bx^{\beta} + Cx^{\gamma} + Dx^{\delta} + \&c. \tag{3.3}$$

where $\alpha, \beta, \gamma, \delta, \&c.$ are any (real) numbers, in ascending or descending order [Friedelmeyer *1993*, 78]. As is well known, Euler had taken for granted the possibility of expanding an arbitrary function y of x as

$$y = A + Bx + Cx^2 + Dx^3 + \&c.$$

and had given (3.3) as an alternative for sceptics, so to speak [Euler *Introductio*, I, § 59; Youschkevitch *1976*, 62-63].

After concluding that the difference Δy of y can be expanded into

$$\Delta y = p\Delta x + \frac{1}{1 \cdot 2} q\Delta x^2 + \frac{1}{1 \cdot 2 \cdot 3} r\Delta x^3 + \frac{1}{1 \cdot 2 \cdot 3 \cdot 4} s\Delta x^4 + \&c. \tag{3.4}$$

---

[25]For instance, Condorcet presents the calculus of finite differences as a fundamental preliminary to the differential calculus, while Lacroix postpones it to the third volume of his *Traité*; both give expansions for a^x by "purely finite" means, but while Lacroix's is based on $a^x \times a^u = a^{x+u}$, Condorcet's is based on $a^{2x} = (a^x)^2$ – in the second edition Lacroix expressly mentioned his preference for the former identity (see below page 264, footnote 19; Fourier, on the other hand, may have been influenced by Condorcet).

[26]There are two surviving manuscripts of the 1789 memoir, one kept at the *Biblioteca Medicea Laurenziana* in Florence, and the other at the *École des Ponts et Chaussées* of Paris. Accounts of the memoir can be found in [Grabiner *1966*, 47-59], [Panza *1985*] and [Friedelmeyer *1993*, 69-131]. I have used them to write this passage and another in section 4.2.1.1 on contact of curves. Later, I was able to make some improvements thanks to photocopies of the Florence manuscript, kindly supplied to me by Marco Panza.

Arbogast calls each of the terms in (3.4) — disregarding the numerical coefficients — *differentials*: $p\Delta x$ the first differential, $q\Delta x^2$ the second differential, and so on. They are given the predictable notation $(dy = p\Delta x, dy^2 = q\Delta x^2, \&c.)$ and then Δx is identified with dx (because $p\Delta x, q\Delta x^2, \&c.$ are differentials, not whole differences), so that it is immediate to conclude that

$$\frac{dy}{dx} = p; \quad \frac{d^2 y}{dx^2} = q; \quad \frac{d^3 y}{dx^3} = r; \quad \&c.$$

$p, q, r, \&c.$ being functions of x, called *differential ratios* ("rapports différentiels") [Friedelmeyer *1993*, 80-81].

An interesting aspect in Arbogast's memoir is his exposition and use of a principle which I will call in this work *Arbogast's principle*: given a series such as

$$\Delta y = \frac{dy}{dx}\Delta x + \frac{1}{1 \cdot 2}\frac{d^2 y}{dx^2}\Delta x^2 + \frac{1}{1 \cdot 2 \cdot 3}\frac{d^3 y}{dx^3}\Delta x^3 + \&c.$$

we can give Δx a value small enough for any of the terms in the series to exceed (in absolute value) the sum of all that follow [Friedelmeyer *1993*, 81]. Arbogast argued for this principle, trying to determine how small Δx had to be. But of course his arguments are flawed (the fundamental flaws amount to using the largest of the terms $\frac{1}{1 \cdot 2}\frac{d^2 y}{dx^2}, \frac{1}{1 \cdot 2 \cdot 3}\frac{d^3 y}{dx^3}, \&c$, of which there may be an infinite number) [Friedelmeyer *1993*, 81-84].

A similar principle (but with a flavour of infinitesimal-neglecting) had already been stated and used by Euler [*Differentialis*, § 122]: in a series $P\omega + Q\omega^2 + R\omega^3 + \&c$, if ω is given a value so small that the terms $Q\omega^2$, $R\omega^3$, etc. become much smaller than $P\omega$, then this first term may be taken for the whole series – this in computations that do not require "the highest rigour". Grabiner calls this "Euler's criterion" [*1981*, 117]. Euler used this to establish the necessary condition $\frac{dy}{dx} = 0$ for a local extreme, without recurring to geometrical considerations [*Differentialis*, II, § 253-254].

It is not surprising that similarly to what Euler had done (or better, according to Friedelmeyer [*1993*, 99]), Arbogast used *his* principle to study local extremes. Apparently he regarded it as one of the most important points in his memoir: in 1800 he made a summary of the unpublished memoir, listing six principles on which it rested, and this was one of them [Grabiner *1966*, 48-49, 54-55]. Arbogast's principle was later used in two developments of the calculus based on power series: those by Lagrange and by Lacroix.[27]

---

[27]It is likely, but not certain, that Lacroix had direct access to the 1789 memoir. He does not mention Arbogast at all while treating the principles of the calculus; he does allude to his memoir in passing in chapter 4 [Lacroix *Traité*, I, 370], but only referring to the similarity between the ways in which Arbogast and Lagrange treated curves - he might know this from elsewhere, namely from Lagrange; while it is very clear that he had read another unpublished memoir by Arbogast, on "arbitrary functions" [Lacroix *Traité*, II, viii, 619], and that he was in contact with Arbogast already in 1794 [*Traité*, III, 543]. It is of course possible that Lacroix read Arbogast's memoir but only in or after 1795, making that reading irrelevant for his development of the principles of the calculus, but in time for the reference in a later chapter.

Lagrange, living in Paris and attending the sessions of the *Académie des Sciences* since 1787, knew Arbogast's memoir. Apparently he was very pleased with it, and in 1797 the only fault he could find in it was that it remained unpublished [Lagrange *Fonctions*, 5].

In 1795 Lagrange was charged with teaching the calculus at the *École Polytechnique*. This was the turning point in which he found the need (and the will) to develop his suggestion of 1772 in detail. A book resulting from these lectures was published in 1797 as *Théorie des Fonctions Analytiques* [Lagrange *Fonctions*].

After some introductory paragraphs (converted into an "Introduction" in the 1813 edition) this book proceeds with a study of the series expansion of $f(x+i)$, where fx is an arbitrary function of x.[28] Lagrange starts by proving that such a series cannot include a fractional power of i, unless x is given certain particular values.[29] The argument is the following: a term of the form $ui^{\frac{m}{n}}$ will have n different values; since $f(x+i)$ and fx must have the same number of values, a series involving the terms fx and $ui^{\frac{m}{n}}$ will have more values than $f(x+i)$ and therefore cannot represent it. The conclusion must be that only integral powers of i may appear in the expansion of $f(x+i)$. No reference is made to the possibility of irrational powers of i. [Lagrange *Fonctions*, 7-8]

Now, since $f(x+0) = fx$, $f(x+i)$ must be equal to fx plus a function of x and i that is zero when $i = 0$. Because of the argument above, this new function must be an integral multiple of i. In other words,

$$f(x+i) = fx + iP \tag{3.5}$$

where P is a function of x and i. But then P is in the same situation as $f(x+i)$, so that calling p the value assumed by P when $i = 0$ and repeating the reasonings above,

$$P = p + iQ \tag{3.6}$$

where Q is a new function of x and i. This can be repeated, so that

$$Q = iq + R, \quad R = ir + S, \quad \text{etc.,} \tag{3.7}$$

and, substituting,

$$f(x+i) = fx + ip + i^2 q + i^3 r + \&c.$$

where $p, q, r, \&c.$ are certain new functions of x. [Lagrange *Fonctions*, 8-9]

The way in which the functions $f, p, q, r, \dots$ relate to each other is explained in the same manner as in [Lagrange *1772a*]: developing $f(x+i+o)$ as $f((x+o)+i)$

---

[28]Like many 18th-century authors, Lagrange only used parentheses around the argument when it involved more than one letter.

[29]Lagrange claims to be the first, as far as he knows, to try to prove this *a priori* [*Fonctions*, 7]. This claim is odd, because Arbogast, as we have seen, did try to prove it, and Lagrange was well aware of this. Unless the fact that Arbogast assumed the binomial expansion prevented his attempt from being *a priori*, to Lagrange's eyes. Be as it may, Lagrange's "proof" is quite different, and much more interesting, than Arbogast's.

and as $f(x + (i + o))$ and equating the resulting series, arriving at

$$f(x + i) = fx + f'xi + \frac{f''x}{2}i^2 + \frac{f'''x}{2 \cdot 3}i^3 + \frac{f^{IV}x}{2 \cdot 3 \cdot 4}i^4 + \&c.$$

where $f'x$ is the first derived function of fx, $f''x$ the first derived function of $f'x$, and so on. fx earns the name *primitive function*, while the *derived functions* $f'x, f''x, f'''x, ...$ are respectively its *first* ("prime"), *second* ("seconde"), *third* ("tierce"), ... functions.

Lagrange also gives a proof of Arbogast's principle, assuming several properties of the function and its power series. In this proof he uses, rather untypically, geometrical language, and considerations close to a limit approach. Given (3.5), (3.6) and (3.7) above, it is enough to prove that i can be given a value small enough that $iP < fx$, or $iQ < p$, or... Now, considering the curve expressed by iP (with i as abscissa), it must of course pass through the origin. Also, unless x assumes one of those particular values mentioned above, the curve must be *continuous* near the origin, so that it approaches the x-axis little by little ("peu à peu") before meeting it, and therefore approaches it by less than any given quantity; it is then enough to take fx as this given quantity; the same argument applies with the curve given by iQ and the quantity p, and so on.[30] Lagrange then comments that this is "one of the fundamental principles of the theory we propose to develop" and that it is tacitly assumed in the differential and fluxional calculi [*Fonctions*, 12]. This suggests that he thought of this principle as a substitute for the neglect of higher-order infinitesimals.

Lagrange did not use Arbogast's principle extensively in [*Fonctions*], at least not in a direct way. But he used it to establish that if $f'z$ is positive from $z = a$ to $z = b$, $b > a$, then $fb > fa$ [*Fonctions*, 45-46], and then used this result to derive what is now called the Lagrange form of the remainder for Taylor's series:

$$\begin{aligned}
f(z + x) &= fz + xf'(z + u) \\
&= fz + xf'z + \frac{x^2}{2}f''(z + u) \\
&= fz + xf'z + \frac{x^2}{2}f''z + \frac{x^3}{2 \cdot 3}f'''(z + u) \\
&\&c
\end{aligned} \qquad (3.8)$$

where in *each* case u is an indeterminate quantity between 0 and x [*Fonctions*, 49]. This he used often, especially in applications to geometry and mechanics; and also, naturally, in the study of maxima and minima [*Fonctions*, 151-154].

---

[30]Grabiner [*1966*, 142] argues that this proof, and particularly the "characterization of the continuity of iP" that Lagrange gives here, can be easily translated into algebra. But then, why did not Lagrange, the algebraist *par excellence*, do so? The fact is that Lagrange does not really *characterize* continuity here; he only uses a *property* of continuity. He did not have an algebraic characterization of continuity – continuity was a fundamentally geometrical property – and when he needed to appeal to continuity he had to resort to geometrical language.

The most marked difference from [Lagrange *1772a*] is the complete absence
of any rapport to differentials or to anything that might remind of them: no correspondence between $f'x$ and $\frac{dfx}{dx}$ is established, because the latter is not even
mentioned. This is an important novelty relative to all other alternative foundations in the 18th century: Lagrange is not trying to justify the differential calculus,
but rather to build afresh a calculus (he would use the expression *calculus of
functions*) that he knows, or hopes, will be equivalent to the differential calculus.

In [Lagrange *Fonctions*] we can see the culmination of a tendency for algebraic formalism that comes from Euler [Fraser *1989*]. While in Euler one can still
notice some remnants of the view that the calculus was concerned with quantities,
in Lagrange the calculus is entirely concerned with expressions (even if he is often
forced to call some of them "quantities" for lack of better words). It is clear, for
instance, that he struggles (not always successfully) to avoid calling i (in $f(x+i)$)
the increment or increase of x, so that instead of Euler's "quantitas variabilis x
accipiat augmentum $= \omega$"[31], we have "à la place de x on met $x + i$"[32].

Moreover, Euler had focused the calculus on functions (which were regarded
as expressions) and had noticed that differential ratios were much more relevant
than differentials themselves; Lagrange took this one step further, abolishing differentials and putting derivatives (derived functions) in the central place of the
calculus.

3.2   The principles of the calculus in Lacroix's *Traité*

We have seen in section 2.1 that Lacroix presented [Lagrange *1772a*] as one of the
main motivations for writing the *Traité*. From the start, it was to be a development
of Lagrange's suggestion.

3.2.1   Dating the Introduction and first two chapters of volume I

Lagrange taught the calculus using the power-series foundation at the *École Polytechnique* in 1795 and 1796, but he only published it in detail [Lagrange *Fonctions*]
in 1797,[33] the same year that Lacroix published the first volume of his *Traité*
(apparently Lacroix's book appeared a little earlier that year than Lagrange's
[Lacroix *Traité*, I, xxx]). Lacroix seems to have attended Lagrange's lectures, but
since he was working on the *Traité* at least since 1787, he probably had already
written its first chapters. This is what he had to say on this in the Preface to the
first volume:

> "L'impression de mon Livre fut commencée en frimaire an 4 (novem-
> bre 1795) et suspendue par des raisons particulières pendant quelques

---

[31]"variable quantity x receives an increase $= \omega$" [Euler *Differentialis*, I, § 112].

[32]"instead of x is put $x + i$" [Lagrange *Fonctions*, 2].

[33]Lagrange taught it again in 1799, from which originated [Lagrange *Calcul*], but that is
irrelevant here.

mois; depuis cette époque Lagrange est revenu sur ses premières idées, à l'occasion d'un Cours qu'il a fait à l'École Polytechnique. J'ai suivi ses leçons avec tout l'intérêt qu'elles devoient inspirer; mais l'etat où étoit mon ouvrage et la marche de l'impression me n'ont permis de profiter que d'un petit nombre de ses remarques que j'ai eu soin de rapporter à leur Auteur."[34] [Lacroix *Traité*, I, xxiv]

In his *Compte rendu [...] des progrès que les mathématiques ont faits depuis 1789* (see appendix B, page 399), Lacroix was even more incisive. Speaking of a passage of chapter 1 with similarities to [Lagrange *Fonctions*], he said

"mais on observera que l'article du traité dont on parle ci dessus était composé, imprimé, et entre les mains de plusieurs personnes, entr'autres du C. Prony, avant que le C. Lagrange fît à l'école polytechnique les leçons qui ont donné naissance à la théorie des fonctions"[35].

There is a problem here. Lagrange gave those lectures for the first time in year 3 [Prony *1795b*]. If the printing of Lacroix's *Traité* started in Frimaire year 4, then Lacroix's claim for priority is false. There are several possibilities:
1 - Lacroix may have just lied, trying to pass off his *Traité* as more original than it really was;
2 - he may have attended Lagrange's lectures on the calculus only in the second year of the *École Polytechnique*, and assumed that in the first year Lagrange had not really taught that subject (according to [Prony *1795b*] Lagrange's course of analysis in 1795 started with arithmetic and covered several topics before finally arriving at the calculus);
3 - Lacroix may have incorrectly remembered the date when the printing started, and correctly remembered that it was before Lagrange's course.

Be that as it may, we can add some evidence corroborating Lacroix's claim of early circulation of part of volume I. Prony indeed had access to it, and cited it in [*1795a*, IV, 548]:

"J'ai donné une règle générale fort simple pour étendre le théorème de *Taylor* à un nombre quelconque de variables; cette matière sera discutée dans l'ouvrage de *Lagrange*, et se trouve aussi exposée avec beaucoup de clarté et de détail dans le traité du calcul différentiel et intégral de Lacroix (tome 1, page 131 et suiv.)."[36]

---

[34]"The printing of my book was started in Frimaire of year 4 (November 1795) and was suspended for personal reasons for a few months; after that time Lagrange returned to his early ideas, with regard to a course that he gave at the École Polytechnique. I followed his lectures with all the attention that they should inspire; but the state in which my work was and the progress of its printing only allowed me to profit from a few of his observations, which I took care in ascribing to their author.

[35]"but it should be noted that the article of the *Traité* mentioned above was composed, printed, and in the hands of several people, among whom citizen Prony, before citizen Lagrange had given at the *École Polytechnique* the lectures that gave rise to the *Théorie des Fonctions*"

[36]"I have given a very simple general rule to extend *Taylor*'s theorem to any number of varia-

[Prony *1795a*, IV] is in the fourth *cahier* of the *Journal de l'École Polytechnique*, referring to the autumn of 1795 but published only in September-October 1796; but this passage can also be found, with precisely the same words, in the version of lecture notes[37] distributed to students in the first year (lecture n.° 30). So we can say that by the end of the first school year of the *École Polytechnique* (late summer or autumn 1795) the first volume of Lacroix's *Traité* was printed at least until page 133 (Taylor's theorem for functions of three or more variables is in pages 131-133).

Considering this, of the possibilities above number 3 seems the least unlikely.

In the following sections we will analyse Lacroix's development of the Lagrangian foundations of the calculus, and we will see internal evidence for its independence from [Lagrange *Fonctions*] (section 3.2.4). A deeper, more philosophical, divergence will be referred to in section 3.2.8. But it is possible to locate at least some of the few "remarques" of Lagrange from which Lacroix profited, as will be seen in section 3.2.5. The conclusion is that the Introduction and chapter 1 predate Lagrange's lectures (or at least Lacroix's attendance of Lagrange's lectures), and chapter 2, in its final form, is posterior.

3.2.2   Functions of one variable

Lacroix starts chapter 1 of the first volume by showing that $f(x+k)$ can be expanded in a power series of k, provided that the function $f(x)$ be rational, exponential, logarithmic or trigonometric. "By analogy", this should happen for all functions; Lacroix promises us that we will see in the following that this analogy is correct [*Traité*, I, 85].

In fact, what he concludes some pages afterwards is somewhat weaker: that we can always expand $f(x+k)$ into a series like $X_0 + X_1k + X_2k^2+$ etc., "si on sait trouver le coefficient de la première puissance de k [that is, how to find the derivative of f], quelque soit la fonction f"[38] [*Traité*, I, 92-93] – which sounds to us like "if every function were differentiable, then every function would be analytic"; but this is not what Lacroix had in mind.

Lacroix's point is that each of these functions X_1, X_2, X_3, etc. can be derived from the previous one (and X_0 from f) by the same procedure, and this procedure is that of deriving X_0 from f. He shows this by comparing $f((x+k)+k')$ with $f(x+(k+k'))$, just like Lagrange had done in [*1772a*] (and as he did in [*Fonctions*]). A power series for the former is obtained from

$$f(x+k) = f(x) + X_1k + X_2k^2 + \text{etc.}$$

---

bles; this topic will be discussed in *Lagrange*'s work, and is also exposed with plenty of clarity and detail in the traité du calcul différentiel et intégral by *Lacroix (vol. 1, pages 131 and following)*."

[37] *Leçons d'Analyse données à l'École Centrale des Travaux Publics, par R. Prony. Première Partie – Introduction à la Mécanique. Première Section – Méthode directe et inverse des différences.* [Éc. Pol. Arch].

[38] "if we know how to find the coefficient of the first power of k [that is, how to find the derivative of f], whatever the function f"

expanding each term in the right side, so that the first becomes

$$\mathrm{f}\left(x\right) + X_1 k' + X_2 k'^2 + \text{etc.,}$$

the second becomes

$$\left(X_1 + X_1' k' + X_1'' k'^2 + \text{etc.}\right) k$$

(where X_1', X_1'', etc. are the functions derived from X_1 as X_1, X_2, X_3, etc. are derived from $\mathrm{f}\left(x\right)$) , the third becomes

$$\left(X_2 + X_2' k' + X_2'' k'^2 + \text{etc.}\right) k^2,$$

and so on. Now, of course

$$\mathrm{f}\left(x + (k + k')\right) = \mathrm{f}\left(x\right) + X_1 \left(k + k'\right) + X_2 \left(k + k'\right)^2 + \text{etc.}$$

Expanding each power of $(k + k')$ and comparing these two power series, Lacroix concludes that

$$X_2 = \frac{X_1'}{2}, X_3 = \frac{X_2'}{3}, X_4 = \frac{X_3'}{4}, \text{etc.}$$

He then adopts the notation $\mathrm{f}'\left(x\right)$ for the coefficient of k in $\mathrm{f}\left(x + k\right)$ (that is X_1); $\mathrm{f}''\left(x\right)$ for the coefficient of k in $\mathrm{f}'\left(x + k\right)$ (that is X_1'); $\mathrm{f}'''\left(x\right)$ for the coefficient of k in $\mathrm{f}''\left(x + k\right)$, etc., obtaining

$$\mathrm{f}\left(x + k\right) = \mathrm{f}\left(x\right) + \frac{\mathrm{f}'\left(x\right)}{1} k + \frac{\mathrm{f}''\left(x\right)}{1 \cdot 2} k^2 + \frac{\mathrm{f}'''\left(x\right)}{1 \cdot 2 \cdot 3} k^3 + \text{etc.}$$

Thus the development into power series is reduced to this recursive process of "*dérivation*": knowing how to go from $\mathrm{f}(x)$ to $\mathrm{f}'(x)$ (whatever f), is enough to get all the coefficients.

This also gives us an idea of the calculus that is "claire et indépendante des notions vagues et paradoxales de l'infini"[39]: the object of the differential calculus is precisely this process of "descendre de la fonction génératrice aux fonctions dérivées"[40] and that of the integral calculus is the inverse process of "remonter de l'une quelconque des fonctions dérivées, à la fonction génératrice"[41] [Lacroix *Traité*, I, 94].

The first term $\mathrm{f}'(x)k$ of the difference $\mathrm{f}(x + k) - \mathrm{f}(x)$ is christened *differential* because it is only "une portion de la différence"[42] and is given the symbol $d\mathrm{f}(x)$. This carries the introduction of the concept of "differentiation": the search for the differentials of quantities [Lacroix *Traité*, 94-96].

Now, for the full introduction of the Leibnizian notation, dx is also required:

---

[39]"clear and independent of the vague and paradoxical notions of infinity"
[40]"descending from the generating function to the derived functions"
[41]"reascending from any one of the derived functions to the generating function"
[42]"a portion of the difference"

"Pour mettre de l'uniformité dans les signes et faire de l'expression $\frac{df(x)}{k}$ un type général qui puisse s'employer quelle que soit la lettre par laquelle on représent la variable d'où dépend la fonction proposée, on écrira dx au lieu de k"[43] [Lacroix *Traité*, I, 95].

Then,

$$f'(x) = \frac{df(x)}{dx}$$

is an immediate *conclusion*.

This means that to obtain the differential $df(x)$ one expands $f(x + dx) - dx$ into a power series and then takes the first term. But is this definition any better than the one by Cousin cited in page 58 above? What kind of object is dx? Trying to explain this, Lacroix uses the expression "hypothetical increment" once, and soon after he elaborates:

"dx n'est, à proprement parler, qu'un signe destiné à retracer la marche qu'on a suivie pour arriver à l'expression de $f'(x)$, et à rappeler qu'on n'a considéré que le premier terme du développement de la différence indiquée; car d'ailleurs on fait toujours abstraction de la valeur de l'accroissement qu'il représente."[44] [Lacroix *Traité*, I, 95-96]

There are some inconsistencies here: dx is just a sign, subordinate to $f'(x) = \frac{df(x)}{dx}$, but it also represents an increment (although a "hypothetical" one, the value of which is never taken into account). I think that Lacroix is struggling here with a lack of appropriate language (or of more sophisticated mathematical concepts). Unlike Lagrange, he wishes to keep differentials, but like Lagrange, he rejects infinitesimals (at least in this section), and wishes to develop a calculus based on functions, not variable quantities. What could then dx be? A later mathematician could tell him that dx could be the identity function $dx : k \mapsto k$ and $df(x)$ the linear function $df(x) : k \mapsto f'(x)k$, so that in fact $df(x)(k) = f'(x) \cdot dx(k)$. But you would really need a *later* mathematician for this.

A slightly later mathematician, Cauchy [*1823*, 13], moved a little in that direction, identifying dx with the differential of the identity function $x \mapsto x$ (by a certain confusion between a function and its value). But Cauchy did not yet have an appropriate language to deal with a functional concept of differential (as opposed to the variable-oriented, Leibnizian one): he defined $df(x)$ as the limit, when α tends to zero, of

$$\frac{f(x + \alpha h) - f(x)}{\alpha} = \frac{f(x + i) - f(x)}{i}h,$$

---

[43]"To introduce uniformity in the symbols and to turn the expression $\frac{df(x)}{k}$ into a general form that may be employed whatever the letter that represents the variable on which the proposed function depends, dx will be written instead of k"

[44]"dx is only, properly speaking, a sign intended to retrace the course followed to arrive at the expression of $f'(x)$, and to recall that only the first term of the development of the indicated difference was considered; besides, abstraction is always made of the value of the increment that it represents."

where h is a *constant* finite quantity and $i = \alpha h$, and therefore his differential of $f(x)$ always involved this constant h (which turned out to be equal to dx). Presumably because h was a constant he did not explicitly draw the conclusion that $df(x)$ was a function of h (or of dx) — $df(x)$ was apparently a function only of x.[45]

As has been seen above, f$'(x)$, f$''(x)$, etc. are sometimes called "derived functions" (an expression taken from [Lagrange *1772a*]), because of the derivation process, but the name that they gain in page 98 (and which will be used throughout the three volumes) is *differential coefficients* ("coefficiens différentiels"). In fact, "derivation" is not a common word at all in [Lacroix *Traité*], but "differentiation" is. After all, this is a treatise on *differential* and integral calculus. It should be noted that this is the first occurrence in print of the name "differential coefficient", which would become very popular in the 19th century, being "adopted in all languages" [Anonymous *1900*; Cajori *1919*, 272]. Lacroix had already used it in 1785, but in a memoir that remained unpublished, and only for partial derivatives (that is, the coefficients in a differential like $dz = p\frac{dz}{dx} + q\frac{dz}{dy}$) (see footnote 1 in page 355 below). This name was probably "on the air": Bossut used it in [*1798*, II, 351], again only for partial derivatives.

In Lacroix's *Traité* the differential notation

$$\frac{du}{dx}, \frac{d^2u}{dx^2}, \frac{d^3u}{dx^3}, \dots$$

and even the Eulerian

$$p, q, r, \dots$$

will also be much more frequent than

$$u', u'', u''', \dots.$$

Often, particularly in differential equations, the differentials $dx, dy, d^2x, d^2y, \dots$ will occur without explicit reference to differential coefficients. Overall this foundation for the calculus is Lagrangian, but much closer to [Lagrange *1772a*] than to [Lagrange *Fonctions*], where differentials have no place.

The results obtained in the Introduction allow easy deductions of the differentials of one-variable algebraic, logarithmic, exponential and trigonometric functions: as has already been noted, it is only necessary to expand f$(x+dx)$ and extract the term with the first power of dx.

3.2.3 Functions of two or more variables

Differentiation of functions of two variables is also inspired by [Lagrange *1772a*], but without resorting to the cumbersome notation employed by Lagrange there

---

[45]This was useful later in establishing higher-order differentials: the differential of $dy = y'h$ was of course $dy' \cdot h = y''h^2 = y'''dx^2$, since h was a constant [Cauchy *1823*, 45]. The alternation between dx constant/variable according to x as independent/dependent variable followed [Cauchy *1823*, 48].

(u',$''$ for modern $\frac{\partial u^3}{\partial x \partial y^2}$). f$(x+h, y+k)$ is expanded in two steps and in two ways (via f$(x + h, y)$ and via f$(x, y + k)$), whence the conclusion is drawn that $\frac{d^2 u}{dx dy} = \frac{d^2 u}{dy dx}$. It is worth mentioning that the notation $\frac{d^2 u}{dy dx}$ is introduced as an abbreviation for $\frac{d\left(\frac{du}{dx}\right)}{dy}$, so that this is to be understood as a differential coefficient, not as a quotient.

The definition of differential as the first-order term in the expanded series of the incremented function is extended to the first order *terms* of $u = $ f(x, y) giving

$$df(x, y) = du = \frac{du}{dx}\, dx + \frac{du}{dy}\, dy. \tag{3.9}$$

The ∂ notation, which had been used occasionally by Legendre [Cajori *1928-1929*, II, 225] is absent, but proper warning is given about the fact that $\frac{du}{dx} dx$ is the differential of u regarding only x as variable and not to be confused with du [Lacroix *Traité*, I, 121, 122-123]. Lacroix was well aware of the existence of notations for partial derivatives: in volume III he mentions several of them, including Euler's $\left(\frac{dz}{dx}\right)$ and $\left(\frac{dz}{dy}\right)$ – but not Legendre's ∂ [Lacroix *Traité*, III, 10-11]. However, he believed that $\frac{dz}{dx}$ and $\frac{dz}{dy}$ are equally clear.

Lacroix used a different kind of parentheses and only for a very special case: if both x and y appear in the expression for u, and at the same time y is regarded as a function of x, then $\frac{du}{dx}$ is the differential coefficient of u taken regarding y as a constant (notwithstanding the supposition that it is a function of x) – a sort of partial derivative; while

$$\frac{d(u)}{dx}$$

is the differential coefficient of u taking in account the supposition that y is a function of x. In such a situation, $u' = \frac{d(u)}{dx}$ [*Traité*, I, 163]; if z is an implicit function of x and y given by an equation $u = 0$, then [*Traité*, I, 174]

$$\frac{d(u)}{dx} = \frac{du}{dx} + \frac{du}{dz}\frac{dz}{dx}.$$

In page 123 Lacroix criticizes the habit of calling $\frac{du}{dx}$, $\frac{du}{dy}$ the first-order *partial differences* of u.[46] The real partial differences of u are f$(x + h, y) - f(x, y)$ and f$(x, y + k) - f(x, y)$, while $\frac{du}{dx} dx$ and $\frac{du}{dy} dy$ should be called its first-order *partial differentials* and $\frac{du}{dx}$ and $\frac{du}{dy}$ its first-order *differential coefficients*.

To find the higher-order differentials of $u = $ f(x, y), Lacroix differentiates (3.9) twice (assuming dx and dy as constant), notices a similarity to the binomial formula and confirms this similarity by an impeccable proof by mathematical

---

[46]This habit can be seen for instance in [Bossut *1798*, II, 351]. Partial differential equations were usually called "equations in partial differences" [Condorcet *1770*; Lagrange *1772b*; Laplace *1773c*; Monge *1771*].

induction[47]: he looks for "la loi qui règne entre deux différentielles consécutives"[48] [*Traité*, I, 125] and confirms the result for the case $n = 1$. The final result is that

$$d^n u = \frac{d^n u}{dx^n} dx^n + \frac{n}{1} \frac{d^n u}{dx^{n-1} dy} dx^{n-1} dx + \frac{n(n-1)}{1} \frac{d^n u}{2} \frac{d^n u}{dx^{n-2} dy^2} dx^{n-2} dy^2 + \text{etc.}$$

that is, $d^n u$ can be obtained by expanding $(dx + dy)^n$ and introducing into each term the corresponding differential coefficient.

A careful argument involving the general term of $f(x+h, y+k)$ allows Lacroix to prove the two-variable version of Taylor's theorem:

$$
\begin{aligned}
f(x + h, y + k) = u & \\
+ \frac{1}{1} & \left\{ \frac{du}{dx} h + \frac{du}{dy} k \right\} \\
+ \frac{1}{1 \cdot 2} & \left\{ \frac{d^2 u}{dx^2} h^2 + 2 \frac{d^2 u}{dxdy} hk + \frac{d^2 u}{dy^2} k^2 \right\} \\
+ \frac{1}{1 \cdot 2 \cdot 3} & \left\{ \frac{d^3 u}{dx^3} h^3 + 3 \frac{d^3 u}{dx^2 dy} h^2 k + 3 \frac{d^3 u}{dxdy^2} hk^2 + \frac{d^3 u}{dy^3} k^3 \right\} \\
\text{etc.} &
\end{aligned}
\tag{3.10}
$$

Functions of more than two variables bring no surprises, and (3.10) is generalized to

$$f(x + h, y + k, \text{etc.}) = u + \frac{du}{1} + \frac{d^2 u}{1 \cdot 2} + \frac{d^3 u}{1 \cdot 2 \cdot 3} + \text{etc.}$$

3.2.4 Differentiation of equations

After the sections on differentiation of (explicit) functions of one, two, and more than two, variables, Lacroix has a large section on differentiation of equations [*Traité*, I, 134-178]. As in [Euler *Differentialis*, I, ch. 9], this is both a manner of dealing with implicit functions and of preparing the way for the treatment of differential equations in the integral calculus.

Here occur two passages that seem independent from Lagrange. The first is about the differentiation of an equation in two variables $u = f(x, y) = 0$ (from which y is to be regarded as an implicit function of x).

In the first edition of the *Traité*, it takes Lacroix almost three and a half pages [*Traité*, 134-138] to arrive at the process to calculate $\frac{dy}{dx}$: calling h the increment of x, he concludes from the fact that the corresponding increment k of y is

$$\frac{y'h}{1} + \frac{y''h^2}{1 \cdot 2} + \frac{y'''h^3}{1 \cdot 2 \cdot 3} + \text{etc.} \tag{3.11}$$

---

[47]But without using the word *induction*: for him it still had the meaning of a generalization drawn by analogy from a number of examples.

[48]"the law that reigns between two consecutive differentials"

that $f(x+h, y+k)$ must have the form

$$f(x,y) + P_1 h + P_2 h^2 + P_3 h^3 + \text{etc.} \tag{3.12}$$

and since $f(x+h, y+k) = f(x,y) = 0$ and h is indeterminate, $P_1 = 0$, $P_2 = 0$, $P_3 = 0$, etc. He then proceeds to show that each of the coefficients in the series (3.12) is derived from the previous one just as in the Taylor series of an explicit function of one variable (invoking arguments analogous to those he had used before), so that $P_1 = u'$, $P_2 = \frac{u''}{1 \cdot 2}$, $P_3 = \frac{u'''}{1 \cdot 2 \cdot 3}$, etc. and therefore $u' = 0$, $u'' = 0$, $u''' = 0$, etc.

To evaluate u', Lacroix uses the fact that

$$f(x+h, y+k) = u + \frac{du}{dx} h + \frac{du}{dy} k + \frac{1}{2} \left(\frac{d^2 u}{dx^2} h^2 + \text{etc.} \right) + \text{etc.}; \tag{3.13}$$

substituting (3.11) for k and disregarding all powers of h other than the first, he gets

$$u' = \frac{du}{dx} + \frac{du}{dy} y' = 0; \tag{3.14}$$

and he still occupies a few more lines arguing that y' in (3.14) is precisely $\frac{dy}{dx}$ (although that is how he had introduced y' for (3.11), three pages earlier), so that naturally $\frac{dy}{dx}$ is obtained by differentiating u as if x and y were independent, putting the result equal to zero, and then solving for $\frac{dy}{dx}$.

By the second edition [*Traité*, 2nd ed, I, 188-90], Lacroix had realized that he did not need to establish the recursive relation between the coefficients in (3.12). It was enough to substitute (3.11) for k in (3.13) to conclude that $\frac{du}{dx} + \frac{du}{dy} y' = 0$, since that is the coefficient of h in the resulting series and all the coefficients should be zero to allow $f(x+h, y+k) = 0$, h being indeterminate.

The way in which Lagrange handles this in [*Fonctions*, 31-32] is a little different (and much simpler than Lacroix's first edition): firstly he notices that $f(x,y)$ may be regarded as a function φx of x only (since y is itself being regarded as a function of x); then, since $\varphi(x+i) = 0$ and i is indeterminate, $\varphi' x$ must also be zero (this is quite similar to Lacroix's second edition); finally, to evaluate $\varphi' x$, Lagrange uses a previously established result to the effect that the derivative of a function of two variables is the sum of the partial derivatives, as well as the chain rule; therefore $\varphi' x$, being the derivative of $f(x,y)$, is equal to $f'(x) + y' f'(y)$ (this is Lagrange's way of writing $\frac{\partial f}{\partial x} + \frac{dy}{dx} \frac{\partial f}{\partial y}$). The conclusion is that

$$y' = -\frac{f'(x)}{f'(y)}.$$

In the same section there is another passage that represents a small original contribution by Lacroix, if we take his word for it [*Traité*, 2nd ed, I, xxi], although he recognizes that similar reasonings appear in [Lagrange *Fonctions*].

In modern terms it would have to do with the inverse function theorem, although for Lacroix (and for Lagrange) it only amounts to knowing what to do to a differential equation on x and y if we want to revert from considering y as a function of x to consider x as a function of y.

In [Euler *Differentialis*] this problem is related to the question of which first differential is set as constant. After giving the method for removing higher-order differentials that was seen in page 56 above, Euler taught how to revert the process, and recover a formula where no first differential is supposed to be constant from another formula with

$$p = \frac{dy}{dx}, q = \frac{dp}{dx}, r = \frac{dq}{dx}, \ldots$$

where dx is set constant. This is not very difficult: if no differential is constant, then

$$dp = \frac{dx\,ddy - dy\,ddx}{dx^2}$$

so that

$$q = \frac{dx\,ddy - dy\,ddx}{dx^3};$$

similarly

$$r = \frac{dx^2 d^3 y - 3dx\,ddy\,ddy + 3dy\,ddx^2 - dx\,dy\,d^3 x}{dx^5};$$

and so forth; it is then enough to substitute these expressions for $p, q, r, \ldots$ [*Diffe-rentialis*, I, § 271-278]. Now, if we want to have dy constant, we just have to put $ddx = d^3 x = \ldots = 0$ [*Differentialis*, I, § 279-280].

A process based on constant differentials was not suitable for [Lagrange *Fonctions*]. Lagrange had to give an alternative approach. But there is a parallelism between this alternative and Euler's process: just as Euler's was a natural consequence of a process to derive a formula where no differential was constant, Lagrange's was a natural consequence of a process to start regarding both variables x and y as functions of a third variable t. It is easy to see how this relates to Euler's approach: if x and y are functions of t, then neither is regarded as an independent variable. Lagrange deduced his process in two different ways. The first is the following [*Fonctions*, 60]: if $y = f(x)$, and x and y are functions of t, then by the chain rule $y' = x' f'(x)$; but if y were simply a function of x, we would have $y' = f'(x)$; so the difference is that $\frac{y'}{x'}$ should replace y'. Similarly,

$$\frac{\left(\frac{y'}{x'}\right)'}{x'} = \frac{y''}{x^2} - \frac{y' x''}{x'^3}$$

should replace y'';[49] and so forth.

---

[49] Lagrange does not explain this second part, but presumably its justification is that from $\frac{y'}{x'} = f'(x)$ comes $\left(\frac{y'}{x'}\right)' = x' f'(x)$ and from $y' = f'(x)$ comes $y'' = f''(x)$, so that we have $\frac{\left(\frac{y'}{x'}\right)'}{x'}$ instead of y''.

Lagrange deduced these formulas in a second way, in the section on applications to mechanics (where he explicitly said that t was *time*). Although this second deduction is in a chapter on applications, it is in fact closer to the basic principles of the Lagrangian calculus.[50] If t becomes $t + \theta$, then x and y become respectively

$$x + \theta x' + \frac{\theta^2}{2} x'' + \frac{\theta^3}{2 \cdot 3} x''' \ \&c. \tag{3.15}$$

and

$$y + \theta y' + \frac{\theta^2}{2} y'' + \frac{\theta^3}{2 \cdot 3} y''' \ \&c. \tag{3.16}$$

However, if we regard y as a function of x, and x becomes i, then y becomes

$$y + i(y') + \frac{i^2}{2}(y'') + \frac{i^3}{2 \cdot 3}(y''') + \&c. \tag{3.17}$$

where $(y'), (y''), (y'''), \ldots$ represent the derivatives of y as a function of x, as opposed to $y', y'', y''', \ldots$ the derivatives of y as a function of t. Now, we have here two expressions for the increment of x, namely i, in (3.17), and $\theta x' + \frac{\theta^2}{2} x'' + \&c.$, in (3.15); and we have two expressions for the development of y, namely (3.16) and (3.17); putting $i = \theta x' + \frac{\theta^2}{2} x'' + \&c.$ in (3.17), comparing with (3.16), and ordering the terms by the powers of θ, we get

$$\theta y' + \frac{\theta^2}{2} y'' + \frac{\theta^3}{2 \cdot 3} y''' + \&c. = (y')x'\theta$$
$$+ \left((y')\frac{x''}{2} + \frac{(y'')}{2} x'^2 \right) \theta^2$$
$$+ \ldots$$

whence we can take

$$(y') = \frac{y'}{x'}, \quad (y'') = \frac{y'' - (y')x''}{x^2} = \frac{y''}{x^2} - \frac{y'x''}{x'^3}$$

and so on [*Fonctions*, 239-241].

Of course, no matter how these formulas are deduced, to change from x to y as independent variable it is enough to take $y = t$, so that $y' = 1$ and $y'' = y''' = \ldots = 0$ and they become

$$(y') = \frac{1}{x'}, \quad (y'') = -\frac{x''}{x'^3}, \quad \text{etc.}$$

Lacroix, the *encyclopédiste*, managed to give three processes. Later, in his *Compte rendu [...] des progrès que les mathématiques ont faits depuis 1789* (see

---

appendix B, page 399) and in [*Traité*, 2nd ed, I, xxi], he claimed that he had felt that Euler's approach was not compatible with the foundation he was trying to implement, so that he decided to substitute new considerations, and that he did this independently from Lagrange. Indeed it seems that the first two processes that he gives are his own. One obvious difference between them and Lagrange's is that these are direct methods, not consequences of methods for introducing a third variable. Lacroix's third process is not so original, but it comes from Euler, not Lagrange.

For the first process [*Traité*, I, 149-150], Lacroix reminds the reader that

$$\left.\begin{array}{l} dy = y'dx \\ dy' = y''dx \\ dy'' = y'''dx \\ \text{etc.} \end{array}\right\} \text{and} \left\{\begin{array}{l} dx = x'dy \\ dx' = x''dy \\ dx'' = x'''dy \\ \text{etc.} \end{array}\right. \tag{3.18}$$

Also, from (3.14) (in page 76 above)[51] there exist M and N such that $M + N\frac{dy}{dx} = 0$, or $M\,dx + N\,dy = 0$, which means that

$$M + Ny' = 0 \quad \text{and} \quad Mx' + N = 0.$$

From this it is immediate that

$$x' = \frac{1}{y'}.$$

(Would it not have been more immediate to conclude this from the first line in (3.18)?) Differentiating this gives

$$dx' = -\frac{dy'}{y'^2}$$

and using the second line in (3.18),

$$x'' = -\frac{y''}{y'^2}\frac{dx}{dy} = -\frac{y''}{y'^3}.$$

Similarly he arrives at

$$x''' = \frac{-y'y''' + 3y''^2}{y'^5}.$$

And so on.

The second way to derive these results is closer to the *first principles* of the foundation Lacroix is following: if h is the increment of x and k is the associated increment of y, then

$$k = \frac{y'h}{1} + \frac{y''h^2}{1 \cdot 2} + \frac{y'''h^3}{1 \cdot 2 \cdot 3} + \text{etc.} \tag{3.19}$$

---

[51] The context of this passage is still the differentiation of an equation f$(x, y) = 0$. Of course in such a situation y is an implicit function of x and x is an implicit function of y.

and

$$h = \frac{x'k}{1} + \frac{x''k^2}{1 \cdot 2} + \frac{x'''k^3}{1 \cdot 2 \cdot 3} + \text{etc.} \tag{3.20}$$

Now, using the method of reversion of series (a purely combinatorial method which had been reported in the Introduction) to obtain a series for h from (3.19), the result is

$$h = \frac{1}{y'}k - \frac{y''}{y'^3}\frac{k^2}{1 \cdot 2} + \left(\frac{3y''^2 - y'y'''}{y'^5}\right)\frac{k^3}{1 \cdot 2 \cdot 3} + \text{etc.}$$

which, being compared with (3.20), confirms the previous results for x', x'', x''', etc. [*Traité*, I, 150-151]

      After reporting these two processes, Lacroix gives a summary of Euler's considerations on differentiation without taking any first differential as constant (and on recovering this situation, that is, given a second-order differential equation in which, say dx is constant, to obtain an equivalent equation in which no first differential is constant). The advantage of doing this is that afterwards we can regard indifferently y as function of x or x as function of y. And it is then possible to make the corresponding first differential constant [*Traité*, I, 151-154].

3.2.5   Particular cases where the Taylor series does not apply

From what has been discussed in the paragraphs above, it is apparent that chapter 1 of the first volume of Lacroix's *Traité* was not influenced by Lagrange's lectures at the *École Polytechnique*. Rather, it presents an independent development of Lagrange's suggestion of 1772.

      However, Lacroix recognized that he had profited somewhat (if not much) from those lectures (see above, page 69). There must be some influence from Lagrange's lectures in [Lacroix *Traité*, I]. And in fact there is, but in chapter 2 (devoted to "the main analytical uses of the differential calculus"[52] — mainly the use of the calculus to develop functions into series, to raise indeterminacies like $\frac{0}{0}$, and to find maxima and minima).

      We have seen above (page 66) that Lagrange gave a *proof*, in [*Fonctions*, 7-8] and probably in his 1795 lectures at the *École Polytechnique*, that the series for $f(x+i)$ cannot include a fractional power of i, unless x is given certain particular values. He also explored those particular cases: if the expression of fx includes a radical that disappears for some particular value of x, then the argument quoted above in page 66 does not apply to that value. For instance, let $fx = (x-a)\sqrt{x-b}$. This function has two values, except when $x = a$ or $x = b$, in which cases it has only one. Because, *in general*, it has two values, so must its development

$$f(x + i) = fx + if'x + \frac{i^2}{2}f''x + \&c. \tag{3.21}$$

---

[52]"Des principaux usages analytiques du Calcul différentiel".

where i is indeterminate, have two values. This happens for $x = a$ because the radical $\sqrt{x - b}$, which disappears in fx, reappears in $f'x, f''x, \ldots$ But it does not happen for $x = b$. In the latter case, (3.21) is faulty ("fautif") – and in fact $f'b = f''b = \ldots = \infty$ (more generally, in such cases $f^n x = f^{n+1} x = \ldots = \infty$, for some integer n). The correct development of $f(x + i)$ for $x = b$ is $(b - a)\sqrt{i} + i^{\frac{3}{2}}$. The fractional powers of i are necessary to give back to the function its double value. [Lagrange *Fonctions*, 32-39]

Lacroix [*Traité*, I, 232-236] reports this latter case, but not the former, where the irrationality disappears only because it is multiplied by an expression that becomes null. In fact, Lacroix even finds it obvious that "toutes les fois que la fonction qu'on voudra développer sera irrationelle en général et que par la substitution d'une valeur particulière de x elle cessera de l'être, alors l'irrationalité tombera nécessairement sur l'acroissement"[53] (i for Lagrange, k for Lacroix). His example, instead of $(x - a)\sqrt{x - b}$, is $b + \sqrt{x - a}$.

He takes the chance also to report Lagrange's argument for the impossibility of an irrational exponent in the expansion of f$(x + k)$. However, he does not call it a "proof", but rather a "more solid foundation than the induction" used in chapter 1.[54]

Lacroix acknowledges Lagrange's authorship of these considerations [Lacroix *Traité*, I, 235], but the one paper by Lagrange [*1776*] associated to this section in the table of contents for [Lacroix *Traité*] is there by mistake: it relates in fact to the previous section. Lacroix almost certainly got this from Lagrange's lectures at the *École Polytechnique* in 1795 or 1796.

It is tempting to wonder if in those lectures Lagrange had not thought of the case in which it is a multiplier of the irrationality that disappears (or maybe he had thought of it but for some reason, say lack of time, chose not to address it in class). For the second edition of the *Traité* Lacroix had of course access to [Lagrange *Fonctions*] and he addressed both cases [*Traité*, 2nd ed, I, 333].

3.2.6 Foundations for algebraic analysis

Before the chapter on the principles of the calculus, [Lacroix *Traité*, I] includes an "Introduction". Its purpose is to give "series expansions of algebraic, exponential, logarithmic and trigonometric functions" by algebraic means, without recourse to the notion of infinity [*Traité*, I, xxiv]. The goal of studying functions by expanding them in series (before presenting the differential calculus) makes clear the intended equivalence to the first volume of [Euler *Introductio*]. But Euler had used infinite quantities quite freely, and Lacroix explicitly avoids them.

---

[53] [Lacroix *Traité*, I, 233]: "Every time the function we want to expand is irrational in general, but ceases to be so by the substitution of a particular value of x, the irrationality must then fall upon the increment".

[54] [Lacroix *Traité*, I, 234]: "Nous offre [...] le moyen de l'établir sur des fondemens plus solides que l'induction dont nous l'avons déduite". "Induction", of course, is here used in the non-mathematical sense – see footnote 47.

Together with chapter 3 (a "digression on algebraic equations"), the Introduction also corresponds to what was known for some time as *algebraic analysis* ("analyse algébrique"). Nowadays this expression is often used by historians to refer to an algebraic conception of analysis, particularly Lagrange's [Fraser *1989*], but also that of the German Combinatorial School [Jahnke *1993*]. However, around 1800 the expression was somewhat ambiguous. It could have that meaning, as in the full title of [Lagrange *Fonctions*], where the principles of differential calculus are declared to be "reduced to the algebraic analysis of finite quantities" (see footnote 24 above). But in the *École Polytechnique* it was used to refer to a section in the syllabus of analysis, composed of aspects of higher algebra that did not use differential or integral calculus: the fundamental theorem of algebra, series expansions of particular functions, algorithms for third- and fourth-degree equations, etc. (see appendices C.2.2 and C.3.1 for details). It is in this latter sense of a subject, not a point of view, that this expression is used here. Lacroix, who did not use this expression much (and may not have been very fond of it) would describe the subject as "l'analyse intermédiaire entre les Élémens d'Algèbre proprement dits, et le Calcul différentiel"[55] [*Traité*, 2nd ed, I, xx]. Cauchy's *Analyse algébrique* [*1821*] transformed radically the meaning of the expression, turning the subject into a pre-calculus study of functions based on a theory of limits, and explicitly rejecting the "generality of algebra".

This Introduction contains material that the modern reader regards as related to the foundations of the calculus, but it must be stressed that in Lacroix's arrangement it comes before the differential calculus and of course before the principles of the calculus are addressed.

The first issue addressed in the Introduction is one of those with a "foundational" character: the concept of function. For Lacroix the content of this word had been going through a progressive enlargement, until at that time it could be defined as follows.

> "Toute quantité dont la valeur dépend d'une ou plusieurs autres quantités, est dite *fonction* de ces dernières, soit qu'on sache ou qu'on ignore par quelles opérations il faut passer pour remonter de celles-ci à la première."[56]

Grattan-Guinness [*1990*, I, 141] compares this definition with the general conception of function used by Dirichlet in 1829 (when he introduced the characteristic function of the rationals). But he also notes that the functions with which Lacroix worked were not that arbitrary: he "often stayed in or around power series in his introduction".

In fact, the example given by Lacroix of a function for which it is not known

---

[55]"the intermediary analysis between the elements of algebra in the strict sense, and the differential calculus"

[56][Lacroix *Traité*, I, 1]: "Any quantity the value of which depends on one or more other quantities is said to be a *function* of these latter, whether or not it is known which operations are necessary to go from them to the former."

which operations are necessary to go from the argument to the corresponding value of the function, is the root of a 5th degree equation: "in the present state of algebra" it was not possible to assign an expression to it. It is doubtful that Lacroix would recognize the characteristic of the rationals as a function. Be that as it may, no function so strange ever occurs in Lacroix's *Traité*.

The introduction of series[57] is justified by two observations: first, that some algebraic functions give rise to them, when one tries to express one such function by an "assembly" ("assemblage") of monomials (it is the case of $\frac{a}{a-x} = 1 + \frac{x}{a} + \frac{x^2}{a^2} + \frac{x^3}{a^3} + $ etc.); second, that some functions, as is the case of the logarithms, sine, and cosine, are not expressible by a limited number of algebraic terms (these functions are called *transcendental*).

This means that series are not studied here for their own sake, rather as *developments* of functions, and therefore it is mainly power series that appear.

This use of series as developments of functions makes it necessary for Lacroix to warn that a series does not always have the *value* of the corresponding function [*Traité*, I, 4]. This entails a discussion on convergence, illustrated by the example of $\frac{a}{a-x} = 1 + \frac{x}{a} + \frac{x^2}{a^2} + \frac{x^3}{a^3} + $ etc., divided into the cases $x < a$, $x > a$ and even $x = a$. It is a careful discussion, making use of the remainder $\frac{x^n}{a^{n-1}(a-x)}$ of $1 + \frac{x}{a} + \frac{x^2}{a^2} + ... + \frac{x^{n-1}}{a^{n-1}}$.

It is here that Lacroix introduces a definition of limit:

> "Dorénavant nous appellerons *limite, toute quantité qu'une grandeur ne sauroit passer dans son accroissement ou son décroissement, ou même qu'elle ne sauroit atteindre, mais dont elle peut approcher aussi près qu'on le voudra.*[58]

So, if a given series has a limit, its value is that limit. But even if the series does not converge, as long as it is the development of some known function it can be used for some purposes as a representation of that function:

> "Si une question nous conduisoit à une série telle que

$$1 + \frac{x}{a} + \frac{x^2}{a^2} + \frac{x^3}{a^3} + \text{etc.}$$

> nous serions en droit de conclure que la fonction cherchée, n'est autre que $\frac{a}{a-x}$; ou si nous découvrions quelques propriétés relatives à une suite de termes tels que $1 + \frac{x}{a} + \frac{x^2}{a^2} + $ etc., nous pourrions affirmer qu'elle appartient à la fonction $\frac{a}{a-x}$. Mais toutes les fois que qu'il s'agira de

---

[57]Lacroix uses the "series" and "sequence" ("suite") interchangeably, but "series" occurs more often and since in the Introduction he is almost always referring to what we call series, I will use only this word here.

[58][Lacroix *Traité*, 1st ed, I, 6]:"Henceforth, we will call *limit, every quantity which a magnitude cannot surpass as it increases or decreases, or even that it cannot achieve, but which it can approach as close as one might wish.*"

la valeur absolue de cette quantité, nous ne saurions employer la suite trouvée par son développement, qu'en ayant égard au reste."[59]

Lacroix duly reports the well-known fact that it is necessary, but not suffici-ent, for a series to have a limit, that its terms be eventually decreasing [*Traité*, I, 9].

He thus seems to use the modern concept of *convergent* (equivalent to "ha-ving a limit") not the one that d'Alembert had used in the articles "Convergent" and "Divergent" of the [*Encyclopédie*]: there d'Alembert had called *convergent* a series the terms of which are always decreasing, and *divergent* one with increa-sing terms.[60] This is in spite of Lacroix citing in the table of contents, associated to the Introduction, a memoir where d'Alembert uses those concepts of conver-gence/divergence (and gives two famous examples of a series that is convergent until the 299th term and divergent from then on and of another that is divergent until the 99th term and convergent from the 100th onwards [d'Alembert *1768*, 175-176]).

But in fact Lacroix never defines *convergent* nor *divergent* explicitly, and in those cases where a series might have decreasing terms but not have a limit, he seems to avoid the words convergence and divergence: for instance, when stating the necessary condition mentioned above, his wording is "pour qu'une série qui est le développement d'une fonction finie, approche continuellement de la vraie valeur, il faut que les termes qui la composent aillent en décroissant"[61]. But at least in one occasion Lacroix *proves* that a series ($e^x = 1 + \frac{x}{1} + \frac{x^2}{1\cdot2} +$ etc.) is *convergent* just by arguing that its terms must be eventually decreasing [Lacroix *Traité*, I, 37].

It is also worth mentioning that Lacroix often speaks of one series being "more convergent" than another, meaning that it converges more rapidly (as, for example, in [*Traité*, I, 42-47], on series for calculating logarithms): convergence is apparently a *practical* issue: it concerns the usefulness of the series as a means to calculate an approximate value of a function; convergence/divergence is about whether it can be used at all for that purpose and the degree of convergence is about how

---

[59][Lacroix *Traité*, I, 7]:"If, while addressing some question, we were led to the series

$$1 + \frac{x}{a} + \frac{x^2}{a^2} + \frac{x^3}{a^3} + \text{etc.}$$

we would be allowed to conclude that the function we were looking for is none other than $\frac{a}{a-x}$; or if we discovered some properties relative to a series of terms such as $1 + \frac{x}{a} + \frac{x^2}{a^2} +$ etc., we would be able to state that they belong to the function $\frac{a}{a-x}$. But whenever the subject is the absolute value of that quantity, we cannot employ the series found by developing it, without taking the remainder into account."

[60]By the article "Série", d'Alembert's ideas seem to have changed a little: a series was then convergent if it approached more and more a finite quantity and, continued to infinity, it would finally become equal to that quantity. That its terms would be decreasing was by then a conse-quence, not the definition.

[61]"for a series that is the development of a finite function to approach continuously its true value, it is necessary that the terms that compose it decrease progressively"

good it is for that purpose. But the status of the series as a representation of the function of which it is a development is not affected by such questions.

There seems to be an odd mixture of rigour and carelessness. This can be seen when Lacroix addresses what I have called in page 3.1.4 Arbogast's principle; Lacroix gives a more general version: given a series of terms like $Ax^\alpha + Bx^\beta + Cx^\gamma + $ etc., where $\alpha > \beta > \gamma > $ etc. or $\alpha < \beta < \gamma < $ etc., it is possible to find some value m to substitute for x such that $Am^\alpha > Bm^\beta + Cm^\gamma + $ etc.

The proposition is stated in this way (and is therefore wrong); it is *proven* (in the first case; the second is analogous) by writing the difference between Am^α and the rest of the series as

$$ m^\alpha \left\{ A - \left(\frac{B}{m^{\alpha-\beta}} + \frac{C}{m^{\alpha-\gamma}} + \text{etc.} \right) \right\}, $$

which *shows* that it increases with m, and therefore that "it is clear" that m can be chosen so as to ensure $Bm^\beta + Cm^\gamma + $ etc. $< Am^\alpha$ [*Traité*, I, 10-12].

However, Lacroix decides to show *how* such a number m can be found. He uses the geometric series (with ratio 2) as a starting point, so that he wants an m that will ensure that each term of the series will be larger than twice the next. Analysing the case of a series $A + Bx + Cx^2 + Dx^3 + ...$, he arrives at the condition $m > \frac{2Q}{P}$, where P, Q are the consecutive coefficients out of $A, B, C, D, ...$ with the largest ratio.

Of course the existence of such a pair P, Q is not assured for every power series, and at the end of the argument just described Lacroix introduces the extra condition that the ratios between consecutive coefficients must have an upper bound [*Traité*, I, 13]. He even gives a counter-example(!):

$$ 1 + \frac{1 \cdot 2}{x} + \frac{1 \cdot 2 \cdot 3}{x^2} + \frac{1 \cdot 2 \cdot 3 \cdot 4}{x^3} \text{ etc.} \tag{3.22} $$

Lacroix used Arbogast's principle in chapter 2 (on analytical applications of the differential calculus), to study maxima and minima, like Euler and Arbogast had done before him and Lagrange was doing more or less at the same time in [*Fonctions*]; and later in chapter 4 to apply the differential calculus to the theory of plane curves (tangents, osculation, areas and arc-lengths), like Arbogast and Lagrange − see section 4.2.1.2. But after the Introduction he did not seem to worry about avoiding situations like (3.22).[62]

Lacroix handles limits still very intuitively, in a way similar to d'Alembert or Cousin (see subsection 3.1.2). Thus, in page 189, adapting the differential calculus to the method of limits, the limit of $p + qh + rh^2 + $ etc., when h vanishes, is p, without further ado. In particular, he does not feel the need for prevention against a counter-example similar to (3.22).

---

[62]In the Introduction there is a situation in which he does verify that the extra condition holds and thus he can use Arbogast's principle [Lacroix *Traité*, 58-59]. This is in a deduction of a power series for the sine.

Infinity $\left(\frac{1}{0}\right)$ is introduced as a "negative" concept: an exclusive limit, a limit that quantities can never reach [Lacroix *Traité*, I, 7, 9-10]. This is the "true metaphysics" that should replace the actual infinity usually employed by mathematicians. But in fact the actual infinite appears every once in a while throughout the three volumes, when Lacroix feels the need or the usefulness of resorting to the Leibnizian calculus. He does think that the way Leibniz presented the calculus was less rigorous than limits or power series [Lacroix *Traité*, I, 193]. But rigour, for Lacroix, seems to be a matter of more or less, rather than yes or no, just like convergence.

3.2.7 Alternative principles for the differential calculus

Chapter 1 ends with a section about alternative foundations for the calculus: d'Alembert's limit approach and Leibniz's infinitesimals. Lacroix does not address Newton's theory, "parce qu'elle tient à la considération du mouvement qui est étrangère à l'analyse et à la géométrie"[63] [*Traité*, I, 194].

Because he has already addressed the theory of limits in the Introduction (see subsection 3.2.6 above), Lacroix only needs to apply it to give limit-based definitions of differential and differential coefficient. But his power-series considerations also play a role here (although a technical one). Given a function u of x, when x becomes $x + h$, u will become $u + ph + qh^2 + rh^3 +$ etc.; therefore, calling k the increment of u, we have

$$\frac{k}{h} = p + qh + rh^2 + \text{etc.}$$

Letting the two increments k and h vanish, the limit of $\frac{k}{h}$ is then p, which is the first differential coefficient of u.

It is clear that Lacroix's purpose in this section is to show that the same results are achieved with the method of limits as with the power-series definition of the differential coefficients. He gives a few examples of how some particular differential coefficients can be deduced using limits, including deducing those of logarithmic and trigonometric functions, without resorting to series expansions.

Leibnizian infinitesimal differentials are also briefly introduced [*Traité*, I, 193-194], in spite of being less rigorous than both limits and power-series, because they are "plus commode[s] dans les applications"[64]. Perhaps Lacroix should have included here a footnote that appears only in chapter 4, when explaining the application of infinitesimals to the study of curves: in that footnote, he quotes Leibniz to the effect that the consideration of a curve as a polygon is an approximation, whose error can be made as small as possible – so that the use of infinitesimals is simply an abbreviation of "Archimedes' style" (i.e., the method of exhaustion), or of the method of limits [*Traité*, I, 423-424].

---

[63]"because it draws on the consideration of motion which is foreign to analysis and geometry"
[64]"more convenient for applications"

This section is typical of Lacroix's *encyclopédiste* approach: to expound all relevant alternative methods or theories (and trying to conciliate them). It is also an essential instance of that approach because in future chapters Lacroix will sometimes need to resort to one or another of those alternative foundations in order to explain some particular method.

This is most marked in chapter 5 of volume I and chapter 5 of volume II. Chapter 5 of volume I is dedicated to analytic and differential geometry in space, and it is essentially based on work by Gaspard Monge. Lacroix had no choice but to follow Monge in speaking, for instance, of envelopes of one-parameter families of surfaces as the limits of their consecutive intersections (where "consecutive" suggests infinitesimal considerations, mixed here with limits). Space curves are regarded mainly as polygons in which three consecutive sides are not coplanar.

Chapter 5 of volume II is dedicated to the method of variations. Lacroix makes no attempt to suit the calculus of variations to the Lagrangian power-series foundation of the calculus, so he presents Lagrange's δ-algorithm in its Leibnizian shape (the rules of δ-differentiation come from those of d-differentiation by plain analogy and $\delta dy = d\delta y$ is justified using infinitesimal considerations).

3.2.8   A criticism of Lagrange

Lacroix's approach to the principles of the calculus, although technically drawn from Lagrange's work, has a fundamental difference in relation to the latter's view on this issue. Lacroix saw in [Lagrange *1772a*] a *more rigorous* and more elegant way to justify the *differential calculus* than the other ways available, but he did not seek to exclude these other views, as they were often useful.

They were useful not only for technical reasons, but also for the insights they allowed. In the preface to the first volume he quoted a letter he had received from Laplace in January 1792, while he was collecting material for the *Traité*:

> "Le rapprochement des Méthodes que vous comptez faire, sert à les éclaicir mutuellement, et ce qu'elles ont de commun renferme le plus souvent leur vraie métaphysique: voilá pourquoi cette métaphysique est presque toujours la dernière chose que l'on découvre."[65]

These words from Laplace certainly mirror the way Lacroix felt about the principles of the calculus.

Lagrange, on the other hand, sought to establish a coherent and comprehensive foundation for the calculus, excluding all alternative views. This included renaming the subject as calculus of functions, and abandoning notations that were evocative of infinitesimals.

---

[65] [Lacroix *Traité*, I, xxiv]: "The reconciliation of the Methods which you are planning to make, serves to clarify them mutually; and what they have in common contains very often their true metaphysics: this is why that metaphysics is almost always the last thing that one discovers". This translation is taken from [Grattan-Guinness *1990*, I, 139]

Lacroix did not comment explicitly on the *fundamentalism*, as it were, of Lagrange; but in volume III, which appeared in 1800 (three years after [Lagrange *Fonctions*]), he did comment, rather disapprovingly, on Lagrange's exclusion of the traditional notations.

In a very lengthy footnote ([*Traité*, III, 10-12]: almost two and a half pages!), Lacroix argues that a change in metaphysics does not necessarily entail a change in notation; that the first two volumes of his *Traité* are proof enough that the traditional notations are compatible with the power-series approach; that Lagrange's ′-notation is not at all convenient for functions of more than two variables; that Lagrange's contributions to analysis using the calculus of functions could be equally obtained using the differential calculus; that the passage from algebra to the differential calculus, as presented in his own *Traité* or in [Lagrange *1772a*], was as simple as the passage from algebra to the calculus of functions; that everyone who had already studied the calculus, reading [Lagrange *Fonctions*] was forced to translate (at least mentally) its results into the usual symbols; and finally, that original notations embarrass students.

His comparison between [Lagrange *Fonctions*] on one side, and [Lacroix *Traité*] and [Lagrange *1772a*] on the other, is particularly suggestive of Lacroix's disappointment with [Lagrange *Fonctions*]. Not a mathematical disappointment, of course: he is very clear about the worth of [Lagrange *Fonctions*], and about the fact that he profited from it; more of a philosophical disappointment, as well as pedagogical.

Chapter 4

Analytic and differential geometry

The two final chapters in volume I comprise a "complete theory of curves and curved surfaces"; that is, not only the "application of the differential calculus to the theory of curves" (and of curved surfaces) − what we now call differential geometry − but also the "purely algebraic part of that theory" − analytic geometry. Lacroix explained the inclusion of analytic geometry by his desire to offer a full set ("ensemble complet") and to relate notions that were usually presented from very different points of view [*Traité*, I, xxv, 327].

Lacroix, a good teacher, divided these chapters according to number of dimensions: chapter 4 is devoted to both analytic and differential geometry on the plane; chapter 5 in space. In this study the main division will be by subject: first analytic geometry, then differential geometry.

The boundaries between analytic geometry and differential geometry are sometimes a little artificial, particularly when talking about the 18th century, an age in which the study of infinite series could be regarded as "purely algebraic". In these two chapters it is often not clear into which of the two subjects a particular passage should be classified. Nevertheless, there is an interesting story to be told about *analytic geometry*, in which Lacroix plays an important role, and that was decisive in the choice for this division.

In these two chapters the influence from Monge is most marked: he was one of the chief authors of the version of analytic geometry that emerged in the late 18th century; and as for chapter 5, the "theory of curved surfaces and curves of double curvature" presented there "is almost entirely due to Monge" [Lacroix *Traité*, I, 435].

Another influence from Monge is in the parallelism that Lacroix tries to draw between analysis and geometry. The purpose of these two chapters is perhaps most fully explained in a draft letter dated 22 Nivose year 3 (11 December 1794), kept at [Lacroix *IF*, ms. 2397][1]:

---

[1] The addressee is not identified, but was probably Regnard, possibly a private pupil, to whom

"Ne croyez pas que les chapitres d'application que je veux intercaler puissent deranger la marche analytique car ils seront isoles du reste, ils ne serviront pas à decouvrir ou à demontrer aucun resultat de calcul. Mais ils seront l'image des chapitres precedens. On les passera si on veut sans nuire à la lecture du reste mais aussi ils reposeront et *amuseront* l'imagination de l'eleve par des *peintures sensibles* des procédés de calcul donnés dans les chapitres précedens.

Ainsi l'analyse et la geometrie ne seront point melées mais cet ouvrage séparé que j'avais commencé avec soin[2] sur l'application de l analyse, se trouvera intercalé par chapitres dans l'autre. Ainsi après les principes du calcul differentiel, on trouvera un traite des proprietes generales des courbes, des courbes a double courbure et des surfaces courbes qu'on lira ou qu'on passera a volonte. On y verra la *peinture* bien complette et bien interessante de ce que c'est que differences partielles."[3]

4.1   Analytic geometry

4.1.1   From "the application of algebra to geometry" to "analytic geometry"

This subsection is based mainly on [Boyer *1956*] and [Taton *1951*, 101-124]. It is an attempt to explain how in a certain sense *analytic geometry* was a novel subject in 1797. Lacroix's presentation of it was one of the very first to take a certain new point of view.

An explanation on terminology is in order here: the expression "application of algebra to geometry" (very much common in the 18th century) will be used

---

Lacroix had been writing, at least since 1789, explaining several issues of mathematics.

[2]Belhoste [*1992*, 568] reads here "avec vous"; but given the teacher-pupil tone of the rest of the letter, this does not sound very convincing (unless of course Lacroix was writing that separate work as lectures for this student). Belhoste also interprets this whole passage as meaning that Lacroix intended to interpose his "descriptive geometry" [Lacroix *1795*] in the *Traité*. I disagree: Lacroix certainly made many references to [Lacroix *1795*] in chapter 5, but what he says here is that a work he had been writing on the application of analysis (to geometry, presumably) was going to be interposed in the *Traité* – that separate work must correspond to chapters 4 and 5.

[3]"Do not think that the chapters of application which I wish to interpose might disturb the analytical course: they will be isolated from the rest, they will not be used to discover or demonstrate any result of calculus. But they will be the image of the preceding chapters. One may pass over them, if one wishes, without hindering the reading of the rest, but they will also rest and *amuse* the imagination of the student through *sensible depictions* of the procedures of calculus given in the preceding chapters.

Thus analysis and geometry will not be mixed, but that separate work which I had begun with care on the application of analysis will be found inserted by chapter in the other. Thus after the principles of the differential calculus, one will find a treatise of the general properties of curves, of curves of double curvature and of curved surfaces, which may be read or passed over as one may wish. One will see there a quite complete and quite interesting *depiction* of what are partial differentials."

for any application of techniques of symbolic algebra in geometry; the much less common expressions "coordinate geometry" and "coordinate methods" will refer to the kind(s) of application of algebra to geometry that used coordinates (not necessarily orthogonal, not necessarily with explicit x- and y-axes), which allowed one to represent the geometrical objects involved by means of equations; "analytic geometry" will be used for a refinement of "coordinate geometry" that sought to be as independent as possible from synthetic (i.e., non-algebraic) geometry. A very simple example of "application of algebra to geometry" which is not "coordinate geometry" is the following, taken from [Bézout *1796*, III]: given the sides of a triangle ABC, to find its height and the lengths of the segments it forms on the basis. That is, we know AB, BC, AC, and wish to know BD, AD, DC. Following

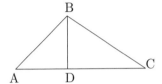

the usual conventions of algebra, we put $BC = a, AB = b, AC = c$, and $CD = x, BD = y$; of course $AD = c - x$. The theorem of Pythagoras gives

$$xx + yy = aa \quad \text{and} \quad cc - 2cx + xx + yy = bb$$

whence

$$x = \frac{1}{2}\frac{(a+b)(a-b)}{c} + \frac{1}{2}c.$$

"Application of algebra to geometry" was an umbrella term for all uses of algebra in geometry, but we can say that its non-coordinate section (which by the 18th century was purely a school subject, not a research topic) focused on the same objects as elementary synthetic geometry: triangles, squares, circles, and so on. "Coordinate geometry", on the other hand, focused on curves and surfaces. In its common form, straight lines and planes were not included in those "curves and surfaces". "Analytic geometry" changed this.

4.1.1.1   From Descartes to Euler

One of the best known *facts* of the history of mathematics is that analytic geometry was invented (or discovered) by the French philosopher and mathematician René Descartes, and that he published this invention (or discovery) in 1637 in [Descartes *Géométrie*]. What is somewhat less well known is that analytic geometry as we know it from school is only a distant relative from what we can find in Descartes' famous book. The object of [Descartes *Géométrie*] was the solution of problems from classical (Greek) geometry.[4] François Viète (1540-1603), in 1591,

---

[4]In the 16th century the possibility of access to ancient Greek mathematical works had increased considerably because of the printing of both original versions and (usually Latin) translations.

had already used symbolic algebra in those problems (by reducing problems to equations). But Descartes went much further in that direction and introduced new algebraic techniques, namely — in a somewhat casual way — the use of coordinates, which allowed one to deal algebraically with curves. Seeking an equation for a curve EC drawn by a certain device, he wrote:

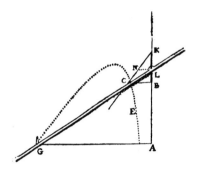

"Je choisis une ligne droite, comme AB, pour rapporter a ses divers poins tous ceux de cette ligne courbe EC, et en cette ligne AB je choisis un point, comme A, pour commencer par luy ce calcul. [...] Aprés cela prenant un point a discretion dans la courbe, comme C, sur lequel je suppose que l'instrument qui sert a la descrire est appliqué, je tire de ce point C la ligne CB parallele a GA, et pourceque CB et BA sont deux quantités indeterminées et inconnuës, je les nomme l'une y et l'autre x. [...] l'equation qu'il falloit trouver est $yy \propto cy - \frac{cx}{b}y + ay - ac$."[5] [Descartes *Géométrie*, 320-322]

However, this was just a new technique: his starting point (geometric problems) and his goal (the geometric *construction* of the solutions) were two thousand years old. Moreover, in Descartes' *Géométrie* no curve is *defined* by an equation; equations are just convenient means to handle curves that are already known; and those curves are not the object of study; they are only auxiliary objects or solutions to loci problems.

When the solution to a problem appeared as an equation, it still had to be reverted to geometry. This led to the rise of a mathematical theory: the "construction of equations" [Bos *1984*]. A process had to be found to construct geometrically

---

This (particularly the publication in 1588 of Commandino's Latin translation of Pappos' *Mathematical Collection*) had given origin to what Bos calls "the early modern tradition of geometrical problem solving" [Bos *2001*, ch. 4].

[5]"I choose a straight line, as AB, to which to refer all its points [i.e. those of the curve EC], and in AB I choose a point A at which to begin the investigation. [...] Then I take on the curve an arbitrary point, as C, at which we will suppose the instrument applied to describe the curve. Then I draw through C the line CB parallel to GA. Since CB and BA are unknown and indeterminate quantities, I shall call one of them y and the other x. [...] the required equation is $y^2 = cy - \frac{cx}{b}y + ay - ac$."[Descartes *Géométrie*, 51-52]

the roots of the equation. This construction was performed by intersecting simpler curves. According to Bos [*1984*, 355], "after 1750 the construction of equations quickly fell into oblivion". It did disappear as a subject of research, but it survived a little longer, although weakened, in school curricula, or at least in textbooks (as for instance in [Lacroix *1798b*, 250-260]).

The next two centuries would witness the gradual transformation of Descartes' coordinate techniques into *analytic geometry*. The first step was the use of those techniques for the study of curves for their own sake, not as auxiliary objects. 1679 saw the posthumous publication of *Varia Opera Mathematica* by the French lawyer Pierre de Fermat (c. 1608-1665). These included an *Ad Locos Planos et Solidos Isagoge*, composed before the publication of [Descartes *Géométrie*], revealing that Fermat had independently created (or discovered) essentially the same techniques. There were some important differences, and Fermat was more interested in the analytic study of curves than Descartes; unlike Descartes, he introduced them through their equations. However, making use of a more cumbersome algebraic notation than Descartes, and being published when Cartesian geometry was already quite popular, Fermat's work on coordinate geometry went largely unnoticed.[6]

Mathematicians in the 17th century who used coordinate methods used them to study old curves; new curves (such as the cycloid) were usually defined by non-algebraic means, which parted them from the "application of algebra to geometry".

According to Boyer, "Fermatian" geometry came into its own only in Newton's *Enumeratio linearum tertii ordinis*, written not later than 1676,[7] revised in 1695 and finally published in 1704 as an appendix to his *Opticks*. Being a study of curves defined by cubic equations in two unknowns, it is "the first instance of a work devoted to the theory of curves as such"[Boyer *1956*, 139].

However, in spite of his contributions to the subject, Newton complained in his *Arithmetica Universalis* (1707) about the mixture of algebra and geometry: "The Ancients did so industriously distinguish them from one another, that they never introduced Arithmetical Terms into Geometry. And the Moderns, by confounding both, have lost the Simplicity in which all the Elegancy of Geometry consists"[8]. Boyer [*1956*, 148] suggests as a solution to the apparent contradiction that Newton recognized the power of algebraic methods in geometry but did not allow them in *elementary* geometry. The view would remain throughout most of the 18th century that the circle and straight line belonged exclusively to the realm of synthetic geometry (the conic sections were perhaps a debatable land): they only appeared as auxiliary lines in coordinate geometry; this had consequences:

> "La faiblesse essentielle d'une telle conception était de négliger ainsi les problèmes élémentaires sur les points et les droites qui, en dehors de leur

---

[6]But it should be mentioned that several of Fermat's works, including the *Isagoge*, had circulated much before, in manuscript form, among the Parisian mathematicians [Boyer *1956*, 82; Bos *2001*, 205-206].

[7]Hence *before* the publication of Fermat's *Opera*. But see previous footnote.

[8]Quoted in [Boyer *1956*, 148].

intérêt propre, permettent de simplifier considérablement la solution de
la plupart des problèmes plus complexes."[9] [Taton *1951*, 102]

The fact that the circle and the straight line were not thoroughly studied in algebraic form made it necessary for propositions from elementary synthetic geometry
to be invoked once and again. Reliance on diagrams was much stronger than it
would later become.

Meanwhile the appearance of the differential and integral calculus had provided mathematicians with a much more powerful tool for the study of curves than
mere algebra. It is only natural that the field known as application of algebra to
geometry had a much slower evolution than the calculus.

"In formalization, infinitesimal analysis had [by the first half of the 18th
century] far outstripped Cartesian geometry [...]. Formulae had been a
natural outgrowth of the algorithms of Newton and Leibniz, but the
coordinate geometry of Descartes and Fermat still leaned heavily upon
auxiliary diagrams"[Boyer *1956*, 170].

This is the explanation for the surprising claim by Boyer that the oldest known
appearance of the formula for the distance between two points dates only from
1731, almost a century after the publication of [Descartes *Géométrie*]. This appearance is to be found in the *Recherches sur les courbes à double courbure* by
the French mathematician Alexis Claude Clairaut (1731-1765).[10] This does not
mean, of course, that previous mathematicians did not use that formula in some
way: it is implicit for example in the equations for a circle or a sphere; and it is a
close relative of the formula for the differential of the arc length $ds = \sqrt{dx^2 + dy^2}$.
But apparently whenever someone needed to calculate a distance, or to write an
expression involving one, the basis for the result was the pythagorean theorem,
not an established formula.

In fact even the passage that Boyer claims to contain the distance formula
for the first time is *not* explicitly about distance. It concerns the deduction of
the equation of a sphere whose centre is not the origin of the coordinates, making
use of the pythagorean theorem. The expression for the radius of such a sphere
is $\sqrt{\overline{x \mp a}^2 + \overline{y \mp b}^2 + \overline{z \mp c}^2}$ [Clairaut *1731*, 98] (the symbol $\mp$ is due to some
uneasiness with the use of signs). Earlier in the same book Clairaut had given the
equation of a sphere with the origin as centre, using quite casually $\sqrt{xx + yy + zz}$
as an expression for its radius [Clairaut *1731*, 8]. Boyer apparently saw a significant
difference between $\sqrt{xx + yy + zz}$ and $\sqrt{\overline{x \mp a}^2 + \overline{y \mp b}^2 + \overline{z \mp c}^2}$. Perhaps more
interesting is the plain fact that in neither occasion is a *distance formula* deduced
for its own sake – it is only equations of spheres that are sought. We will see
below that the formula for the distance from a point to the origin appeared in

---

[9]"The essential weakness in such a conception was the neglect of the elementary problems on
points and straight lines which, besides their own interest, allow one to simplify considerably the
resolution of most of the more complex problems."

[10]For this claim, see [Boyer *1956*, 168-170].

[Euler *Introductio*], but the general distance formula would not appear *explicitly* until the late 18th century.

The publication of [Euler *Introductio*] in 1748 was a big step in the direction towards *analytic geometry*. The purpose of this book was to develop those parts of algebra necessary for the study of calculus, and its second volume was devoted to coordinate geometry. It is very much relevant that the usefulness of coordinate methods was now related to the calculus; quite a different situation from that when Descartes used them to solve problems from classical geometry. It is also quite telling that [Euler *Introductio*, II] has only one chapter (out of 28) dedicated to the "construction of equations".

Moreover the principal object of study in [Euler *Introductio*] is the *function*, something which happens for the first time. Thus, in its second volume coordinate geometry is a method for the study of functions. Each curve is associated with a function but, more importantly, each function can be represented by a curve. The functional approach allows Euler to start by giving a short general theory of curves, instead of starting by the conic sections, as was usual (although conics play a fundamental role in the introduction of several aspects of curves); it also allows him to include a chapter on transcendental curves. He also strives to give a thoroughly analytic treatment of conic sections: they are called "second-order lines", and their study is based upon the general second-order equation on two unknowns; they are defined by their equations, not as sections of cones, nor as planar geometrical loci (as was often the case: we will see two examples below, in section 4.1.1.2).

However, Euler's coordinate geometry still relied heavily (according to later standards) on diagrams and elementary synthetic geometry. An example of this is his deduction of the equation of a circle of centre C and radius $AC = a$, AB being the axis and A the origin of the abscissas; the abscissa is $AP = x$ and the ordinate is $PM = y$. Then $PM^2 = AP \cdot PB$ and $PB = 2a - x$, so that the equation is $y^2 = 2ax - x^2$ [Euler *Introductio*, II, §64]. The main result used here is a well-known property of the circle given in Euclid's *Elements*, VI, 13.

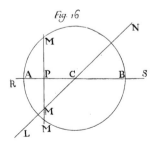

Fig. 16

It may be worth mentioning that the above deduction of the equation of the circle appears as a detail in an example about "complex lines": Euler finds the equations of the circle and the straight line in the figure and multiplies them, obtaining a "third-degree complex equation". The chapter is on the "classification of algebraic curves by order".

A general equation for the straight line had been obtained previously [*Introductio*, II, §39], in a chapter on "change of coordinates", also in an incidental manner. Boyer [*1956*, 182] says of Euler's treatment of the straight line equation that it "is characteristic for its generality, but it is startlingly abbreviated".

The length of a straight line from a point in space to the origin of coordinates is given as $\sqrt{xx + yy + zz}$, without any justification (geometrical or otherwise), but apparently the only use for this formula is to provide the equation of the sphere [Euler *Introductio*, II, Appendix § 10, 14]. On the plane, at least in three occasions [Euler *Introductio*, II, § 127, 139, 396] $\sqrt{xx + yy}$ appears as the distance to the origin (also without an explicit justification); but they are somewhat incidental: in the first two of these occasions its sole use lies in recognizing ellipses with equal axes as circles; and the third is related to conversion of polar to rectangle coordinates.

In the decade after the publication of [Euler *Introductio*], appeared two important treatises on algebraic curves, pointing in the same analytic direction: [Cramer *1750*] and [Goudin & du Séjour 1750]. They have a common characteristic, that makes their treatments of curves seem even more general than Euler's: while in the latter's work there are separate chapters for second, third and fourth-order lines, and many properties of general curves are only studied afterwards, in [Cramer *1750*] and [Goudin & du Séjour 1750] those lines are not more than interesting examples.

However analytic these three works are, each of them is a "study of higher plane curves, rather than an analytic geometry in the modern sense"[Boyer *1956*, 198]. But they represented what for some time seemed the definitive aspect of the subject of coordinate geometry; until Lagrange made an important suggestion for a somewhat new approach in 1773 (see section 4.1.1.3).

4.1.1.2 Two traditional elementary accounts: Bézout and Cousin

[Euler *Introductio*, II], [Cramer *1750*], or [Goudin & du Séjour *1756*] do not seem to represent accurately the version of coordinate geometry dominant in the second half of the 18th century for educational purposes. A good example of the standard, not-too-difficult, educational version of the subject at that time is more likely to be found in the third part of Bézout's *Cours de Mathématiques* [Bézout *1796*, III], which is dedicated to algebra and contains a section on the application of algebra to geometry, pages 289-488 (almost half of the volume, in fact)[11].

It was translated into English in the United States in 1820 instead of the corresponding section in one of Lacroix's textbooks.[12] The reasons for this choice

---

[11] According to [Boyer *1956*, 272] its "treatment of analytic geometry is typical of the time about 1775".

[12] Lacroix had published a textbook *Traité élémentaire de trigonométie et d'application de l'algèbre à la géométrie* [*1798b*], combining in one volume these two subjects; [Lacroix & Bézout

were that "analytical geometry[13] [had until then] made no part of the mathematics taught in the public seminaries of the United States", and was to have little time allotted and to be taught "in many instances [to students] at an age not sufficiently mature for inquiries of an abstract nature" (although this book was intended "for the use of the students of the University at Cambridge, New England"!); so "it was thought best to make the experiment with a treatise distinguished for its simplicity and plainness" [Lacroix & Bézout *1826*, iii].

Simple and plain it is. It is also much more *old fashioned* than [Euler *Introductio*, II] (and incredibly more elementary). Because the results of operations on geometrical magnitudes can be given either in numbers or in *lines*, the first few pages are dedicated to the "geometrical construction of algebraic quantities": from the construction of $\frac{ab}{c}$ (a fourth proportional) to that of $\frac{a\sqrt{b+c}}{\sqrt{d+e}}$ [Bézout *1796*, III, 289-303]. Then comes a long section [Bézout *1796*, III, 304-360] on the use of equations to solve geometric problems, *without using coordinates*. These problems range from inscribing a square in a given triangle to questions about volumes of simple solids. An example of this was seen in page 91.

Coordinates are finally introduced for the study of "curved lines in general, and conic sections in particular". The first example [Bézout *1796*, III, 361-372] is that of a curve defined by the property that its ordinate is a mean proportional between its abscissa and the complement of the abscissa in a given segment; after plotting the curve, Bézout deduces that it is a circle (using the defining property and, of course, Pythagoras' theorem) and proves a couple of properties about it. The only change of coordinates considered is a change of origin, from an end point of a diameter to the centre of the circle.

But the example of the circle is just an introduction. Apparently the main (or sole) purpose of coordinate geometry is the study of the conic sections [Bézout *1796*, III, 372-456]. Each one is defined by the respective property of the distances between its points and its foci (to express algebraically those distance properties, right triangles are always invoked, of course). Various properties are found or stated and proven (including ways of drawing the curves and equations for their tangents). Some changes of coordinates are given, but each is particular to a conic section, and their purpose is to be able to reduce any second-degree equation (in two unknowns) to a conic section (and thus to construct that equation, a deployment referred to in the preface [Bézout *1796*, III, ix]).[14]

After some examples [Bézout *1796*, III, 456-482], the deduction of a few trigonometric formulae [Bézout *1796*, III, 482-488] closes the volume.

1826] was a combined translation of Lacroix's trigonometry and Bézout's application of algebra to geometry.

[13]By the 1820's the expression "analytic(al) geometry" had already become popular enough to be used in the "advertisement" to this American translation. Its author seems to use it as synonymous with "application of algebra to geometry"

[14][Lacroix & Bézout *1826*] closes just after the study of the conic sections, so that it does not include the construction of equations. It is unlikely that this is due to the obsolescence of the subject, since an 1829 French edition of Bézout's *Cours* (Paris: Bachelier) still includes that section.

Of course a comparison between [Euler *Introductio*, II] and [Bézout *1796*, III, 289-488] is unfair for several reasons. One of them is that the former was part of an introduction to the calculus, while the latter was part of a general mathematical education for naval personnel.

Closer in aim to [Euler *Introductio*, II] (and to [Lacroix *Traité*, I, ch. 4]) was the first chapter of the introduction to [Cousin *1796*]. This chapter, entitled "Application de l'Algébre à la Géométrie", is one of the main additions in [Cousin *1796*], when compared to [Cousin *1777*]. It is much less elementary than [Bézout *1796*, III, 289-488], but not more modern in tone. This can be seen from the start, in the following sentence, characterizing the way to solve problems by applying algebra to geometry:

> "Tout se réduit à se procurer des équations: & comme la Géométrie
> ne nous offre pour cela que des triangles semblables ou des triangles
> rectangles; il ne s'agit que de former triangles semblables ou des triangles
> rectangles, au moyen de quelque construction simple que la nature du
> problême indique."[15] [Cousin *1796*, I, 1]

As in [Bézout *1796*, III], the first examples have little or nothing to do with coordinate geometry. However, here the pre-coordinate section is much shorter [Cousin *1796*, I, 1-6]. Two equations for the circle are deduced (from the radius as hypotenuse of a right triangle): $y = \pm\sqrt{r^2 - x^2}$ and hence, if $r = 1$, $\sin m^2 + \cos m^2 = 1$; but this is done without explicit reference to coordinates, and its purpose is not to study the circle, but rather to develop several trigonometric formulas. In the second example Cousin, without any recourse to coordinates, arrives at various formulas relating angles, sides, and area in a generic triangle.

Once again as in [Bézout *1796*, III], coordinates are introduced for the study of conic sections. These have definitions equivalent to those in [Bézout *1796*, III], but in an even more geometrical language: instead of speaking of distances, their points are the intersections of circles, or of circles and straight lines (in the case of the parabola) [Cousin *1796*, I, 6-9].

The properties of the conic sections are then studied [*1796*, I, 9-20] (including tangents, asymptotes, and infinite branches). Formulas are given for a general change of coordinates (in a very unclear way), and they are used to prove that any second-order curve is a conic section [*1796*, I, 10-12].

Unlike Bézout, Cousin considers curves of any order (although in practice he does not go beyond the third order) [*1796*, I, 20-27]. The questions asked about them have to do with their centres, diameters, and infinite branches. He also considers curved surfaces in a short section (dealing mainly with solids of revolution) [*1796*, I, 27-30].

The chapter closes with a tiny and very awkwardly placed section on geometrical loci [*1796*, I, 30-31] and another on "construction of determinate equations"

---

[15]"It all comes down to search for equations: & since Geometry does not offer for that but similar triangles or right triangles; it amounts to form similar triangles or right triangles, by means of some simple construction indicated by the nature of the problem."

[*1796*, I, 31-36]. It is somewhat mysterious what possible use this last section could have in a treatise on differential and integral calculus published in 1796.

4.1.1.3 The analytic program for elementary geometry: Lagrange and Monge

Three-dimensional coordinate geometry had a much slower development than its planar counterpart. The first major accounts of it [Clairaut *1731*; Euler *Introductio*, II, appendix] date from the 18th century, about a century after the appearance of the subject. This was partly due to the facts that the space is much harder to visualize than the plane, and that therefore space synthetic geometry is much more difficult than plane synthetic geometry. Coordinate geometry as it was before the end of the 18th century, relying heavily on diagrams and on frequent use of elementary synthetic geometry, was also much more well adapted to the plane than to space.

It should be no surprise, then, that it was in relation to three-dimensional geometry that further algebrization took place. Nor is it surprising that Lagrange was involved in that.

[Lagrange *1773b*] was the first published suggestion for a really algebrized geometry. In that memoir Lagrange studied several properties of a generic tetrahedron: the areas of its faces, its height, volume, inscribed and circumscribed spheres, centre of gravity, etc. He regarded tetrahedra as the equivalent in solid geometry of triangles in plane geometry; but he had noticed that while triangles had always been an object of the geometers' closest attention, on tetrahedra only a handful of the many possible problems had been solved [Lagrange *1773b*, 661]. However, this was not really the motivation behind this memoir: however useful the results obtained might be

> "elles serviront principalement à montrer avec combien de facilité et de succès la méthode algébrique peut être employée dans les questions qui paraissent être le plus du ressort de la Géométrie proprement dite, et les moins propres à être traitées par le calcul."[16] [Lagrange *1773b*, 662]

We can see that there is a sense of novelty here. Lagrange feels the need to explain the spirit and the method of the memoir: "Ces solutions sont purement analytiques et peuvent même être entendues sans figures"[17] [*1773b*, 661]. The memoir is in fact devoid of diagrams. Using rectangular coordinates for the significant points of the tetrahedron,

> "tout se réduit à une affaire de pur calcul, et il est très-facile de déter-
> miner la valeur des lignes qu'on veut connaitre, puisqu'il ne faut que

---

[16]"they will serve mainly to show how easily and how successfully the algebraic method can be employed in those questions that most seem to fall within the scope of Geometry proper, and appear the least suitable to be dealt with by calculation."

[17]"These solutions are purely analytic and can even be understood without figures."

prendre la somme des carrés des différences des coordonnées qui répondent aux deux extrémités de chaque ligne proposée."[18] [Lagrange *1773b*, 662]

This is a much more explicit statement of the distance *formula* than Clairaut's (see above page 94) and more general than Euler's (see above page 96). But the main innovation is that here it is a fundamental tool throughout.

A typically analytic passage in this memoir is that in which Lagrange seeks the height of the tetrahedron. Its summit being the origin of the coordinates, he takes a generic point in the base plane, with coordinates s, t, u, so that the distance between the point and the summit is $\sqrt{s^2 + t^2 + u^2}$; he then minimizes it, making its differential equal to zero, and combines the equation $u = l + ms + nt$ of the base plane, arriving at the result $\frac{l}{\sqrt{1+m^2+n^2}}$ [*1773b*, 670-672]. There is little geometrical reasoning involved here. But there is another interesting aspect in this passage: Lagrange has to resort to differential calculus, probably because perpendicularity had not yet been properly expressed in algebraic form.

An algebraic treatment of perpendicularity in space would be published by Monge [*1785a*]. This is a memoir on evolutes that contains important aspects of analytic geometry, pointing in a direction very similar to the one suggested in [Lagrange *1773b*]. A version of Monge's memoir was submitted to the Paris Academy of Sciences in 1771 (thus before the publication of [Lagrange *1773b*]), but it is not clear whether that version already included those aspects of analytic geometry − a preliminary manuscript of 1770 did not [Taton *1951*, 114]. On the paternity of this conception of analytic geometry, Lacroix would later say:

> "Lagrange a donné, dans les Mémoires de l'Académie de Berlin (année 1773), une Théorie des Pyramides, qui est un chef-d'œuvre dans ce genre; mais Monge est, je crois, le premier qui ait pensé à présenter sous cette forme l'application de l'Algèbre à la Géométrie."[19] [Lacroix *Traité*, I, xxvi]

In [*1785a*, 524-527] Monge seeks the equation of the normal plane to a space curve; for this he needs the equation of the plane perpendicular to a given straight line that passes through a given point on that straight line. Starting from two equations defining the straight line, he projects it on the three coordinate planes (by eliminating each of the variables in turn); removes the constant terms so as to have a parallel through the origin; determines the cosines of the angles between this parallel and the three coordinate axes; using these and a little trigonometry he arrives at the relation between the distances from the origin to the point where the plane intersects the parallel and the points where the plane intersects the axes; this

---

[18]"it all amounts to an affair of pure calculation, and it is very easy to determine the value of the lines we wish to know, since it is enough to take the sum of the squares of the differences between the coordinates that correspond to the extremities of each proposed line."

[19]"Lagrange gave, in the Memoirs of the Berlin Academy (year 1773), a Theory of Pyramids which is a masterpiece in this genre; but it was Monge, I believe, the first who thought of presenting under this form the application of Algebra to Geometry."

relation gives a proportion between the coefficients in the plane's equation (which is the same as that between the coefficients α, β, γ of y, x, z in the equations of the projections); it only remains to force it to pass through the point with coordinates x', y', z', which is easily done putting its equation in the form

$$\alpha[z - z'] + \beta[y - y'] + \gamma[x - x'] = 0.$$

Next, to determine the distance from a point to a straight line, Monge just has to determine the (equation of the) plane that is perpendicular to the straight line and passes through the point, intersect this plane with the straight line (which gives a point of coordinates x, y, z), and take the distance between this and the original point (which had coordinates x', y', z'): $\sqrt{(x - x')^2 + (y - y')^2 + (z - z')^2}$ [Monge *1785a*, 527-528]. This is later applied in finding the radius of curvature of a space curve in a given point.[20]

What is interesting here is that these results are in a research paper. They are only auxiliary tools, not the subject of the paper; but their explanation makes it clear that the reader was not supposed to have seen them (or similar ones) before.

In the next few years Monge published a few more memoirs on differential geometry, where he kept using elementary geometry in this *analytic* fashion.[21]

This new algebrized version of elementary solid geometry would be systematized by Monge in 1795, in his lectures at the newly founded École Polytechnique. Monge was quite influential in the setting up of the curriculum of the École, and he managed to include a course in "analysis applied to geometry" that addressed differential geometry ("a branch of science which only Monge could teach" [Taton *1951*, 40]), and also those purely algebraic solutions for elementary geometrical problems that he had been using in his research memoirs (as well as the algebra and calculus necessary for these applications)[22].

Monge supplied his students with notes containing the applications of analysis to geometry given in these lectures (*Feuilles d'Analyse appliqué à la Géométrie*). The first edition of these notes, printed in 1795, was never published as a volume and is very rare. I have only consulted the second edition, published in 1801, but according to Taton [*1951*, 121] the differences regarding analytic geometry between the first and the second edition amount only to insignificant details (that is, differences in *text*; information provided in [Belhoste & Taton *1992*, 292-301] implies a stronger association of the first edition to the contemporary course

---

[20] As the distance between that point and its corresponding straight line in the developable surface.

[21] This included what according to Boyer [*1956*, 205-206] was perhaps the first explicit appearance of the point-slope equation of the straight line: $y - y' = a(x - x')$, where a is the tangent of the angle between the straight line and the abscissa axis and x', y' are the coordinates of a given point on it [Monge *1781*, 669].

[22] An abridged syllabus of this course is in [Langins *1987a*, 130-131]. Of course, there is no guarantee that Monge really followed this syllabus. One serious possibility is that he may have taught only the geometrical applications, while others (Hachette, Malus, Dupuis) taught algebra and the calculus [Langins *1987a*, 78]. See also section 8.2.

on descriptive geometry, which in *teaching practice* may have been a significant difference – see below).

The introduction on analytic geometry in [Monge *Feuilles*] is composed by its first 14 pages (leaves nᵒˢ 1 − 3*bis*). It opens with a short paragraph on the equation of the straight line on the plane. The coordinates involved are x and z, and that is for a good reason: the objects of study in the introduction are planes and straight lines in space, but the former's traces and the latter's projections on the coordinate planes (especially the vertical ones) are fundamental tools (in fact the coordinate planes are called "plans rectangulaires des projections" [Monge *Feuilles*, nᵒ 1-iii]).

The style is very concise. Immediately after that opening paragraph Monge attacks several problems, such as finding equations for a straight line parallel to a given straight line, or perpendicular to a given straight line, a given plane, or two given straight lines; and the calculation of angles between planes and/or straight lines of distances between points, a point and a plane, or the shortest distance between skew straight lines.

On two occasions differential calculus is used [*Feuilles*, nᵒˢ 1-iii; 2-i]. In both passages the purpose of this use is to minimize distances in order to express perpendicularity (the first is very similar to [Lagrange *1773b*, 671] − see above page 100). But an algebraic alternative is given, based on the "known fact" that if a plane is perpendicular to a straight line, then their respective traces and projections are also perpendicular[23] [Monge *Feuilles*, nᵒ 2-i].

This geometry is very much algebraized, but it is not easy to understand how purely algebraic it was in practice. In 1795 Monge taught descriptive geometry to the same students and he tried to associate the two courses [Belhoste & Taton *1992*, 295]. This association is well illustrated by the fact that the problems solved algebraically in [Monge *Feuilles*, nᵒˢ 1-3 bis] are precisely the same (and almost in the same order) that were treated in lectures 1-5 and 8 of his course of descriptive geometry [Monge *Stéréotomie*, 11-12; *1992*, 292-293]; in fact, each of the leaves of that preliminary section in the first edition of [Monge *Feuilles*] has an indication for the corresponding diagram in the lecture notes of descriptive geometry [Belhoste & Taton *1992*, 295-297]. Moreover, on several occasions the reader was required to supply some basic geometrical reasoning (or to have some previous knowledge of space geometry), particularly on how to operate with projections: one example is the known fact ("on sait que...") about perpendicularity quoted in the paragraph above.

But of course it would have been impossible to dispense with all geometrical reasoning in the setting up of analytic geometry. Its purpose was to derive algebraic formulas to be used *subsequently* instead of synthetic geometry; in the deduction of those formulas, formulas previously obtained might be preferred to geometrical reasonings, but the occasional recourse to the latter was unavoidable.

---

[23]Perpendicularity on the plane had been swiftly taken care of in the opening paragraph, using the fact that, in $x = az + b$, a is the tangent of the angle between the straight line and the z-axis.

It must also be noticed that in the second edition (whose text, as has already been mentioned, seems to be almost unaltered) the association with descriptive geometry is no longer apparent: at least, there is no indication for external dia- grams, and no internal diagrams replacing them – which means zero diagrams for this introduction on analytic geometry (the whole [Monge *Feuilles*, 2nd ed] has only 10 figures). Thus, diagrams were not regarded as indispensable to the reader.

The reason for this entanglement is that Monge did never see analytic ge- ometry as a replacement for synthetic (or descriptive) geometry; rather, he saw these two as distinct ways of expressing the same objects. Each had its own ad- vantages (the "evidence" of descriptive geometry, the "generality" of analysis) and they should be cultivated simultaneously and in parallel [Monge *1795*, 317]. In later years Monge (quoted by Olivier [*1843*, vi]) would go as far as to claim that if he were to rewrite [*Feuilles*], it would have two columns with the same results: one in analysis, and the other in descriptive geometry.

4.1.2 Lacroix and analytic geometry

Lacroix was familiar with Monge's algebraic approach to geometry much before the latter's lectures at the *École Polytechnique* (and more than the common reader of Monge's memoirs on differential geometry). In fact, it was one of the topics they addressed in their correspondence, when Lacroix was in Rochefort and Besançon; Taton [*1951*, 119-120] cites a letter from 1789, kept in [Lacroix *IF*, ms 2396], where Monge answers a problem proposed to him by Lacroix, about the minimum distance between two straight lines. We have seen above (page 100) a reference in the preface to Lacroix's *Traité* to his belief on Monge's priority on "analytic geometry".

In the preface to the first volume of his *Traité*, Lacroix gave a statement of his adherence to the analytic program for geometry:

> "En écartant avec soin toutes les constructions géométriques, j'ai voulu faire sentir au Lecteur qu'il existoit une manière d'envisager la Géo- métrie, qu'on pourroit appeler *Géométrie analytique*, et qui consisteroit à déduire les propriétés de l'étendue du plus petit nombre possible de principes, par des méthodes purement analytiques, comme Lagrange l'a fait dans sa Méchanique à l'égard des propriétés de l'équilibre et du mouvement."[24] [Lacroix *Traité*, I, xxv]

Apparently, in this passage Lacroix is even introducing the new name "analytic geometry", inspired by Lagrange's "analytic(al) mechanics"[25], instead of the old

---

[24]"In carefully avoiding all geometric constructions, I would have the reader realize that there exists a way of looking at geometry which one might call *analytic geometry*, and which consists in deducing the properties of extension from the smallest possible number of principles by pu- rely analytic methods, as Lagrange has done in his mechanics with regard to the properties of equilibrium and movement". This translation is taken from [Boyer *1956*, 211].

[25]That Lagrange's *méchanique analytique* has been translated as *analytical mechanics* while

"application of algebra to geometry". The expression "analytic geometry" had occurred before: it seems to have been used for the first time, in 1709, by the French
mathematician Michel Rolle (1632-1719); there were a few other occurrences in
the 1770s and 1780s, but not associated to the "analytic program" of Lagrange
and Monge [Boyer *1956*, 155, 215-216]. However, these were isolated occurrences.
Lacroix may not have been aware of them or, if he was aware, he did not feel they
were enough to have given a definite meaning to the expression: it was available
to be used for the new kind of coordinate geometry.

   Lacroix never wrote a work bearing the expression "analytic geometry" in the
title. The textbook [Lacroix *1798b*] in which he included the subject was called
"Traité élémentaire de trigonométrie rectiligne et sphérique, et d'application de
l'algèbre à la géométrie": the old name surviving. But the chapter on "application
of algebra to geometry" contains more than what Lacroix had proposed to call
"analytic geometry".[26] Its first sections are concerned with the use of "algebraic
operations to combine several theorems of geometry so as to deduce their consequences" [Lacroix *1798b*, 83]; this is non-coordinate algebraic geometry, similar to
that seen in the works of Bézout and Cousin (section 4.1.1.2), and in the style of
the example given in page 91. There are also a few small sections on the construction of equations. It is true that the bulk of it is in fact *analytic geometry*; but it
seems that conceptually, *analytic geometry* was only a part (although the major
part) of the *application of algebra to geometry*.

   There is another possible explanation, given by Boyer [*1956*, 217], for the
absence of the phrase "analytic geometry" in the title of [Lacroix *1798b*]: Lacroix
might have avoided it because of the confusion that existed at the time as to the
distinction(s) between *analysis* and *synthesis*. In fact, on a later text about that
distinction, he wrote that

> "L'exactitude du langage semblerait demander qu'on prévint l'équivoque
> occasionnée par les divers sens dans lesquels se prend le mot *analyse*, et
> que pour cela on désignât autrement l'emploi du signe arbitraire [i.e.,
> of algebraic symbolism]."[27] [Lacroix *1805*, 2nd ed, 232]

But *what* other designation could be adopted? After discussing briefly the possibilities of *logistics* and *calculus*-"calcul" ("too vulgar" − especially as it would
bring along the word "calculators", easily confused with "arithmeticians"), Lacroix
concludes that

> "le changement de dénomination est peu important en lui-même dès que
> l'on conçoit nettement la différence des procédés; et par cette différence

---

analytic geometry is more common in English than *analytic*al *geometry* is just an unfortunate
miscoincidence.
   [26]That chapter and an appendix on analytic geometry in space, together comprise more than
two thirds of the book.
   [27]"Exactitude in language would seem to demand that the ambiguity which is caused by the
different meanings in which the word *analysis* is taken be avoided, and that therefore the use of
arbitrary signs [i.e., of algebraic symbolism] be designated differently."

on saura toujours bien quand une *analyse* méritera véritablement ce
nom, ou ne sera qu'une *synthèse* réduite en calcul."[28] [Lacroix *1805*,
2nd ed, 233]

It is worth noticing that Lacroix repeated in 1810 (in the preface to the second
edition) his suggestion for the name "analytic geometry" [*Traité*, 2nd ed, I, xxxvii].
He had *not* changed his mind.

The phrase "analytic geometry" would be used for the first time in the title of
a work in 1804, in the second edition of a textbook by Frédéric-Louis Lefrançois;
that title was *Essais de géométrie analytique*; the title of the first edition (1801)
had been *Essais sur la ligne droite et les courbes du second degré*.[29]

It is important to examine the relationship that Lacroix proposed between
analytic geometry and synthetic geometry. In the preface to the *Traité*, still referring to the chapters on geometry, Lacroix stated very clearly that his insistence on
the "advantages of algebraic analysis" did not mean a criticism of either synthesis
or geometrical analysis. He just thought that geometrical considerations and algebraic calculations should be kept apart as much as possible; and that its respective
results "s'éclairassent mutuellement, en se correspondant, pour ainsi dire, comme
le texte d'un livre et sa traduction"[30]. This is remarkably similar to Monge's views
mentioned above, and to Monge's practice when teaching at the *École Polytechnique*. The letter quoted in the beginning of this chapter shows that in 1794 Lacroix
already had this conception. It certainly is a very important conception in Lacroix's *Traité*, not only in the two final chapters of volume 1, but also in several
passages in volume 2.

4.1.2.1 Analytic geometry on the plane in Lacroix's *Traité*

In 1797 analytic geometry (in the new sense) had not yet been applied to the plane
– with the sole exception of the short opening paragraph of [Monge *Feuilles*]. It
was up to Lacroix to do this, systematically, for the first time. As Boyer [*1956*, 211]
puts it (speaking of both [Lacroix *Traité*] and [Lacroix *1798b*]: "Here Lacroix did
for two dimensions what Lagrange and Monge had done for three-space"; he even
finds it "probably fair to speak of the new program as 'analytic geometry in the
sense of Lagrange, Monge and Lacroix'". At least some of their contemporaries
had a similar perspective, as can be seen by the title of a book published in

---

[28]"the change in denomination is not very important in itself, as long as the difference between
the processes is clearly understood; and by that difference it will always be known when an
analysis is really worthy of that name, or is just a *synthesis* reduced to calculus".

[29]Both Taton [*1951*, 135] and Boyer [*1956*, 220] wrongly ascribe this little priority to Jean-
Baptiste Biot. Biot published in 1802 a *Traité analytique des courbes et des surfaces du second
degré*; he changed the title of this work in the second edition (1805) to *Essai de géométrie
analytique, appliqué aux courbes et aux surfaces du second degré*. Boyer had the excuse that
he apparently did not see the first edition and assumed it had the same title as the second
[Boyer *1956*, 273]; but Taton [*1951*, 132] gave all these (and more) bibliographic details.

[30]"should serve for mutual clarification, corresponding, so to speak, to the text of a book and
its translation". This translation is taken from [Boyer *1956*, 212].

1801 by Louis Puissant: *Recueil de diverses propositions de géométrie résolues ou démontrées par l'analyse algébrique, suivant les principes de Monge et de Lacroix*[31] [Taton *1951*, 132].

Lacroix did include several diagrams, but usually their role is purely illustrative. Apart from a few exceptions (particularly those related to graphical representation), they could be omitted with only a pedagogical loss, not a logical one.

The first few pages of chapter 4 are taken up by a short introduction to rectangular coordinates and an extensive study of fundamental formulae for straight lines and distances [Lacroix *Traité*, I, 327-332].

The usual form of the equation for a straight line will be $y = ax + b$; this form is thoroughly explored: a is the tangent of the angle between the line and the abscissa axis, b the ordinate at the origin, $-\frac{b}{a}$ the abscissa at the origin. Much attention is given to negative coordinates. The equation of a straight line that passes through the points that have coordinates α, β and α', β' is easily found combining $\beta = a\alpha + b$ with $\beta' = a\alpha' + b$; the equation of the straight line that passes through the point with coordinates α, β and is parallel to $y = a'x + b'$ is almost immediately given as $y - \beta = a'(x - \alpha)$ because $y - \beta = a(x - \alpha)$ is the general equation of the lines satisfying the first condition and the coefficient a' gives the second.

A slightly unnecessary geometrical *intrusion* occurs apropos of perpendicularity: similar triangles are invoked to justify that $-\frac{1}{a}$ is the slope coefficient of a straight line perpendicular to $y = ax + b$; it would have been more *algebraic* to say that that is the cotangent of the angle which has a as tangent, as in [Monge *Feuilles*, n° 1-i].

To justify the distance formula $\sqrt{(\alpha' - \alpha)^2 + (\beta' - \beta)^2}$ Lacroix invokes a right triangle. It could not have been otherwise. But once these formulas have been established, it takes Lacroix only six lines (and no diagram) to deduce a formula for the distance of a point to a straight line [*Traité*, I, 332].

The equation of the circle is explicitly derived from the distance formula, much further along, in the section on osculation of curves [*Traité*, I, 392].

Of course all of these preliminary results are quite elementary. Also its substance was not really new. But this form of exposition was. Boyer [*1956*, 213-214] stresses as novel the "continued emphasis upon the almost automatic application of formulas[, making] the subject resemble an algorithm, in which independent reference to the geometrical properties of figures is dispensed with".

Afterwards, Lacroix included those preliminary considerations in his [*1798b*] and subsequently several textbooks were published that also contained them: Taton [*1951*, 132-133] lists six books on the new *analytic geometry* between 1801 and 1809, not including Monge and Hachette's *Application d'algèbre à la géométrie* of 1802. Because of this, in 1810 Lacroix was able to remove this preliminary section

---

[31] *Collection of several propositions of geometry solved or demonstrated by algebraic analysis, following the principles of Monge and Lacroix.*

from the second edition [*Traité*, 2nd ed, I, xxxvii].

The rest of the plane analytic geometry is not so original: [Euler *Introductio*, II], [Cramer *1750*] and [Goudin & du Séjour *1756*] provided versions of coordinate geometry of algebraic curves beyond the straight line and circle that would fit well in an *analytic geometry* (these three works are cited in the *Traité*'s table of contents for chapter 4).

Right after the preliminaries on the straight line comes a section in which Lacroix addresses the graphical representation of algebraic curves, in the case where it is possible to *solve* the equation in y (that is, to turn it into several expressions such as $y = f(x)$ – the roots of the equation). There is one instance in which plotting by joining points is recommended [Lacroix *Traité*, I, 336-337]; but the main tool is the study of the roots of $f(x)$: they show the number of branches of the curve, which of them are infinite, etc. Points of the curve with remarkable characteristics ("particularités remarquables") are called *singular points* (including cases in which the partial derivatives at the point are not null). Several kinds of singular points are introduced: multiple points, inflexion points, conjugate (i.e. isolated) points, nodes and cusps ("points de rebroussement"). Lacroix strives to give analytical characterizations of these singular points (speaking of multiple values, situations in which certain coefficients are null, etc.); but that is not always feasible, as when introducing inflexion points, where he appeals to the graph of an example curve [*Traité*, I, 339].

Next comes transformation of coordinates. This is a very powerful tool. It allows Lacroix to give a short study of second-order curves (without any diagram), and briefly indicate how the same could be done for third-order curves [*Traité*, I, 345-351]. It also gives a means to find centres and diameters of curves [*Traité*, I, 351-353].

Transformation of coordinates also provides a "very elegant means" to determine the tangent to a curve in a given point M: M being the origin of the new coordinates u, t (which will be oblique), and the u axis being parallel to the x axis, one tries to get a t axis that will be tangent to the curve. Imagining first that it cuts the curve in some point m besides the origin, one *approaches* m and M until they are the same; since there will be two null values of t at the same time, the new equation of the curve will be divisible by t^2 when $u = 0$ [*Traité*, I, 353-355].[32]

Similar considerations on divisibility of a transformed equation by powers of t give algebraic characterizations of multiple points and inflexion points.

This section finishes with a few considerations on the number of possible intersections between two algebraic curves of given degrees, and the number of points necessary to determine a curve of a given degree (and, in a footnote, a statement of Cramer's paradox).

---

[32]To be more precise, it will be divisible by t^{n+1}, where n is the largest integer by which it would be divisible in general (that is, the multiplicity of that point). This procedure can be found in [Cramer *1750*, 460-464] and [Goudin & du Séjour *1756*, 77-78]. Transformation of coordinates are fundamental tools in these books.

Next comes a section on the "application of the expansion of functions into series to the theory of curves". It might seem that such a section should be classified as differential, rather than analytic, geometry, since it involves power-series expansions; but in the context of late 18th-century mathematics it is an application of *algebraic analysis*. Therefore we will examine it here, although an important passage on tangents will be postponed to the section on differential geometry (4.2.1.2).

Lacroix takes up again an example he had given in chapter 2, to illustrate a (non-differential) method by Lagrange [*1776*, §2-5] for obtaining "convergent" series. From the equation

$$ax^3 + x^3y - ay^3 = 0$$

he had obtained [Lacroix *Traité*, I, 229-230] four power series:

$$y = x + \frac{x^2}{3a} - \frac{x^4}{81a^3} + \frac{x^5}{243a^4} \text{ etc.} \tag{4.1}$$

is "more convergent as x is smaller"; while

$$y = -a - a^4x^{-3} - 3a^7x^{-5} - 12a^{10}x^{-9} - 55a^{13}x^{-12} \text{ etc.,} \tag{4.2}$$

$$y = a^{-\frac{1}{2}}x^{\frac{3}{2}} + \frac{1}{2}a - \frac{3}{8}a^{\frac{5}{2}}x^{-\frac{3}{2}} + \frac{1}{2}a^4x^{-3} \text{ etc., and} \tag{4.3}$$

$$y = -a^{-\frac{1}{2}}x^{\frac{3}{2}} + \frac{1}{2}a + \frac{3}{8}a^{\frac{5}{2}}x^{-\frac{3}{2}} + \frac{1}{2}a^4x^{-3} \text{ etc.} \tag{4.4}$$

are convergent for large values of x. (4.1) gives $y = x$ as tangent to the curve at the origin (we will see how in section 4.2.1.2); (4.2)-(4.4) give the asymptotes $y = -a$, $y = a^{-\frac{1}{2}}x^{\frac{3}{2}}$ and $y = -a^{-\frac{1}{2}}x^{\frac{3}{2}}$. Asymptotes correspond to infinite branches of the curve, and this is explored by Lacroix, including a classification in hyperbolic and parabolic branches: the former have straight lines as asymptotes, as in the hyperbola; the asymptotes of the latter are (generalized) parabolic curves. But Lacroix does not spend an awful lot of time on this. He mentions that Euler and Cramer had used the number and nature of infinite branches to classify third- and fourth-order curves into genera, but "ces détails, plus curieux qu'utiles, sortent entièrement du plan que je me suis proposé"[33] [Lacroix *Traité*, I, 368].

Analytic geometry seems to be concerned almost exclusively with algebraic curves. Lacroix includes a section on transcendental curves, but only after having introduced differential geometry (it is the penultimate section of chapter 4), and it mixes analytic and differential considerations. He favours differential equations over (non-differential) transcendental ones: for instance, he gives a differential equation between the coordinates of the cycloid, but not a non-differential one, because it would involve an inverse sine.

---

[33]"those details, more curious than useful, entirely depart from the plan I have proposed myself".

Besides the cycloid, only the logarithmic and the spirals are dealt with. The study of spirals brings the only really relevant aspect for analytic geometry in this section: polar coordinates, with formulas for transformation of polar into rectangular coordinates and vice-versa.

4.1.2.2 Analytic geometry in space in Lacroix's *Traité*

The three-dimensional version of the *new* analytic geometry had already been presented, in [Monge *Feuilles*] (see section 4.1.1.3). But Lacroix's presentation has significant differences in exposition. Lacroix explains the basics of coordinate geometry in space carefully, not assuming a previous knowledge of descriptive geometry, as Monge apparently had done. It is true that Lacroix refers occasionally to his own textbook on descriptive geometry [*1795*] to justify certain reasonings; but overall his exposition is much more self-contained than Monge's – Lacroix's references to his [*1795*] seem sometimes superfluous. And when both he and Monge explain the same thing, Lacroix is more detailed and clearer.

Lacroix starts by introducing projections in space, the three coordinate planes, and their intersections (the three coordinate axes).

Then come two pages on how a first-degree equation corresponds to a plane (culminating on the equations of its intersections with the coordinate planes) – the closest to this one can find in [Monge *Feuilles*] is contained in problem II, which occupies half a page. The equation of any plane will be presented as

$$Ax + By + Cz + D = 0$$

for reasons of symmetry. It must always be kept in mind that any one of the constants may be regarded as equal to 1, or determined by particular conditions [Lacroix *Traité*, I, 438]. Monge [*Feuilles*, n° 1-ii,iii] had given similar considerations.

A straight line is characterized by the intersection of any two planes that contain it, but a clear preference is given to those that are perpendicular to the coordinate planes, so that none of their equations contains all the three coordinates (and of course, such that they represent the projections of the line).

Having established the equations of the plane and the straight line, Lacroix proceeds to solve several problems, most of them similar to those found in [Monge *Feuilles*]: for example, to determine the plane that passes through three given points; or to find the equation of a plane perpendicular to a given straight line. In this second example, the *known fact* to which Monge had appealed to, and that was quoted in page 102 above, is also invoked, but here a clear reference is given to [Lacroix *1795*, 24].

Although the problems are very similar, the solutions are not always the same. For example, to determine the angle between two planes, Monge [*Feuilles*, n°s 2-iv, 3-i] asks to conceive a perpendicular to one of the planes lowered from any point on the other, and a perpendicular to this other plane lowered from the foot of the first perpendicular; it is obvious ("il est évident") that the quotient of the second perpendicular divided by the first is the cosine of the angle between the planes (of course Monge is thinking here of the *lengths* of the *segments* of the lines determined by the planes). In the next problem, to determine the angle between two straight lines, he applies the formula just obtained to two planes perpendicular to the given lines [Monge *Feuilles*, n° 3-i,ii].

Lacroix's solution is simpler to follow, requiring less geometrical reasoning: he had just deduced the distance formula and the equation of the sphere; to determine the angle between two straight lines he intersects them with a sphere; the distance between (two of) the intersections will be the chord of the angle, from which the cosine is easily derived. Next, to determine the angle between two planes he only has to calculate the cosine of the angle between two straight lines perpendicular to the planes [*Traité*, I, 444-446]. The cosine of this later angle is

$$\frac{A'A + BB' + CC'}{\sqrt{(A^2 + B^2 + C^2)(A'^2 + B'^2 + C'^2)}}$$

(where the planes are given by the equations $Ax + By + Cz + D = 0$ and $A'x + B'y + C'z + D' = 0$), so that it is immediate to conclude that if the planes are perpendicular we will have

$$AA' + BB' + CC' = 0 \qquad (4.5)$$

(naturally this is to be found also in [Monge *Feuilles*]).

The preliminary section on planes and straight lines finishes with two formulas derived using differential calculus: one on the minimum distance between two straight lines and the other on a straight line perpendicular to a given plane and through a given point (which of course also amounts to a minimum distance). It is interesting to note that Lacroix decided not to use the purely algebraic solution to the former problem that Monge had given to him in 1789 (see page 103 above).

The second (and final) section on analytic geometry in space is entitled "On second-order curved surfaces" [Lacroix *Traité*, I, 448-465]. But it contains a little more than that, since to study properly those surfaces it is convenient to simplify their general equation

$$\left. \begin{array}{r} Ax^2 + By^2 + Cz^2 + 2Dxy + 2Exz + 2Fyz \\ +2Gx + 2Hy + 2Kz \\ -L^2 \end{array} \right\} = 0. \qquad (4.6)$$

This is done by transformation of coordinates, which of course has to be discussed previously.

This approach to the study of quadric surfaces came from chapter 5 in the appendix to [Euler *Introductio*, II], "the first unified treatment of the subject" [Boyer *1956*, 189]. That chapter is the sole item cited in the table of contents of Lacroix's *Traité* for this section. But it must be noted that the formulas given by Lacroix for the transformation of coordinates are not those given by Euler (which were non-symmetric and involved the sines and cosines of the angles between the old and the new axes); instead he uses formulas given for the first time in a paper by Lagrange "sur l'attraction des sphéroïdes elliptiques" [Lagrange *1773a*, 646-648]. He reports Lagrange's derivation of those formulas: if the origin remains the same (it is easy to translate it afterwards), the most general form for the old coordinates in terms of the new is

$$x = \alpha t + \beta u + \gamma v$$
$$y = \alpha' t + \beta' u + \gamma' v$$
$$z = \alpha'' t + \beta'' u + \gamma'' v.$$

But of course the distance to the origin remains the same, that is, $t^2 + u^2 + v^2 = x^2 + y^2 + z^2 = (\alpha t + \beta u + \gamma v)^2 + (\alpha' t + \beta' u + \gamma' v)^2 + (\alpha'' t + \beta'' u + \gamma'' v)^2$, whatever the values of t, u, v, whence

$$\left.\begin{array}{l} \alpha^2 + \alpha'^2 + \alpha''^2 = 1 \\ \beta^2 + \beta'^2 + \beta''^2 = 1 \\ \gamma^2 + \gamma'^2 + \gamma''^2 = 1 \end{array}\right\} \text{ and } \left\{\begin{array}{l} \alpha\beta + \alpha'\beta' + \alpha''\beta'' = 0 \\ \alpha\gamma + \alpha'\gamma' + \alpha''\gamma'' = 0 \\ \beta\gamma + \beta'\gamma' + \beta''\gamma'' = 0. \end{array}\right. \quad (4.7)$$

These conditions allow us to determine six of the nine constants involved. The other three are dependent on the particular transformation. [Lacroix *Traité*, I, 451-452]

But just prior to this Lacroix [*Traité*, I, 450-451] also presents a different derivation for a set of similar formulas: given

$$At + Bu + Cv = 0$$
$$A't + B'u + C'v = 0 \quad (4.8)$$
$$A''t + B''u + C''v = 0$$

as the equations in the new coordinates for the old coordinate planes (y,z, x,z, and x,y, respectively) and since the coordinates of a point are equal to its distances to the coordinate planes, it follows from a formula obtained previously that

$$x = -\frac{At + Bu + Cv}{\sqrt{A^2 + B^2 + C^2}}$$
$$y = -\frac{A't + B'u + C'v}{\sqrt{A'^2 + B'^2 + C'^2}} \quad (4.9)$$
$$z = -\frac{A''t + B''u + C''v}{\sqrt{A''^2 + B''^2 + C''^2}}.$$

Now, in each of the equations (4.8) there is one superfluous constant; therefore it is possible to put

$$A^2 + B^2 + C^2 = 1$$
$$A'^2 + B'^2 + C'^2 = 1 \quad (4.10)$$
$$A''^2 + B''^2 + C''^2 = 1$$

so that (4.9) become

$$x = -At - Bu - Cv$$
$$y = -A't - B'u - C'v$$
$$z = -A''t - B''u - C''v.$$

Also, because the coordinate planes are perpendicular, from (4.5) we have

$$AA' + BB' + CC' = 0$$
$$AA'' + BB'' + CC'' = 0$$
$$A'A'' + B'B'' + C'C'' = 0.$$

These formulas had been deduced by Monge [*1784-1785*, 28; *1784a*, 112-114], although without any reference to (4.9): (4.10) had been chosen "to simplify the expressions" (but which expressions?).

Lacroix then combines both sets of formulas, arriving at several results, including that $\alpha, \beta, \gamma, \alpha', \beta', \ldots$, taken with opposite signs, give the cosines of the angles between the old and the new coordinate planes.

Although Lacroix does not cite either of the memoirs by Lagrange or Monge in the table of contents for this section, he does cite their names in the text, apropos of further calculations for the determination of the constants in particular transformations. He refers the reader to Lagrange's *Méchanique analitique* and quotes (and praises) a few formulas that can be found in [Monge *1784a*].

Of these two procedures, Lagrange's is certainly shorter. But Monge's, at least in Lacroix's version, seems clearer and it is a fine example of *analytic geometry*: it is algebraic, but the calculations, while not requiring diagrams to be understood, can be given geometrical meanings (perpendicularity, distance of a point to a plane) − like "the text of a book and its translation" (see pages 89 and 105 above). Lagrange does use a distance formula at the start, but apart from that he − typically − compares coefficients.

Returning to (4.6), using a translation of the origin followed by a rotation of the axes, Lacroix reduces it to

$$A't^2 + B'u^2 + C'v^2 - L'^2 = 0$$

which gives the second-degree surfaces that have a centre. Lacroix then studies them by giving particular signs − or eliminating − each coefficient, and then cutting plane sections and analysing the resulting second-degree curves. Recognizing that the transformation of coordinates he had done is not always possible, Lacroix returns to (4.6) for a second, more general one, in order to study the second-degree surfaces that do not have a centre. The result is

$$Ax'^2 + By'^2 + Cz'^2 + 2K'z' = 0$$

and a similar study follows.

Lacroix does not report Euler's taxonomy (elliptic hyperboloid, etc.); he seems more concerned with recognizing conic, cylindric, and revolution surfaces.

He also dedicates only one short article to asymptotes of second-degree surfaces, something to which Euler had given considerably more attention, connected as it was with the question of part(s) of the surface going to infinity. Similarly, while Euler had dedicated his whole final chapter to intersections of surfaces, Lacroix has one article (half a page) on this.

This section — and analytic geometry — finishes with another short article, on "polar coordinates" in space. Only a few formulas are presented, but Lacroix manages to introduce two different systems: the first corresponds to what we call spherical coordinates, while the other, "more symmetrical", uses the three angles π, ψ, φ between the radius vector and the coordinate axes, so that

$$x = r\cos\pi, \quad y = r\cos\psi, \quad z = r\cos\varphi$$

(where, of course, r is the distance of the point from the origin); clearly there is one unnecessary coordinate: as it happens,

$$\cos^2\pi + \cos^2\psi + \cos^2\varphi = 1.$$

Both systems had been introduced by Lagrange: the first in [*1773a*, 626-627]; the second in his *Méchanique analitique* [Taton *1951*, 127].

4.2 Differential geometry

4.2.1 Differential geometry of plane curves

4.2.1.1 Differential geometry of plane curves in the 18th century

Differential calculus developed in part from techniques used in the 17th century to study certain properties of curves [Pedersen *1980*]. It is only natural that the most prominent of its applications in its initial period was precisely the study of those properties of curves.

The first textbook on the differential calculus [l'Hôpital *1696*] is also a textbook on differential geometry of plane curves, as can be seen from its full title: *Analyse des infiniment petits pour l'intelligence des lignes courbes*[34]. It can also be seen from its table of contents, where the titles of seven chapters, out of ten, refer explicitly to curves. L'Hôpital teaches how to use the differential calculus to find the tangents of curves, their points of inflexion and cusps, their evolutes and radii of curvature (called "radii of the evolute"), the caustic curves generated by reflection, those generated by refraction, envelopes of families of curves, and a few more things.

L'Hôpital [*1696*, 3] puts as a postulate that a curve be considered as a polygon with an infinite number of sides, each of them infinitely small. To find a tangent it is enough to prolong one of these infinitely small sides (this is in fact his *definition* of tangent [*1696*, 11]; see figure below). Given a curve AM by an equation between x (AP) and y (PM), if we wish to draw the tangent MT, we should conceive another ordinate, mp, infinitely close to PM, so that $Pp = MR = dx$ and $Rm = dy$; the triangles mRM and MPT are similar, so that $dy.dx :: MP.PT$, and therefore the subtangent PT is equal to $\frac{ydx}{dy}$. The subtangent is information enough to draw ("mener") the tangent.

---

[34] *Analysis of the infinitely small, for the understanding of curved lines.*

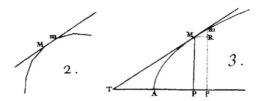

The treatment of curvature in [l'Hôpital *1696*, ch. 5] is attached to the theory of evolutes and involutes: given a curve BDF, one is asked to conceive a string $ABDF$ wrapped against it, fixed in F and extended to A; keeping the string taut and unwrapping it, the point A describes a new curve AHK; BDF is then the *evolute* of AHK;[35] the straight portions of the string AB, HD, KF are the *radii of the evolute*.

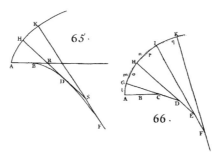

Regarding the curve BDF as a polygon with infinitely small sides (BC, CD, DE, EF), AHK can be seen as composed of infinitely small arcs of circle (AG, GH, HI, IK), the centres of those circles being the points of the evolute (C, D, E, F).[36] This means that the radii of the evolute are tangent to the evolute and normal to the involute. It also means that curvature can be measured, since "la courbure des cercles augmente à proportion que leurs rayons diminüent"[37] [*1696*, 73].

[l'Hôpital *1696*] soon became the standard account of the application of differential calculus to the theory of curves. It remained standard for a long time. The corresponding sections in [Bézout *1796*, IV] follow l'Hôpital closely, and the table of contents for the section on the "use of differential calculus to find the tangents of Curves, their inflexions and their retrogressions" in [Lacroix *Traité*, I, xxxii] contains one single item: precisely [l'Hôpital *1696*]. As we will see below, Lacroix did have two more sources for this section − and he mentions them in the

---

[35] AHK is usually called an *involute* (in French: développante) of BDF, but l'Hôpital does not seem to give it any particular name.

[36] An interesting remark is that between the curve and any of these circles it is impossible to pass another circle [l'Hôpital *1696*, 73]. It is interesting because Lagrange will use this property as a definition of contact.

[37] "the curvature of the circles increases proportionally to the decrease of their radii"

text − but they were then very recent and not yet published.

Nevertheless, there were competing approaches to calculate these curve-re-lated quantities. One of them, of course, was the method of limits. D'Alembert, in the article "Différentiel" of the [*Encyclopédie*], gave a famous example of the calculation of the ratio between the ordinate and the subtangent of a parabola as the limit of the ratios between the ordinate and the subsecants. Cousin [*1796*] calculated tangents, inflexion points, cusps, evolutes and radii of curvature in a chapter dedicated to the method of limits, even before introducing the differential calculus.

There were also *algebraic* approaches, regarded as belonging to the applica-tion of algebra to geometry, rather than the application of differential calculus to geometry. Algebraic methods were sometimes regarded as more appropriate for the study of algebraic curves. That is the point of view in a book entitled *Usages de l'Analyse de Descartes, Pour découvrir, sans le secours du Calcul Différentiel, les Proprietés, ou Affections principales des Lignes Géométriques de tous les Or-dres*[38] [Gua de Malves *1740*] – "lignes géométriques" referring in fact to algebraic curves; a similar stand is found in [Cramer *1750*] and [Goudin & du Séjour *1756*]. We saw above (page 107) Lacroix use one of those algebraic methods, taken from those books: an application of transformation of coordinates. But those methods could only be justified with recourse to either infinitesimal or limit-oriented argu-ments. In the case of the tangent method used by Lacroix, its justification involves a point *approaching* another until both are the same, so that in fact this is not very distant from the method of limits.

An interesting case, as usual, is that of Euler. [Euler *Differentialis*] does not include any applications to geometry. But some of the problems that could be treated as such are studied in [Euler *Introductio*, II]; in an *algebraic* fashion, of course. The process to find tangents is the following: given a curve nMm, its equation in x and y, and a point M in it with abscissa $AP = p$ and ordinate $PM = q$, we translate the origin of the coordinates to M, and call t, u the new coordinates; the new equation for the curve is found simply by substituting $p + t$ for x and $q + u$ for y; but since the curve passes through the new origin, the new equation cannot have an independent term, so that it is of the form

$$0 = At + Bu + Ct^2 + Dtu + Eu^2 + Ft^3 + Gt^2u + Htu^2 + \&c.$$

Now, taking "very small" values of t, u will be also very small, but t^2, tu and u^2 will be even smaller, t^3, t^2u, tu^2, u^3, etc. much smaller even, and so on. Thus, all these terms can be omitted, and

"remanebit ista aequatio $0 = At + Bu$, quæ est æquatio pro Linea recta $M\mu$ per punctum M transeunte, atque indicat hanc rectam, si punctum

---

[38] *Uses of Descartes' Analysis, To find, without the aid of the Differential Calculus, the main Properties, or Affections of the Geometrical Lines of all Orders.*

m ad M proxime accedat, cum Curva congruere."[39] [Euler *Introductio*,
II, §288]

We recognize here the use of *Arbogast's principle*, or rather of "Euler's criterion"
(see page 65 above).

Later in the same book, Euler neglects only the terms of third and higher
orders to obtain an *osculating parabola*, the vertex of which coincides with the
infinitely small arc Mm. Because he wants to measure the curvature of a curve, he
decides that it is equal to the curvature of the osculating parabola at its vertex.
But a parabola is not the ideal figure to help measure curvature: the circle is,
because it has the same curvature at every point and because this curvature is
inversely proportional to its radius. So what he wants is an *osculating circle*. The
way to define this is through the parabola: the osculating circle is the circle that
shares its osculating parabola with the curve at the given point. The radius of this
circle is the *osculating radius* or *radius of curvature* of the curve [*Introductio*, II,
§304-310].

So, Euler's *algebraic* method rests on a mixture of naïve limits ("m approaches M") and the neglect of higher-order infinitesimals. But it contains a fruitful
idea, typical of him: to take advantage of a power-series form of the equation of
the curve.

This idea was expanded by Arbogast in his 1789 memoir on the principles
of the calculus (see section 3.1.4), where he developed a theory of osculation.
Arbogast considered two curves with one common point M: one of the curves was
given, while on the other certain conditions were to be determined according to
how "intimately" it should touch the former. Their expressions should be put in
power-series form, so that we would have as equation for the given curve

$$y' = y + \frac{dy}{dx}\Delta x + \frac{d^2 y}{1 \cdot 2\, dx^2}\Delta x^2 + \frac{d^3 y}{1 \cdot 2 \cdot 3\, dx^3}\Delta x^3 + \&c.$$

and for the one to be specified

$$u' = u + \frac{du}{dt}\Delta t + \frac{d^2 u}{1 \cdot 2\, dt^2}\Delta t^2 + \frac{d^3 u}{1 \cdot 2 \cdot 3\, dt^3}\Delta t^3 + \&c.^{40}$$

Now, in order to have them meet at the point M with coordinates x, y, we make
the first terms equal, that is, we put $u = y$ and $t = x$; and in order to have y'
and u' correspond to the same ordinate, we also put $\Delta t = \Delta x$. So now we have

---

[39]"it will remain this equation $0 = At + Bu$, which is an equation of a straight line $M\mu$ passing
through the point M, and that indicates that if m approaches M this straight line will coincide
with the curve."

[40]Two remarks on notation: y' and u' are not derivatives, of course — y' stands for $y(x + \Delta x)$
and u' for $u(t + \Delta t)$; also, the difference in coordinates (x, y for one curve and u, t for the
other) is related to the usual 18th-century conflation between symbols for variables and for their
values: x, y represent the coordinates of one fixed point (M) and x', y' represent the (values of)
coordinates assumed by the first curve — and could not be used for a different curve.

$u' = y + \frac{du}{dt}\Delta x + \frac{d^2 u}{1\cdot 2\, dt^2}\Delta x^2 + \frac{d^3 u}{1\cdot 2\cdot 3\, dt^3}\Delta x^3 + \&c$ [Arbogast *1789*, § 47-48]. Putting in addition $\frac{du}{dt} = \frac{dy}{dx}$, this curve is tangent to the other one [*1789*, § 50].

Why is it so? Arbogast argued that the two curves do not intersect other than in M, at least not in the range of abscissas from $x - \Delta x$ to $x + \Delta x$; this for a small Δx – small enough for $\frac{d^2 y}{dx^2}\Delta x^2$ and $\frac{d^2 u}{dt^2}\Delta x^2$ to be greater than the sums of the remaining terms in the series. That is, he used "Arbogast's principle" where Euler had used "Euler's criterion"; but notice that Arbogast's argument is much more algebraic – instead of having the two curves "coincide" infinitesimally (or in the limit), he argued that y' is greater or smaller than u' according to whether $\frac{d^2 y}{dx^2}$ is greater or smaller than $\frac{d^2 u}{dt^2}$ [*1789*, § 50].

There are other advantages in Arbogast's theory. One is that the touching curve does not need to be a straight line. Another is its adaptation to osculation. Tangency is called *first-order contact*. If in addition we put $\frac{d^2 u}{1\cdot 2\, dt^2} = \frac{d^2 y}{1\cdot 2\, dx^2}$ we get a more intimate contact, called *second-order contact*; and so on. Of course this gives a much more elegant way of defining the osculating circle than Euler's resort to the osculating parabola: it is just a circle with a second-order contact. In this way tangency and curvature are united under the same theory.

However, this union was not so novel: the idea of orders of contact, and the names "first-order", "second-order", etc., had already been presented by Lagrange [*1779*, art. III]: given a curve with equation $V = 0$, for another curve to have a first-order contact with it, it would have to satisfy (at the point of contact) the equations $V = 0$ and $dV = 0$; for a second-order contact, it should satisfy in addition $d^2 V = 0$; and so on. But Lagrange had defined first-order contact by the meeting of two points of intersection, second-order contact by the meeting of three points of intersection, third-order contact by the meeting of four points of intersection, and so on. Moreover, these definitions were perfunctory: he had not given a justification based on them for the equations that the contacting curve had to satisfy (nor any other justification). Arbogast's theory of osculation can thus be seen as a justification of Lagrange's.

A justification of his theory of the contact of curves by comparison of coefficients in power-series expansions was ideal for Lagrange, and he adopted it and improved upon it in [*Fonctions*]. Lagrange starts the chapter on applications to geometry by adopting a definition of tangent line inspired by the ancient (Greek) geometers: "une ligne droite est tangente d'une courbe, lorsqu'ayant un point commun avec la courbe, on ne peut mener par ce point aucune autre droite entre elle et la courbe"[41]. He contrasts this definition with those used in the 17th and 18th centuries: secants of which the two points of intersection are united; prolongation of an infinitely small side of the curve seen as a polygon with infinite sides; and the direction of the movement by which the curve is described.[42] The methods based

---

[41]"a straight line is tangent to a curve when, having a point in common with the curve, it is not possible to draw any other straight line between them"

[42]The first of these definitions is the one used by himself in [*1779*]. We have also seen it being used by Lacroix (following Cramer and Goudin and du Séjour) in a context of analytic geometry

on these definitions were general and simple, but lacked the evidence and rigour of the ancient proofs. Fortunately there was now his theory of analytic functions. [Lagrange *Fonctions*, 117-118]

To apply his definition of tangency, Lagrange considers a curve with equation $y = f(x)$, a different one – supposed to be tangent to the first – with equation $y = F(x)$, and a third one – the one that will be proved not capable of being drawn between the first two – with equation $y = \varphi(x)$.[43] These curves are supposed to intersect at a point of coordinates x and $f(x) = F(x) = \varphi(x)$. He then examines what happens close to that point, that is, when the abscissa is $x + i$. For this he considers the differences between the ordinates of the first curve and the other two curves:

$$D = f(x + i) - F(x + i),$$
$$\Delta = f(x + i) - \varphi(x + i).$$

Expanding the functions using the Lagrange remainder (3.8), these differences become

$$D = i(f'(x) - F'(x)) + \frac{i^2}{2}[f''(x + j) - F''(x + h)],$$

$$\Delta = i(f'(x) - \varphi'(x)) + \frac{i^2}{2}[f''(x + j) - \varphi''(x + k)]$$

(where h, j and k are indeterminate quantities between 0 and i)[44]. Now, if $f'(x) = F'(x)$, D reduces to

$$\frac{i^2}{2}[f''(x + j) - F''(x + h)]$$

and as long as $f'(x) \neq \varphi'(x)$, D will be less than Δ for values of i small enough: it is sufficient to take values of i small enough for $f'(x) - \varphi'(x)$ to be larger than $\frac{i}{2}[\varphi''(x+j) - F''(x+h)]$.[45] This means that if two curves have the same derivative at a common point, then no curve with a different derivative can pass between them [Lagrange *Fonctions*, 118-120]. In other words, they are tangent.

Likewise, if two curves have the same first and second derivatives at a common point, then no curve with a different first or second derivative can pass between

---

(page 107 above); it is also used by Euler in the passage above, and it is implicit in the limit-oriented works, like those by d'Alembert and Cousin referred to above. The second definition was the one commonly used in differential calculus; its use by l'Hôpital was mentioned above. The third definition had fallen somewhat in disuse after the end of the 17th century, except possibly in the English method of fluxions.

[43]In fact the equations are $y = fx$, $q = Fp$, and $s = \varphi r$, although the coordinate axes are the same. That is because Lagrange also conflates variables and their values; taking the same abscissa for the three curves is done by taking $r = p = x$, and the curves intersect at that point if $s = q = y$.

[44]In fact Lagrange calls all three of them j, but he remarks that j may take different values in $f''(x + j), F''(x + j)$ and $\varphi''(x + j)$.

[45]Of course some regularity is needed for this argument, namely that φ'' and F'' be bounded in a neighbourhood of x. On a different note, there is a printing error here: $\frac{i}{2}[f''(x+j) - F''(x+j)]$ instead of $\frac{i}{2}[\varphi''(x + j) - F''(x + j)]$; this was later corrected (at least in the *Œuvres* printing [Lagrange *Fonctions*, 2nd ed, 187]).

them; and so on. Thus we have different degrees of contact or osculation [*Fonctions*, 120-122; 127-128].

Lagrange applies this theory to tangent circles: in the general equation of the circle $(x - a)^2 + (y - b)^2 = c^2$, which can be written as $F(x) = y = b + \sqrt{c^2 - (x - a)^2}$ there are three indeterminate constants, a, b, c. Two of them (say, a and b) can be determined by putting $F(x) = f(x)$ and $F'(x) = f'(x)$; this leaves one indeterminate constant (c), which means that for each value of c there is a circle of radius c tangent to the curve $y = f(x)$. But if in addition we put $F''(x) = f''(x)$, c is determined and there will be no other circle between $(x - a)^2 + (y - b)^2 = c^2$ and the curve. This is the *osculating circle*, or *circle of curvature*, and c is the *radius of curvature* [*Fonctions*, 124-127].

It is interesting to remember here that Lagrange had used Arbogast's principle to derive Lagrange's remainder (page 67 above), so that his theory of osculation not only seems to owe something to Arbogast's, but is also an indirect use of Arbogast's principle.

There is more in the section on plane differential geometry in [Lagrange *Fonctions*] than just a theory of contact of curves, but this theory is one of the main driving forces there, along with the connection between envelopes ("courbes enveloppantes") and singular solutions (a connection that had been revealed in [Lagrange *1774*]; see section 6.1.3.3).

4.2.1.2 Differential geometry of plane curves in Lacroix's *Traité*

We saw in page 108 that Lacroix, in order to study the curve $ax^3 + x^3y - ay^3 = 0$, expands it into the series

$$y = x + \frac{x^2}{3a} - \frac{x^4}{81a^3} + \frac{x^5}{243a^4} \text{ etc.,}$$

"convergent" for small values of x. The discussion on how Lacroix concluded from that series that $y = x$ is tangent to the curve was then postponed. We will see it now, because it is an introduction to his plane differential geometry.

Lacroix invokes his version of Arbogast's principle (which we saw in section 3.2.6): we can take values of x small enough to make the rest of the series ($\frac{x^2}{3a} - \frac{x^4}{81a^3} + \frac{x^5}{243a^4}$ etc.) even smaller than x. This means that the curve differs as little as we may wish from the straight line $y = x$, and a very small portion of it around the origin will become undistinguishable ("se confondra sensiblement") from that straight line.

But Lacroix also presents another (and more interesting) argument in favour of $y = x$ being tangent to the curve at the origin: it is impossible to draw another straight line through the origin passing between them. The argument is very similar to Lagrange's but it uses Arbogast's principle directly: denoting now by y' the ordinate of the straight line, so that $y' = x$ is its equation,[46] the difference between

---

[46] Apparently this is once again the conflation between symbols for variables and for their

that ordinate and the ordinate of the curve is

$$y - y' = \frac{x^2}{3a} - \frac{x^4}{81a^3} \text{ etc.}$$

He then considers another straight line through the origin, with equation $y'' = Ax$, so that the difference between the ordinates of the two straight lines is

$$y'' - y' = (A - 1)x$$

which can be made larger than $\frac{x^2}{3a} - \frac{x^4}{81a^3}$ etc. by taking a value of x small enough [Lacroix *Traité*, I, 363].

Lacroix then comments that "il est facile de voir qu'on peut prendre pour le caractère de la tangente, l'impossibilité de faire passer une autre droite entr'elle et la courbe"[47] to conclude that in fact $y = x$ is the tangent at the origin [*Traité*, I, 364]. Having already used the "two intersection points becoming one" characterization of the tangent (page 107 above), he could not adopt, as Lagrange did, the "no straight line between..." property as *the definition*. But he could use it as *a* working definition (a "caractère"). He did not attempt to prove the equivalence between the two characterizations.

When finally addressing directly the use of the differential calculus to study curves [*Traité*, I, 369], Lacroix returns to the considerations he had made apropos of $ax^3 + x^3y - ay^3 = 0$, but this time in a general form, introducing local coordinates (in that example he had only studied the tangent at the origin).

It is necessary here to remark some notational peculiarities that Lacroix introduces at this point: he decides to distinguish the coordinates x', y' from x, y, the former referring to points of the curve under study, and the latter to any points on the plane. We have seen Arbogast and Lagrange make similar distinctions, but Lacroix seems clearer and more systematic.

If x', y' are the coordinates of the point M through which we want to pass a tangent, when x' becomes $x' + h$, y' becomes $y' + k$. h, k will be regarded as new coordinates, the origin being M. By Taylor's theorem

$$k = \frac{dy'}{dx'}\frac{h}{1} + \frac{d^2y'}{dx'^2}\frac{h^2}{1 \cdot 2} + \frac{d^3y'}{dx'^3}\frac{h^3}{1 \cdot 2 \cdot 3} + \text{etc...}$$

or more simply,

$$k = ph + qh^2 + rh^3 + \text{etc.} \tag{4.11}$$

The argument given above (more precisely the "no straight line between..." version) is repeated to conclude that $k' = ph$ is tangent to the curve at M.[48] It remains

---

values. But Lacroix only makes the distinction that is useful (and in fact necessary), that of the ordinates: he is aware that he is comparing different ordinates for the same abscissas, so that the latter remain plainly x.

[47]"it is easy to see that we can take as the character of the tangent, the impossibility of passing another straight line between it and the curve"

[48]k' instead of k presumably because it is an ordinate not belonging to the curve, but this is not very consistent with y' for the curve.

to return to the original coordinates: this is done by substituting $x - x'$ for h and $y - y'$ for k', so that the tangent at M is given by

$$y - y' = p\,(x - x') \qquad \text{or} \qquad y - y' = \frac{dy'}{dx'}(x - x').$$

The sign of the second term of $k = ph + qh^2 + rh^3 +$ etc. can be used to study the concavity of the curve at M: the difference between the ordinates of the curve and of the tangent is $k - k' = qh^2 + rh^3 +$ etc.; h can be given values small enough for qh^2 to surpass $rh^3 +$ etc. and therefore for the sign of $k - k'$ to be the same as the sign of q; if q (or $\frac{d^2y'}{dx'^2}$) is positive, the curve is above the tangent "immediately before and after M", so that its convexity is turned to the abscissa axis; if q is negative, the opposite is the case [*Traité*, I, 368]. A similar discussion had already occurred in the previous section about $ax^3 + x^3y - ay^3 = 0$, and there the association between *inflexion point* and $q = 0$ had been noted (as it is noted further ahead, in more detail, when Lacroix studies singular points).

Lacroix observes that the role of the differential calculus here is auxiliary: it could be replaced by any other process that would give the development of k (as in fact had been the case in the previous section).

He also remarks that Arbogast was the first person who presented under this point of view the application of the differential calculus to the theory of curves; Lagrange was also led to it by his way of viewing the calculus [Lacroix *Traité*, I, 370].

Lacroix then spends a few pages exploring this: for instance, he teaches how to determine a tangent to a given curve with the condition that it passes through a given point not on the curve, or parallel to a given straight line; and how to calculate the subtangent, the normal and the subnormal.

Asymptotes are treated as limits of the tangent (as the point of tangency moves away from the origin).

The study of cases in which certain terms of (4.11) are null or infinite permits to characterize singular points: for instance, there is an inflexion point when the first non-null term (after ph) is of odd degree (or in certain situations in which $\frac{d^2y'}{dx'^2}$ is infinite) [*Traité*, I, 377-378].

Naturally, Lacroix presents the theory of osculation of Lagrange and Arbogast, and it deserves its own section (under the title "Théorie des osculations des courbes") [*Traité*, II, 388-401], which also includes a treatment of curvature. Lacroix starts his presentation, however, by considering only the simplest osculating curves for each degree, that is, the parabolic curves

$$k' = ph, \qquad k'' = ph + qh^2, \qquad k''' = ph + qh^2 + rh^3, \qquad \text{etc.}$$

The second of these curves passes between the given curve and the first of these; the third passes between the given curve and the second of these; and so on (this, of course, is proved using Arbogast's principle). Clearly, the first of these curves is the tangent straight line; the second is called the osculating parabola (in the singular)

and has a second-order contact with the given curve (unless $r = 0$, in which case it
has at least a third-order one); but all these curves (except apparently the tangent)
are called osculating parabolas (first osculating parabola: $k'' = ph + qh^2$; second
osculating parabola: $k''' = ph + qh^2 + rh^3$, etc.).

Probably the only reason for this introduction to osculation is pedagogical.[49]
What is important, mathematically, is the general concept of osculating curves:
adopting the same local coordinate system as in the study of tangents, an arbitrary
curve that passes by M, having its ordinate named K, has an equation of the form

$$K = Ph + Qh^2 + Rh^3 + Sh^4 + \text{etc.}$$

Putting $P = p$, this curve will have the same tangent as the given curve, and a
first-order contact with it; if in addition $Q = q$, the contact is of second order; and
so forth.

The most obvious (and useful) example is the circle. Lacroix considers a
circle $(x - \alpha)^2 + (y - \beta)^2 = a^2$, and determines the three arbitrary constants
α, β, a by the conditions of passing by M and having a second-order contact. This
is the *osculating circle* and no other circle can pass between this and the given
curve. Since the circle has a uniform curvature, and this curvature is inversely
proportional to its radius, the osculating circle is used to estimate the curvature
of the given curve: for this "on compare la courbe à son cercle osculateur, de même
qu'on la compare à sa tangente, pour connoître la direction vers laquelle tendroit
à chaque instant le point qui la décriroit"[50] [*Traité*, I, 396]. Because of this the
radius of the osculating circle is also called *radius of curvature*.

Lacroix defines the evolute of the curve as the curve formed by the centres
of all the osculating curves (and having, thus, coordinates α, β). He then proves
that the radii of the osculating circles are tangent to the evolute. He also alludes
to the relations between the behaviour of the evolute and singular points of the
involute (a topic favoured by l'Hôpital), but does not dwell long on them.

The next section is on transcendental curves, and it has already been mentioned briefly in the section on analytic geometry (page 108 above), since Lacroix
mixes analytic and differential considerations there.

For some reason, it is in this section that Lacroix calculates the differentials
of the arc-length of a curve and of the area under a curve. For this, he uses a
consequence of Arbogast's principle that he had given in the Introduction and
which amounts to a pinching theorem: given three "expressions"

$$
\begin{array}{ccccccccc}
A & + & Bx & + & Cx^2 & + & Dx^3 & + & \text{etc.} \\
A' & + & B'x & + & C'x^2 & + & D'x^3 & + & \text{etc.} \\
A'' & + & B''x & + & C''x^2 & + & D''x^3 & + & \text{etc.}
\end{array}
$$

---

[49]Lagrange had also given these simplest osculating curves, but only as a comment, after
having dealt with the general theory [*Fonctions*, 129-130].

[50]"the curve is compared to its osculating circle, in the same way that it is compared to its
tangent to get to know the direction towards which would tend in each instant the point that
would describe it"

such that the values of the second are always between those of the first and those of the third, if $A = A''$, then also $A = A'$.

To prove this, he gives x a value small enough for A, A', A'' to be larger than the rest of the respective series, and thus represents those series by $A + \delta$, $A' + \delta'$, $A'' + \delta''$. Now, the differences between the second and the first, and between the third and the second are $A' - A + \delta' - \delta$ and $A'' - A' + \delta'' - \delta'$, respectively, and if $A = A''$, the latter is $A - A' + \delta'' - \delta'$; these differences must have the same sign. Now, if $A' = A + d$ or $A' = A - d$, with positive d, then those differences reduce to $d + \delta' - \delta$ and $-d + \delta'' - \delta'$; but it is possible to take $\delta, \delta', \delta''$ smaller than d (presumably by taking x even smaller than before) so that the signs of the differences are those of $d, -d$, and thus not the same. The conclusion is that $A = A'$ [*Traité*, I, 58-60].

The applications of this to area and arc-length are almost obvious. Lacroix considers a curve DM, with abscissa $x = AP$ and ordinate $y = PM$, and an increment of the abscissa, $h = PP'$, small enough for the curve not to have any inflexion between its ordinates PM and $P'M'$ (that is, for the function y of x to be monotonic in the interval PP').[51] If the area $ADMP$, which is a function of

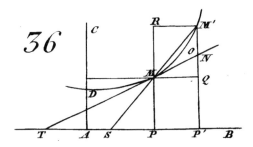

x, is represented by s, then its increment $PMM'P'$ (corresponding to h) is

$$\frac{s'h}{1} + \frac{s''h^2}{1 \cdot 2} + \frac{s'''h^3}{1 \cdot 2 \cdot 3} + \text{etc.}$$

But "it is easy to see" that this increment is comprised between the rectangles $PP' \times PM$ and $PP' \times P'M'$, that is between

$$yh \qquad \text{and} \qquad h\left(y + \frac{y'h}{1} + \frac{y''h^2}{1 \cdot 2} + \text{etc.}\right)$$

so that $s' = y$, or $ds = ydx$ [*Traité*, I, 416-417].

---

[51]It was common belief in the 18th century that all functions were piecewise monotonic. [Lagrange *Fonctions*, 155-156] for instance, has a similar assumption (also in a proof that the ordinate is the derivative of the area).

As for the arc-length z, Lacroix inserts its increment between the segment of straight line MM' and the sum $MN + NM'$ (where MN is tangent to the curve DM), to arrive at $z' = (1 + y'^2)^{\frac{1}{2}}$, or $dz = \sqrt{dx^2 + dy^2}$ [*Traité*, I, 414-416].

This section finishes with an interesting article [*Traité*, I, 418-419] on the characterization of curves using equations other than those between rectangular or polar coordinates; for instance, using an equation between the radius of curvature and the arc length. Struik [*1933*, 114-115] locates here the origin of discussions on intrinsic coordinates, soon to be taken up by Ampère and Carnot.[52]

The last section of chapter 4 has a purpose very typical of Lacroix: to present alternative points of view, namely an application of the method of limits to find tangents and osculating lines and the Leibnizian consideration of curves as polygons.

Consider an equation between x, y and three arbitrary constants, so that it represents a family of curves; we can specify the curves by subjecting them to pass through three particular points. Now imagine that these three points are in a curve of which we have an equation on x' and y', and that their abscissas are equally distanced: they are, say, $x', x' + h, x' + 2h$. After having obtained the respective conditions on the arbitrary constants (as power series on h), we make the three points approach each other (that is, h tend to 0) until they are only one and we have a second-order contact [*Traité*, I, 419-421].

Adopting the Leibnizian approach, and interpreting curves as polygons with infinitely small sides, then two curves have a contact if they have a certain number of common sides (say n), and therefore they must have the same differentials, up to order n (and of course that is a contact of order n) [*Traité*, I, 425].

But there is a different way, still in Leibnizian terms, to characterize the osculating circle, and that Lacroix thinks is "trop élégante et trop féconde"[53] to be omitted. This alternative way amounts to characterizing the centre of the osculating circle as the intersection of two normals to the curve infinitely close [*Traité*, I, 426].

It is "féconde" indeed because, giving also a characterization of the evolute as formed by all such intersections, and the evolute being tangent to the radii of curvature, it entails the more general consideration of envelopes of one-parameter families of curves. Lacroix does not use the word *envelope*, but the concept is there; his wording of the problem is: "trouver l'équation de la courbe qui en touche une infinité d'autres d'une nature donnée et assujetties à se succéder suivant une certaine loi"[54] [*Traité*, I, 427].

Lacroix's treatment of envelopes, as the wording above suggests, is the tra-

---

[52] Actually, Struik only locates it there tentatively: he uses the second edition of Lacroix's *Traité*, posterior to Ampère and Carnot's works; he recognizes the origin in Lacroix because of Ampère's acknowledgment. The change in this article introduced in the second edition amounts to only one sentence, where Lacroix cites Ampère and Carnot.

[53] "too elegant and too fruitful"

[54] "to find the equation of the curve that is tangent to an infinity of other [curves] of a given nature and subject to follow one another according to some law"

dicional Leibnizian one, with just a little amount of very naïve limit-oriented language: for two intersecting curves in the family (Lacroix has an example with circles), "it is evident" that the point of intersection will be the closer to the envelope as the two curves are closer to each other (that is, as the parameter varies less from one to the other); and if those two curves are made to coincide, their point of intersection will coincide with the points where they touch the envelope. This means that the envelope is formed by the successive intersections of the curves in the family.[55]

Thus, if the family is given by an equation $V = 0$ in x, y, and a parameter α, we should differentiate this equation relative to α, assuming that as α becomes $\alpha + d\alpha$ (as we pass from one curve to the next) the coordinates x, y remain unchanged (because we want the common point between the two curves). Since we want an equation of the envelope (where y should be a function of x) we take V as a function of x and α; differentiating $V = 0$ relative to α gives

$$\frac{dV}{d\alpha} + \frac{dV}{dy}\frac{dy}{d\alpha} = 0,$$

but as we had assumed that $\frac{dy}{d\alpha} = 0$, this becomes

$$\frac{dV}{d\alpha} = 0.$$

Eliminating α between this and $V = 0$ gives the equation of the envelope [*Traité*, I, 427-429].

But Lacroix also gives another justification for this procedure, without resorting to "successive intersections": at each point the envelope is tangent to one of the curves in the family, and therefore has the same coordinates and the same differential as that curve; it must therefore satisfy

$$V = 0 \qquad \text{and} \qquad \frac{dV}{dx}dx + \frac{dV}{dy}dy = 0$$

once the corresponding value of α has been substituted. Now, since x and y in fact vary with α, they are functions of α; differentiating $V = 0$ under this assumption gives $\frac{dV}{dx}\frac{dx}{d\alpha} + \frac{dV}{dy}\frac{dy}{d\alpha} + \frac{dV}{d\alpha} = 0$, or

$$\frac{dV}{dx}dx + \frac{dV}{dy}dy + \frac{dV}{d\alpha}d\alpha = 0;$$

but since $\frac{dV}{dx}dx + \frac{dV}{dy}dy = 0$, we have $\frac{dV}{d\alpha} = 0$ [*Traité*, I, 429].

Lacroix then remarks that the process of variation of constants is a very important one in analysis, and very fruitful in geometry, and he ends the chapter by applying it to the study of *roulettes*: the curves produced by the movement of a determinate point on a curve that rolls over the perimeter of another (the cycloid is the most important example of a *roulette*).

---

[55] This characterization of envelopes can be seen for instance in [l'Hôpital *1696*, ch. 8].

4.2.2   Differential geometry of surfaces and "curves of double curvature"

4.2.2.1   Differential geometry of surfaces and "curves of double curvature" in the 18th century

It was mentioned in section 4.1.1.3 that the development of three-dimensional coordinate geometry was slower than that of planar coordinate geometry. Differential geometry needs a background of coordinate geometry, so that state of affairs reflected on spatial differential geometry [Taton *1951*, 148]. It is revealing that the origins of partial differentiation are related to the study of parameterized families of (plane) curves, instead of the study of surfaces [Engelsman *1984*]. The only problem in spatial differential geometry that appears to have been seriously studied in the early stages of the differential calculus is that of geodesics[56] on a surface. In 1698 Johann Bernoulli gave a geometrical solution: the osculating planes to the curve should be perpendicular to the tangent planes to the surface [Coolidge *1940*, 324]; in 1728-1729 he and Euler gave solutions in the form of differential equations [Eneström *1899*]. But apparently this did not lead into further studies on spatial differential geometry.

The first major analytic (both algebraic and differential) study of space geometry was [Clairaut *1731*]. Its second chapter is dedicated to the application of the differential calculus to curves of double curvature[57], but does not go beyond tangents and normals. A tangent to a curve can be found prolonging an infinitely small side of the polygon-curve, or intersecting two planes perpendicular to the vertical coordinate planes and passing through the tangents of the corresponding projections of the curve. Either way, Clairaut's goal is not to determine the equations of the tangent, but rather to calculate the subtangent (the length of the projection into the horizontal coordinate plane of the segment of tangent between that same plane and the point of tangency). The tangent plane to a surface in a given point is determined by two of its straight lines, namely the tangents to the sections of the surface through the given point that are parallel to the vertical coordinate planes [Clairaut *1731*, 49]. Curiously, this is only a lemma, and it does not occur to Clairaut to calculate the equation of a tangent plane. The use for this lemma is to deduce geometrical properties of the normal line to a surface at a point, so that it is possible later to determine the equation of the curve generated by the intersections of the horizontal coordinate plane with all those normals through the points of a curve in the surface [*1731*, 57-58]. The third chapter of [Clairaut *1731*], dedicated to applications of the integral calculus, consists in calculations of arc lengths, areas of surfaces, and volumes.

[Euler *Introductio*, II] contains an appendix on surfaces, but it is very uninte-

---

[56] In the sense of "shortest path between two points".

[57] Henri Pitot had used the name "curves of double curvature" for space curves in 1724, but it was Clairaut [*1731*] who established it as standard. It was used throughout the 18th century. Neither Pitot nor Clairaut seemed to have in mind first curvature and torsion when using the name [Struik *1933*, 100-101].

resting as far as differential geometry is concerned: the only problem tackled there that might be regarded as part of the subject is tangency between surfaces, and it is dealt with algebraically. Tangency is interpreted as a coincidence of two intersections, so that enquiring whether two surfaces are tangent (and where, in case they are) involves searching for double roots of equations expressing intersection [*Introductio*, II, appendix, § 139-142]. There is no attempt to adapt directly the power-series method for plane curves (page 115 above). There is a distinct process for the search of tangent planes that uses it, but in a *planar* way: the tangent plane to a point in a surface can be defined in the same way as in [Clairaut *1731*]; this involves calculating tangents to two plane curves, so that that method can be used. In neither process is the equation of a tangent plane actually written.

Euler's great contributions to differential geometry in space came later. In [*1760*] he addresses for the first time the problem of curvature of surfaces. He calculates the osculating radius for an arbitrary plane section through a given point, then concentrates on normal sections; taking one as the "principal section", he shows how to determine the osculating radius of any section using that of the principal and the angle between them. He then notices that the normal sections that give the largest and smallest radii make a right angle and arrives at the formula

$$r = \frac{2fg}{f + g - (f - g)\cos 2\varphi}$$

for an osculating radius r, where f is the largest osculating radius, g the smallest, and φ the angle between the sections that give r and f.

In 1770 Euler wrote another important article, where he studied developable surfaces (surfaces that can be unfolded onto a plane). There Euler tried to show that every developable surface is a ruled surface (that is, composed of straight lines) but, according to [Coolidge *1940*, 331], without much success. In that article Euler did show that the tangents to a space curve form a developable surface [Struik *1933*, 104]. He also gave a set of conditions for a surface to be developable, for which he represented the coordinates x, y, z of a point on the surface as functions of two variables t, u. However, this idea was not followed before Gauss, in the 19th century.

As we can see, in mid 18th century the differential geometry of surfaces and space curves was not a very dynamic subject; but then appeared Gaspard Monge, and it was set in motion. In an analytical age, Monge combined a knowledge of analysis with a deep geometrical intuition. Speaking of a memoir in which Monge took up Euler's theory of developable surfaces, Struik [*1933*, 106] said that

> "the formulas always follow the dynamics of geometrical development, so that the integration of a partial differential equation becomes the gradual building up of a geometrical system in space. Nobody except Lie ever equalled Monge in that direction".

Monge's first article on differential geometry was a "Mémoire sur les développées, les rayons de courbure, et les différens genres d'inflexions des courbes à double courbure"[58] [Monge *1785a*], submitted to the Paris Academy of Sciences in 31 August 1771, and which has already been mentioned in section 4.1.1.3. In that memoir Monge expands the theory of evolutes to the space.

The first third of the memoir contains the geometrical exposition of his study (with many infinitesimal considerations). If at each point of a curve (plane or of double curvature) we take its normal plane, the normal planes through two infinitely close points will meet along a straight line. These straight lines form a developable surface, nowadays called the *polar developable*[59]. There is an infinity of envelopes of straight lines normal to the given curve, they are all in the polar developable, and he calls them the *evolutes* of the curve. In the case that the given curve is plane (and only in that case), one of its evolutes is plane (it is the usual evolute). The polar developable of a plane curve is a cylinder erected upon the plane evolute. Unfortunately, unless the given curve is plane, its centres of curvature do not form one of its evolutes.[60] He also characterizes developable surfaces as composed by a system of straight lines[61] and introduces the concept of *edge of regression* of a developable surface (other than cylinders and cones): the curve formed by the intersections of consecutive generating lines.[62] The edge of regression of the polar developable of a curve is composed of the centres of curvature of the curve (and therefore is *not* an evolute[63]).

Monge then "applies analysis" to these considerations. After some preliminaries of analytic geometry, he calculates the equations of the normal plane to a given curve through a given point, of its polar developable, of the edge of regression of this polar developable (and the radii of curvature of the curve), of a curve formed by folding a straight line on a surface, and of an arbitrary evolute.

He then addresses points of inflexion. There are two types of inflexion: in a simple inflexion, the curve is locally planar, that is, three consecutive "elements" (sides of the polygon-curve) are in the same plane; in a double inflexion, the curve is locally linear, that is, two consecutive elements are in a straight line. A simple inflexion can be recognized because the polar developable behaves locally like a

---

[58]"Memoir on the evolutes, the radii of curvature, and the different kinds of inflexion of curves of double curvature"

[59]Monge occasionally refers to it as the "surface des pôles [de la courbe]", because those straight lines are seen as axes through the centres of the osculating circles, and their points are poles of those circles; but he usually calls it "surface of the evolutes", rather than "surface of the poles".

[60]For each point in a curve, the radius of curvature is the radius of an evolute, but for two consecutive points in a space curve, the radii of curvature are radii of *different* evolutes. In fact, there is an important exception to this rule: when the curve is a line of least or greatest curvature of a surface, its centres of curvature do form an evolute. Monge implicitly reported this in [*1781*, 690], stating that the normals are tangent to that curve, but apparently he never recognized explicitly that it is an *evolute*. Lagrange [*Fonctions*, 183], on the other hand, was quite explicit, and Hachette cited him in a footnote in [Monge & Hachette *1799*, 357].

[61]That is, they are ruled surfaces. But he does not attempt to prove this in general. The case of polar developables is immediate from the definition.

[62]In the case of cones it can be said that the edge of regression is the vertex. In the case of cylinders, the generating lines are all parallel.

[63]In fact, the edge of regression is such that none of its tangents meet the curve. There is, however, the exception mentioned in footnote 60.

cylinder[64]; a double inflexion happens when the radius of curvature is 0 or ∞.

In 1775 Monge submitted to the Paris Academy a memoir on developable surfaces [Monge *1780*], where he proposed to simplify and amplify Euler's work on the subject, and where, naturally, he reworked some ideas from [Monge *1785a*]. A developable surface is one that, supposed flexible and inextensible, can be applied on a plane, so as to touch it without gaps or duplication. The obvious examples are cones and cylinders. The *application* process can be thought of the other way around – a plane being wrapped on the surface – and that is the path that Monge follows. He imagines the wrapping as consisting of an infinity of rotations along straight lines tangent to the surface. These tangent straight lines must belong to the surface, and two consecutive ones must be coplanar. If they are all parallel, we have a cylinder; if they all meet in one point, we have a cone; but in the general case they will meet along a curve (that is, they have an envelope) which is the *edge of regression*.

This gives two characterizations of any developable surface: first, it is formed by the tangents to some space curve; second, at each point it contains one of its tangents, and two consecutive tangents are coplanar [Monge *1780*, 383-385]. Using this he arrives in three different ways at the differential equation for a developable surface

$$\delta\delta z \cdot \mathrm{d}\mathrm{d}z = (\delta\mathrm{d}z)^2$$

(where δ stands for partial differentiation relative to x and d relative to y). The second of those characterizations also gives a distinction between developable surfaces and general ruled surfaces: a surface may be composed of straight lines, but such that two consecutive ones are not coplanar (which is the case of skew surfaces).

In the second section of the memoir, Monge applies this to the theory of shadows and penumbrae: if a light source and an opaque body are given as surfaces, then both the shadow and penumbra are delimited by developable surfaces.[65]

Monge then gives a few analytical applications [Monge *1780*, 423-426], and finishes the memoir with a study of ruled surfaces [Monge *1780*, 427-440].[66] He gives the third-order partial differential equation for ruled surfaces and shows that developable surfaces are a particular case of them.

Monge's work on differential geometry soon generated disciples, and the first two of them had been students of his at Mézières. Charles Tinseau (1749-1822) submitted two memoirs to the Paris Academy shortly after leaving Mézières in 1771 [Tinseau *1780a*; *1780b*].[67] The first is a collection of problems revolving around

---

[64]Two consecutive generating lines are parallel.

[65]In the special case that the light source is a point, the penumbra does not exist and the shadow is delimited by a cone, circumscribed to the opaque body and with vertex at the light source.

[66]He does not use the expression *ruled surface* ("*surface reglé*"). This last section of the memoir is ostensibly only about skew surfaces ("surfaces gauches"): surfaces composed of straight lines, but such that no two consecutive ones are coplanar.

[67][Tinseau *1780a*, 593] has an indication of having been submitted in 1774, but according to

Monge's differential geometry, the kind of simple problems that the creators of theories often do not bother to solve. In [Tinseau *1780a*, 593-594] we find, apparently for the first time [Taton *1951*, 119], the determination of an *equation* for the tangent plane to a surface. Unfortunately, Tinseau was not a master of notation, and choosing x, y, z for the coordinates of the point of tangency (and of the surface) and π, φ, ω for the coordinates of the plane, the equation obtained is

$$(x - \pi) \times dy \times \left(\frac{dz}{dx}\right) \cdot dx + (y - \varphi)dx \times \left(\frac{dz}{dy}\right) \cdot dy - (z - \omega)dxdy = 0.^{68}$$

The second memoir deals with quadratures and cubatures of ruled surfaces.

Jean-Baptiste Meusnier (1754-1793) was also a student of Monge at Mézières, from 1774 to 1775. In 1776 he submitted to the Paris Academy his only work on mathematics, a memoir on the curvature of surfaces [Meusnier *1785*]. There he derives Euler's results in a different way and improves upon them. Meusnier takes as *element* of curvature a small portion of a torus: he rotates a small arc of circle, tangent to the tangent plane, around an axis that is parallel to the tangent plane; this is done under such conditions that the resulting torus will have the same first and second differentials that the surface at the touching point. The radius of the arc of circle r, and the distance from the touching point to the axis ρ are called the radii of curvature. He then proves that r and ρ correspond to Euler's maximum and minimum osculating radii, and Euler's results follow. But he also addresses the curvature of non-normal sections, arriving at what is still called *Meusnier's theorem*[69] [Meusnier *1785*, 490-491].

Meusnier also interprets the signs of r and ρ in terms of convexity and concavity, noting for instance that when those signs are different some sections are concave and others convex [*1785*, 490-500]. He also proves that the only (presumably curved) surface that has both equal radii of curvature everywhere is the sphere; and determines a condition for minimal surfaces: that the radii of curvature are "equal" with opposite signs [*1785*, 500-504]. This allows him to find two examples, the twisted helicoid and the catenoid, the only minimal surfaces that were known for a long time [Struik *1933*, 107]. [Meusnier *1785*] is a remarkable piece, especially being the author's single mathematical work.

Meanwhile, Monge kept working on differential geometry, and including considerations of differential geometry in memoirs on other subjects. In fact, one of the main themes of his mathematical work (since its beginning) was the association of differential equations in three variables with families of surfaces sharing a common form of "generation" (usually they are generated by the movement of a

---

Taton [*1951*, 76] the correct date is 7 December 1771. [Tinseau *1780b*] has no date but, also according to Taton [*1951*, 76], appears to be contemporary of the former memoir.

[68]Which is finite: it is possible to divide each term by $dxdy$; apparently Tinseau admitted only the partial differentials $\left(\frac{dz}{dx}\right) dx$, $\left(\frac{dz}{dy}\right) dy$, not the partial differential ratios $\left(\frac{dz}{dx}\right)$, $\left(\frac{dz}{dy}\right)$.

[69]The curvature of a non-normal section that intersects the tangent plane in a straight line a is the same as that of the section made by the same plane in a sphere tangent to the surface and having as radius the radius of curvature of the normal section through a.

curve; sometimes as envelopes of other surfaces). This is perhaps explained best in [Monge *1784a*, 85-86]: a "finite" equation in three variables may refer to a family of surfaces by including arbitrary elements (particular values of which give rise to specific surfaces in the family), which may be constants or (more commonly) functions (as is the case with the family of surfaces of revolution around a fixed axis – the arbitrary functions represent the coordinates of the revolving curve); we can eliminate those arbitrary elements (constants or functions) between the "finite" equation and its differentials, so that such a family of surfaces can be represented by a differential equation, which expresses only the mode of generation. For more on this see sections 6.1.3.2 (pages 202 ff.) and 6.1.3.4.

Monge managed to include a section on differential geometry of surfaces [*1781*, 685-699] even in a memoir on the problem of minimizing the work done in the transport of rubble (he studies the problem on the plane and in space, the latter case essentially as a theoretical exercise [Taton *1951*, 297]). In that memoir Monge addresses the issue of curvature, not following precisely either Euler or Meusnier. Instead, he considers the normal straight lines to the surface, and asks when do two consecutive such normals intersect. The answer is that for each point the normal only intersects the consecutive normals in two directions, and these directions are orthogonal. Of course these correspond to the principal directions of curvature, and the curvature of the surface along one of them is established as the curvature of the sphere with centre in the corresponding intersection of the normals. Following these directions from point to point in the surface, *lines of least, or greatest, curvature* are formed.

As has already been mentioned (page 101 above), Monge taught differential geometry at the *École Polytechnique* from 1795, and from (or rather, for) that teaching resulted [Monge *Feuilles*], the first textbook on differential geometry.

Most of [Monge *Feuilles*] is composed of studies of particular families of surfaces.[70] For each family Monge seeks a differential equation and an equation "in finite quantities". Naturally, as the text proceeds other aspects are introduced and studied from these equations. Those families are ordered by the complexity of the differential equations that arise: first-order linear, first-order non-linear, second-order, and third-order. But he manages to introduce them naturally through other means: for instance those that have first-order linear equations are the cylindrical and conical surfaces, surfaces of revolution and those generated by the movement of a horizontal straight line that stays horizontal and always intersects a given (static) vertical line.

Interspersed are a few chapters dealing with more general aspects: tangent planes and normal straight lines [Monge *Feuilles*, n° 4-i,ii]; envelopes of families of surfaces [Monge *Feuilles*, n° 7]; developable surfaces [Monge *Feuilles*, n°s 13-iv - 15-iii]; curvature of surfaces [Monge *Feuilles*, n°s 17-iv - 19-i; and evolutes, radii of curvature and inflexions of curves of double curvature [Monge *Feuilles*,

---

[70] As Taton [*1951*, 210] puts it, these studies take up a score ("une vingtaine") of chapters out of about twenty-five ("quelque vingt-cinq") in the differential part of [Monge *Feuilles*].

n°s 32-34].[71]

[Monge *Feuilles*] is then a reformulation and systematization of previous work, containing also a few new results.

[Lagrange *Fonctions*] also includes a section on spatial differential geometry. It essentially attempts to address the same questions as its planar counterpart (contact and curvature of curves, evolutes, contact and curvature of surfaces). For more advanced studies Lagrange refers the reader to Monge's works [Lagrange *Fonctions*, 168, 184, 187]. Occasionally it is apparent that Lagrange's fundamentalist approach was not very well suited for advanced differential geometry. For instance, he briefly mentions developable surfaces, giving their equation and characterizing each of them as the "intersection continuelle"[72] of a family of planes, so that we can conceive that any of these planes "supposé flexible et inextensible, s'applique et se plie sur la surface"[73].

4.2.2.2 Differential geometry of surfaces in Lacroix's *Traité*

Apart from a few considerations on tangency and contact, closer to Lagrange, we will see that Lacroix essentially follows Monge in his account of the differential geometry of surfaces.

In the first few pages Lacroix relates vertical sections with partial series expansions:

$$z + \frac{dz}{dx}\frac{h}{1} + \frac{d^2z}{dx^2}\frac{h^2}{1\cdot 2} + \frac{d^3z}{dx^3}\frac{h^3}{1\cdot 2\cdot 3} + \text{etc.}$$

for a section parallel to the x, z plane, and

$$z + \frac{dz}{dy}\frac{k}{1} + \frac{d^2z}{dy^2}\frac{k^2}{1\cdot 2} + \frac{d^3z}{dy^3}\frac{k^3}{1\cdot 2\cdot 3} + \text{etc.}$$

for a section parallel to the y, z plane. Other vertical sections are obtained by making the ratio $\frac{k}{h}$ constant in

$$z + \frac{dz}{dx}h + \frac{dz}{dy}k + \frac{1}{2}\left(\frac{d^2z}{dx^2}h^2 + \frac{d^2z}{dx\,dy}hk + \frac{d^2z}{dy^2}k^2\right) + \text{etc.}$$

Of course this series is (3.10); the equality of mixed differential coefficients expresses the fact that to go from a point of coordinates x, y to a point of coordinates $x + h, y + k$ one may use the first series first (to go to $x + h, y$, along the section parallel to the x, z plane) and then the second (to to go to $x + h, y + k$, along the section parallel to the y, z plane), or the second series first (to go to $x, y + k$, along the section parallel to the y, z plane) and then the first (to to go to $x + h, y + k$, along the section parallel to the x, z plane), and that the two results must coincide – as Lacroix [*Traité*, I, 467] puts it, it expresses "la continuité de la surface"[74].

---

[71]This last chapter was absent from the first edition [Taton *1951*, 219].
[72]"continued intersection"
[73]"supposed flexible and inextensible, is applied and folded onto the surface"
[74]"the continuity of the surface"

There is some discussion of the contact of two surfaces using these series expansions. If they have a common point with coordinates x', y', z' and series expansions

$$z' + ph + qk + \frac{1}{2}(rh^2 + 2shk + tk^2) + \text{etc.}$$

and

$$z' + Ph + Qk + \frac{1}{2}(Rh^2 + 2Shk + Tk^2) + \text{etc.},$$

a first-order contact will happen when $p = P$ and $q = Q$; a second-order contact when in addition $r = R$, $s = S$ and $t = T$; and so on. This easily gives the equation

$$z - z' = p(x - x') + q(y - y')$$

for the tangent plane [*Traité*, I, 467-468].

But an alternative way is given for finding this equation – a "translation into analysis" of a construction from [Lacroix 1795]: the tangent plane through a point with coordinates x', y', z' can be determined by the tangents to the sections parallel to the vertical coordinate planes; these tangents have equations

$$z - z' = \frac{dz'}{dx'}(x - x'), \qquad y - y' = 0$$

and

$$z - z' = \frac{dz'}{dy'}(y - y'), \qquad x - x' = 0;$$

representing the equation of the tangent plane by $z - z' = A(x - x') + B(y - y')$, it follows that

$$A = \frac{dz'}{dx'} = p \qquad \text{and} \qquad B = \frac{dz'}{dy'} = q.$$

Interestingly, Lacroix feels the need to argue that this plane is in fact tangent to the surface, and not only to the two sections. This is so not only because the result is the same as in the power-series argument above (and therefore, by *transitivity*, because of Arbogast's principle); but also because it carries a coincidence between the first-order differentials of the surface and the plane, and consequently a coincidence of their points "immediately around" the point of tangency [*Traité*, I, 470-471]. As we move into this section, the power-series foundation gives way to infinitesimal considerations.

Lacroix addresses osculating spheres next [*Traité*, I, 471-472]. Using the conditions for first-order contact, he finds that all spheres tangent to a surface at a point M have their centres in the normal line through M. Trying next to use the conditions for second-order contact poses a problem: he has three more equations to satisfy and only one constant left to determine, so instead of $r = R$, $s = S$ and $t = T$, he takes $Rh^2 + 2Shk + Tk^2 = rh^2 + 2shk + tk^2$; putting this as an equation in $\frac{k}{h}$, he manages to find an osculating sphere (for each value of $\frac{k}{h}$). However, this osculation only happens along one direction indicated by $\frac{k}{h}$, that is along one normal section to the surface.

Having an expression for the radii of curvature of normal sections through a given point, Lacroix determines their maximum and minimum. Conceiving a transformation of coordinates such that the new horizontal coordinate plane is the tangent plane and the tangency point is the new origin, Lacroix shows that the directions of maximum and minimum curvature are perpendicular. It remains to establish Euler's relation between the radius of curvature of an arbitrary normal section, the maximum and minimum values of those radii, and the angles between the arbitrary section and those of maximum and minimum radii [*Traité*, I, 473-478].

Naturally, Lacroix reports also Monge's consideration of intersection of normals: the two directions in which this can happen, how they correspond to directions of maximum and minimum curvature, and the formation of lines of curvature. This is done briefly [*Traité*, I, 478-480] and referring to equations obtained previously. Even briefer is an argumentation for the possibility of obtaining the conditions for surface contact from the "coincidence of their consecutive points" [*Traité*, I, 480-481].

Also very brief is the reference to a surface with a complete second-order contact: it is Meusnier's torus [Lacroix *Traité*, I, 482]. However, Meusnier's results are mostly absent. Meusnier's theorem is not given, and the only mention of concavity and inflexion of surfaces (and their relation to the signs of radii of curvature) appears in a short footnote in the section on space curves [*Traité*, I, 519].

The rest of the section on surfaces [*Traité*, I, 482-504] appears in the subject index under the general heading "*Surfaces courbes*, leur génération"[75] (but there are also particular headings for many articles included there) [*Traité*, III, 575]. Of course this reflects Monge's views on the study of families of surfaces – with a nuance: Monge seemed to prefer generation by movement of a (usually straight) line – at least for the simpler families –, while Lacroix gives preference to envelopes.

In the text Lacroix does not announce that he is to address the generation of surfaces: instead he says that he wishes to follow the same order here as in the chapter on plane curves, so that after having dealt with tangency, second-order contact and curvature, he should address envelopes. Lacroix speaks of a "surface formée par les intersections successives d'une infinité d'autres d'une nature donnée"[76] – these other surfaces sharing a general equation with an arbitrary constant m [*Traité*, I, 482]. Maybe this language is too *infinitesimal*, so he tries to be precise: for two very close values of m, the resulting surfaces must intersect along a line; imagining these intersections to become closer, they "détermineront un espace dont la surface que nous cherchons sera la *limite*"[77] [*Traité*, I, 482]. He

---

[75]"*Curved surfaces*, their generation"

[76]"surface formed by the successive intersections of an infinity of others of a given nature"

[77]"will determine a space, the *limit* of which is the surface that we seek"

decides to use the name *limit* for the envelope: as in the planar case, the word *envelope* is absent[78].

This is presented through examples. If the generating surfaces are planes, all of them with a common point, we get of course a conical surface; if they are planes, all perpendicular to a given one, then the result is a cylindrical surface; a sequence of spheres with colinear centres gives a surface of revolution; a more complicated case is that of an "annular surface": generated by a sequence of spheres (first of constant radius and such that their centres form a plane curve [*Traité*, I, 488-489]; later the general case, but only briefly [*Traité*, I, 497, 501]).

Let us look at an example: conical surfaces [*Traité*, I, 483-486]. Lacroix starts with the equation

$$f(n)(x-\alpha) + n(y-\beta) + (z-\gamma) = 0$$

of the generating planes (α, β, γ are the coordinates of the common point to all these planes – that is, the vertex; the equations of two of these planes must differ only by two parameters, but we can put one as function of the other); differentiation (on the surface, so to speak, so that x, y, z remain constant as they represent the common points between one plane and the next) gives

$$f'(n) = -\frac{y-\beta}{x-\alpha}$$

so that $n = \psi\left(\frac{y-\beta}{x-\alpha}\right)$, for some function ψ, and therefore

$$-f\left[\psi\left(\frac{y-\beta}{x-\alpha}\right)\right] - \frac{y-\beta}{x-\alpha}\,\psi\left(\frac{y-\beta}{x-\alpha}\right) = \frac{z-\gamma}{x-\alpha}$$

which can be simplified to

$$\frac{z-\gamma}{x-\alpha} = \varphi\left(\frac{y-\beta}{x-\alpha}\right) \tag{4.12}$$

where φ is an undetermined function. This is the general non-differential equation of conical surfaces. Eliminating φ' between the first-order differentials of this equation yields

$$z - \gamma = p(x-\alpha) + q(y-\beta)$$

(where p, q are such that $dz = p\,dx + q\,dy$).

The function φ can also be determined, particularizing the conical surface: for instance by forcing it to pass through a given curve, or by imposing it to circumscribe a given surface. This had been a favorite theme of Monge in his early work (see below pages 202 ff.).

---

[78] Although Monge [*Feuilles*, n° 7-i] had already used it in this sense, applied to surfaces. Lagrange [*Fonctions*] spoke of "courbes enveloppantes" and "surfaces enveloppantes".

Addressing the general case: given an equation $V = 0$ in x, y, z and m to represent a family of surfaces, the equation of the *limit surface* comes from the elimination of the parameter m between $V = 0$ and $\frac{dV}{dm} = 0$. If, instead of eliminating it, one assigns a particular value to m, these two equations give a curve, along which the corresponding generating surface and the limit surface are tangent. These curves, which are also the intersections of the successive generating surfaces, are called, following Monge, *characteristics* [Lacroix *Traité*, I, 490-491].

A special case is that in which the generating surfaces are planes: the limit surface is then called a *developable surface*. The general equation of developable surfaces comes from the elimination of m between

$$z - m = x\,\varphi(m) + y\,\psi(m) \qquad \text{and} \qquad -1 = x\,\varphi'(m) + y\,\psi'(m).$$

Eliminating φ and ψ by differentiation gives

$$rt - s^2 = 0$$

(where r, s, t are the second-order differential coefficients: $dp = r\,dx + s\,dy$ and $dq = s\,dx + t\,dy$). The characteristics of a developable surface also produce a curve by successive intersections, and that curve is the *edge of regression* [*Traité*, I, 494-495]. Lacroix [*Traité*, I, 496-497] also pays some attention to the determination of φ and ψ, given particular conditions (partly because of the problem of shadows and penumbrae).

In the final pages of the section [*Traité*, I, 498-504] surfaces are studied as composed by lines (that is, generated by the movement of lines – straight lines or curves in space), instead of as envelopes of other surfaces. The simplest example is once again that of conical surfaces: if α, β, γ are the coordinates of the vertex, the equations of the straight lines that compose the surface are

$$y - \beta = a\,(x - \alpha), \qquad z - \gamma = b\,(x - \alpha).$$

Putting $b = \varphi(a)$ gives once again the equation (4.12).

This point of view allows Lacroix to characterize developable surfaces as formed by straight lines with consecutive intersections, and skew surfaces as formed by straight lines that do not intersect consecutively.[79]

4.2.2.3   Differential geometry of "curves of double curvature" in Lacroix's *Traité*

As everyone else in the 18th century, Lacroix takes any space curve to be the intersection between two surfaces.[80] This seems particularly adequate for a geometry based on projection planes. Given the equations $F(x, y, z) = 0$ and $f(x, y, z) = 0$ of

---

[79]Of course one has to allow here for some sloppiness in language: cylinders are developable surfaces, despite the fact that their straight lines do not intersect; instead, they are parallel, which should be mentioned by Lacroix as an alternative to intersection.

[80]Which is not correct in general. [Coolidge *1940*, 136] gives the example of any non-planar curve with prime order.

the two surfaces that intersect in the curve, by eliminating for instance x between them we get the projection of the curve in the horizontal coordinate plane; but this equation is also that of the cylinder erected upon that projection; if we eliminate one of the other variables, we get another projection and another cylinder. The curve can be studied using two of those projections, and it is the intersection between those cylinders [Lacroix *Traité*, I, 504].[81]

This idea can be found already in [Clairaut *1731*, 1-3], but it is easy to see how appealing it should be to Lacroix (and Monge), who appreciated a parallelism (as it were) between descriptive geometry on one side and analytic and differential geometry on the other.

However, besides being incorrect (which Lacroix does not seem to have been aware of) it is not a very fruitful idea, and Lacroix does not insist much on it. He uses it to give the equations of the tangent to a curve at a given point: the projections of the tangent must also be tangent to the projections of the curve; combining them,

$$y - y' = \frac{dy'}{dx'}(x - x') \quad \text{and} \quad z - z' = \frac{dz'}{dx'}(x - x') \tag{4.13}$$

are the equations of the tangent at the point with coordinates x', y', z'.

But he quickly moves on to another approach, that of power series: given a curve with coordinates x', y', z', we can take two of them, for instance y' and z', as functions of the third (in this case x'). Then, when x' becomes $x' + h$, y' and z' become

$$y' + \frac{dy'}{dx'}\frac{h}{1} + \frac{d^2y'}{dx'^2}\frac{h^2}{1 \cdot 2} + \text{etc.} \quad \text{and} \quad z' + \frac{dz'}{dx'}\frac{h}{1} + \frac{d^2z'}{dx'^2}\frac{h^2}{1 \cdot 2} + \text{etc.}$$

while for an osculating line with coordinates x, y, z, when x becomes $x + h$, y and z become

$$y + \frac{dy}{dx}\frac{h}{1} + \frac{d^2y}{dx^2}\frac{h^2}{1 \cdot 2} + \text{etc.} \quad \text{and} \quad z + \frac{dz}{dx}\frac{h}{1} + \frac{d^2z}{dx^2}\frac{h^2}{1 \cdot 2} + \text{etc.}$$

For a first-order contact it is enough to put $y = y'$, $z = z'$, $\frac{dy}{dx} = \frac{dy'}{dx'}$ and $\frac{dz}{dx} = \frac{dz'}{dx'}$ when $x = x'$. In the case of a straight line, this gives the same equations as in (4.13).

This approach can also be followed to study the contact between a curve and a surface. If x, y, z are the coordinates of the surface, when x becomes $x + h$ and

---

[81] A very simple example can be given to show that this is not always so: take the helix $x = \cos z, y = \sin z$; its vertical projection generates the cylinder $x^2 + y^2 = 1$, and its projection onto the plane x, z generates the cylinder $x = \cos z$; however, the intersection of those two cylinders is the *double* helix $x = \cos z, y = \pm \sin z$.

y becomes $y + k$, z becomes

$$z + \frac{dz}{dx}h + \frac{dz}{dy}k$$

$$+ \frac{1}{2}\left\{ \frac{d^2z}{dx^2}h^2 + 2\frac{d^2z}{dxdy}hk + \frac{d^2z}{dy^2}k^2 \right\}$$

$$+ \text{etc.}$$

But because we want to study the contact of this surface with a curve (with similar conventions as above), not only should we put $x = x'$, $y = y'$, and $z = z'$, but also the increment k of y' must be equal to $\frac{dy'}{dx'}\frac{h}{1} + \frac{d^2y'}{dx'^2}\frac{h^2}{1 \cdot 2} +$ etc. Substituting will give a series of the form $z' + Ph + Qh^2 + Rh^3 +$ etc. Then $P = \frac{dz'}{dx'}$ gives a first-order contact; this and $Q = \frac{d^2z'}{dx'^2}$ gives a second-order contact; and so on. An obvious example is the osculating plane to a curve.

But this is enough as a demonstration of the power-series theory of contact. Lacroix intends to present Monge's results about space curves, and so in the rest of the section he regards curves of double curvature as polygons where three consecutive sides are not coplanar.

This infinitesimal approach gives very easily the equations of tangents and osculating planes, and the expression $\sqrt{dx'^2 + dy'^2 + dz'^2}$ for the differential of arc-length.

The bulk of the section is dedicated to what is essentially an account of Monge's work on evolutes of space curves and polar developables [Monge *1785a*] (see page 127 above). There are only a few differences in the presentation: Lacroix had already introduced developable surfaces (and their edges of regression) in the previous section; he chooses to study the evolutes of a plane curve in space, and then those of curves of double curvature, instead of taking the former as a particular case of the latter; he adopts the unfortunate name *radii of curvature* for the radii of the evolutes, and calls *absolute radius of curvature* the shortest one (which Monge had called simply *radius of curvature*) [Lacroix *Traité*, I, 512-513].

Lacroix repeats Monge's mistake of stating that the centres of curvature only form an evolute in the case of a plane curve, forgetting the case of the lines of curvature of a surface (see footnote 60, page 128).

This section (and the chapter, and the volume) finishes somewhat abruptly with a short comment on inflexions. Lacroix mentions two kinds of inflexions of space curves: the first happens when the radius of curvature of the polar developable changes sign; and the second when the absolute radius of curvature changes sign. But

> "Cette matière demanderoit pour être traité avec exactitude et clarté, quelques détails, dans lesquels je ne puis entrer maintenant; il me suffit d'avoir mis le lecteur sur la voie de ces recherches, dont l'application d'ailleurs n'est pas fréquente."[82] [Lacroix *Traité*, I, 519]

---

[82]"To address this matter with exactitude and clarity would demand certain details in which

Why could he not enter in those details? They do not require integral calculus, so the reason is not one of order. One possible reason (but this is pure conjecture) is that Lacroix, knowing that [Lagrange *Fonctions*] was about to appear, hurried to print his first volume; having already failed to publish it in 1795, he might wish to secure his proper place in the chronology of calculus books authors. Or perhaps he really did not see those details as too important; in the second edition he did not add much, and what he did add was motivated by a work by Lancret that had appeared in the meantime.

On the whole, the two sections on spatial differential geometry in Lacroix's *Traité* seem to have offered around 1800 a more accessible introduction to the subject than the more specialized [Monge *Feuilles*], and a far more suitable one for contemporary research than the corresponding sections in [Lagrange *Fonctions*] (which were somewhat marred by the author's *fundamentalist* approach to the calculus).

I cannot enter at this point; I am content with having shown to the reader the path for these researches, which anyway do not often have applications."

Chapter 5

Approximate integration and conceptions of the integral

5.1 Conceptions of the integral and approximate integration in the 18th century

5.1.1 Conceptions of the integral

It is well known that one of the first *innovations* introduced by the Bernoulli brothers on the Leibnizian differential calculus was the answer to "what is $\int y\,dx$?". Leibniz originally meant this to be the *ſum* of the infinitesimally narrow rectangles of sides y and dx ($\int$ is a typical 18th-century italic *s*) – and therefore the area under the curve represented by y. However, he later adopted the name *integral*, coined by Johann I Bernoulli but first proposed in print by his brother Jacob, suggestive of a different definition for the operation represented by $\int$: simply the inverse operation of differentiation [Bos *1974*, 20-22; Boyer *1939*, 205].

This was the definition adopted in the first account of the integral calculus, the *Lectiones Mathematicæ de Methodo Integralium* (*Mathematical Lectures on the Method of Integrals*), written by Johann Bernoulli in 1691-1692 for the use of the Marquis de l'Hôpital, but published only in 1742:

> "Vidimus in præcedentibus quomodo quantitatum *Differentiales* inveniendæ sunt: nunc vice versa quomodo differentialium *Integrales*, id est, eæ quantitates quarum sunt differentiales, inveniantur, monstrabimus."[1]
> [Joh. Bernoulli *Integralium*, 387]

---

[1]"We have seen before how to find the *Differentials* of quantities: now, reversely, we will show how to find the *Integrals* of the differentials, ie, those quantities of which they are the differentials."

At least on one occasion this difference in approaches gave Leibniz an advantage over Johann Bernoulli, namely when the issue of differentiating an integral relative to a different variable occurred to the latter in 1697: trying to solve a problem involving a one-parameter family of ellipses, he was not able to advance when faced with the need to differentiate, relatively to the parameter a, an integral of the form $\int X(x,a)\,dx$. Returning to the original view of the integral as a sum, and remembering that "the sum of the differences of the parts is equal to the difference of the sums of the parts", Leibniz provided the answer:[2] $\int d_a X(x,a)\,dx$ [Engelsman *1984*, 41-46].

But Bernoulli's definition gained ground and was widely adopted throughout the 18th century [Boyer *1939*, 239, 278], consistently with an increasing formalism in mathematics. In the 1710's, Nicolaus I Bernoulli, nephew of Johann and Jakob, discovered the equality of mixed second-order differentials, and derived from that Leibniz's result on differentiation under the integral sign [Engelsman *1984*, 105-107].[3] This derivation made sense under Bernoulli's definition of integral and was adopted by Euler [Engelsman *1984*, 128-131].

Another situation in which the conception of the integral as a sum was useful at first occurred in the calculus of variations. In 1744 Euler published *Methodus inveniendi lineas curvas maximi minimive proprietate gaudentes*[4], the first book on that subject. There, in order to study conditions under which the curve *amnz* would extremize $\int Z\,dx$ (Z being a function of the abscissa $x = AH, AI, \ldots, AZ$, the ordinate $y = Aa, Hh, \ldots, Zz$ and $p = \frac{dy}{dx}$), Euler regarded $\int Z\,dx$ as an infinite sum of terms $Z\,dx$ corresponding to the infinitely close abscissas $AH, AI, \ldots, AZ$.[5] Introducing an infinitesimal increment nv to the ordinate Nn and putting $dZ = M\,dx + N\,dy + P\,dp$, he derived the "Euler-Lagrange equation" $N - \frac{dP}{dx} = 0$, a fundamental result [Fraser *1985*, 156-158; *1994*, 104-105].

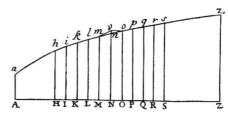

But in 1755 the nineteen-year old Lagrange discovered another approach for

---

[2]Here in an anachronistic notation, where d_a represents differentiation relative to a.

[3]A simple (although perhaps not very faithful) rendition of that derivation could be: put $y = \int X(x,a)\,dx$, so that $d_x y = X(x,a)\,dx$; from $d_x d_a y = d_a d_x y$ comes $d_x d_a \int X(x,a)\,dx = d_a X(x,a)\,dx$; integration (on x) gives $d_a \int X(x,a)\,dx = \int d_a X(x,a)\,dx$. The original is in a very geometrical language [Engelsman *1984*, 202-203].

[4]*Method to find the curved lines which enjoy a property of maximum or minimum*

[5]The figure from Euler's *Methodus* reproduced here may be a little misleading: it is a representation *only* of x and y, and particularly of the succession of their values; the integral under consideration does not correspond to the area under the curve *amnz*, nor to anything pictured there.

the calculus of variations, based on the introduction of a new operator δ. Using this he was able to derive the "Euler-Lagrange equation" without having to regard the integral as a sum and avoiding any appeal to geometry [Fraser *1985*, 155, 160-162]. This new approach was adopted by Euler and "quickly became standard" [Fraser *1994*, 103].[6]

Naturally, when Euler wrote his very influential treatise on the integral calculus, he used Bernoulli's definition:

> "functio, cuius differentiale est $= X\,dx$, huius vocatur integrale, et praefixo signo $\int$ indicari solet: ita ut $\int X\,dx$ eam denotet quantitatem variabilem, cuius differentiale est $= X\,dx$."[7] [Euler *Integralis*, I, § 7]

In this book Euler referred to the conception of "integrale tanquam summa omnium differentialium" ("integral as the sum of all the differentials") as "parum idoneo" ("too little appropriate"), no more reasonable than considering a line as composed of points[8] [*Integralis*, I, § 11] (but see also page 152 for some compromise on these principles). The idea of $P = \int X\,dx$ as a solution to the differential equation $dP = X\,dx$ introduced fewer complications (or at least confined them to the principles of the differential calculus).[9]

It was also more elegant, because it gave a unified definition for integration of functions and integration of equations:

> "Calculus integralis est methodus, ex data differentialium relatione inueniendi relationem ipsarum quantitatum: et operatio, qua hoc praestatur, integratio vocari solet."[10] [Euler *Integralis*, I, § 1]

Whether the given relation was in the form $dP = X\,dx$ or in a more complicated form (say, a third-degree, second-order differential equation) was, from the conceptual point of view, irrelevant. Thus, when Euler divided the integral calculus in two parts, and [Euler *Integralis*] in two "books", the first referred to functions of only one variable and the second to functions of two or more variables [*Integralis*, I, § 13-14, p. 16]; moreover, the further division of the first book was between a first part for first-order and a second part for higher-order problems; only then was the first part of the first book (corresponding to the first volume) divided between

---

[6] A more serious challenge was posed by Euler's "isoperimetric rule"; Lagrange was able to derive it without resorting to integral-as-sum considerations only in 1806. It is almost certainly not a coincidence that isoperimetric problems were neglected in the meantime [Fraser *1992*].

[7] "the function, whose differential is $= X\,dx$, is called its integral, and is usually indicated by the sign $\int$ in front of it: that is $\int X\,dx$ denotes the variable quantity whose differential is $= X\,dx$."

[8] Arguably, this is an incorrect analogy, since the rectangles $X \times dx$, while infinitesimal, have as many dimensions as $\int X\,dx$; points, however, have one dimension less than lines.

[9] The conception of the integral as sum also carried – in theory at least – the danger of more frequent appearances of infinitely large quantities of the form $\int y$, where y is finite [Bos *1974*, 22].

[10] "The integral calculus is the method for finding the relation between quantities, from a given relation between their differentials: and the operation thus manifested is usually called integration."

a first section for "integration of differential formulas" and a second section for "integration of differential equations" [*Integralis*, I, § 17-20, pp. 16-17, 251].

The extent to which the conception of integration as inverse of differentiation was successful in the 18th century can be assessed by looking at works based on limits.

The method of limits was naturally related to the Greek method of exhaustion; the first example given by Cousin in his chapter on the method of limits is precisely that of the area of the circle as the limit of the areas of inscribed regular polygons [*1777*, 17-19; *1796*, I, 84-85]; Newton had *proved* that the area under a curve *abcdE* is the limit of the sum of the areas of the inscribed parallelograms *AKbB, BLcC, CMdD*, etc., or of the circumscribed ones *AalB, BbmC, CcnD*, etc. (lemmas II and III, section I, book I of [Newton *Principia*]); l'Huilier gives this same result (with the same argument) as an example of a limit situation [*1786*, 9-10].

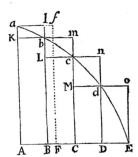

However, Cousin and l'Huilier's examples were just that – introductory examples, and explicitly about areas.[11] The integral calculus proper is introduced by Cousin as the "inverse method of limits": "remonter des limites des rapports entre les différences, au rapport même des quantités"[12] [Cousin *1777*, 56, 72; *1796*, I, 128, 150]. L'Huilier gives a similar definition and explicitly rejects the association between sums and integrals (the idea of integral as *limit* of sums simply does not seem to have occurred to him) [*1786*, 32, 143-144].

Of course, the association between integrals and sums, even if nearly always rejected, was never forgotten. Bézout gave the same definition of integral calculus as everyone else:

> "Il s'agit ici de revenir des quantités différentielles, aux quantités finies dont la différentiation a produit celles-là: la méthode qui enseigne comment se fait ce retour, s'appelle le *Calcul intégral.*"[13] [Bézout *1796*, IV, 97]

---

[11]As for [Newton *Principia*], it was a very explicit attempt at writing in a synthetic and geometric style – soon very old fashioned; in his other writings it is the inverse relationship between fluxions and fluents that we see [Bos *1980*, 54-60; Boyer *1939*, 190-202, 206; Guicciardini *2003*, 78-84, 100-102].

[12]"to reascend from the limits of the ratios of the differences to the ratio itself of the quantities"

[13]The American translation of this passage is not quite literal: "The method known by the

Nevertheless, being an orthodox Leibnizian (see section 3.1.1), he accepted also the infinite-sum interpretation of the integral:

> "Pour indiquer l'intégrale d'une différentielle, nous nous servirons de la lettre $\int$ que nous mettrons devant cette quantité: cette lettre équivaudra à ces mots *somme de*, parce que *intégrer*, ou *prendre l'intégrale*, n'est autre chose que sommer tous les accroissements infiniment petits que la quantité a dû prendre pour arriver à un état fini déterminé."[14] [Bézout *1796*, IV, 98-99]

However, by the end of the 18th century the association between integrals and sums seems to only occur either by pedagogical reasons (as to motivate the symbol $\int$) or to be rejected.[15]

5.1.2 Constants of integration, particular integrals, and definite integrals

An aspect that is important in conceptions of the integral is the treatment of arbitrary constants, and how they relate to definite integrals. This aspect becomes more relevant as the concept of function becomes central in analysis: the *definite integral*, in the sense of the integral "evaluated from $x = a$ to $x = b$" (suggesting symmetrical roles for both endpoints), is a kind of *quantity* not particularly well-suited to be expressed as a function.[16]

Let us see how the issue is addressed in [Euler *Integralis*].

Let X be a function of x. Its integral (or rather the integral of $X\,dx$) is also a function of x; but of course it must contain an arbitrary constant: if $X\,dx$ is the differential of P, then it is also the differential of $P + C$, whatever the constant C; the "complete integral" of $X\,dx$, $\int X\,dx = P + C$ is thus an indeterminate function of x; however, if C is somehow determined, we have a "particular integral" [*Integralis*, I, § 31-39]. C (and therefore $P + C$) can be determined "from the nature of the question" (but since the purpose of [Euler *Integralis*] is to treat integration *in genere*, Euler warns that those constants will generally remain indeterminate).

---

name of *Integral Calculus* is the reverse of the Differential Calculus. It has for its object to ascend from differential quantities to the functions from which they are derived" [Bézout *1824*, 74]. Note that the word "function" is only defined three paragraphs below.

[14]"To indicate the integral of a differential, the letter $\int$ is written before this quantity; this letter is equivalent to the words *sum of*, because, *to integrate*, or *take the integral*, is nothing but to sum up all the infinitely small increments which the quantity must have received, to arrive at a determinate, finite state." [Bézout *1824*, 75]

[15]Except in the few situations in which it was technically unavoidable (see footnote 6).

[16]The difference $F(b) - F(a)$ is obviously not a function, nor even a value of the function F. It could be argued that it is a value of the two-variable function $F(u) - F(v)$, but in the 18th century that would go against the *obvious* idea that the integral – definite or indefinite – of a function of x must also be a function of x. Anyway, it will be seen below that, contrary to intuition, definite integrals (or their equivalent) were not commonly (especially before the 1770's?) evaluated automatically as the difference between two values of the antiderivative – although, of course, their calculations can be easily interpreted in that way.

A condition of the form "posito $x = a$ fiat $y = b$"[17] is quite enough for such a determination. The "simplest" determination, and in fact the one that Euler apparently prefers in the few examples he gives, amounts to asking that the integral "euanescat, posito $x = 0$"[18] [*Integralis*, I, § 35, § 64, § 128].[19]

What we would call a definite integral corresponds to the situation in which we have a particular integral *and then* compute it for a specific value of x. Chapter VIII (of the first section of the first part of "book" I), "de valoribus integralium quos certis tantum casibus recipiunt"[20] is precisely dedicated to such situations. Euler calculates specific values of (particular) integrals which are not expressible in terms of elementary functions: the first problem addressed in that chapter is:

"Integralis $\displaystyle\int \frac{x^m\,dx}{\sqrt{1-xx}}$ valorem, quem posito $x = 1$ recipit, assignare,

integrali scilicet ita determinato, ut evanescat posito $x = 0$."[21] [Euler *Integralis*, I, § 330]

which amounts to $\displaystyle\int_0^1 \frac{x^m\,dx}{\sqrt{1-xx}}$, for integer m (in fact, separately for even and odd m).

Notice the asymmetry between the equivalent to limits of integration. It does not seem completely obvious that an integral, supposed to vanish at $x = a$ and calculated at $x = b$, differs only in sign from the same integral calculated at $x = a$ and supposed to vanish at $x = b$.

We would expect definite integrals to appear in a form closer to that of an evaluation "from $x = a$ to $x = b$" in a different situation: calculation of areas under curves or calculation of other geometrical magnitudes expressible by integrals and having naturally two endpoints. There also definite integrals would more naturally appear as *objects* (rather than as particular values of other objects). Such calculations do not occur in [Euler *Integralis*], which addresses no applications of integral calculus other than purely analytical ones. We then turn our attention to [Bézout *1796*, IV].

To calculate the area contained between the curve $ALMm$ and the abscissa axis AP, Bézout considers the curve as a polygon with infinitely small sides Mm;

---

[17]"let $x = a$ make $y = b$"

[18]"vanish, when x is set $= 0$"

[19]Often Euler *forgets* to include the constant of integration. Sometimes this is because C was previously set $= 0$ for the same or a similar integral. When that is not the case it might be interpreted as an implicit setting of $C = 0$, particularly if that would make the integral vanish for $x = 0$; this interpretation is weakened before a list of integrals such as in [*Integralis*, I, § 77-78], all lacking a constant of integration, and having different values for $x = 0$. Whatever the case, often the integral is afterwards calculated for a specific value of the variable; this is what happens, for instance, in the title of [Euler *1774a*], which includes the expression "casu quo post integrationem ponitur $z = 1$" ("when after the integration z is set $= 1$").

[20]"on the values that integrals receive in certain cases"

[21]"To assign the value that the integral $\int \frac{x^m\,dx}{\sqrt{1-xx}}$ takes when $x = 1$, naturally this integral being determined so that it vanishes when $x = 0$."

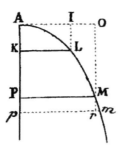

then the differential of the area is the trapezium $PpmM = \frac{PM+pm}{2} \times Pp = \frac{y+(y+dy)}{2} \times dx = y\,dx + \frac{dy\,dx}{2} = y\,dx$ (because $dy\,dx$ is infinitely smaller than $y\,dx$) [Bézout 1796, IV, 114-116; 1824, 85-86].

But Bézout remarks that $PpmM$ is the differential both of the area APM reckoned from A and of any other area such as $KPML$ reckoned from a point K. The solution to distinguish these cases is to determine the constant C accordingly: if $\int y\,dx = Y + C$, we must calculate Y for $x = AK$ and set C so that $Y + C = 0$ at that point [1796, IV, 115-118; 1824, 85-87]. In other words, the definite integral "from K to P" is notoriously absent. Instead, what we see here is something very similar to what we saw above in Euler: the determination of a "particular integral" (although Bézout does not use this expression).

Curiously, soon after the publication of the third and last volume of the first edition of [Euler *Integralis*] (1770) Euler started(?) to speak of integration "from $x = a$ to $x = b$". The "fourth volume" in the second edition of [Euler *Integralis*] is in fact a posthumous collection of memoirs on the integral calculus (mostly reprints, but including a few unpublished memoirs that had been presented to the St. Petersburg Academy). Some are about or at least contain what we call calculations of definite integrals. In those memoirs we watch an interesting oscillation in language. [Euler *1774a*] has terminology similar to what we have already seen: "post integrationem ponitur $z = 1$"[22]; "integrale euanescat posito $z = 0$" [Euler *1774a*, 122-123]. An apparently previous memoir, [Euler *1771*], seems to be the first (at least in order of presentation to the Academy) to speak of a situation in which "integratio a valore $x = 0$ vsque ad $x = 1$ extendatur"[23] [Euler *1771*, 78].

But it is another memoir, presented only in 1774 [Euler *1774b*], that seems to most clearly show the evolution in language. It addresses calculation of definite integrals, with substitution of variables. Euler starts by speaking of an integral vanishing for $z = 0$, and then setting $z = 1$; however, he quickly introduces a geometrical argument for $\int \frac{(z-1)dz}{lz}$ (under those conditions) to be not much larger than $\frac{1}{2}$, that is the area between the curve $y = \frac{z-1}{lz}$ "a termino $z = 0$ vsque

---

[22]"after the integration z is set $= 1$"
[23]"integration extends from $x = 0$ till $x = 1$"

ad terminum $z = 1$ extensa"[24] to be only slightly larger than the triangle with vertices on the origin, on the abscissa axis for $z = 1$ and on the curve for $z = 1$ (for which $y = 1$); by the third page he is introducing (for another integral) a change of variable $z = x^{\frac{1}{i}}$ (i an infinite number!), calculating the new limits of integration[25] and speaking of integration "a termino $x = 0$ vsque $x = 1$" [Euler *1774b*, 260-262].[26] Notice how integration between two endpoints appears associated to a geometrical visualization of the integral.

The next step was the introduction of a notation for this. The earliest occurrences of such a notation seem to be associated with changes in direction of integration: in [Euler *1775*, 387], a couple of changes of variables lead from $\int \frac{e^{-x}\,dx}{\sqrt{x}} = \sqrt{\pi}$ (the integral vanishing for $x = 0$, and being set $x = \infty$ after the integration) to

$$-\int \frac{dz}{\sqrt{-lz}} \left[\begin{array}{cc} \text{a} & z = 1 \\ \text{ad} & z = 0 \end{array} \right] = \sqrt{\pi}$$

whence, permutating the limits of integration,

$$\int \frac{dz}{\sqrt{-lz}} \left[\begin{array}{cc} \text{a} & z = 0 \\ \text{ad} & z = 1 \end{array} \right] = \sqrt{\pi}.$$

[Euler *1776*, 298] has a similar argument to prove that $\int \frac{x^{p-1}\,dx}{\sqrt[n]{(1-x^n)}^{n-q}}$, "extended from $x = 0$ till $x = 1$", being equal to $-\int y^{q-1}dy(1-y^n)^{\frac{p-n}{n}}$ "from $y = 1$ till $y = 0$", is also equal to

$$\int \frac{y^{q-1}dy}{\sqrt[n]{(1-y^n)}^{n-p}} \left[\begin{array}{cc} \text{ab} & y = 0 \\ \text{ad} & y = 1 \end{array} \right].$$

Finally, the name *definite integral* was introduced by Laplace [*1779*, 209]:

"je nomme *intégrale définie*, une intégrale prise depuis une valeur déterminée de la variable jusqu'à une autre valeur déterminée."[27]

The context is that of a method to reduce the solution of a linear finite difference equation to that of a linear differential equation; for that Laplace uses definite integrals on a new variable. In that memoir Laplace also used occasionally the expression "indefinite integral", without feeling the need for a definition [Laplace *1779*, 275].

---

[24]"extending from the limit $z = 0$ till the limit $z = 1$"

[25]Which happen to be the same numerically: $z = 0 \to x = 0$ and $z = 1 \to x = 1$.

[26]There is at least one precedent for this sort of thing: in [Euler *Integralis*, I, §304] Euler speaks of the formula $\frac{x\,dx}{\sqrt{1-x^3}}$ in the interval $x = 1 - \omega$ to $x = 1$; he introduces the change $x = 1 - z$, so that the new bounds are $z = 0$ and $z = \omega$. The context is that of approximating integrals (see section 5.1.3).

[27]"I call *definite integral*, an integral taken from a determinate value of the variable until another determinate value."

By the end of the 18th century this special concept of a definite integral was not yet standard enough to appear in every major treatise of integral calculus. It is absent, for instance, from [Cousin *1796*].

It is also absent from [Lagrange *Fonctions*], but that is hardly surprising. In fact, it would not fit very well in Lagrange's scheme: Lagrange spoke not of integrals, but of *primitive functions*, that is, antiderivatives; naturally, he was much more comfortable with the conception described above (on [Euler *Integralis*, I]) of determining the arbitrary constant when necessary (as for calculation of areas [Lagrange *Fonctions*, 156]), thus obtaining a particular primitive function which then might be calculated for specific values of the variable.[28] Moreover, Lagrange tried to base the calculus upon a small set of concepts; the definite integral was an unnecessary object, which would spoil the economy of [Lagrange *Fonctions*].

[Bossut *1798*] does contain a chapter on definite integrals (entitled "Intégration entre des limites données: Comparaison de certaines intégrales pour des intervalles aussi déterminés"[29] [*1798*, I, 415-431]). However, Bossut is only interested in giving an introduction to Euler's works on the subject.

We will see in section 5.2.3 that definite integrals did appear in Lacroix's *Traité* and that they seem to fit well in Lacroix's conceptions of the integral.

5.1.3   Series integration and approximate integration

5.1.3.1   Series integration

Integration by means of series was a fundamental procedure since the earliest times of the integral calculus. It was particularly important in the development of the Newtonian "inverse method of fluxions", and remained a traditional practice in the "English school" [Chabert *1999*, 434]. Its relevance lay at least as much in the fact that a power series gave a very convenient representation of a quantity (for instance, being easily integrable term by term), as in its approximative qualities [Bos *1980*, 54-56; Boyer *1939*, 190, 192].

The first section of [Euler *Integralis*, I] includes two chapters dedicated to series integration: chapter III addresses power series and chapter VI addresses trigonometric series. In both chapters the basic idea is to integrate term by term. There is not an openly declared purpose in these integrations, so that it all seems like a pure exploration of the infinite-series form. Often Euler already has a finite expression for the integral, so that this looks like a means to obtain a series expan-

---

[28]This could be particularly cumbersome in the calculus of variations, where one tries to find the function y of x for which "la fonction primitive de $f(x, y, y', y'' \ldots)$, fût un *maximum* ou un *minimum*, en supposant que cette fonction soit nulle lorsque x aura une valeur donnée a, et qu'elle devienne un *maximum* or a *minimum* lorsque x aura une autre valeur donnée b" ("the primitive function of $f(x, y, y', y'' \ldots)$ is a *maximum* or a *minimum*, supposing that that function is null when x has a given value a, and that it becomes a *maximum* or a *minimum* when x has a different given value b") [Lagrange *Fonctions*, 201].

[29]"Integration between given limits: Comparison of certain integrals for intervals also determined"

sion for such an expression. Only occasionally does the issue of practical usefulness openly arise: an example occurs when Euler addresses the formula $dy = \frac{dx}{x\sqrt{1-xx}}$; he already knows that $y = l\frac{1-\sqrt{1-xx}}{x}$, but he integrates it by series, arriving at

$$y = \left(\frac{1}{xx} + \frac{2}{3x^4} + \frac{2 \cdot 4}{3 \cdot 5x^6} + \frac{2 \cdot 4 \cdot 6}{3 \cdot 5 \cdot 7x^8 + \text{etc.}} \right) \sqrt{1 - xx};$$

the problem is that for the series to converge it is necessary to have "$x > 1$" (that is, $|x| > 1$), and in that case $\sqrt{1 - x^2}$ is imaginary, so that the series is useless [*Integralis*, I, § 168-169].

Similarly, in the section dedicated to second-order ordinary differential equations (first section of the second part of "book I"[30]) there are two chapters (VII and VIII) devoted to solutions by infinite series.[31] It is essentially the method of undetermined coefficients that is used there, along with many particular tricks and strategies. The limitations in generality of these methods can be seen by the fact that chapter VII is only about integration by series of the equation $ddy + a\,x^n\,y\,dx^2 = 0$; chapter VIII is about integration by series of "other" second-order ordinary differential equations, but to avoid too much complicated calculations Euler sticks to linear equations $ddy + M\,dx\,dy + Ny\,dx^2 = X\,dx^2$ (actually, he sticks to $xx(a + b\,x^n)ddy + x(c + e\,x^n)dx\,dy + (f + g\,x^n)y\,dx^2 = 0$). Again, the relation between this and approximations is not made explicit. On the contrary, the title of chapter XII, "De aequationum differentio-differentialium integratione per approximationes"[32] (where series only appear as a last resource — see below) suggests a distinct subject.

A different situation can be seen in [Bézout *1796*, IV]. There series integration is addressed in a section entitled "De la maniere d'intégrer par approximation, & quelques usages de cette Méthode"[33] [*1796*, IV, 145-164]:

> "L'art d'intégrer par approximation, consiste à convertir la quantité proposée, en une suite de monomes dont la valeur aille continuellement en diminuant; chaque terme s'intégre alors aisément, & il suffit d'en prendre un certain nombre, pour avoir une valeur suffisante de l'intégrale."[34] [Bézout *1796*, IV, 145]

Bézout's discussion revolves around finding series expansions that, once integrated, converge quickly enough. This is accompanied by an *ad hoc* evaluation

---

[30] The second part of "book I" corresponds to volume II.

[31] For some reason, there is no such chapter in the section on first-order ordinary differential equations.

[32] "On the integration of differentio-differential equations by approximation"

[33] "On the mode of integrating by approximation and some uses of that method" [Bézout *1824*, 106-119]

[34] "The art of integrating by approximation, consists in converting the proposed quantity into a series of simple quantities whose value continually diminishes; each term is then easily integrated and it is sufficient to take a certain number of them, in order to obtain an approximate value for the integral" [Bézout *1824*, 106].

of errors: calculating the length of an arc of a circle of diameter 1 "by means of its versed sine $AP\,[=x]$", i.e. calculating $\int \frac{dx}{2\sqrt{x-x^2}}$, he arrives at

$$x^{\frac{1}{2}}\left(1 + \frac{1}{6}x + \frac{3}{40}x^2 + \frac{5}{112}x^3 \text{ \&c.}\right);$$

the fact that x is always smaller than 1 (the diameter) guarantees that the terms of the series decrease, and that the smaller x is, the faster they decrease; for $x = 0.01$ each term is more than a hundred times less than the preceding, so that Bézout is happy in taking the hundredth part of $\frac{5}{112}(0.01)^3$ to judge the error committed by confining to the first four terms [*1796*, IV, 146-148; *1824*, 106-108].

5.1.3.2 Euler's "general method" for explicit functions

[Euler *Integralis*, I] also addresses approximation of integrals — only not, at least not explicitly, in the chapter dedicated to series integration. He does so in chapter VII (of the first section), entitled "Methodus generalis integralia quaecunque proxime inueniendi"[35].

The general method given by Euler is (in its simpler form) an approximation by rectangles, but introduced in a quite un-geometrical way [*Integralis*, I, § 297]: we want to approximate $y = \int X dx$, knowing in some way that y takes the value b for $x = a$;[36] if x increases by an extremely small ("valde parva") quantity α, X will increase very little, so that it may be regarded as constant; X being constant, we (would) have $y = Xx + \text{Const.}$; because of the initial conditions, $b = Xa + \text{Const.}$, so that $\text{Const.} = b - Xa$ and consequently

$$y = b + X(x - a)$$

(a convoluted argument to introduce the first rectangle without appealing to geometrical intuition); now, dropping the assumption of constant X, when $x = a + \alpha$ it will be $y = b + \beta$; these values serve as new initial conditions, from which we arrive at

$$y = b + \beta + X(x - a - \alpha),$$

X being again assumed as constant (in fact a new one, its value for $x = a + \alpha$); repeating this process, and calling $A, A', A'', A''', \text{etc.}$, and $b, b', b'', b''', \text{etc.}$ the values of X and y, respectively, for $x = a, a', a'', a''', \text{etc.}$ (where the differences $a' - a, a'' - a', a''' - a'', \text{etc.}$ are extremely small), we will have

$$\begin{aligned} b' &= b + A(a' - a) \\ b'' &= b' + A'(a'' - a') \\ b''' &= b'' + A''(a''' - a'') \end{aligned}$$
$$\text{etc.}$$

---

[35]"[A] general method to find all integrals approximately"

[36]Notice the *initial conditions*, and how Euler seems to have in mind more a particular integral than a definite integral.

or, substituting,

$$b' = b + A(a' - a)$$
$$b'' = b + A(a' - a) + A'(a'' - a')$$
$$b''' = b + A(a' - a) + A'(a'' - a') + A''(a''' - a'')$$
etc.;

this process is supposed to be continued until x is reached, that is, until the value for which we wish to calculate the integral is reached; but in two out of three examples given this value remains undetermined (that is, x remains a variable), unlike the initial values a, b, which are always given in the usual form "euanescat posito $x = a$" (b is always 0) [Euler *Integralis*, I, § 305-316].

For formula's sake, the penultimate value for x is represented by $'x$, and the corresponding value of X by $'X$, so that the integral is approximated by

$$b + A(a' - a) + A'(a'' - a') + A''(a''' - a'') \ldots + {'X}(x - {'x}) \qquad (5.1)$$

[Euler *Integralis*, I, § 301].

This is followed by a few interesting considerations. Firstly, Euler revisits the idea of integral as a sum (and even that of a line as an aggregate of points). What in the beginning of the book had been qualified as "little appropriate" (see page 143) is now tolerable, as long as it is well explained: integration can be attained by summation approximately, but not exactly, unless the differences $a' - a$, $a'' - a'$, $a''' - a''$, etc. are infinitely small, that is, null; hence the elongated S as the symbol for integration, and even the alternative name *summation*, are acceptable [Euler *Integralis*, I, § 302].

The other considerations have to do with the errors committed in the approximation. Since at the beginning of the first interval $X = A$ and at its end $X = A'$, it seems more convenient to use some value between A and A',[37] instead of A as above; this might suggest taking the (arithmetical) mean between A and A', but Euler does not do that yet – he will later take the arithmetic mean between two estimates of y given by an improved version of the method (see below); in the meantime he finds useful to give an estimate of y by excess and another by defect: the true value of y should be contained between two "limits" ("bounds") given by an estimate that takes the initial value of X for each interval, that is, as before,

$$b + A(a' - a) + A'(a'' - a') + A''(a''' - a'') \ldots + {'X}(x - {'x}) \qquad (5.1)$$

and another taking the final value of X for each interval,

$$b + A'(a' - a) + A''(a'' - a') + A'''(a''' - a'') \ldots + X(x - {'x}). \qquad (5.2)$$

This is not accompanied by any explicit imposition of monotonicity.[38] Euler just seems to assume that, for each interval, taking the initial value of X gives an

---

[37]"Medium quoddam inter A et A'": "some mean between A and A'", as in arithmetical, geometrical or some other mean?

[38]According to [Grabiner *1981*, 149], Euler did impose monotonicity: "first, he [Euler] said, assume that the function is always increasing or always decreasing on the given interval". I cannot locate any such passage in Euler's text.

estimate by defect and taking the final value gives an estimate by excess, or vice-versa — or at least that this happens most frequently ("plerumque"); and that in some way the sums of the interval estimates will maintain their excess or defect character[39] [*Integralis*, I, § 303].

The last of these remarks is a warning about the importance of the rate of change of the integrand function. The rate of change of $\frac{1}{\sqrt{1-x^2}}$ increases and tends to infinity as x approaches 1, so that putting $a' - a = a'' - a' = a''' - a'' = \ldots$ will not be appropriate; the length of the intervals must decrease as the rate of change of X increases [Euler *Integralis*, I, § 304].

Euler later gives an improved version of this method: it is not really true in general that $\int X\,dx = X(x - a)$, as was assumed for each interval, but an integration by parts gives $\int X\,dx = X(x - a) - \int P(x - a)dx$, where $dX = P\,dx$; assuming P to be constant in the first interval, we get $b + A(a' - a) - \frac{1}{2}B(a' - a)^2$ (where B is the value of P for $x = a$), which is a better approximation than the one used above, namely $b + A(a' - a)$; this can be continued, as it is not really true in general that $\int P(x - a)dx = \frac{1}{2}P(x - a)^2$ (P is not constant), but rather $\int P(x-a)dx = \frac{1}{2}P(x-a)^2 - \frac{1}{2}\int Q(x - a)^2 dx$ (where $dP = Q\,dx$); and so on. This leads to the formula

$$y = b + X(x - a) - \frac{1}{2}P(x - a)^2 + \frac{1}{6}Q(x - a)^3 - \frac{1}{24}R(x - a)^4 + \text{etc.} \qquad (5.3)$$

that is, to the Bernoulli series for $y = \int X\,dx$ around x (equivalent to the Taylor series for b around x). The improvement comes from substituting (5.3) for the above linear approximations; that is, the term $A'(a' - a)$ in (5.2) is replaced by $A'(a' - a) - \frac{1}{2}B'(a' - a)^2 + \frac{1}{6}C'(a' - a)^3 - \text{etc.}$, $A''(a'' - a')$ is replaced by $A''(a'' - a') - \frac{1}{2}B''(a'' - a')^2 + \frac{1}{6}C''(a'' - a')^3 - \text{etc.}$, and so on, where $B', B'', \ldots$ are the values of $P = \frac{dX}{dx}$ for $a', a'', \ldots$, $C', C'', \ldots$ are the corresponding values of $Q = \frac{dP}{dx}$, and so on [Euler *Integralis*, I, § 317].

A similar improvement can be made of (5.1) using the Taylor series for y around a,

$$y = b + A(x - a) + \frac{1}{2}B(x - a)^2 + \frac{1}{6}C(x - a)^3 + \frac{1}{24}D(x - a)^4 + \text{etc.} \qquad (5.4)$$

The final formula in this chapter is the arithmetic mean between these two improved "bounds" for y, in the case that the differences $a' - a, a'' - a', \ldots$ are all equal (to some α) [Euler *Integralis*, II, § 322]:

$$y = b + \alpha(A + A' + A'' + \ldots + X) - \frac{1}{2}\alpha(A + X) + \frac{1}{4}\alpha^2(B - P)$$

$$+ \frac{1}{6}\alpha^3(C + C' + C'' + \ldots + Q) - \frac{1}{12}\alpha^3(C + Q) + \frac{1}{48}\alpha^4(D - R) \qquad (5.5)$$

etc.

---

[39] Because the initial values will (almost) always give excesses or (almost) always defects?

Apparently both this method and the whole first section of [Euler *Integralis*] were outside the mainstream development of numerical analysis. A recent "history of algorithms" [Chabert *1999*] mentions four methods for approximate quadratures up to Euler's times: Gregory's formula, Newton's three-eighth rule, Newton-Cotes formulas (including Simpson's rule), and Stirling's correction formulas (for Newton-Cotes formulas) [Chabert *1999*, 353-363]. None of these methods are mentioned in [Euler *Integralis*] and Euler's "general method" is absent from [Chabert *1999*]. Similar remarks can be made for the older (and less clearly organized) [Goldstine *1977*]. Euler's "general method" seems to have been more influential in the development of *pure* mathematics; partly via [Lacroix *Traité*].

5.1.3.3   Euler's "general method" for differential equations

What has just been said about the influence of Euler's "general method" applies only to its first version, on integration of *functions*. Euler returns to this method in the second section of [*Integralis*, I], to find approximate solutions of first-order ordinary *differential equations*. This second appearance of the method is the subject of a section in [Chabert *1999*, 374-378] (the only one about differential equations prior to the 19th century), and according to Goldstine [*1977*, 285] it "is basically responsible for the present-day methods". However, neither [Chabert *1999*] nor [Goldstine *1977*] acknowledge the fact that Euler's method for approximation of solutions of differential equations is merely an adaptation of his "general method" for approximation of integrals.[40]

The differential equation whose solution is to be approximated is of the form $\frac{dy}{dx} = V$, where V is a function of both x and y, subject to the initial condition that $y = b$ when $x = a$ (that is, the only difference from the situation above is the substitution of $V(x,y)$ for $X(x)$). Now, we can calculate the value A of V for $x = a$ and $y = b$; if ω is very small, we can assume V to be constant between $x = a$ and $x = a' = a + \omega$, for which we will have $y = b' = b + A(x - a)$; with these new conditions we can calculate a new value A' for V; proceeding like this we will generate, as above, three (finite) sequences,

$$\begin{array}{c|cccccc} x & a, & a', & a'', & a''', & a^{IV}, & \ldots \\ y & b, & b', & b'', & b''', & b^{IV}, & \ldots \\ V & A, & A', & A'', & A''', & A^{IV}, & \ldots \end{array}$$

the middle one giving the desired approximate solution [Euler *Integralis*, I, § 650].

Of course there are differences between this and the corresponding method for integrals of functions. Although the solution is here still made up of products such as $A'(a'' - a')$, we cannot associate them to rectangles, since the constant A' no longer represents an ordinate (a side of a rectangle), but rather a slope.

Much more importantly, in the former case we had a polygonal approximation which had (at least) as many points in common with the true function X as the

---

[40]Tournès [*2003*, 458-463] indicates several geometrical antecedents of this method in its version for differential equations.

number of elements in the sequence $a, a', a'', \ldots, x$. Here, on the other hand, the only point in which it is guaranteed that the slope V is accurately evaluated is the initial point. This is so because the calculation of each of $A', A'', A''', \ldots$ involves the previous approximated value of y $(b', b'', b''', \ldots)$. And of course the errors accumulate from one interval to the next, as Euler admits [*Integralis*, I, § 652].

For the same reason it would seem pointless to give a different approximation using the (estimated) final values of V for each interval, that is, something equivalent to (5.2).[41]

However, the relationship between the two methods is undeniable, and the fact that the former was more developed (and developable) than the latter may be a good sign of which one was prior.

In fact, Euler [*Integralis*, I, § 656] expressly invokes the appropriate articles in section I to justify the use of the Taylor series (5.4) also for differential equations (the Bernoulli series (5.3) is not applicable since in this case $X, P, Q, R, \ldots$ cannot be calculated without knowing the final value of y). It is (5.4) that is then used in the two examples of this chapter [Euler *Integralis*, § 661-662].

In the second volume of [Euler *Integralis*][42] a similar method is developed for second-order differential equations (in chapter XII of the first section, entitled "De aequationum differentio-differentialium integratione per approximationes"[43]). However, in this case Euler pays very little attention to the intervals beyond the first one. Given an equation in x, y, p, q, where $dy = p\,dx$ and $dp = q\,dx$, q may be seen as a function V of x, y, p; if the initial conditions are that $y = b$ and $p = c$ when $x = a$, and if V is taken as constant $(= F)$ between $x = a$ and $x = a + \omega$ (ω being very small), then at $x = a + \omega$ Euler concludes that $p = c + F\omega$ and $y = b + c\omega$; Euler remarks that this can be repeated for further small intervals as in the methods above, but he does not do it [*Integralis*, II, § 1082]. What he does do is to improve upon the method by regarding not V as constant, but rather $\frac{dV}{dx}$, similarly to what he had done for integration of functions: integration by parts gives $p = c + V(x-a) - \int (x-a)dV$; putting $dV = P\,dx + Q\,dy + R\,dp = (P + Qp + RV)dx$, and taking $P + Qp + RV$ as constant, gives $p = c + F(x-a) - \frac{1}{2}(P + Qc + RF)(x-a)^2$ and $y = b + c(x-a) + \frac{1}{2}F(x-a)^2 - \frac{1}{6}(P + Qc + RF)(x-a)^3$ (where P, Q, R are calculated at $x = a$) [Euler *Integralis*, II, § 1094].

It must be mentioned that in this chapter the word "series" occurs, albeit quite timidly: in case a power of $x - a$ appears in $P + Qp + RV$, this cannot be taken as constant; in that case a truncated series[44] approximation is used, of the form $p = c + A(x-a)^\lambda$; $y = b + c(x-a) + \frac{A}{\lambda+1}(x-a)^{\lambda+1}$ [Euler *Integralis*, II, § 1094, 1098].

---

[41] Although the arithmetic mean between these *upper* and *lower* estimates was used, namely by Carl Runge (1856-1927), to obtain an improved method [Chabert *1999*, 381-387].

[42] This second volume constitutes the second part of the first "book", dedicated to higher-order ordinary differential equations. It is divided into two sections: the first on second-order equations and the second on third- and higher-order equations.

[43] "On the integration of differentio-differential equations by approximation".

[44] "Seriei initium" ("beginning of a series").

In the third volume of [Euler *Integralis*], dedicated to partial differential equations, there is no chapter devoted to series integration or approximate integration.

5.1.3.4   Other methods for differential equations

In spite of Goldstine's quote above about Euler's "general method" being the ancestor of (nearly all?) the modern methods, other methods can be found in the 18th century.

An important motivation for approximation of differential equations was astronomy: the motion of celestial bodies is too complicated for rigorous solutions to be achievable (because of multiple-body gravitational influences). But approximate values are easily accessible, and can be improved using an adaptation of Newton's approximation method for numerical equations: one takes the initial approximate value plus an undetermined quantity, which should be very small; then the terms involving the square and higher powers of this undetermined quantity are neglected, resulting in a linear differential equation; by integrating this linear equation (which is much easier), a new approximate value is obtained; and the procedure is repeated with this new approximate value. Versions of this method are found in works by d'Alembert on lunar theory [d'Alembert *1754-1756*, I, 31-34; Tisserand *1894*, 60-62], by Euler on the three-body problem and by Lagrange [*1766*, 110] on the satellites of Jupiter [Wilson *1994*, 1049]; at least hints at this method were also present in Clairaut's earlier work on lunar theory [Tisserand *1894*, 51-56]. Gillispie [*1997*, 48] says that Laplace attributed this method to d'Alembert, but in fact what Laplace [*1772b*, 267] attributes to d'Alembert is the use of indeterminate coefficients for the integration of the linear differential equations involved in the method. The method itself "se presenta naturellement aux Géomètres, qui résolurent les premiers le Problème des Trois-corps"[45] [Laplace *1772b*, 268], which would include not only d'Alembert but also Clairaut and Euler (d'Alembert [*1754-1756*, I, xxxv] himself referred to this as "Méthodes connues"[46]).

This method had problems, particularly in the case of the Moon (where it introduced undesirable "arcs of circle" – terms containing integer powers of angles instead of sines and cosines of angles, which are incompatible with the fact that the Moon orbits the Earth and therefore its distance remains bounded) and in the case of a planet with more than one satellite (where it mixed first-order terms in the second-order solutions). D'Alembert [*1754-1756*, I, 34-37] noticed the former difficulty and gave a means to avoid it, and Lagrange overcame the latter difficulty by "an elaborate algebraic process" [Wilson *1994*, 1049]; nevertheless, Laplace proposed a new method – also of successive approximations – consisting "à faire varier les constantes arbitraires dans les intégrales approchées, et à trouver ensuite par l'intégration, leurs valeurs pour un temps quelconque"[47] [Laplace *1772b*, 268].

---

[45]"[had] appeared naturally to the Geometers who first solved the three-body problem"
[46]"known methods"
[47]"in varying the arbitrary constants in the approximate integrals and then determining their values for a given time by integration." [Gillispie *1997*, 48]

Later Laplace [*1777*] simplified this method of variation of arbitrary constants [Gillispie *1997*, 70]; there he summarized it in a rule: one should solve approximately the differential equation in the traditional way, and then erase the terms containing "arcs of circle" and at the same time replace the arbitrary constants with variables subject to certain differential condition equations [Laplace *1777*, 381].

Lagrange was naturally quite sympathetic to techniques of variation of constants (see sections 6.1.2.3, 6.1.4.1 and 6.1.4.2), but in this particular case he thought that Laplace's method rested on a "metaphysics" that was not satisfying; besides, it failed in cases in which an arbitrary constant occurred within the argument of a sine, cosine, or exponential [Lagrange *1783*, 227]. He thus presented his own method of variation of constants [Lagrange *1781*, § 25-27; *1783*], introducing corrections to Laplace's condition equations [*1783*, § 3-5].

An entirely different method for approximating solutions of differential equations, using continued fractions, was also proposed by Lagrange [*1776*]. Given a differential equation in x and y, Lagrange's method consisted in finding a first approximation ξ of y for very small x (ξ should be of the form ax^α); substitute $y = \frac{\xi}{1+y'}$ in the given equation, resulting in a new equation in x and y'; and repeat these steps, so that

$$y = \cfrac{\xi}{1 + \cfrac{\xi'}{1 + \cfrac{\xi''}{1 + \cfrac{\xi'''}{1 + \ddots}}}}$$

The method of series had "l'inconvénient de donner des suites infinies lors même que ces suites peuvent être représentées par des expressions rationnelles finies"[48]; a continued fraction, on the other hand, would stop whenever the solution was finite and rational [Lagrange *1776*, 301]. This method, however, was not much pursued in the 18th century [Chabert *1999*, 373].[49]

5.1.3.5 Two accounts in the 1790's: Cousin and Bossut

To conclude this section on series and approximate integration, it remains to look at how this subject is treated in important treatises at the end of the 18th century.

[Bézout *1796*, IV] (not really an important treatise, but rather a standard elementary textbook) has been seen above to conflate approximate integration with series integration, but also to be more practical than [Euler *Integralis*]. Naturally for its level, it does not address approximations of solutions of differential equations.

---

[48]"the inconvenience of giving infinite series even when such series can be represented by finite rational expressions"

[49][Lagrange *1776*] is nevertheless an important work, namely for the (pre-)history of Padé approximants [Brezinski *1991*, 137-139]. Also from that memoir Lacroix extracted a method for expanding functions into series, which he reported in chapter 2 and used in chapter 4 of [Lacroix *Traité*, I] (see page 108).

Cousin [*1777* 446-455; *1796* II, 30-40] uses two methods to approximate integrals: undetermined coefficients to find a series for the integral; and what is probably a version of Euler's "general method". Starting from Taylor's theorem around two different points, corresponding to Bernoulli series (5.3) and Taylor series (5.4), Cousin decides to divide the interval between x and a into several small subintervals, all of the same length Δa; then apply both formulas to each subinterval, so that he gets two estimates for the integral $y = \int X dx$, corresponding to (5.1) and (5.2) but with full series for each subinterval instead of just a linear polynomial; and finally take the arithmetic mean between these two estimates (which Euler, as we have seen, preferred not to do).

A little afterwards Cousin [*1777*, 484-508; *1796*, II, 59-77] returns to the application of infinite series to differential equations, namely to separate variables, but this seems to be equivalent to chapters VII and VIII of the first section of [Euler *Integralis*, II], and approximation appears far from the point.

[Bossut *1798*, I] includes three chapters on approximation of integrals, in the first part of the integral calculus. In chapter XII, "Méthodes pour intégrer par approximation les Formules qui ne peuvent l'être en rigueur"[50] [*1798*, I, 432-456], the goal is to express integrals as infinite series. Bossut uses continued division, the binomial formula, the method of undetermined coefficients, and Bernoulli series. Although there is no attempt at evaluation of errors, there is much more concern with the practical issues of convergence than in [Euler *Integralis*].

Chapter XIII, "Suite: Autres méthodes pour l'approximation des Intégrales"[51] [Bossut *1798*, I, 457-471] is more geometrical. Firstly, Bossut presents a version of Euler's "general method", in a geometrical guise: his idea is to consider the integral as the area under a curve, and to approximate it by trapezia; the result is thus the average between (5.1) and (5.2) that Euler did not calculate (but Bossut's reasoning is closer to Bézout's calculation of areas − see page 146 above). In the rest of the chapter Bossut interpolates curves and integrates the resulting polynomials.

Chapter XIV [Bossut *1798*, I, 472-484] treats only of applications of previous methods to the calculation of the arc-length of ellipses.

Volume 2 of [Bossut *1798*] contains two small chapters on approximate solutions of differential equations, one for first-order and another for higher-order equations [*1798*, II, 197-205, 282-293]. Both deal in fact with finding series solutions (namely using undetermined coefficients), the latter being a summary of chapters VII and VIII in the first section of [Euler *Integralis*, II]. A scholion at the end of the former suggests that approximate solutions be calculated along small subintervals and then added together, but this is the only vague reference to Euler's "general method" within the context of differential equations.

---

[50]"Methods to integrate by approximation those formulas that cannot be [integrated] exactly".
[51]"Continuation: Other methods for the approximation of integrals"

5.2 Approximate integration and conceptions of the integral in Lacroix's *Traité*

5.2.1 Integration (of explicit functions) by series

In the chapter on integration of functions of one variable, Lacroix dedicates a section to "intégration par les séries"[52] [*Traité*, II, 66-88].

Its beginning is very typical, with a remark that, if a function has been expanded into series, then it is easy to integrate it, because it is enough to integrate each of the monomials that compose the series. Lacroix explores this, giving several examples taken from [Euler *Integralis*]. But slightly more than half of the section [*Traité*, II, 77-88] is taken up with a summary of a memoir by Lagrange on series expansion of elliptic integrals [Lagrange *1784-1785*]. There are also references to integration by series in the section on integration of logarithmic and exponential functions and especially in the section on integration of trigonometric functions (namely a long passage on $\int dz \, (1 + n \cos z)^m$ [*Traité*, II, 118-133]).

Clearly there can be two different purposes in integration by series, as in fact is the case for any use of infinite series (see also section 3.2.6): it can be used to facilitate (or to enable) the operation of integration,[53] which is but a useful instance of the use of a series as a representation of a function; or it can be used to "parvenir à des valeurs approchées des intégrales dont on ne peut obtenir l'expression algébrique"[54] [Lacroix *Traité*, II, 73].

This latter purpose, however, only appears in the eighth page of this section, and it is never deeply explored. It brings along the issue of convergence,[55] which Lacroix addresses in his down-to-earth manner: he suggests the importance of having several series expansions for the same integral, so that it may be possible to use the one that is convergent for the relevant value of x.

In fact, Lacroix remarks the inconvenience that integration by series does not always give (any) convergent series, and that divergent series do not give approximations [*Traité*, II, 135]. This motivates a distinct section, on a "méthode générale pour obtenir les valeurs approchées des intégrales"[56] [*Traité*, II, 135-160] – Euler's "general method".

5.2.2 Euler's "general method"

Lacroix's derivation of the method is not the same as Euler's but it is not terribly original either. The main difference is that Lacroix takes full advantage of Taylor

---

[52]"integration by series"

[53]Reducing it either to the integration of expressions of the form ax^n or to their differentiation, in the case of the method of undetermined coefficients.

[54]"arrive at approximate values of integrals whose algebraic expression is not obtainable"

[55]This issue had already appeared, apropos of an expansion for $\int \frac{x^m dx}{x^n + a^n}$ [Lacroix *Traité*, II, 68-69]. Apparently Lacroix always *preferred* convergent series.

[56]"general method to obtain approximate values of integrals"

series as a well-established tool. In this context this might be evocative of Cousin, were it not for the overall importance of Taylor series in Lacroix's *Traité*. Lacroix starts by considering the Taylor series

$$Y_1 = Y + Y'\frac{(a_1 - a)}{1} + Y''\frac{(a_1 - a)^2}{1 \cdot 2} + Y'''\frac{(a_1 - a)^3}{1 \cdot 2 \cdot 3} + \text{etc.} \qquad (5.6)$$

$$Y_2 = Y_1 + Y_1'\frac{(a_2 - a1)}{1} + Y_1''\frac{(a_2 - a_1)^2}{1 \cdot 2} + Y_1'''\frac{(a_2 - a_1)^3}{1 \cdot 2 \cdot 3} + \text{etc.} \qquad (5.7)$$

and so on,

where Y, Y', Y'', etc. are the values of $y = \int X dx$, $\frac{dy}{dx} = X$, $\frac{d^2y}{dx^2} = \frac{dX}{dx}$, etc. at $x = a$; Y_1, Y_1', Y_1'', etc. are the values of the same expressions at $x = a_1$; Y_2, Y_2', Y_2'', etc. the same at $x = a_2$; and so on. But unlike Cousin he follows Euler in taking only linear polynomials: supposing that the quantities a_1, a_2, a_3, etc. are chosen so that the second and higher powers of $a_1 - a$, $a_2 - a_1$, $a_3 - a_2$, etc. may be neglected "sans erreur sensible"[57], the following approximations result

$$Y_1 = Y + Y'(a_1 - a)$$
$$Y_2 = Y_1 + Y_1'(a_2 - a_1)$$
$$Y_3 = Y_2 + Y_2'(a_3 - a_2)$$
etc.;

these may be combined, giving

$$Y_n = Y + Y'(a_1 - a) + Y_1'(a_2 - a_1) + Y_2'(a_3 - a_2) \ldots + Y_{n-1}'(a_n - a_{n-1}) \qquad (5.8)$$

as an approximation for the value Y_n of $\int X dx$ for $x = a_n$. This approximation will be "the more exact" as the quantities a, a_1, a_2, etc. are closer to one another [Lacroix *Traité*, II, 136-137].

Now for a second estimate. If the process were to start from a_n instead of a, that is, to follow the sequence $a_n, a_{n-1}, a_{n-2}, \ldots, a_1, a$ instead of $a, a_1, a_2, \ldots, a_{n-1}, a_n$, the first step would consist in the Taylor series

$$Y_{n-1} = Y_n - Y_n'\frac{(a_n - a_{n-1})}{1} + Y_n''\frac{(a_n - a_{n-1})^2}{1 \cdot 2} - \text{etc.}$$

Proceeding with series for Y_{n-2}, Y_{n-3}, etc., neglecting higher powers of $a_n - a_{n-1}$, $a_{n-1} - a_{n-2}$, etc., and combining the results gives

$$Y = Y_n - Y_n'(a_n - a_{n-1}) - Y_{n-1}'(a_{n-1} - a_{n-2}) \ldots - Y_1'(a_1 - a) \qquad (5.9)$$

which of course is the same thing as

$$Y_n = Y + Y_1'(a_1 - a) + Y_2'(a_2 - a_1) \ldots + Y_{n-1}'(a_{n-1} - a_{n-2}) + Y_n'(a_n - a_{n-1}). \qquad (5.10)$$

---

[57]"without noticeable error"

That is, (5.8) uses the initial value of the function at each interval, and (5.10) uses the final value (they are precisely the same as, respectively, (5.1) and (5.2)) [Lacroix *Traité*, II, 138-139].

Further ahead in the same section Lacroix gives a geometrical interpretation for these approximations [*Traité*, II, 143-144]: if the curve BMZ represents the function X (AP being the abscissa axis) and AP, AP', AP'', AP''', etc. are respectively equal to a, a_1, a_2, a_3, etc., then $Y'(a_1 - a) + Y_1'(a_2 - a_1) + Y_2'(a_3 - a_2) +$ etc. is represented by the polygon $PMRM'R'M''R''$, etc. and $Y_1'(a_1 - a) + Y_2'(a_2 - a_1) + Y_3'(a_3 - a_2) +$ etc. is represented by the polygon $PSM'S'M''S''M'''$, etc. To have an approximation of the value of the integral since the origin of the abscissas one must add a first term Y, equal to the area $ACMP$. It must be stressed that Lacroix gives this simply as a geometrical *illustration* of results already obtained "from analysis" (cf. pages 89 and 105).

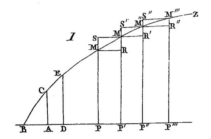

The rest of Lacroix's account of the method *itself* (how best to use it; examples) follows Euler closely (although somewhat shortened). For instance, Lacroix reports Euler's advice against taking the differences $a_1 - a$, $a_2 - a_1$, $a_3 - a_2$, etc. all equal; instead, they should be smaller where X varies most [*Traité*, II, 145].

Lacroix also reports Euler's improved method, and in fact it occurs more naturally here: it is enough not to neglect the second and higher powers of $a_1 - a$, $a_2 - a_1$, $a_3 - a_2$, etc. Taking these differences to be all equal (to some α) gives the estimates

$$Y_n = Y + (Y' + Y_1' + Y_2' \ldots\ldots + Y_{n-1}')\frac{\alpha}{1}$$
$$+ (Y'' + Y_1'' + Y_2'' \ldots\ldots + Y_{n-1}'')\frac{\alpha^2}{1 \cdot 2} \qquad (5.11)$$
$$+ (Y''' + Y_1''' + Y_2''' \ldots\ldots + Y_{n-1}''')\frac{\alpha^3}{1 \cdot 2 \cdot 3}$$
$$+ \text{etc.}$$

and

$$Y_n = Y + (Y_1' + Y_2' + Y_3' \ldots\ldots + Y_n')\frac{\alpha}{1}$$

$$- (Y_1'' + Y_2'' + Y_3'' \ldots\ldots + Y_n'')\frac{\alpha^2}{1 \cdot 2} \qquad (5.12)$$

$$+ (Y_1''' + Y_2''' + Y_3''' \ldots\ldots + Y_n''')\frac{\alpha^3}{1 \cdot 2 \cdot 3}$$

$$- \text{etc.}$$

According to Lacroix, in case "none of the coefficients $X, \frac{dX}{dx}, \frac{d^2X}{dx^2}$, etc. changes sign in the interval from $x = a$ til $x = b$"[58], the true value of Y_n is between these two estimates[59], and a better approximation is given by their arithmetic mean (5.5) [*Traité*, II, 147-148].

But Lacroix does much more than just report Euler's method, and his additions and remarks make this one of the most interesting sections in his *Traité*. We will look at that additional work by Lacroix in the next paragraphs and in section 5.2.3.

We saw above that Euler was not very clear about the monotonicity of the function whose integral was to be approximated: he did not explicitly assume it, yet his argument for (5.1) and (5.2) to be bounds for the true value of the integral makes sense only if the function is monotonic.

Lacroix, on the contrary, was very clear about that. Included in his geometrical interpretations of the method is a sort of counterexample: There is no

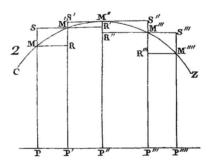

reason to assume that either of the polygons $PMRM'R'M''S''M'''S'''P''''$ or

---

[58]See below a discussion about this condition.

[59]This is not always true. For a very simple example, take $X = \frac{x^2}{2}$, consider only one subinterval, from $x = 0 = a$ to $x = 1 = a_1$, and truncate after the term with α^2: (5.11) will give $0 + 0 \cdot \frac{1}{1} + 0 \cdot \frac{1^2}{2} = 0$ and (5.12) will give $0 + \frac{1^2}{2} \cdot \frac{1}{1} - 1 \cdot \frac{1^2}{2} = 0$; yet $\int_0^1 \frac{x^2}{2}dx = \frac{1}{6}$. An example with less simple calculations, but where the truncation is less artificial, is $X = \sin x$, with $a = 0$ and $a_1 = \frac{\pi}{4}$; truncation is indispensable, because otherwise both (5.11) and (5.12) will give infinite series; truncating after α^2, (5.11) gives $\frac{1}{32}\pi^2 \approx .30843$, and (5.12) gives $\frac{1}{8}\sqrt{2}\pi - \frac{1}{64}\sqrt{2}\pi^2 \approx .33727$; however, $\int_0^{\frac{\pi}{4}} \sin x\, dx = -\frac{1}{2}\sqrt{2} + 1 \approx .29289$.

$PSM'S'M''R''M'''R'''M''''P''''$ is smaller (or larger) than the curvilinear area $PMM'M''M'''M''''P''''$ [*Traité*, II, 144].

He also gives a sufficient condition: $\int X\,dx$ is included between the values given by (5.8) and (5.10) *if* X "conserve le même signe et varie dans le même sens"[60] [Lacroix *Traité*, II, 139].

Lacroix proves this by examining one of the subintervals, namely the first, between a and a_1. Dividing it further with a "great number" of intermediary values $\alpha_1, \alpha_2, \alpha_3, \ldots, \alpha_m$ of x, then (5.8) and (5.10), which in this case were $Y+Y'(a_1-a)$ and $Y+Y_1'(a_1-a)$ respectively, become

$$Y + Y'(\alpha_1 - a) + Y'(\alpha_2 - \alpha_1) \ldots \ldots + Y'(a_1 - \alpha_m) \qquad (5.13)$$

and

$$Y + Y_1'(\alpha_1 - a) + Y_1'(\alpha_2 - \alpha_1) \ldots \ldots + Y_1'(a_1 - \alpha_m). \qquad (5.14)$$

But if we consider the values $y_1', y_2', \ldots y_m'$ of X corresponding to the values $\alpha_1, \alpha_2, \ldots \alpha_m$ of x, then we can have better approximations of Y_1, one of which is

$$Y + Y'(\alpha_1 - a) + y_1'(\alpha_2 - \alpha_1) \ldots \ldots + y_{m-1}'(\alpha_m - \alpha_{m-1}) + y_m'(a_1 - \alpha_m). \ (5.15)$$

Now, if X is, for instance, always increasing between a and a_1, then Y', y_1', $y_2', \ldots y_m'$, Y_1' is an increasing progression, and it is clear that (5.15) is between (5.13) and (5.14). Finally (using a very interesting argument on which we will comment below), "comme on peut concevoir que la [série (5.15)] soit aussi près qu'on voudra de la vraie valeur de Y_1, en imaginant un nombre suffisant de termes intermédiaires"[61], the conclusion must be drawn that that true value of Y_1 is in fact between $Y + Y'(a_1 - a)$ and $Y + Y_1'(a_1 - a)$ [Lacroix *Traité*, II, 140].

Apparently mysterious is the condition that X should keep the same sign (always positive or always negative): this condition does not seem to be used at all in the proof. An explanation may lie in the concept of "increasing": there are plenty of examples in Lacroix's *Traité* where expressions like $x < a$ clearly mean, in modern terms, $|x| < |a|$ (see for instance section 3.2.6); in this very section there is a passage which reinforces this view of "greater" and "less" referring to the absolute *size* of magnitudes (see footnote 65 below); this would entail that, say, $-2, -1, 0, 1, 2$ was *not* an increasing progression, since $0, 1, 2$ was increasing but $-2, -1, 0$ was decreasing.[62]

As we have already seen, when later on Lacroix reports Euler's improved method he gives as a sufficient condition that none of the coefficients $X, \frac{dX}{dx}, \frac{d^2X}{dx^2}$, etc. change sign for the true value of Y_n to be included between (5.11) and (5.12). This is clearly a generalization of the simpler proposition whose proof we have

---

[60]"keeps the same sign and varies in the same direction"

[61]"since it is possible to conceive the [series (5.15)] as close as one may wish to the true value of Y_1 by imagining enough intermediary terms"

[62]Lagrange, on the other hand, in a passage equivalent to that referred to in footnote 65, decided to have $-1 > -2$, but he had to state this explicitly [Lagrange *Fonctions*, 46].

just examined, but the proof is not generalizable (the minus signs in (5.12) and the fact that, say, $\frac{dX}{dx}$ may be always positive and $\frac{d^3X}{dx^3}$ always negative thwarts argumentation involving the monotonicities – which may be in opposing directions – of sequences appearing in the formulas). In fact, this latter proposition is wrong – see footnote 59 above. Lacroix's correction of Euler thus fails for the improved method.

But Lacroix did not just impose extra conditions for (5.8) and (5.10) (and (5.11) and (5.12), albeit wrongly) to be upper and lower bounds. He felt the need for bounds that would be general, that would not require conditions of monotonicity. This was possibly his motivation for remarking that one can divide the interval into portions where the function is increasing and portions where it is decreasing[63], and treat them separately [*Traité*, II, 140]. This *might* have given something like

$$Y+Y'(a_1-a)+Y_1'(a_2-a_1)+\ldots+Y_{i-1}'(a_i-a_{i-1})+Y_{i+1}'(a_{i+1}-a_i)+\ldots+Y_n'(a_n-a_{n-1})$$

as a lower bound, in case the function is increasing from $x = a$ till $x = a_i$ and decreasing thenceforward; however, Lacroix did not derive any explicit result from that remark. Instead, he found much simpler (but also quite uninformative) expressions for bounds that do not require monotonicity in a passage from Lagrange's derivation of the remainder for Taylor series [Lagrange *Fonctions*, 46]: calling M the greatest value[64] that X takes between $x = a$ and $x = b$ and m the smallest value of X in the same interval,[65] the difference $Y_b - Y_a$ between the values of $\int X dx$ for $x = a$ and $x = b$ is contained between $M(b - a)$ and $m(b - a)$.

These bounds are a straightforward result from a lemma which will be discussed below: if X is always positive between $x = a$ and $x = b$, then $Y_b - Y_a$ is also positive. This means that, since $M - X$ and $X - m$ are by definition positive, the differences between the values of $\int (M - X)dx$ and $\int (X - m)dx$ for $x = b$ and $x = a$, are also positive; that is, $Mb - Y_b - (Ma - Y_a)$ and $Y_b - mb - (Y_a - ma)$ are positive, whence $mb - ma < Y_b - Y_a < Mb - Ma$ [Lacroix *Traité*, II, 141].

This result for itself has of course very little use, but Lacroix also gives an improvement: if $X = PQ$, M and m are the greatest and smallest values of P, and it is possible to calculate $Z = \int Q dx$, then $mZ_b - mZ_a < Y_b - Y_a < MZ_b - MZ_a$. He later uses this to prove that $\frac{1}{\sqrt{1+r}} \int \frac{N du}{\sqrt{r+u^2}} < \int \frac{N du}{\sqrt{r+u^2} \cdot \sqrt{1+ru^2}} < \int \frac{N du}{\sqrt{r+u^2}}$ (all the integrals taken from $u = 0$ till $u = 1$) [Lacroix *Traité*, II, 152-153].

---

[63]He assumed, as usual at the time, that every function is piecewise monotonic.

[64]As usual at the time, Lacroix did not distinguish between a maximum and a least upper bound. Similarly, there was no distinction between positive and nonnegative, and the symbol $<$ might sometimes be interpreted as meaning $\leq$.

[65] In case X takes negative values somewhere in the interval, m must be the "greatest" of these – that is, the greatest in absolute value, what we would still call the smallest. Similarly, if X only takes negative values, then M must be the "smallest", not the "greatest" value [Lacroix *Traité*, II, 142].

5.2.3 "On the nature of integrals, and on the constants that must be added to them"

When Lacroix published the first edition of his *Traité élémentaire du calcul...* he kept this section on the "general method to obtain approximate values of integrals" virtually unchanged [Lacroix *1802a*, 284-309]. A very interesting detail is that in the table of contents of [Lacroix *1802a*] – which unlike that of [Lacroix *Traité*] contains titles of *subsections* – we find the following subsection of this section: "De la nature des intégrales, et des constantes qu'il faut y ajouter"[66] [*1802a*, xxxviii]. Indeed Lacroix had included, in a section supposedly devoted to approximate integration, some conceptual remarks about that object called *integral*.

But should not such remarks appear before, at the start of the integral calculus, that is at the beginning of the second volume? In that apparently more suitable context Lacroix pays remarkably little attention to foundational or conceptual issues: the integral calculus is simply the inverse of the differential calculus, so that its purpose is, given X, to find y such that $\frac{dy}{dx} = X$, and this is done by reversing the rules of differentiation [*Traité*, II, 1-2].

A certain lack of care in writing this passage (as if it was not terribly important?) can be seen in the fact that the names *primitive* or *integral* for the function y are introduced *only* in a footnote – a not very large footnote (by Lacroix's standards) whose purpose is to explain the origin of the notation $\int X\,dx$ for y: $\int$ for the infinite *ſum* of the infinitely small increments $X\,dx$, according to Leibniz's views. The name *integral* is then predominantly used throughout the volume, without further ado.

Not even the issue of arbitrary constants receives much attention. It is only introduced when dealing with the first example of a rational function ($\int ax^n dx = \frac{ax^{n+1}}{n+1} + B$ because $d(Ax^m + B) = mAx^{m-1}dx$), *not* when speaking of integrals in general. For its arbitrariness, the reader is referred to the first volume.

This almost exclusive referral to the principles of the differential calculus is consistent with what Lacroix had said in the general preface at the beginning of the first volume:

> "Lorsque les principes du Calcul différentiel sont bien établis, le Calcul intégral, qui en est l'inverse, n'offre plus qu'une collection de procédés analytiques, qu'il suffit d'ordonner de manière à en faire appercevoir les rapports."[67] [Lacroix *Traité*, I, xxvii]

It is also consistent with the usual approach to the integral calculus at the end of the 18th century (see section 5.1.1).

After the small and perfunctory introduction to the integral calculus which we have just discussed, Lacroix occupies over a hundred pages with "procédés

---

[66]"On the nature of integrals, and on the constants that must be added to them"

[67]"Once the principles of the differential calculus are well established, the integral calculus, which is its inverse, offers but a collection of analytical procedures, which is enough to order so as to make perceive their connections."

analytiques", that is, the integration of rational and irrational functions, series, and logarithmic, exponential, and trigonometrical functions. And then, in the section dedicated to approximate integration, he returns to conceptual issues.

First, Lacroix timidly introduces what we may interpret as limit considerations, and without pausing he substantiates Leibniz's original concept of the integral as an infinite sum of infinitesimals:

"Ces valeurs [approchées de $\int X dx$] seront d'autant plus exactes que les quantités a, a_1, a_2 seront plus voisines les unes des autres. En regardant les différences $a_1 - a$, $a_2 - a_1$, $a_3 - a_2$, comme infiniment petites, les quantités $Y'(a_1-a)$, $Y_1'(a_2-a_1)$, $Y_2'(a_3-a_2)$, etc. seront ce que devient la différentielle $X dx$, lorsqu'on fait successivement $x = a$, $x = a_1$, $x = a_2$, etc. C'est sous ce point de vue que l'on conçoit l'intégrale comme la somme d'un nombre infini d'élémens, égaux aux valeurs consécutives que prend la différentielle par les divers changemens qu'éprouve la variable x."[68] [Lacroix Traité, II, 137]

This is followed by a reference to the footnote on the notation $\int X dx$ at the beginning of the volume.

But what Lacroix subsequently uses from this passage is the naïve limit approach, not the infinitesimal one. In the chapter dedicated to the calculus of variations he would remark that

"il faut se rappeler qu'une intégrale peut être envisagée (n°. 470 [the article quoted above]), comme la limite des sommes d'un nombre indéfini d'élémens"[69] [Lacroix Traité, II, 686].

We have seen already (page 163) that Lacroix uses the property of the integral Y_1 being the limit of the approximating sum (5.15) to prove that (5.13) and (5.14) are bounds for its true value. A naïve limit argument is also used to prove that, if X is always positive between $x = a$ and $x = a_n$, then $Y_n - Y$ is also positive: for this difference we may give the approximate equation

$$Y_n - Y = Y'(a_1 - a) + Y_1'(a_2 - a_1) \ldots + Y_{n-1}'(a_n - a_{n-1}),$$

the right side of which is clearly positive if all the coefficients Y', Y_1', etc. are positive (which is an obvious consequence of X being positive); but it is possible to take the elements of the sequence $a, a_1, a_2, \ldots, a_n$ as close together as necessary to "porter ainsi le degré d'exactitude de l'équation ci-dessus, aussi loin qu'on le jugera à propos"[70]; the conclusion follows.

---

[68]"These [approximate] values [of $\int X dx$] will be the more exact as the quantities a, a_1, a_2 are closer to one another. Regarding the differences $a_1 - a$, $a_2 - a_1$, $a_3 - a_2$, as infinitely small, the quantities $Y'(a_1 - a)$, $Y_1'(a_2 - a_1)$, $Y_2'(a_3 - a_2)$, etc. will be the result of putting successively $x = a$, $x = a_1$, $x = a_2$, etc. in the differential $X dx$. It is from this point of view that the integral is conceived as the sum of an infinite number of elements, equal to the consecutive values which the differential receives through the varying changes experienced by the variable x."

[69]"one must remember that an integral may be viewed (n°. 470 [the article quoted above]), as the limit of the sums of an indefinite number of elements"

[70]"thus carry the degree of exactness of the above equation as far as deemed fit"

This same lemma can be found in [Lagrange *Fonctions*, 45-46], but in a different context (an important step in the derivation of the remainder for Taylor series), and with a different proof: Lagrange invokes Arbogast's principle to say that one can take i small enough for $f(a+i) - f(a) = if'(a) + \frac{i^2}{2}f''(a) +$ etc. to be positive, provided that $f'(a)$ is positive; dividing the interval from a to b into subintervals of length i and applying this argument also to $f(a+2i) - f(a+1)$, $f(a+3i) - f(a+2i)$, etc., he concludes that $f(b) - f(a) = f(a+i) - f(a) + f(a+2i) - f(a+1) +$ etc. is positive, if $f'(z)$ is always positive from $z = a$ till $z = b$. It is interesting to notice that Lacroix could have used Lagrange's proof, or at least a close adaptation − he had used Arbogast's principle before (see sections 3.2.6 and 4.2.1.2) and we have seen that this section starts with Taylor series; but instead he gave the above limit argument.

Of course these two proofs are of results ostensibly related to approximations − a subject which suggests the issue of convergence and hence of limits. What then has this to do with general conceptions of the integral? Well, first of all, whatever the subject of the section, these are proofs in which the integral − the true value of the integral − is represented as the limit of a sum.

Perhaps more importantly, in this section there are three articles, which have not yet been discussed, whose relation to the subject of approximations is, to say the least, not at all obvious. Those three articles address arbitrary constants of integration, the distinction between primitive functions and integrals, the distinction between definite and indefinite integrals – issues notoriously overlooked in the beginning of the volume – and a geometrical illustration of these considerations.

The first of those articles [Lacroix *Traité*, 137-138] is the one which, as mentioned above, was reproduced in [*1802a*, 287-288] with the title "On the nature of integrals, and on the constants that must be added to them". It occurs immediately after the passage quoted above suggesting limit- and infinitesimal-based approaches. Lacroix proposes to explain how the integral $\int X \, dx$ differs from a "given primitive function" (what Euler called a "particular integral"): if we assign a value to x, that of a "given primitive function" becomes perfectly determined (i. e., it is a function only of x); according to Lacroix, the same does not happen to the integral, because the same operation (the assignment of a value to x) only determines where the series Y, $Y'(a_1 - a)$, $Y_1'(a_2 - a_1)$, $Y_2'(a_3 - a_2)$, etc. should end, not where it should start: "la somme de cette série restera encore indeterminée tant qu'on n'aura rien statué sur la valeur de x, à laquelle doit répondre son premier terme et sur celle de ce premier terme"[71] [*Traité*, II, 138]. Although this is not the clearest explanation one could wish for,[72] it shows that for Lacroix the sum (or limit of sums) approach is not limited to approximation of definite integrals or of particular integrals; it also refers to indefinite integrals, by allowing the first term in the series to remain indeterminate.

---

[71]"the sum of this series will remain indeterminate while one has not pronounced about the value of x to which corresponds its first term nor about the value of this first term"

[72]It must have been clear enough for the textbook writer Jean-Guillaume Garnier, who reproduced it almost word for word in [Garnier *1812*, 108].

Lacroix's conclusion from this (also not exposed too clearly) is that the integral $\int X\,dx$ "est une fonction de x, dont la valeur se trouve renfermée entre deux limites qui sont indeterminées"[73]. These limits are independent of the constant of integration: if $\int X\,dx = P + const.$ and A and B are the values of P for $x = a$ and $x = b$ respectively, then the difference between the respective values of $\int X\,dx$, that is $A + const.$ and $B + const.$, is $A - B$.[74] According to Lacroix, this difference is nothing else than the sum of some of the terms in the series (5.8), namely those from the one corresponding to $x = a$ up to the one corresponding to $x = b$ (notice once again the indeterminacy of the first term in the global series, which might start before $x = a$).

The determination of the constant of integration (by forcing the integral to take a certain value for a specified value of x) corresponds to the determination of one term in the series, "Y, par exemple", from which one proceeds to form the other terms.[75] After this the integral becomes a *primitive function*, which only requires the specification of x for its complete determination.

The second of the three articles mentioned above occurs, strangely enough, four pages afterwards [*Traité*, II, 142-143]. It addresses mainly issues of terminology, introducing the terms *indefinite integral* (what he had been calling simply integral, the general value of $\int X\,dx$, which must contain an arbitrary constant to be complete) and *definite integral* (the result of giving a determined value to the variable, after having determined the constant of integration), and the expression "to take the integral $\int X\,dx$ from $x = a$ till $x = b$" (to calculate the difference between the corresponding values of the integral).

Lacroix attributes these names, rather vaguely, to "the Analysts"; presumably this is a reference to [Laplace *1779*]. The names *definite integral* and *indefinite integral* were by then rare enough for Cajori [*1919*, 272] to attribute their introduction to Lacroix himself.[76]

It must be said that these names do not occur often in the rest of the second volume (and not at all in this section; apparently the next occurrence is in the chapter on the calculus of variations [*Traité*, II, 685]); they, or rather "definite integral", only becomes frequent in the third volume, where Lacroix reports the works of Euler and Laplace that bear on definite integrals [*Traité*, III, 392-418, 445-529]. In two articles there Lacroix uses Euler's notation for the limits of integration

---

[73]"is a function of x whose value is enclosed between two indeterminate limits"

[74]*Sic*; not only is this not corrected in the errata as it is repeated in [Lacroix *1802a*, 288] and [Lacroix *Traité*, 2nd ed, II, 134] (but, curiously, it appears as $B - A$ in [*1802a*, 2nd ed, 303] and subsequent editions). One can only assume that Lacroix is only concerned here with the *absolute* difference. Nevertheless, as we have seen above, he speaks further ahead of this difference as $Y_b - Y_a$.

[75]Lacroix is quite clear about Y being completely independent from the other terms, so that what this means must be that one proceeds from the corresponding specified value of x.

[76]In this same year (*an* VI $\approx$ 1798) "indefinite integral" made a fleeting appearance in [Bossut *1798*, I, 415], but "definite integral" does not seem to have accompanied it there.

("lefthand" limit at the top) [*Traité*, III, 446-447, 475]:

$$\int \frac{x^{m-1} + x^{n-m-1}}{1 + x^n} \, dx \begin{bmatrix} x = 0 \\ x = 1 \end{bmatrix}.$$

But naturally the concept of definite integral occurs without need for mention of the name. It is clearly present, for instance, still in the section on approximation, in "the integral $\int \frac{dx}{\sqrt{1-x}}$, from $x = 0$ until $x = 1 - \delta$" [*Traité*, II, 145]. Similar expressions appear in the chapter on areas, volumes, etc., particularly when using double (repeated) integrals to calculate volumes [*Traité*, II, 195-197]. Integral splitting occurs very casually (for instance in [*Traité*, II, 152]); it may have been a motivation for one the few uses of Euler's notation [*Traité*, III, 447]:

$$\int \frac{x^{m-1}}{1 + x^n} \, dx \begin{bmatrix} x = 0 \\ x = \inf \end{bmatrix} - \int \frac{x^{m-1}}{1 + x^n} \, dx \begin{bmatrix} x = 0 \\ x = 1 \end{bmatrix} = \int \frac{x^{m-1}}{1 + x^n} \, dx \begin{bmatrix} x = 1 \\ x = \inf \end{bmatrix}.$$

Of course this would not be as obvious in a context of particular integrals/primitive functions.

Finally, the third of the certainly non-*approximative* articles [*Traité*, II, 143] gives a geometrical illustration (it is followed by the geometrical interpretation of the approximation method mentioned above): if the curve BCZ represents the function X, the integral $\int X dx$ may be regarded as representing "a variable portion" of the area under it. This portion may be indeterminate – doubly indeterminate, in fact – while its limits are arbitrary; but once the outmost abscissas are fixed – for instance AD and AP – it is determined – $DEMP$.

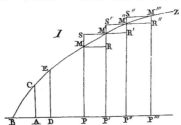

What can we make of Lacroix's section on the "general method" for approximation of integrals? Is it really just about approximation of integrals? I hope the preceding paragraphs will convince the reader that that section has another subject: the "nature of integrals".

Judith Grabiner [*1981*, 150-152] has concluded that that section was an important source of inspiration for Cauchy's theory of the integral: not only was it the most probable means through which Cauchy knew Euler's "general method", but also "Lacroix had picked out the key property of the definite integral – the integral is the limit of sums – and used it in a proof" (two proofs, actually), and had implied, "though not saying explicitly, that for any piecewise monotonic function approximating sums can be found that are arbitrarily close to the function's integral" (a reference to his remark about treating separately the portions where the function is increasing and those where it is decreasing). But she adds that

> "the technical similarities in their treatments of the definite integral cannot dispel the differences in points of view between Cauchy and his predecessors. For Euler and Lacroix, approximation by sums is just one property of the integral, related to little else in the theory of the integral calculus. For Cauchy, it became the defining property. For Euler and Lacroix, the integral is the antiderivative, whose value can be approximated by sums." [Grabiner *1981*, 152]

Also, "as usual, Lacroix had not intended to do anything new; in elaborating Euler's work, his goal was to present, explain, and clarify" [Grabiner *1981*, 150].

Lacroix's intentions regarding originality are not completely clear. In the general Preface of the *Traité* he does suggest that there are some details that belong to him [*Traité*, I, xxviii]; in later writings he claimed priority for some of those details (see section 10.1.1). But *this* detail is not among them. Apparently he did not see it as important enough. Perhaps because it still went against the prevailing tendencies in analysis?

Clearly, the differences between Lacroix's remarks on the "nature of integrals" and Cauchy's theory of the integral are huge. Lacroix did not give the limit of sums as *the definition* of definite integral; he did not question the existence of integrals; he did not prove that the limit of the approximating sums is independent of the mode of partition of the interval; and, more importantly, his remarks occupy a modest place in the structure of his integral calculus. It could not be otherwise: the purpose of Lacroix's *Traité* is to report the calculus as it was in the end of the 18th century and to prepare its readers to understand the research done in that area; and the integral calculus at that time was almost exclusively based on the conception of the integral as antiderivative.

However, there is enough evidence to say that for Lacroix, approximation by sums was *not* just another property of the integral, "related to little else in the theory of the integral calculus". It is true that it was not its *defining property*, but it was a property that allowed him to explore "the nature of integrals", and to explain the concepts of *indefinite integral*, *primitive function*, and *definite integral*. This may be regarded as "related to little else" in the integral calculus, in the sense that it had few technical consequences (if any), but such conceptual considerations would certainly be quite relevant for the intended readers – training mathematicians. It is also quite interesting to notice how Lacroix used this material in his first course of analysis at the *École Polytechnique*: of the 5 articles from this section mentioned in the summary of that course (see page 408), only two are about the approximation method (and one of these two is the geometrical illustration of the method and the other also includes the interpretation of the integral as a sum of infinitesimals or limit of sums); the other three are concerned with the distinctions between integral and primitive function and between definite and indefinite integral, with the determination of constants of integration and with the geometrical interpretation of integrals.

I believe that in the passage quoted above Grabiner fails to take full account of a fundamental distinction between Cauchy's and Lacroix's approaches. Cauchy wanted *one definition* for each concept from which all the results concerning that concept had to stem; Lacroix, on the other hand, thought that a concept could be seen from several perspectives, and that different aspects of that concept might be better illuminated from different perspectives.[77]

Thus for Lacroix the integral is the antiderivative *and* it is a limit of sums.

Another aspect of this that must be mentioned is its Leibnizian genealogy. It was remarked in section 5.1.1 that the Leibnizian idea of the integral as a sum of infinitesimals had never completely disappeared in the 18th century. Euler had established a connection between his "general method" of approximation and that idea, by allowing the differences between the abscissas used to be infinitely small (see page 152). However, he left that connection as an unconsequential remark.

What Lacroix did here, apart from improving on Euler's method itself, was to seriously pursue that connection, and give it a more solid ground. Believing that the correct interpretation of the infinitesimal method lies in taking it as an abbreviation for the method of limits [Lacroix *Traité*, I, 423-424], it should not be difficult for Lacroix to make the leap to the integral as "limit of sums", in order to provide a good explanation, an acceptable interpretation, of the Leibnizian infinite sum of infinitesimals.

Why did he do it? Probably for two reasons: firstly, because the encyclopedic character of his *Traité* demanded some acknowledgement of the original Leibnizian approach to the integral; but also because it seemed a worthwhile perspective: it made the integral a more concrete object, a better understandable one.

This *concreteness* helps us also to explain the puzzling location of Lacroix's remarks on the "nature of integrals". If we look at chapter 1 of [Lacroix *Traité*, II], we see 135 pages of formalistic, *algebraic* integration, based on the integral as antiderivative, followed by 21 pages of approximation and conceptual remarks.[78] For those first 135 pages, and indeed for most of the integral calculus, the quick definition of integral as antiderivative and the matter-of-fact reference to arbitrary constants were quite enough. The perspective of the integral as a limit of sums appears in a section which has a different *flavour*: an integral whose *value* is approximated is something more *concrete* than an antiderivative; and, very importantly, the derivation of formulas (5.8) and (5.10) is quite distinct from the formal manipulation of series and other expressions that can be seen in those 135 initial pages.

---

[77]Grabiner is of course well aware of Lacroix's "eclectic view" of the concepts of the calculus, but explains it on purely technical grounds: "Lacroix, like most mathematicians of the time, wanted to show how to solve problems; therefore his *Traité* included whatever techniques were applicable to this end" [Grabiner *1981*, 79-80]. This interpretation of Lacroix's motivations, while not at all wrong, is in my view too restrictive.

[78]And then 5 final pages on integration of higher-order differentials, which in the second edition of the *Traité* constitute a section, but in the first edition are included in this section on the "general method" of approximation.

In fact, what may be surprising is the occurrence at all of these conceptual remarks, and the fact that they appear so *early*: it would be conceivable for them (particularly the definition of definite integral) to appear in the chapter on calculation of lengths, areas, and volumes, or in the chapter on calculus of variations (a subject naturally related to definite integrals), or yet in the third volume, which is where definite integrals are effectively used.[79] Their occurrence in the first chapter is of course a consequence of their connection to the method of approximation, but this is not a full explanation: what depends on that method may appear at any time after the method. The location of these remarks in the chapter on integration of functions also reflects, in my opinion, a more general significance than they would have if they appeared only where they are more directly relevant.

To summarize: Euler's "general method" for approximation of integrals provided Lacroix with the chance of exploring the "nature of integrals" in an original way: referring back to the Leibnizian conception of the integral as a sum of differentials, but reinterpreting this in terms of limits. Given that the dominant approach at the time was that of the integral as antiderivative, the encyclopedic character of Lacroix's *Traité* would not allow this to be more than a detail (at least if evaluated lengthwise); but it was also this encyclopedic character that allowed this *detail* to appear at all. And how irrelevant could be to a training mathematician a *detail* which explained the "nature of integrals"?

5.2.4   Approximation of solutions of differential equations

Approximation of solutions of differential equations does not provide such interesting conceptual reflections. Or rather, it does, but in an incredibly fleeting way (see below). Lacroix mainly reports several methods, divided into first-order differential equations, second-order differential equations, and a combination of successive substitutions with integration of "first-degree" differential equations. All of this is in the chapter on ordinary differential equations: in section 5.1.3 we saw no attempts to approximate partial differential equations in the 18th century (apparently there were none), and we do not see them in Lacroix's *Traité*.

In the section on approximate solutions of first-order differential equations [*Traité*, II, 284-296], Lacroix is more inclined than in the chapter on integration of explicit functions to match series integration with approximate integration:

> "Après avoir épuisé les moyens connus pour intégrer une équation différentielle, il faut chercher à la résoudre par approximation, c'est-à-dire, à en tirer la valeur de y en x, au moyen d'une série."[80] [Lacroix *Traité*, II, 284]

---

[79]It is true that in the examples of the use of the approximation method Lacroix uses, if not the name *definite integral*, at least the idea of integration "from $x = a$ till $x = b$". But of course he did not have to: in the same context Euler had stuck to particular integrals.

[80]"After having exhausted all known means of integrating a differential equation, we must try to solve it by approximation, that is, to extract from it the value of y as a series in x."

Naturally, he starts by undetermined coefficients [Lacroix Traité, II, 284]: if we know that $y = b$ when $x = a$ (a and b constants), we can put $x = a + t$, $y = b + u$, and $u = At^\alpha + Bt^{\alpha+1} + Ct^{\alpha+2} +$ etc., substitute in the differential equation (choosing α appropriately) and solve for A, B, C, etc.

Next Lacroix considers Taylor-series expansions. He uses them in deriving the series

$$Y_1 = Y + Y'\frac{(a_1 - a)}{1} + Y''\frac{(a_1 - a)^2}{1 \cdot 2} + Y'''\frac{(a_1 - a)^3}{1 \cdot 2 \cdot 3} + \text{etc.} \tag{5.16}$$

equivalent to (5.6), except in that now the coefficients $Y', Y'', Y''', \ldots$ depend not only on a but also on Y (since $\frac{dy}{dx}$ depends on x and y). Euler's "general method" is a natural consequence, but Lacroix is extremely brief about it: he mainly remarks that what had been said in the articles leading to (5.8), (5.9), and (5.11-5.12) also applies here – in the latter case minding that the coefficients also depend on $\frac{dy}{dx}$ and its differentials; and it also seems that he has in mind formulas more complicated than (5.11-5.12), involving differences $a_1 - a, a_2 - a_1, \ldots$ not all equal, and in the second case probably with Y, not Y_n, on the left side. It is not completely clear whether Lacroix excludes from this the use of the average between (5.11) and (5.12) (or rather between its correspondents), which had appeared in the case of explicit functions [Traité, II, 148]; but his implicit reference to (5.9) instead of (5.10) ("revenir de cette valeur $[Y_n]$ à celle de Y"[81] [Traité, II, 286]) suggests that what Lacroix had in mind for the differential-equation equivalents of (5.9) and (5.12) was situations in which the initial conditions refer to a_n, Y_n and it is a *left-hand* value of y that is approximated.

Here occurs a most interesting remark, although also very casual (it is the fleeting conceptual remark announced above):

> "Ce qui précède fait voir que les équations différentielles du premier ordre à deux variables sont toujours possibles, c'est-à-dire, qu'on peut toujours assigner des valeurs soit rigoureuses, soit approchées de la fonction qu'elles déterminent"[82] [Lacroix Traité, II, 287].

(This opens an article which is somewhat out of place: Lacroix argues that the "possibility" of first-order differential equations may also be shown by geometrical considerations – by presenting a construction for those equations; details of the construction will be given in section 6.2.3.2.) Is this not a (very crude) attempt at an existence theorem? Of course, one must not exaggerate its relevance: it is very far from Cauchy's results of the 1820's [Cauchy 1981]; and it is even much less developed than Lacroix's considerations on integrals of explicit functions using similar approximations (section 5.2.3). But Lacroix's concern with showing an

---

[81]"to return from this value $[Y_n]$ to that of Y"

[82]"The preceding shows that the differential equations of first order are always possible, that is, that one can always assign values, either rigorous or approximate, to the function which they determine"

existence that most people around 1800 took for granted is noteworthy.[83] Lacroix may have been inspired by a similar remark by Leibniz: having constructed a polygon approximating a certain transcendental curve starting at an arbitrary point $1C$, Leibniz concluded

> "Et sic habebitur polygonum $1C2C3C$ &c. lineae quaesitae succeda-
> neum, seu *linea Mechanica Geometricae vicaria*; simulque manifeste
> cognoscimus, possibilem esse Geometricam per datum punctum $1C$
> transeuntem, cum sit *limes, in quem* tandem polygona continue *ad-*
> *vergentia evanescunt.*"[84] [Leibniz *1694*, 374]

Still, Lacroix does not mention Leibniz in connection to this subject (either possibility/existence or approximation in general), nor does he cite this memoir in the table of contents. A probable indirect influence, motivating the concern with possibility, is Clairaut's claim for the impossibility of some differential equations in three variables (see section 6.1.3.1), as well as Monge's denial of that impossibility; in fact, this denial opens with a short remark [Monge *1784c*, 502] on the possibility of every first-order differential equation in two variables, based on the argument that using the equation one can always find the slope of the tangent to the (integral) curve – Lacroix's remark is very likely an elaboration of Monge's.

Next, Lacroix shows his awareness of insufficiencies in the Taylor series (5.6) and (5.7) and in the formulas derived from them. But he reduces them to situations in which some differential coefficient of the function y of x becomes infinite, and solves those insufficiencies by considering more general power series – with non-integer exponents – as he had done in chapter 2 of the first volume, extracting a method for obtaining those series from a memoir by Lagrange on continued fractions (see pages 108 and 157).

Lacroix does not dwell on Lagrange's method for obtaining those power series, since he already had done so [*Traité*, I, 220-230]; but he does dwell on Lagrange's use of it for obtaining continued fractions (see page 157). In fact, this takes up about two thirds of the section on approximation methods for first-order differential equations [*Traité*, II, 288-296]. However, it would be wrong to conclude from the number of pages that this is the most important method for approximation. It might need more pages to be explained, but was probably less relevant: it is not taken up for second-order differential equations, and it is dropped off from [Lacroix *1802a*].

---

[83]Concerning the influence of Lacroix's *Traité*, it is also noteworthy that Cauchy's first existence theorem derived from the same method of approximation [Cauchy *1981*, 39-66]. Gilain [*1981*, xxiv-xxv, xxxiii] compared Cauchy's work with Lacroix's *Traité*, but because he used only the second edition of the latter he missed Lacroix's connection between the analytical version of this method and the "possibility" of differential equations.

[84]"And thus we will have a polygon $1C2C3C$ &c. replacing the required curve, that is, *a mechanical curve in place of the geometrical one*; at the same time we clearly perceive that the geometrical [curve], passing through a given point $1C$, is possible, since it is the limit into which the continually *converging* polygons finally *vanish.*"

The section on approximation methods for second-order differential equations [*Traité*, II, 349-364] is essentially an account of chapters VII and VIII of the first section of [Euler *Integralis*, II] – that is, power-series developments for $ddy + a\,x^n\,y\,dx^2 = 0$ and $xx(a + b\,x^n)ddy + x(c + e\,x^n)dx\,dy + (f + g\,x^n)y\,dx^2 = 0$.

Euler's "general method" is mentioned also for second-order equations, but only in a short article [*Traité*, II, 351], remarking that what was said about its use for first-order equations also applies here, except that now in series such as (5.16) the second term is arbitrary, since a second-order differential equation leaves the first differential coefficient undetermined; one must then have as initial condition not only the value of y but also that of $\frac{dy}{dx}$, for $x = a$.

This is accompanied by an article on the construction of second-order equations [*Traité*, II, 351-352], entirely analogous to the one on first-order equations mentioned above, and which has little to do with approximation (see section 6.2.3.2).

Lacroix includes one final section on approximate integration of differential equations, namely on the use of integration of "first-degree" (that is, linear) differential equations to obtain successive approximate solutions of non-"first-degree" differential equations [*Traité*, II, 394-408]. That is, this section deals with the methods used in obtaining approximations of planetary orbits (see page 156 above). However, Lacroix does not mention that motivation for these methods. The only hint is when he refers the reader seeking further details to the "excellens Mémoires d'Astronomie-physique de Lagrange et de Laplace"[85] [*Traité*, II, 407]. Lacroix is clearly not interested in astronomy (not in the *Traité*, that is – "un ouvrage consacré uniquement à l'Analyse et à la Géométrie"[86] [*Traité*, II, 299]), but rather simply in mathematical methods that happened to have originated from astronomical problems. This idea is reinforced by his closing sentence saying that he had had as only goal in this section to "rattacher à l'ensemble des méthodes du Calcul intégral"[87] several procedures which had thus far always been expounded isolated – isolated, one gathers, from pure analysis.

Lacroix [*Traité*, II, 394-397] introduces the traditional method through the example

$$\frac{d^2y}{dx^2} + y + \alpha y^2 = b,$$

where α is very small; neglecting α yields the first-degree equation

$$\frac{d^2y}{dx^2} + y = b,$$

whose integral is

$$y = b + p\,\cos x + q\,\sin x;$$

---

[85]"excellent memoirs of physical astronomy by Lagrange and Laplace"
[86]"a work solely devoted to analysis and geometry"
[87]"restore to the collection of methods of integral calculus"

putting $Y = b + p \cos x + q \sin x$, $y = Y + \alpha y'$, substituting in the original equation, and neglecting α^2 and α^3, yields

$$\frac{d^2 y'}{dx^2} + y' = -Y^2 \left[= -b^2 - 2b(p \cos x + q \sin x) - (p \cos x + q \sin x)^2 \right];$$

from which a second approximate value for y is obtained; and so on. Next [*Traité*, II, 398] he remarks that this ammounts to assuming

$$y = Y + \alpha Y' + \alpha^2 Y'' + \alpha^3 Y''' + \text{etc.,}$$

obtaining Y, Y', Y'', etc. from

$$\frac{d^2 Y}{dx^2} + Y = b, \quad \frac{d^2 Y'}{dx^2} + Y' = -Y^2, \quad \frac{d^2 Y''}{dx^2} + Y'' = -2YY', \quad \text{etc.}$$

After two iterations Lacroix has something of the form

$$y = A + (B + Cx + Dx^2) \cos x + (E + Fx) \cos 2x + G \cos 3x$$
$$+ (B' + C'x + D'x^2) \sin x + (E' + F'x) \sin 2x + G' \sin 3x,$$

which he says is only an approximate value in case x is very small [*Traité*, II, 400] – a reference to the "arcs of circle", i. e. to the powers of x higher than zero which appear in the coefficients of the sines and cosines; while if one had a result of the form

$$y = A_1 + B_1 \cos \beta x + C_1 \cos \gamma x + \text{etc.}$$
$$+ B_1' \sin \beta x + C_1' \sin \gamma x + \text{etc.,}$$

"et que les coefficiens $A_1, B_1, B_2, \ldots B_1', B_2'$, etc. formassent une suite convergente"[88], the fact that the sine and cosine are bounded would assure the convergence of the expression for y. Thus Lacroix presents as motivation for the avoidance of "arcs of circle" the fact that they make convergence harder to achieve, not any astronomical considerations.

Notice the twofold mistake above: the sequence $A_1, B_1, B_2, \ldots B_1', B_2', \ldots$ does not even make sense; and if we assume that it is a typo for $A_1, B_1, C_1, \ldots B_1', C_1', \ldots$, we still have to face the fact that Lacroix should be asking for $B_1, C_1, \ldots$ and $B_1', C_1', \ldots$ to be *two* convergent sequences. This is only the first of a series of strange mistakes in this section. The following ones become even stranger when we notice that Lacroix was following [Lagrange *1783*, § 1-4] closely – where these mistakes do not occur.

Thus Lacroix [*Traité*, II, 400-403] reports Lagrange's method of variation of constants: assuming

$$y = P + P'x + P''x^2 + P'''x^3 + \text{etc.,}$$

---

[88]"and the coefficients $A_1, B_1, B_2, \ldots B_1', B_2'$, etc. formed a convergent sequence".

where P, P', P'', P''', etc. contain only exponentials, sines and cosines of multiples of x, along with the arbitrary integration constants p, q, etc., differentiation yields

$$\frac{dy}{dx} = \frac{dP}{dx} + P' + \left(\frac{dP'}{dx} + 2P''\right) x + \left(\frac{dP''}{dx} + 3P'''\right) x^2 + \text{etc.},$$

$$\frac{d^2y}{dx^2} = \frac{d^2P}{dx^2} + 2\frac{dP'}{dx} + 2P'' + \left(\frac{d^2P'}{dx^2} + 4\frac{dP''}{dx} + 6P'''\right) x + \text{etc.},$$

and so on – here occurs the second mistake: in the equations above Lacroix writes the differentials dy, dy^2 on the left-hand sides instead of the differential coefficients; now, in order to have the equation free of powers of x, the coefficients in these series must be null, and we must have

$$y = P, \quad \frac{dy}{dx} = \frac{dP}{dx} + P', \quad \frac{d^2y}{dx^2} = \frac{d^2P}{dx^2} + 2\frac{dP'}{dx} + 2P'', \quad \text{etc.}$$

– and another mistake: d^2y instead of $\frac{d^2y}{dx^2}$ in the third equation (but a correct $\frac{dy}{dx}$ in the second one); and for this to make sense (if $y = P$, $\frac{dy}{dx}$ should certainly not be $\frac{dP}{dx} + P'$) it is necessary to regard the arbitrary constants as variables, to differentiate accordingly and to determine them so as to verify the equations above. Here occurs yet another mistake: Lacroix seems to forget the "etc." in the list of constants "p, q, etc." which he had given, and writes

$$dy = \frac{dP}{dx}dx + \frac{dP}{dp}dp + \frac{dP}{dq}dq;$$

he proceeds using only p and q in the next formulas (the corresponding formulas in [Lagrange 1783, § 3] have the appropriate &c.'s), although also repeating the list "p, q, etc."; of course this might be simply dismissed as sloppy language, but it is uncharacteristically sloppy for Lacroix, and culminates an uncharacteristic sequence of typos/mistakes. This section seems to have suffered from a very poor editorial job.

After extending this method to systems of equations [*Traité*, II, 403-406], Lacroix comments on the "arcs of circle" being terms from power series expansions of sines and cosines, so that Lagrange's method really amounts to replacing those series with the original functions; he then mentions a method by Trembley which uses this idea, by grouping the terms so as to form recognizable series – but which entails calculations too long to be included in Lacroix's *Traité*.

This section finishes with a footnote (slightly over a page in size), where Lacroix [*Traité*, II, 407-408] reports the first version of Lagrange's method of variation of constants, following [Lagrange 1781, §25-26] (in the first edition Lacroix *forgets* to mention [Lagrange 1781] in the table of contents – which may be why in the second edition it receives a "N.B." [Lacroix *Traité*, 2nd ed, II, xvi]).

Chapter 6

Types of solutions of differential equations

This chapter deals with several aspects of differential equations relating to *types of solutions* (complete, general, particular, and singular integrals or solutions), as opposed to *methods of solution*. That is, the subject here is not so much the processes for solving differential equations, as the conceptions about what kind of object a final solution might be. For this reason, the word "solution" will be used here in the sense of *answer*, but not in the sense of *process for obtaining an answer*.

6.1 Types of solutions of differential equations in the 18th century

It has been seen in section 5.1.1 that Euler tended not to distinguish conceptually integrating functions from solving differential equations. Thus, his definitions of *complete* and *particular* integral (from the general preface to [*Integralis*]) applied to both situations:

> "Integrale *completum* exhiberi dicitur, quando functio quaesita omni extensione cum constante arbitraria representatur. Quando autem ista constans iam certo modo est determinata, integrale vocari solet *particulare*."[1] [Euler *Integralis*, I, § 36]

In these definitions, the phrase "arbitrary constant" should not be taken too literally: Euler had mentioned a few articles earlier the possibility of the function

---

[1]"A *complete* integral is said to be presented when the required function is represented in all its extension with an arbitrary constant. When, on the other hand, this constant has already been determined in some way, the integral is usually called a *particular* one."

y being "defined by a relation between second-order differentials", in which case
it would involve *two* arbitrary constants [*Integralis*, I, § 33]; and the possibility of
y being a function of two variables x and t, in which case it also would seem to
involve an "arbitrary constant", but apparently one for each value of t – that is, in
fact an arbitrary function of t [*Integralis*, I, § 34].

Given these definitions, it is easy to conclude that "integrale ergo comple-
tum omnia integralia particularia in se complectitur"[2] [Euler *Integralis*, I, § 38].
Naturally this applies both to integrals of explicit functions and to solutions of
differential equations, which is confirmed at the beginning of a chapter on "particular integration of differential equations": a particular integral of a differential
equation must be contained in its complete integral [*Integralis*, I, § 540].

The following sections are partly dedicated to the story of how this *neat*
scheme got complicated. The first threat that it faced was the occurrence of singular solutions, that is, solutions not contained in the complete integral. But further
complications appeared in the case of partial differential equations when Lagrange
[*1774*] introduced a distinction between complete and general integrals, that is
between integrals containing arbitrary constants and integrals containing arbitrary functions.

6.1.1   Terminological complications

A modern reader faces additional difficulties when trying to understand the work
of 18th-century mathematicians on this subject, because of the use of different
terminologies, including sometimes the use of the same name for different objects.

Until around 1770 everything was simple: as above, "complete" integrals (or
synonymously "general" integrals [Laplace *1772a*]) opposed to "particular" integrals. The first complication arose when Laplace [*1772a*, 344] decided to distinguish "particular integrals" (contained in the general integral) from "particular
solutions" (not contained in the general integral). Rather confusingly, Lagrange
[*1774*] used the name "particular integrals" for what Laplace had called "particular solutions"; as for what Euler and Laplace had called "particular integrals",
Lagrange used the term "incomplete integral" [*1774*, § 1, § 13]. Even more confusingly, there are a few (fortunately only a few) situations in which "particular
integral" seems to refer to any solution which does not contain the necessary arbitrary elements to be complete, regardless of being contained or not in the complete
integral; for instance, Lagrange in a letter to Euler dated 1769, complimenting the
latter on his "methodes [...] pour reconnoitre si une integrale particuliere peut être
comprise dans l'integrale générale"[3] [Euler & Lagrange *Correspondance*, 464]; or
Trembley [*1790-91*], who seems to usually employ the expression "particular integral" to refer to both particular instances of the complete integral and singular
solutions, but who when addressing the subject of singular solutions refers to

---

[2]"thus the complete integral embraces in itself all the particular integrals"
[3]"methods to recognize whether a particular integral might be contained in the general integral"

"intégrales particulières proprement dites"[4], to be distinguished from "incomplete integrals" [Trembley *1790-91*, 4]. Later Lagrange [*Fonctions*, 69] introduced the adjective "singular"[5], which eventually solved the confusion by displacing the word "particular" from names for singular solutions.[6]

A different complication occurs with partial differential equations, because of Lagrange's [*1774*] distinction between "complete" and "general" integrals, using terms that until then had been synonymous. As will be seen below, not everyone (not even Lagrange!) followed this terminological distinction in the late 18th century. That is, more often than not "complete integral" of a first-order partial differential equation still meant an integral with one arbitrary function [Lagrange *1779*, 153; Monge *1784b*, 120; Legendre *1787*, 338].

An attempt has been made in this chapter to follow the original terminologies when reporting the work of 18th-century mathematicians. Therefore, say "particular integral" will be used when speaking of Euler or Laplace with the same meaning as "incomplete integral" when speaking of Lagrange. There is however one important exception: "particular integral" in the sense of [Lagrange *1774*] – that is, with the meaning of singular solution – would be too confusing, so that in the following sections it was replaced by "singular integral" (both when speaking of Lagrange or of other authors that followed his terminology). Confusion arising from conflicting uses of the expression "complete integral" is a necessary risk: the choice of which kind of integral to name *complete* is an important conceptual clue.[7]

6.1.2 Singular solutions

6.1.2.1 Euler and Clairaut

Euler was well aware of the existence of what is now known as *singular solutions* of differential equations. This existence had been noted in two works that had appeared in 1736,[8] one of which was his *Mechanica* [Euler *1736*]. In its second

---

[4]"particular integrals properly so called"

[5]He probably used this adjective because Taylor [*1715*], when encountering for the first time a singular solution, had remarked that it was "singularis quædam solutio", which may be translated as "a certain unique solution" – "unique" either in the sense of *only one (of its kind)* or of *remarkable*.

[6]But not immediately: in the 1820's the syllabi of the *École Polytechnique* still used Laplace's term "particular solutions" [Gilain *1989*, 112, 116, 120, 126, 130], while Cauchy, in his lectures there, changed from following that in 1819/1820 and 1821/1822 [Gilain *1989*, 61, 67] to speaking of "singular integrals" in 1823/1824, 1827/1828 and 1829/1830 [Gilain *1989*, 73, 85, 93].

[7]Except for authors (possibly influenced by Laplace [*1772a*]) who seemed to prefer "general integral" as the principal term: the syllabi of the *École Polytechnique* from 1817 to 1830 spoke of "general integrals" of *ordinary* differential equations [Gilain *1989*, 108-130], and so did Cauchy in his lectures [Gilain *1989*, 56-94]. But among the authors studied here Laplace and Condorcet [*1765*, 3, 67] were the only ones with that preference.

[8]Brook Taylor had encountered one before that, but he does not seem to have noticed its significance [Taylor *1715*, 26-27].

volume Euler not only gives two examples of equations with singular solutions,[9] but he also gives a rule to find such solutions: if V is a function of u and T is a function of t such that $T = 0$ for $t = 0$, then the equation

$$\frac{dt}{T} = V\,du$$

is satisfied both by

$$t = 0 \quad \text{and} \quad \int \frac{dt}{T} = \int V\,du;$$

moreover, even if T is not null for $t = 0$, $T = 0$ is a solution (since it implies $dt = 0$) [Euler *1736*, II, §335].

The other work published in 1736 which mentions singular solutions is [Clairaut *1734*, 209-213]. Investigating a curve MON with two branches, each

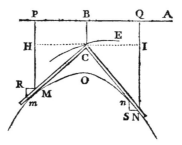

of which is tangent to one of the two arms of a sliding set square MCN whose vertex describes a given curve EC, Clairaut arrives at

$$x\,\Pi u - u\,\Pi u = y - \Phi u \quad \text{and} \quad \frac{dy}{dx} = \Pi u$$

(where x, y are the coordinates AP, PM of MON; $u, \Phi u$ are the coordinates AB, BC of the given curve EC; and Πu is used to express the fact that $\frac{dy}{dx}$ is a function of u). Differentiating, he gets $dx\,\Pi u + x\,\Delta u\,du - du\,\Pi u - u\,\Delta u\,du = \Pi u\,dx - \Xi u\,du$ (where $d\Pi u = \Delta u\,du$ and $d\Phi u = \Xi u\,du$); happily $dx\,\Pi u$ cancels out and all that remains is divisible by du, so that $x\,\Delta u - \Pi u - u\,\Delta u = -\Xi u$, whence the solution

$$x = \frac{\Pi u + u\,\Delta u - \Xi u}{\Delta u} \quad \text{and} \quad y = \frac{(\Pi u)^2 - \Xi u\,\Pi u + \Phi u\,\Pi u}{\Delta u}.$$

An interesting issue, which Clairaut remarks, is that this process does not involve integration, although it is easy to think of a different process that would: to solve

---

[9] $x = a$ for $ds = \dfrac{-dx\sqrt{b}}{\sqrt{a^2 b - bx^2 - a^2 \int P\,dx}}$ and $k^2 u = (k^2 + 1)x$ for $\dfrac{(k^2+1)dx - k^2 du}{\sqrt{(k^2+1)x - k^2 u}} = \dfrac{\pm du}{\sqrt{u}}$ [Euler *1736*, §335, §300].

$\frac{dy}{dx} = \Pi u$ for u, substitute the result in $x\,\Pi u - u\,\Pi u = y - \Phi u$ and integrate the resulting differential equation; the problem is that this integral would inevitably include an arbitrary constant which is absent from the solution obtained above; so we have two non-equivalent solutions, and the one obtained by differentiation would seem to be less general than the one obtained by integration. However, argues Clairaut, the only step in the former that may cause a loss of generality is the division by du, which might be zero; and in the case of $du = 0$, that is $u = a$ for some constant a, we would have only $x\,\Pi a - a\,\Pi a = y - \Phi a$, the equation of a straight line (an arm of the set square, in fact). Calculating two examples ($\Pi u = \Phi u = u$, and $\Pi u = \frac{u}{a+u}$ and $\Phi u = 0$), he concludes that integration leads only to the straight line solution, while the solution he is after is not obtainable by integral calculus. He closes the subject (which is not the central topic of the memoir) with the statement that, more generally, any equation of the form

$$\frac{d(\Phi xy)}{\Phi xy} = \text{some function of } x, y, dx, dy$$

has the solution $\Phi xy = 0$, besides the one obtained by integration (Φxy is Clairaut's notation for a function of the two variables x, y).

This is explained more clearly in [Euler 1756]. There Euler addresses these two interrelated paradoxes: that some differential equations are more easily solvable by further differentiation than by the normal methods of integral calculus, and that some differential equations are satisfied by finite equations which are not contained in their complete integral.

For the first paradox Euler gives four examples, the first of which is that of, given a point A, to find a curve such that all the normals taken from it to A have the same length a. This gives the differential equation

$$y\,dx - x\,dy = a\sqrt{dx^2 + dy^2}, \tag{6.1}$$

which it takes two pages to solve by setting the differentials free of the square root:

$$aa\,dy - xx\,dy + xy\,dx = a\,dx\sqrt{xx + yy - aa}, \tag{6.2}$$

and separating the variables by substituting $y = u\sqrt{aa - xx}$:

$$\frac{du}{\sqrt{u-1}} = \frac{a\,dx}{aa - xx} \tag{6.3}$$

to finally arrive at the solution

$$y = \frac{n}{2}(a + x) + \frac{1}{2n}(a - x) \tag{6.4}$$

(n is the arbitrary constant; this is an equation of all the straight lines at distance a from the origin). Instead of this, it is much easier to differentiate, after putting (6.1) in the form

$$y = px + a\sqrt{1 + pp}$$

(where $dy = p\,dx$); this allows us to cancel out $p\,dx$, and the remaining terms are divisible by dp; this division and a few algebraic manipulations lead to the solution

$$xx + yy = aa; \tag{6.5}$$

while the case $dp = 0$ quickly gives

$$y = nx + a\sqrt{1 + nn}, \tag{6.6}$$

(where once again n is the arbitrary constant; this also gives all the straight lines at distance a from the origin). Not only is this much easier, Euler remarks, but it can also be applied to equations such as

$$y\,dx - x\,dy = a\sqrt[3]{dx^3 + dy^3}, \tag{6.7}$$

whose variables cannot be separated. The other three examples are very similar: after being differentiated, the only terms that are not multiples of dp are two instances of $p\,dx$ which cancel each other.

The same examples can be given for the second paradox. For instance, in the first example the normal procedures give only the solution (6.4), which clearly does not include the circle of radius a (6.5). Euler includes another example (it is in fact the first in the text), where the singular solution is found in a more immediate way: given the equation

$$x\,dx + y\,dy = dy\sqrt{xx + yy - aa}, \tag{6.8}$$

"it is evident" that $xx + yy - aa = 0$ is a solution, although it is not contained in the complete integral $\sqrt{xx + yy - aa} = y + c$ (of course the same immediate reasoning could be applied to (6.2) or (6.3)).

Euler's *explanation* for these two paradoxes relies heavily on the form of the examples, more precisely on the forms such as (6.3) and (6.8): the equation

$$V\,dz = Z\,(P\,dx + Q\,dy),$$

where z, P, Q, V are functions of x, y and Z is a function of z, accepts the solution

$$Z = 0,$$

since this implies $z = \text{Const.}$, that is, $dz = 0$ (a variant of the rule he had given in [Euler *1736*]). As for the first paradox, Euler simply argues that the cases in which it occurs are precisely those in which the second occurs, and that those solutions found by differentiation instead of integration are the ones that are not comprised in the complete integral.

So, Euler had already studied the phenomenon of what are nowadays called singular solutions. Yet, he never gave any special name to these solutions [Rothenberg *1908*, 325, 344]. Moreover, in [Euler *Integralis*] he refused them the status of

integrals; referring to them, he wrote: "Etiamsi scilicet omnia integralia sint eius-
modi valores, qui aequationi differentiali satisfaciant, tamen non vicissim omnes
valores, qui satisfaciunt, sunt integralia"[10] [Euler *Integralis*, I, §546]; even if the paradox had already been *explained*, these solutions were *anomalies*. They were also
tricky: when one is not capable of finding a complete integral, particular integrals
are very useful, but there is the danger of getting instead those solutions which are
not integrals at all. In [Euler *1764*, § 34-35], he attributes a wrong result to the
existence of a singular solution[11], which caused Condorcet to say that "M. Euler a
remarqué [...] que ces solutions particulières non comprises dans l'équation générale
ne pouvaient être employées à la solution des problèmes"[12] [Condorcet *1770-1773*,
13-14]. This negative view of singular solutions motivates the study of the distinction between them and particular integrals [Euler *Integralis*, I, §546]; the sole
object of Euler's researches on singular solutions in [*Integralis*] is to find criteria
for this distinction [Rothenberg *1908*, 341-344]: for instance, in the case $dy = \frac{dx}{Q}$,
$x = a$ is a particular integral if it makes not only $Q = 0$ but also $\int \frac{dx}{Q} = \infty$
[Euler *Integralis*, I, §547]; or, for $y = X$ (where X is a function of x) to be a particular integral of $Pdx = Qdy$, it is necessary, when substituting $y = X + \omega$, that ω
appears with an exponent greater or equal to 1 (in absolute value) [Euler *Integralis*,
I, §565].

An inconvenience in Euler's work on the subject is that, as we have seen,
it was highly dependent on the *forms* of the solutions. For instance this last rule
(Euler's most general) required the proposed particular integral to be in the form
$y = X(x)$. However, it was quite fruitful, being adapted by Laplace and later used
also by Lagrange [*Fonctions*].

6.1.2.2 Laplace

[Laplace *1772a*] was a turning point in several respects. First of all, it introduced
a name for those solutions not comprised in the complete integral (or general
integral, as Laplace calls it): *particular solutions* [Laplace *1772a*, 344].

It also addressed the issue for the first time without relying on the specific forms of the solutions. To determine whether a certain solution $\mu = 0$ of a
differential equation $dy = p\,dx$ is a particular integral, Laplace considers a curve
HCM representing $\mu = 0$, and another curve LCN obtained by determining the
arbitrary constant in the general integral $\varphi = 0$ of $dy = p\,dx$ with the condition
that it should pass through a given point C of HCM. In case $\mu = 0$ is a particular
integral, HCM and LCN are one and the same curve; if for any abscissa P the
points M and N in the two curves do not coincide, then $\mu = 0$ is a particular

---

[10]"Although obviously all integrals are values such that they satisfy a differential equation,
still on the other hand not all values that satisfy [it] are integrals."

[11][Blanc *1957*, xx] presumes that some real mistake had slipped into Euler's reasonings. He is
very critical of the whole memoir [Euler *1764*].

[12]"M. Euler has remarked [...] that those particular solutions not contained in the general
equation could not be employed in the solution of problems"

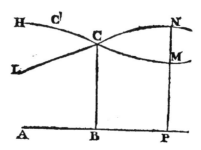

solution. To compare $y' = PM$ with $Y' = PN$ without knowing $\varphi = 0$, Laplace uses Taylor's theorem:

$$y' = y + \frac{\alpha\,\delta y}{\delta x} + \frac{\alpha^2}{1\cdot 2}\cdot\frac{\delta^2 y}{\delta x^2} + \frac{\alpha^3}{1\cdot 2\cdot 3}\cdot\frac{\delta^3 y}{\delta x^3} + \&c.$$

$$Y' = y + \frac{\alpha\,dy}{dx} + \frac{\alpha^2}{1\cdot 2}\cdot\frac{d^2 y}{dx^2} + \frac{\alpha^3}{1\cdot 2\cdot 3}\cdot\frac{d^3 y}{dx^3} + \&c.$$

(where $y = BC$, $\alpha = BP$, δ represents differentiation on the curve HCM, and d represents differentiation on the curve LCN). The conclusion is that $\mu = 0$ is a particular integral if not only it is a solution to $dy = p\,dx$ but also

$$\frac{\delta y}{\delta x} = \frac{dy}{dx}, \quad \frac{\delta^2 y}{\delta x^2} = \frac{d^2 y}{dx^2}, \quad \frac{\delta^3 y}{\delta x^3} = \frac{d^3 y}{dx^3}, \quad \&c.$$

Examining possible power expansions (with positive exponents) for $p = \frac{dy}{dx}$ (but in fact concentrating only on the term with the smallest exponent, n), Laplace reduced the differential equation $dy = p\,dx$ (satisfied by $\mu = 0$) to the form $d\mu = \mu^n \cdot h\,dx$ (where h is a function of x and μ), arriving at two conclusions: one, that if $n \geq 1$ then $\mu = 0$ is a particular integral, and if $n < 1$ then $\mu = 0$ is a particular solution (a development of one of Euler's criteria, seen above) [Laplace 1772a, 347-350]; and two, that $\mu = 0$ is a particular solution if and only if it makes $\left(\frac{dd\mu}{dx^2}\right) + p\left(\frac{dd\mu}{dxdy}\right) + \left(\frac{dp}{dx}\right)\left(\frac{d\mu}{dy}\right)$ infinite [Laplace 1772a, 350-351].

Laplace also gave two methods to find all the particular solutions of a given differential equation $dy = p\,dx$. The first method related to integrating factors: let β be the integrating factor of the equation, so that $\beta(dy - p\,dx)$ is an exact differential; if $\mu = 0$ is a particular solution of $dy = p\,dx$, then μ is a function of x and y, and y is also a function of x and μ, so that the integral of $\beta(dy - p\,dx)$ can be put in the form $\Psi(x, \mu) + C$ (C an arbitrary constant); but whatever value we attribute to C, the condition $\mu = 0$ cannot make $\Psi(x, \mu) + C$ vanish (otherwise $\mu = 0$ would be a particular integral) – and the same applies to its differential $\beta(dy - p\,dx)$; while $\mu = 0$ must make $dy - p\,dx = 0$; therefore $\mu = 0$ must make

β infinite; the conclusion is that the particular solutions are factors of $\frac{1}{\beta} = 0$
[Laplace *1772a*, 352].

The second method could be used without knowing the integrating factor: if μ is a function of x and y, then $\mu = 0$ is a particular solution of $dy = p\,dx$ if and only if μ is a common factor of

$$p + \frac{\left(\frac{ddp}{dx\,dy}\right)}{\left(\frac{ddp}{dy^2}\right)} \quad \text{and} \quad \frac{1}{\left(\frac{dp}{dy}\right)}$$

(as in Euler, parentheses indicate partial differentiation) [Laplace *1772a*, 355].

Laplace also extended these methods to second-order equations [*1772a*, 357-365] and to partial differential equations (on three variables) [*1772a*, 365-370], which was a multiple novelty, as until then only singular solutions of first-order ordinary differential equations had been considered.

6.1.2.3 Lagrange

But the major breakthrough in the theory of singular solutions was the memoir [Lagrange *1774*] "Sur les intégrales particulières des équations différentielles"[13]. Lagrange was able to explain them not as exceptions, but rather as natural outcomes of the complete integrals [Lagrange *1774*, §7].

For this Lagrange explored the relation between a differential equation

$$Z = 0$$

where Z is a function of x, y and $\frac{dy}{dx}$, and its complete integral

$$V = 0$$

where V is a function of x, y and of an arbitrary constant a which does not appear in $Z = 0$: differentiation of $V = 0$ gives something like

$$\frac{dy}{dx} = p$$

where p is a finite function of x, y and a; $Z = 0$ must be the result of eliminating a between $V = 0$ and $\frac{dy}{dx} - p = 0$.[14] Now, the process of elimination of a is not dependent on the constancy of a; so what if a were a variable? Since that would mean

$$dy = p\,dx + q\,da,$$

---

[13]"On the particular [i.e. singular] integrals of differential equations"

[14]This idea of conceiving a differential equation as the result of the elimination of arbitrary constants between a finite equation and its differentials had already been given by Fontaine, but without connection to singular solutions (see section 6.1.4.1).

we would need

$$q \, \mathrm{d}a = 0$$

to ensure $\mathrm{d}y = p \, \mathrm{d}x$ and that we arrive at the same result, namely $Z = 0$; for this either $\mathrm{d}a = 0$ (that is, a is in fact a constant) or

$$q = 0.$$

Thus elimination of a between $q = \frac{\mathrm{d}y}{\mathrm{d}a} = 0$ and $V = 0$ provides a finite equation which satisfies $Z = 0$ and does not contain an arbitrary constant: according to Lagrange, this will be a singular integral[15] of $Z = 0$ [Lagrange *1774*, § 2-4].

As an example, let us look at the differential equation (6.1), that is (with a slight change in notation)

$$y \, \mathrm{d}x - x \, \mathrm{d}y = b\sqrt{\mathrm{d}x^2 + \mathrm{d}y^2}, \tag{6.9}$$

whose complete integral is

$$y = \frac{a}{2}(b + x) + \frac{1}{2a}(b - x) \tag{6.10}$$

([Lagrange *1774*, § 1,6] used $y - ax - b\sqrt{1 + a^2} = 0$, that is, (6.6)). Differentiation of (6.10) relative to a gives

$$\frac{\mathrm{d}y}{\mathrm{d}a} = \frac{b + x}{2} - \frac{b - x}{2a^2},$$

and elimination of a between $\frac{b+x}{2} - \frac{b-x}{2a^2} = 0$ (that is, $a^2 = \frac{b-x}{b+x}$) and (6.10) gives the singular integral $x^2 + y^2 = b^2$. This solution, although not contained in the complete integral (6.10) (that is, it does not represent a determination of the arbitrary constant a), is obtainable from it by this process of elimination.

This can be carried to higher orders: as differentiation of $V = 0$ with a constant gives $\frac{\mathrm{d}y}{\mathrm{d}x} = p$, further differentiations give

$$\frac{\mathrm{d}^2 y}{\mathrm{d}x^2} = p', \quad \frac{\mathrm{d}^3 y}{\mathrm{d}x^3} = p'', \quad \dots$$

So a second-order differential equation $Z' = 0$ satisfied by $V = 0$ must "be formed by combination" of $V = 0$, $\frac{\mathrm{d}y}{\mathrm{d}x} = p$, and $\frac{\mathrm{d}^2 y}{\mathrm{d}x^2} = p'$; a third-order equation $Z'' = 0$ satisfied by $V = 0$ must "be formed by combination" of $V = 0$, $\frac{\mathrm{d}y}{\mathrm{d}x} = p$, $\frac{\mathrm{d}^2 y}{\mathrm{d}x^2} = p'$ and $\frac{\mathrm{d}^3 y}{\mathrm{d}x^3} = p''$; and so on (in all cases a being eliminated). But if a is variable, then it is necessary for $V = 0$ to satisfy $Z' = 0$ that not only $\frac{\mathrm{d}y}{\mathrm{d}a} = 0$, but also $\frac{\mathrm{d}^2 y}{\mathrm{d}x \, \mathrm{d}a} = 0$

---

[15]As long as certain conditions apply: that at least one of x, y appears in $\frac{\mathrm{d}y}{\mathrm{d}a} = 0$ [Lagrange *1774*, § 4] and that not all of $\frac{\mathrm{d}y}{\mathrm{d}a}, \frac{\mathrm{d}^2 y}{\mathrm{d}x \, \mathrm{d}a}, \frac{\mathrm{d}^3 y}{\mathrm{d}x^2 \mathrm{d}a}, \dots$ are zero (see below).

(i. e., that $dp = p'dx + q'da = p'dx$); for $V = 0$ to satisfy $Z'' = 0$ that additionally $\frac{d^3y}{dx^2da} = 0$; and so on [Lagrange *1774*, §8-11].

If, however, all the differentials $\frac{dy}{da}, \frac{d^2y}{dx\,da}, \frac{d^3y}{dx^2da}, \ldots$ are zero, then a is a constant and the solution at hand is in fact an "incomplete integral" [Lagrange *1774*, §13].

[Lagrange *1774*, §14-15] also gives a method to find the singular integral of a first-order equation $Z = 0$, without knowing the complete integral. His proof assumes that no transcendental functions occur in Z – but he argues that "it is not difficult to be convinced" that it also applies whatever the nature and form of Z. Further assuming that $Z = 0$ has been delivered of fractions and radicals, so that the same happens to

$$dZ = A\,d.\frac{dy}{dx} + B\,dy + C\,dx$$

(Z is a function of x, y, and $\frac{dy}{dx}$), Lagrange concludes (using the above fact that at least one of the quantities $\frac{dy}{da}, \frac{d^2y}{dx\,da}, \frac{d^3y}{dx^2da}, \ldots$ is nonzero) that a singular integral will make $A = 0$; since $Z = 0$ implies $dZ = 0$, we have on one hand $B\,dy + C\,dx = 0$, and on the other $A\frac{d^2y}{dx^2} + B\frac{dy}{dx} + C = 0$, whence $\frac{d^2y}{dx^2} = -\frac{B\frac{dy}{dx}+C}{A}$; thus a singular integral makes

$$\frac{d^2y}{dx^2} = \frac{0}{0}.$$

A simpler situation occurs when Z is such that $B\,dy + C\,dx = 0$ always; in that case, of course, the condition for a singular integral is simply $A = 0$. The importance of this special situation is that it is the case for the equations of the form $y - \frac{dy}{dx}x + f.\frac{dy}{dx} = 0$ (f being an arbitrary function); the examples given by Clairaut and Euler fall within this category [Lagrange *1774*, §16-17].[16]

The study of singular integrals of second-order equations is very similar, but somewhat complicated by two facts: one, that the complete integral of a second-order equation contains two arbitrary constants a, b instead of just one a; in order to use the conditions mentioned above, and since a and b are both arbitrary, Lagrange puts $b = f.a$, f being an arbitrary function[17]; the conclusion is that a singular integral to a second-order equation $Z' = 0$ with complete integral $V = 0$ is obtained by elimination of a, b, and $\frac{db}{da}$ from

$$V = 0, \quad \frac{dy}{dx} - p = 0, \quad \frac{dy}{da} + \frac{dy}{db}\cdot\frac{db}{da} = 0 \quad \text{and} \quad \frac{d^2y}{dx\,da} + \frac{d^2y}{dx\,db}\cdot\frac{db}{da} = 0 \quad (6.11)$$

(where a and b are treated as variables) [Lagrange *1774*, §27-29]. The other complicating fact is that the process to obtain a singular (finite) integral involves as an

---

[16] $y - \frac{dy}{dx}x + f\left(\frac{dy}{dx}\right) = 0$ is nowadays called *Clairaut's equation*.

[17] Remarkably *arbitrary* for this period, particularly considering Lagrange's view of functions as analytic expressions (see section 6.1.3.2).

intermediate step to obtain a singular first-order solution (whose integral, including an arbitrary constant α, is the requested singular finite integral); this, being a first-order differential equation, may in turn have a singular integral, which may or may not be a (singular) solution of the second-order equation [Lagrange *1774*, § 30-31].

A more interesting extension of this theory of singular integrals is the one to partial differential equations. This involved a new definition for *complete integral* of a partial differential equation. In Euler's conception, such a complete integral was analogous to a complete integral of an ordinary differential equation, simply with the arbitrary constant(s) replaced by arbitrary function(s) [Euler *Integralis*, I, § 34; III, § 33, § 37-38, § 249]. In a paper on first-order partial differential equations published in the Berlin Memoirs for 1772, Lagrange (still following Euler's terminology) had noticed that "a particular solution [i.e., one without the necessary arbitrary function] which contains two arbitrary constants is sufficient to permit the derivation of the complete solution [i.e., with an arbitrary function]" [Engelsman *1980*, 14]. He pursued this in [Lagrange *1774*, § 39]: if V is a function of x, y, and z involving two arbitrary constants a and b, and if differentiation of $V = 0$ yields $\mathrm{d}z = p \, \mathrm{d}x + q \, \mathrm{d}y$, then a and b may be eliminated from

$$V = 0, \quad \frac{\mathrm{d}z}{\mathrm{d}x} - p = 0 \quad \text{and} \quad \frac{\mathrm{d}z}{\mathrm{d}y} - q = 0,$$

resulting in a differential equation $Z = 0$. Lagrange then adopts $V = 0$ as the *complete integral* of $Z = 0$ – that is, the complete integral of a first-order equation in three variables must contain two arbitrary constants (instead of an arbitrary function).

Now, if a and b are regarded as variables, the differential of $V = 0$ will become

$$\mathrm{d}z = p \, \mathrm{d}x + q \, \mathrm{d}y + r \, \mathrm{d}a + s \, \mathrm{d}b,$$

so that to obtain $Z = 0$ it is necessary to have

$$r \, \mathrm{d}a + s \, \mathrm{d}b = 0. \tag{6.12}$$

A singular integral arises analogously to the case of ordinary differential equations by taking

$$r \left(= \tfrac{\mathrm{d}z}{\mathrm{d}a} \right) = 0 \quad \text{and} \quad s \left(= \tfrac{\mathrm{d}z}{\mathrm{d}b} \right) = 0$$

and combining with $V = 0$ [Lagrange *1774*, § 40-41].

There is, however, one other type of solution: $\frac{\mathrm{d}z}{\mathrm{d}a} = 0$ and $\frac{\mathrm{d}z}{\mathrm{d}b} = 0$ is not the only way of satisfying (6.12); if one assumes for instance b to be a function of a, namely $b = \phi a$, (6.12) becomes

$$\frac{\mathrm{d}z}{\mathrm{d}a} + \frac{\mathrm{d}z}{\mathrm{d}b} \phi' a = 0;$$

the result of eliminating a between this equation and $V = 0$ will also be a solution, one which includes an arbitrary function (and which therefore corresponds to

Euler's complete integral). Because of that arbitrary function, argues Lagrange, this solution is "beaucoup plus général que l'intégrale complette $V = 0$"[18], so that he calls it precisely *general integral* [Lagrange *1774*, § 47][19]

For the geometrical interpretations of all this, see section 6.1.3.3. For more on complete and general integrals, see section 6.1.4.2.

Between [Lagrange *1774*] and [Lacroix *Traité*] there appeared a few more works devoted to or touching upon singular solutions: [Trembley *1790-91*], [Legendre *1790*] and [Lagrange *Fonctions*]. However, they did not bring any dramatic innovations, and will only be mentioned along with their treatment in Lacroix's *Traité*. Suffice to remark here that none of them was mentioned in the other main treatises on the calculus published in the 1790's [Cousin *1796*; Bossut *1798*].

6.1.3 Geometrical connections

Like every other branch of the calculus, differential equations had geometrical beginnings. The French amateur mathematician Florimond de Beaune (1601-1652) is often credited with initiating the subject by proposing a few problems to determine curves given properties of their subtangents – the first *inverse tangent* problems.

Also like every other branch of the calculus, differential equations were affected by the tendency for algebraization of mathematics throughout the 18th century. The problems, although often inspired by more concrete fields (mainly mechanics), became more abstract and geometry was usually invoked only for illustration, for helping in visualization. A good example is the study of singular solutions, whose geometrical counterparts help to understand the relation between types of solutions, even though their derivation is purely algebraic (sections 6.1.2.3 and 6.1.3.3).

But even in the *age of analysis* geometrical considerations played more important roles in certain aspects of the development of differential equations, and namely in the study of their solutions. Gaspard Monge studied differential equations in three variables, interpreting their solutions as surfaces (section 6.1.3.4). And the biggest and most famous challenge to the *rule of analysis* came also from this subject: could the arbitrary functions involved in solutions of partial differential equations be so arbitrary as to include not only functions defined by analytic expressions, but also those defined by the coordinates of a curve drawn "by the free stroke of the hand" (section 6.1.3.2)? Some of the supporters of this "return to geometry" revived in their arguments an old concept – the *construction of differential equations* – which requires some explanation (section 6.1.3.1).

---

[18]"much more general than the complete integral $V = 0$"

[19]Before [Lagrange *1774*] "general integral" had been simply an alternative name for "complete integral": we have seen above Laplace using it in that sense.

6.1.3.1 Construction of differential equations until c. 1750

Henk Bos has called attention to the importance of the concept of construction in 17th-century *analytic geometry* (or rather, "application of algebra to geometry" – see section 4.1.1) [Bos *1984; 1986; 2001*] and in the early history of differential equations [Bos *1986; 2004*]. At a time when new curves were being introduced in mathematics (such as the cycloid and the logarithmic), and the use of algebra for the study of curves was also very recent, it was not obvious when a curve was sufficiently *known*. Only gradually did equations become sufficient *representations* of loci; therefore a geometrical problem was not fully solved simply by having an equation (either algebraic or differential) corresponding to the solution: a geometrical *construction* for that equation was also demanded (although there was no consensus on the best methods for construction). Naturally, the need for such a construction was particularly felt when the solution equation involved a *new* curve (such as a transcendental one) – it was a fundamental factor in the *legitimation* of that new curve.

This changed around the turn of the 17th to the 18th century, with mathematicians' "habituation" to algebraic (and certain types of transcendental) equations and their consequent acceptance as sufficient representations of curves. The construction of algebraic equations slowly died out, and disappeared (except as a school subject) around 1750.

Bos [*2004*] suggests that the construction of differential equations had a similar fate. As for differential equations in two variables, this indeed appears to be the case: there are not many traces of their construction in the latter half of the 18th century.[20]

In the early 18th century, the most natural way to construct a differential equation was to integrate it first, and then to construct the resulting finite equation; when an algebraic integral could not be achieved, some quadrature or rectification had to be assumed. The only method for integration known in those early days was separation of variables, and Johann (I) Bernoulli gave a simple construction (described in [Montucla & Lalande *1802*, 174-175]) for the separated equation $Y\,dy = X\,dx$, which required drawing curves representing the areas $\int Y\,dy$ and $\int X\,dx$. Clairaut remarked in [*1740*, 293] that when variables in differential equations are separated, "on peut toujours ou les intégrer, ou au moins les

---

[20]It is true that in [Euler *Integralis*, II] (published in 1769) there are two chapters which refer to construction of ordinary differential equations: chapters 10 and 11 of the first section, respectively "de constructione aequationum differentio-differentialium per quadraturas curvarum" ("on the construction of differentio-differential [i.e., second-order differential] equations by quadratures of curves") and "de constructione aequationum differentio-differentialium ex earum resolutione per series infinitas petita" ("on the construction of differentio-differential [i.e., second-order differential] equations from their required solution by infinite series"). But one would seek in vain for geometrical constructions in those chapters. Rather, Euler seems to refine problems and techniques which had appeared in the context of construction of differential equations (namely solving an equation assuming certain quadratures or rectifications – see below), but which appear devoid of geometrical meaning. Deakin [*1985*] finds integral transforms in chapter 10.

construire, puisque la difficulté est réduite à la quadrature des Courbes"[21].

There were other methods, seen as alternatives to analytical integration. In 1694 Johann Bernoulli published a short paper entitled "modus generalis construendi æquationes differentiales primi gradus"[22] [Joh. Bernoulli *1694*], where he tried to address precisely the construction of differential equations which he could not integrate – that is, whose variables he could not separate. Given an equation $\frac{dy}{dx} = m$ (m of course being a "quantity made up of x, y, and constants"), the first step in Bernoulli's method is to construct an infinite number of curves m = constant (for very close values of m) – the isoclines or, as Bernoulli called them, the "directrices"[23]; Bernoulli assumed that these were algebraic curves (i.e., that m is an algebraic function) and hence relatively easy to construct. Then it was enough to connect these curves by small straight lines having the corresponding slopes.

The approximative nature of this method is evident. The same is true for other methods of this period; for instance, Tournès [*2003*, 461-463] identifies a polygonal approximation in a construction given by Leibniz [*1694*]. However, this kind of graphical approximation was soon dropped in favour of analytical or numerical methods [*Traité*, II, 296; Montucla & Lalande *1802*, III, 175] – such as the ones mentioned in section 5.1.3; Euler's "general method" is a very clear example of an analytical version of a polygonal method. Graphical approximation only regained importance in the 19th century [Tournès *2003*].

A very interesting illustration of the loss of general relevance of the concept of construction involves the Italian mathematician Vincenzo Riccati. Riccati published in 1752 a treatise in which he proved that all first-order (ordinary) differential equations conceivable at the time could be constructed using tractional motion [Tournès *2003*, 477; *2004*].

> "However, the work of Vincenzo Riccati was neither celebrated nor influential. [...] The book probably arrived too late, at the end of the time of construction of curves, at the moment when geometry was giving way to algebra" [Tournès *2004*, 2742].

It may also be noted that although Cousin included a section on construction of equations in the introductory chapter on "application of algebra to geometry" of his Traité [Cousin *1796*, I, 31-36], he did not do the same for construction of differential equations. Of course in 1796 the relevance of construction of algebraic equations was purely pedagogical; but Cousin seems to have thought that construction of differential equations lacked even that relevance.

But the construction of differential equations in three (or more) variables seems to have a somewhat more complicated history, appearing with some regularity in arguments on *possibilities*. Clairaut [*1740*, 307-311] wanted to show that there

---

[21]"it is always possible to integrate them, or at least to construct them, since the difficulty is reduced to the quadrature of curves"

[22]"general method for constructing first-order differential equations"

[23]"directing [curves]"

are differential equations in three variables which not only cannot be integrated, but also cannot be constructed.[24] The former impossibility had an algebraic proof: elimination of an integrating factor μ and of $\frac{d\mu}{dx}, \frac{d\mu}{dy}, \frac{d\mu}{dz}$ between certain condition equations gave

$$N\frac{dP}{dx} - P\frac{dN}{dx} + M\frac{dN}{dz} - N\frac{dM}{dz} - M\frac{dP}{dy} + P\frac{dM}{dy} = 0 \qquad (6.13)$$

as a necessary condition for the integration of $M\,dx + N\,dy + P\,dz = 0$ to be possible. As for the latter impossibility: suppose that the surface expressed by $dz = \omega\,dx + \vartheta\,dy$ is constructed, PN is a section on it, perpendicular to the x-axis AP, and QN is another section, perpendicular to the y-axis AQ; suppose also that pn and $q\nu$ are sections parallel and infinitely close to PN and QN; they must

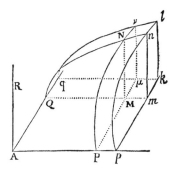

intersect in l, so that $z + \omega\,dx + \vartheta\,dy + \frac{d\vartheta}{dx}\,dx\,dy + \frac{d\vartheta}{dz}\,\omega\,dx\,dy$ must be equal to $z + \vartheta\,dy + \omega\,dx + \frac{d\omega}{dy}\,dy\,dx + \frac{d\omega}{dz}\,\vartheta\,dy\,dx$, that is,

$$\frac{d\vartheta}{dx} + \omega\frac{d\vartheta}{dz} = \frac{d\omega}{dy} + \vartheta\frac{d\omega}{dz} \qquad (6.14)$$

must hold – and this is the same condition as (6.13), with $\omega = -\frac{M}{P}, \vartheta = -\frac{N}{P}$. Clairaut concludes from this that problems whose solution depends on $M\,dx + N\,dy + P\,dz = 0$ are impossible unless (6.13) is verified.

In the second half of the 18th century, geometrical arguments lost much ground. But we will see in section 6.1.3.2 that arguments involving constructions of partial differential equations played a relevant role in another discussion on *legitimation* – the legitimation of so-called "discontinuous" functions.

6.1.3.2 The controversy on vibrating strings and arbitrary functions

One of the most famous controversies in 18th-century mathematics was the one opposing Euler to d'Alembert over which functions could be admitted in the so-

---

[24]He assumed that any possible integral was composed of a single equation, and that any possible construction led to a surface. Later, Monge would be more flexible (see section 6.1.3.5).

lution to the partial differential equation $\frac{\partial^2 y}{dt^2} = c^2 \frac{\partial^2 y}{dx^2}$; and more generally on whether the arbitrary functions appearing in the general solutions to partial differential equations could be arbitrary enough as to include "discontinuous" ($\simeq$ non-analytic) ones.

First, let us examine the concepts of "function", "continuous function", and "discontinuous function". When Euler published his *Introductio in Analysin Infinitorum* in 1748, he defined function as an "analytic expression":

> "Functio quantitas variabilis, est expressio analytica quomodocunque composita ex illa quantitate variabili, et numeris seu quantitatibus constantibus."[25] [Euler *Introductio*, I, § 4]

Analytic expressions were composed by algebraic operations, and by (some) transcendental ones, such as exponentiation, logarithms, and others "quas Calculus integralis suppeditat"[26] [Euler *Introductio*, I, § 6], and their most general form was supposed to be power series:

> "num vero per hujusmodi terminorum $[A + Bz + Cz^2 + \&c.]$ seriem infinitam [Functio quaelibet ipsius z] exhiberi possit, si quis dubitet, hoc dubium per ipsam evolutionem cujusque Functionis tolletur."[27] [Euler *Introductio*, I, § 59]

However, this definition should not be taken too literally. Giovanni Ferraro [*2000*] has analysed Euler's concept of function and has noticed two levels in it: a formalized level corresponding to the definition of function as an analytic expression involving variables and constants; and an intuitive level, corresponding to an "idea of dependence or relation between variables" (a "functional relation") [Ferraro *2000*, 111-112], present in explanations, applications, or other more informal contexts. Ferraro does not see these two levels as contradictory, partly because in 18th-century mathematics "a definition did not necessarily exhaust the defined notion"[28]; and partly because, he argues, an analytic expression or formula was the proper way for expressing a "functional relation" within analysis (how could one *calculate* without an analytic expression?) – but not necessarily in geometry or mechanics [Ferraro *2000*, 112-113].

One of those more informal contexts is the preface to [Euler *Differentialis*], published in 1755, where we find an explicit characterization of the "functional relation" aspect of the concept of function; just after giving a physical example of

---

[25] "A function of a variable quantity is an analytic expression composed in whatever way from that variable quantity and numbers or constant quantities."

[26] "furnished by integral calculus"

[27] "if anyone doubts whether in fact [any function of z] may be displayed by an infinite series of such terms $[A + Bz + Cz^2 + \&c.]$, this doubt will be eliminated by the very development of each function."

[28] In fact Ferraro speaks only of Euler's mathematics; but one might remember here Lacroix's *encyclopedic* views, particularly his exploration of the nature of integrals *not* from their ostensive definition (see section 5.2.3).

dependence involving four variables (amount of gunpowder, angle of fire, range of shot, and length of time before the bullet hits the ground) Euler proceeds:

> "Quae autem quantitates hoc modo ab aliis pendent, ut his mutatis etiam ipsae mutationes subeant, eae harum functiones appellari solent; quae denominatio latissime patet, atque omnes modos, quibus una quantitas per alias determinari potest, in se complectitur. Si igitur *x* denotet quantitatem variabilem, omnes quantitates, quae utcunque ab *x* pendent, seu per eam determinantur, eius functiones vocantur"[29]. [Euler *Differentialis*, vi]

This has often been regarded as a "new", "general" definition of function [Youschkevitch *1976*, 69-70]. But it should be remarked that Ferraro [*2000*, 111] finds examples of the "functional relation" aspect already in [Euler *Introductio*] – as in fact Youschkevitch [*1976*, 69] himself had already found; and it must also be noticed that all the functions studied in [Euler *Differentialis*] are within the scope of the older definition – analytic expressions. This later observation and the fact that Euler did not refer back to this definition when he eventually started talking about "discontinuous" functions leads Lützen [*1983*, 356] to a conclusion analogous to Ferraro's: Euler thought of this definition as equivalent to the older one.

Later, Euler would try to expand the realm of functions. Grattan-Guinness [*1970*, 6] and Youschkevitch [*1976*, 64] both date the start of that expansion to Euler's definition of "discontinuous curves" in the second volume of his *Introductio*. But as a matter of fact, while that definition may be seen as establishing a terminology ("continuous" vs. "discontinuous") that would later be applied to functions, it reinforces the idea that a function must be expressible by *one* formula – curves which do not follow one single law are *ipso facto* not expressed by one single function:

> "Linea scilicet curva *continua* ita est comparata, ut ejus natura per unam ipsius *x* Functionem definitam exprimatur. Quod si autem linea curva ita sit comparata, ut variæ ejus portiones *BM*, *MD*, *DM*, &c., per varias ipsius *x* Functiones exprimantur; [...] hujusmodi lineas curvas *discontinuas* seu *mixtas* & *irregulares* appellamus: propterea quod non secundum unam legem constantem formantur, atque ex portionibus variarum curvarum continuarum componuntur."[30] [Euler *Introductio*, II, § 9]

---

[29]"Those quantities that depend on others in this way, so that if the latter change they also change, are called functions of the latter; this denomination applies very broadly, and comprises all the manners in which one quantity may be determined by others. If, therefore, *x* denotes a variable quantity, all quantities which depend on *x* in whatever manner, or are determined by it, are called functions of *x*"

[30]"A continuous curved line is [one] so arranged that its nature is expressed by one definite function of *x*. Whereas if a curved line is so arranged that several of its parts *BM*, *MD*, *DM*, &c., are expressed by several functions of *x*; [...] we call curved lines of this kind *discontinuous* or *mixed* and *irregular*: on account that they are not formed according to one constant law, but rather composed from parts of several continuous curves."

However, it was about that time that Euler started speaking of functions not corresponding to analytic expressions. This happened in his first contribution to the vibrating-string controversy, in a very matter-of-fact way: two arbitrary functions had to be determined; having described an appropriate curve, "soit réguliere, contenüe dans une certaine équation, soit irreguliere, ou méchanique, son appliquée quelconque PM fournira les fonctions, dont nous avons besoin pour la solution du Problême"[31] [Euler *1748*, §XXII].

It was only almost twenty years later that Euler [*1765a*] explained more or less clearly his "continuous" and "discontinuous" functions. He did this recurring once again to a correspondence between curves and functions: given a function y of x, it is always possible to describe a curve with abscissa a and ordinate y; and in turn, given a curve, its ordinates produce ("exhibent") a function of its abscissas. He now considered a curve to be continuous if its points follow a certain "law or equation" (no longer simply a "function", since this word had now a broader sense than in his *Introductio*), and discontinuous otherwise – and discontinuous curves provide ("suppeditant") discontinuous functions.

Euler was very careful in explaining that the law of continuity does not mean connectedness of trace: a hyperbola is a continuous curve, in spite of its two branches, since it is defined by *one* equation. In discontinuous curves, Euler included those drawn "libero manus tractu"[32] and mixed curves (that is, those composed of several parts, such as the perimeter of a polygon) [Euler *1765a*, § 1-3]. It should be remarked that in practice Euler's discontinuous functions were almost always functions corresponding to mixed curves.[33] Thus Euler's (and generally 18th-century's) "continuous" functions broadly correspond to modern analytic functions, while most "discontinuous" functions would now be called piecewise analytic. Naturally the 18th-century meanings (vague as they are) will be used in the rest of this section.

Let us now turn to the controversy on the vibrating string. This dealt with whether discontinuous functions could be allowed in solutions to the vibrating-string problem, or other problems translated into partial differential equations. It did *not* deal with the concept of function; that is, there was no disagreement between Euler and d'Alembert on what a function was[34], but rather on what curves or functions could be treated by analysis.

D'Alembert first treated the problem of the vibration of a stretched string,

---

[31]"either regular – contained in a certain equation – or irregular or mechanical, its arbitrary ordinate PM will furnish the functions which we need for solving the problem"

[32]"by the free stroke of the hand"

[33]With one possible exception, striking but very isolated: according to Youschkevitch [*1976*, 71] (following Truesdell) Euler [*1765c*, § 39] introduced pulse functions (different from zero only at one point); Lützen [*1982*, 197-198] disagrees with the implication in Youschkevitch's text that those were delta functions; for myself, I am not completely convinced that Euler thought he was talking about *functions* at all.

[34]That is, d'Alembert effectively accompanied Euler in this evolution: in [d'Alembert *1747*], the concept of function is the same as in [Euler *Introductio*]; while [d'Alembert *1780*] is a memoir on discontinuous functions.

fixed at both ends, in a couple of memoirs published in the Berlin Academy volume
for 1747 [d'Alembert *1747*]. Calling y the displacement of a point of the string, so
that y is equal to an unknown function $\varphi(t, s)$ of the time t and of the arc length s
of the string from one end to that point, we have $dy = p\,dt + q\,ds$, $dp = \alpha\,dt + v\,ds$
and $dq = v\,dt + \beta\,ds$, where p, q, α, v and β are other unknown functions of t
and s; d'Alembert established then that $\alpha = \beta\frac{2aml}{\theta^2}$, that is, in modern notation,
something of the familiar form $\frac{\partial^2 y}{\partial t^2} = c^2\frac{\partial^2 y}{\partial s^2}$; choosing a convenient time unit so
that $\theta^2 = 2aml$ $(c = 1)$, d'Alembert arrived at

$$y = \Psi(t + s) + \Gamma(t - s)$$

and, because of the boundary conditions $y = 0$ for $s = 0$ or $s = l$ (the total length
of the string) whatever t,

$$y = \Psi(t + s) - \Psi(t - s),$$

and Ψ is periodic with period $2l$ (the fact that the string only went from $s = 0$
to $s = l$ did not restrict the domain of the function Ψ). If, in addition, the string
starts vibrating from the taut position ($y = 0$ for $t = 0$), then Ψ must be an
even function, or as d'Alembert puts it, "Ψs doit etre une fonction de s dans
laquelle il n'entre que des puissances paires, lorsqu'on l'aura reduite en serie"[35]
[d'Alembert *1747*, 217]. A more general solution depends on the initial form of
the string, given by a function $\Sigma = \Psi s - \Psi(-s)$, and on the initial velocity of
each point of the string, also given by another function σ of s; the plain definition
$\Sigma = \Psi s - \Psi(-s)$ leads to the conclusion that Σ must be an odd function of s,
that is, "oú il n'entre que des puissances impaires de s"[36]; otherwise the problem
is impossible – one cannot find $y = \Psi(t + s) - \Psi(t - s)$ [d'Alembert *1747*, 231].

Thus we see that d'Alembert naturally assumed these functions to have
power-series expansions – in the terminology of [Euler *Introductio*], they were (simply) functions; in slightly later terminology, they were continuous functions.

But Euler expressed a different opinion, in a memoir which appeared in two
versions: the Latin original in 1749 in the *Nova acta eruditorum*, and a French
translation in 1750 in the volume for 1748 of the Berlin Academy [Euler *1748*].
Euler's analysis is very similar to d'Alembert's, arriving at the equation

$$y = f:(x + t\sqrt{b}) + \varphi:(x - t\sqrt{b}),$$

where f and φ are arbitrary functions subject to $\varphi : -t\sqrt{b} = -f : -t\sqrt{b}$ and
$\varphi : (a - t\sqrt{b}) = -f : (a + t\sqrt{b})$ (a is the length of the string). But there is a
significant difference: the curves which the string describes need not be "regular",
because "la premiere vibration dépend de notre bon plaisir, puisq'on peut, avant
que de lacher la corde, lui donner une figure quelconque"[37] [Euler *1748*, §III].

---

[35]"Ψs must be a function of s with only even powers, once expanded into a series"

[36]"with but odd powers of s"

[37]"the first vibration depends on our goodwill, since we may give the string any shape whatsoever, before releasing it"

This had consequences for the arbitrary functions f and φ: a passage has already been quoted above (page 197), where Euler considers these functions furnished by the ordinate of an appropriate curve, even if it is "irregular". Thus he effectively introduces the consideration of discontinuous functions, even if not calling them so.

D'Alembert did not agree, and he *rectified* Euler, "de crainte que quelques lecteurs ne prennent mal le sens de ses paroles"[38]: he insisted that in the equation $y = \Sigma$ of the initial curve, Σ must be an odd function of s, with period $2l$; otherwise "le problème ne pourra se résoudre, au moins par ma methode, et je ne say même s'il ne surpassera pas les forces de l'analyse connuë"[39] [d'Alembert *1750*, 358].

Part of the discussion dealt with Euler's construction of the extension of the initial-shape curve (d'Alembert's $y = \Sigma$): the string corresponds only to a section of this extended curve, but both Euler and d'Alembert agreed that (what we would call) the domain of the function Σ should extend both ways indefinitely. Now, Euler [*1748*, § XXI] simply took the initial shape AMB of the string and copied it alternately on each side of the axis, so as to ensure the necessary periodicity and oddness. But for d'Alembert, it was necessary not only that AMB would be

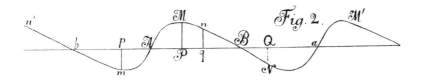

continuous, but also such that the curve $\dots n'bmAMBNaM' \dots$ thus constructed would be continuous.

The controversy proceeded for decades, in multiple publications. To analyse d'Alembert's later argumentation, suffice to mention two memoirs. One is in the first volume of his *Opuscules* [d'Alembert *1761*, 15-42]. D'Alembert's most important argument (expounded in several ways) is that, for the vibrating-string equation $\frac{ddy}{dx^2} = \frac{ddy}{dt^2}$ to be satisfied, the radius of curvature of the initial curve cannot "jump"; in modern terms, at every point the right-hand and left-hand second-order derivatives of the initial-shape function must coincide. Now, the simple act of pulling the string at one point (so as to release it and make it vibrate) introduces one such *forbidden* shape: two straight lines making a finite angle.[40]

Almost twenty years later, d'Alembert would refine this argument, in a memoir on discontinuous functions [d'Alembert *1780*].[41] Considering cases where

---

[38]"fearing that some readers might misunderstand the meaning of his words"

[39]"the problem cannot be solved, at least by my method, and I am not sure whether it does not surpass the power of known analysis"

[40]Moreover, even if the initial shape of the string properly speaking is smooth, the curvature in A and B should also be null; otherwise there is a "jump" in the curvature of the extended curve.

[41]Which was an answer to Monge's stand, rather than Euler's (see below).

these result from the junction of continuous functions (that is, which correspond to Euler's mixed curves),[42] d'Alembert imposes the modern-looking condition that, to appear in solutions to n-th order differential equations, the left-hand and right-hand derivatives, up to n-th order, must be equal at the points of discontinuity (not in these words, of course, but rather: if the discontinuity is such that φz becomes ("devienne") Δz at $z = $ a, then it is necessary that $\frac{d^n \varphi z}{dz^n} = \frac{d^n \Delta z}{dz^n}$, $\frac{d^{n-1} \varphi z}{dz^{n-1}} = \frac{d^{n-1} \Delta z}{dz^{n-1}}$, and so on, all these equalities considered at $z = $ a [d'Alembert *1780*, 307]).[43] It is true that now d'Alembert admits discontinuous functions in solutions to differential equations, but this is hardly an agreement with Euler, even a partial one: the obvious discontinuous solution for the vibrating string (the angle) is still out of the question. Moreover, this argument by d'Alembert is at odds with the *global* way of thinking typical of Euler – d'Alembert [*1780*, 307] requires that "pour toutes les valeurs possibles de z, l'équation différentielle aura rigoureusement lieu"[44].

Many years before this refinement, Euler [*1765b*] had dismissed d'Alembert's objections, using three arguments:

1 – He implied an equivalence between the differential equation $\left(\frac{ddy}{dt^2}\right) = cc\left(\frac{ddy}{dx^2}\right)$ and the integral equation $y = \Gamma : (x + ct) + \Delta(x - ct)$, "qui contient la solution du probleme"[45] [Euler *1765b*, § 44]; that is, Euler replaced the original differential equation with a new, functional one.[46]

2 – He assumed that the "jumps" in the radius of curvature could occur only in isolated points (once again, the curve would be piecewise analytic), and therefore would be of no consequence: "quoiqu'on y commette quelque erreur, cette erreur n'affectera qu'un seul élément, et sera par conséquent sans aucune conséquence, étant toujours infiniment petite"[47] [Euler *1765b*, § 47]; a fine example of the *global* way of thinking that d'Alembert would contradict.

3 – Finally, in order to remove any objection to the second argument, he argued that "on n'auroit qu'à emousser infiniment peu les angulosités [...] et par cela même, qu'on n'auroit changé qu'infiniment peu la figure [...], toutes les conclusions qu'on en tire, demeureront toujours les mêmes"[48] [Euler *1765b*, § 46].

---

[42]It is not clear whether d'Alembert could conceive of any other kind of discontinuous functions.

[43]Condorcet [*1771*, 69-71] and Laplace [*1779*, 299-302] imposed a less strict condition: the functions and the derivatives up to order $n - 1$ were forbidden to have "jumps", but the n-th derivative was not. Note however that Laplace admitted stronger discontinuities in physical (rather than "geometrical") solutions, using an argument similar to Euler's number 3 below [*1779*, 302]. See also section 9.5.4.

[44]"for all possible values of z, the differential equation will take place strictly"

[45]"which contains the solution of the problem"

[46]Lützen [*1982*, 19] sees this as an anticipation of the most common technique in the 20th century for obtaining generalized solutions to differential equations, although this technique consists in replacing the differential equation with different types of integral equations.

[47]"although some error is made there, it will affect only one element, and will therefore be of no consequence, as it will be infinitely small"

[48]"it will be enough to blunt infinitely little the angularities [...] and because the figure [...] will

Euler also saw any objections possible to the second and third arguments as similar to the objections against the infinitesimal calculus, and therefore wrong, since "aujourd'hui ces doutes sont entierement dissipés"[49] [Euler *1765b*, § 48] – a wonderfully optimistic point of view (see section 3.1).

Several other mathematicians expressed their opinions on this issue during the latter half of the 18th century (see for instance footnote 43 above). A curious one was that of Lagrange, combining some of d'Alembert's scruples with the generality of Euler's solution. In fact, the first major work of the young Joseph-Louis Lagrange was on the "nature and propagation of sound" [Lagrange *1759c*]. Lagrange [*1759c*, § 15] agreed with d'Alembert that the differential and integral calculus concerned only "fonctions algébriques", whose values are necessarily "liées ensemble par la loi de continuité", so that d'Alembert's and Euler's solutions, as it was deduced by them, was only applicable when the initial shape of the string was a continuous curve. Finding this insufficient, Lagrange decided to analyse the problem in a different way: he first considered a weightless string loaded with a finite number m of bodies; and then he put $m = \infty$, arriving at Euler's solution (applicable to discontinuous curves). That is, in modern terms Lagrange proceeded to a passage to the limit, which "is valid only subject to hypotheses essentially the same as those necessary to justify the direct use of appropriate differentiations and integrations" [Truesdell *1960*, 263].[50]

Lagrange was not very coherent on this issue. In a second memoir on the same subject, Lagrange came closer to Euler's stand, urging on the need to employ discontinuous functions [Lagrange *1760-61a*, § 5]. Later (in the 1760's) he would change his mind and support d'Alembert [Truesdell *1960*, 279]; while in the end of his life (in the second edition of his *Mécanique Analytique*, 1811) he would return to his initial stand, and acknowledge that Monge's work (see below) had led to the general acceptance of discontinuous functions [Truesdell *1960*, 295].

But what is most important for us to remark here is Lagrange's longest stand, similar to d'Alembert's. In both [*Fonctions*, 1] and [*Calcul*, 6] Lagrange defines function as an "expression de calcul" – a very similar phrase to that of [Euler *Introductio*, I, § 4]; it is a very important characteristic of Lagrange's two books on the "calculus of functions" that any function "is given by a *single* analytical expression" [Fraser *1987*, 40-41].

Although the issue of discontinuous functions appeared with the controversy on the vibrating string, it was of course more general. Euler's memoir [*1765a*], cited above as his first *clear* mention of discontinuous functions, was a defence of the need to consider these in the integral calculus of several variables (which was then relatively recent – about 20 years old – and on whose novelty Euler insisted; this memoir falls within Demidov's "first period" in the history of partial differential equations [Demidov *1982*, 326]). Euler [*1765a*, § 6] recognized that

---

have been changed only infinitely little, all the conclusions drawn will remain the same"

[49]"nowadays those doubts are entirely dispelled"

[50]It is interesting to compare this to Lagrange's later attempt to avoid infinitesimals and limits by recurring to... infinite series (section 3.1.4).

discontinuous functions could not be admitted in the part of infinitesimal analysis which had been treated chiefly until then – namely the calculus of functions of one variable; however, in the "new" integral calculus, which treated functions of two or more variables, discontinuous functions were *indispensable*, since arbitrary functions took the place of arbitrary constants in "common" calculus (cf. page 180 above), and arbitrary functions could be discontinuous [Euler *1765a*, § 18].

The fact that mathematicians did not really know how to work with discontinuous functions was of course a problem. Euler might have this in mind when he incited all geometers to gather their forces in cultivating multivariate analysis [Euler *1765b*, § 32]. Lützen [*1983*] speaks of "Euler's vision of a general partial differential calculus for a generalized kind of function", a vision which was only fulfilled in the 20th century, especially through the theory of distributions. This "vision" did not develop at all during the "age of rigorization of analysis" (most of the 19th century) because of the restriction of differentiation to differentiable functions [Lützen *1982*, 14, 24-25]. But neither did it develop in the pre-Cauchy era, in spite of a growing consensus on the acceptability of discontinuous functions. Euler, for one, did not do much more than what has been mentioned above.[51] As Fraser [*1989*, 326] puts it, "[Euler's] notion of a general function was never incorporated into the analytical theory presented in his mid-century textbooks, and indeed was at odds with its basic direction" – Grattan-Guinness [*1970*, 6] appropriately called Euler's correspondence between arbitrary functions and curves a "return to geometry", but returning to geometry was not exactly the main current in late 18th-century mathematics.

Nevertheless, in the late 18th century there was one important mathematician "returning to geometry" in the study of partial differential equations: Gaspard Monge.

At least from 1771 Gaspard Monge showed a deep interest in something that would be a major theme in his work: the classification of surfaces in families, each corresponding to a certain partial differential equation with two independent variables, and to a certain form of generation. In November 27th that year he presented to the *Académie des Sciences de Paris* a memoir [Monge *1771*] related to that theme, which is in part a defense of discontinuous functions/curves. In spite of a report signed by Bossut, Vandermonde, and d'Alembert supporting its publication in the *Savans Étrangers*, it remained unpublished until [Taton *1950*].[52]

---

[51] Euler even found himself a serious objection to his geometrical correspondence between curves and arbitrary functions: integrating $\left(\frac{ddz}{dy^2}\right)+aa\left(\frac{ddz}{dx^2}\right)=0$ he arrived at $z = f : (x+ay\sqrt{-1})+F : (x - ay\sqrt{-1})$; what could an abscissa like $x + ay\sqrt{-1}$ mean, not even he had any idea [Euler *Integralis*, III, § 301; Ferraro *2000*, 128-129]. Nevertheless, Ferraro [*2000*, 130] exaggerates when he says that the objects that Euler called discontinuous functions "substantially differed from effective functions since only the latter could be manipulated and, therefore, accepted as solutions to a problem"; the vibrating-string controversy shows that Euler did accept discontinuous functions as solutions, and strived to be able to manipulate them.

[52] In fact only the second part of that memoir was devoted to this; the first part was on the integration of a certain kind of linear partial differential equation. However, since that first part is lost (its contents can only be guessed from the report and a letter by Monge), and the two

[Monge *1771*] starts from an analogy with ordinary differential equations: just as the complete integral of an n-th order ordinary differential equation has to contain n arbitrary constants, so too the complete integral of an n-th order partial differential equation has to contain n arbitrary functions – a consensual idea then, as [Lagrange *1774*] was still some years away. Therefore, the complete solution of such an equation corresponds not to a surface, but to a class of surfaces sharing some property. The determination of those arbitrary functions corresponds to the specification of a particular surface in the class; and the most natural way to do this is to subject the surface to pass through n specific curves. This would be a very important subject of research for Monge.

The issues in this memoir are this specification in the case of certain first-order equations, and the claim that the specifying curve does not need to be continuous.

Monge gives three examples, all with first-order equations (involving therefore only one arbitrary function). In the first example he gives two proofs that every horizontal cylindrical surface (that is, a surface generated by a horizontal straight line that slides along some curve – continuous or discontinuous – keeping always the same direction) has as differential equation (where δ refers to partial differentiation relative to x, and ∂ to partial differentiation relative to y)

$$\frac{\delta z}{dx} + a\frac{\partial z}{dy} = 0. \tag{6.15}$$

For the first proof, Monge notes that the vertical planes passing through the generating straight line as it moves are all parallel, so that they have as equation $y = ax - \beta$, where β is constant for each plane but varies from plane to plane; so the surface is such that if one makes $ax - y = $ constant, the result is a horizontal straight line, that is, $z = $ const., or $dz = 0$; therefore the equation of the surface is $z = \varphi(ax - y)$, where the arbitrary function φ depends on the curve along which the generating line slides, and is "assignable" or not according to whether the curve is continuous; finally, $z = \varphi(ax - y)$ always gives $\frac{\delta z}{dx} + a\frac{\partial z}{dy} = 0$, no matter φ. For the second proof, Monge assumes for simplification that $a = 1$, i.e. that the generating line makes angles of $45°$ with the x and y axes; he then considers a tangent plane to the surface, remarking that its intersection with the xy plane also makes angles of $45°$ with the x and y axes, and examines right triangles formed by that plane and planes parallel to the vertical coordinates planes; these are similar to infinitesimal right triangles whose legs are $\delta z, dx$ and $\partial z, dy$, which leads him to the desired conclusion that $\frac{\delta z}{dx} + \frac{\partial z}{dy} = 0$. The second example, with similar proofs, is that of a surface of revolution around the z axis, whose differential equation is

$$y\frac{\delta z}{dx} - x\frac{\partial z}{dy} = 0$$

---

parts are quite independent, we may as well refer to the surviving second part as *the* memoir [*1950*, 48; *1951*, 280].

(so that its finite equation is $z = \varphi(x^2 + y^2)$. The final example is that of a conical surface with vertex at the origin, of which Monge only gives the finite equation[53]

$$z = x\,\varphi\left(\frac{x}{y}\right)$$

and no proof, claiming that it is analogous to the preceding ones.

As corollaries, Monge states the possibility of "constructing" these equations (either the differential or the finite ones), subject to a condition such as that putting $y = \Delta(x)$ will make $z = \Psi(x)$ (the projections of the specifying curve). For instance, in the first example, it is enough to construct a space curve with those projections, to take a horizontal straight line whose projection is $y = ax - b$, and to slide it along the curve. He also insists on the general validity of these constructions, even when either or both of these functions $\Delta(x)$ and $\Psi(x)$ are "discontinuous", that is, not "de nature à être exprimés par des équations"[54] [Monge 1771, 50]; in that case the arbitrary function involved in the finite equation for the class of surfaces is not "expressible analytically" – but Monge does not seem to think that that situation might affect the validity of both the finite and the differential equations.

Reading this, one is led to wonder what precisely Monge meant by "discontinuous". Both of his "proofs" above assume differentiability, of course: the first in a direct way, so as to go from $z = \varphi(ax - y)$ to $\frac{\delta z}{dx} + a\frac{\partial z}{\partial y} = 0$; the second through the existence of a tangent plane in any point of the surface. D'Alembert would challenge these assumptions in [1780]. Although he does not mention Monge, it seems clear for us[55] who did d'Alembert have in mind: his main example is the equation $z = \varphi(ax - y)$ and its relation to $\frac{dz}{dx} + \frac{a\,dz}{dy} = 0$ – if φ changes form at $z = \text{a}$, $\frac{dz}{dx} = -\frac{a\,dz}{dy}$ does not (necessarily) hold at $ax - y = \text{a}$; moreover, the finite angle described by the generating line in such a case thwarts the existence of a tangent plane to the surface at those points [d'Alembert 1780, 302-303, 305-307]. Apparently Monge never replied to d'Alembert (in 1780 he was no longer very much concerned with this issue). But from his wording in [Monge 1771] it seems that the fundamental characteristic of discontinuous curves or functions was that they were not "expressible analytically" – they were objects of geometry, rather than analysis; but their smoothness was always taken for granted.[56]

For some time in the 1770's Monge kept working on the determination of arbitrary functions. One very likely reason for [Monge 1771] not having been published is that Monge soon wrote three others which superseded it. In [Monge 1770-1773], he gives more general procedures for the determination of the arbitrary functions

---

[53] In a letter to Condorcet dated 2nd September 1771 (published by Taton [1947, 979-982]), he had given all of these equations plus $\frac{x\,\delta z}{dx} + \frac{y\,\partial z}{dy} = z - a$ for a conical surface.

[54] "of such a nature as to be expressed by equations"

[55] Not so for the general 18th-century reader who did not know the manuscript of [Monge 1771].

[56] A different possibility is that, similarly to Arbogast (see page 207 below), he assumed something like piecewise continuity and could work with two tangent planes at a point of discontinuity. But I do not see any suggestion of this in his words.

given appropriate conditions, and for their geometrical construction. Most of the examples involve two arbitrary functions, and the last one involves an indeterminate number of arbitrary functions, so that they correspond to second- and higher-order equations. However, the differential equations themselves do not play any role. [Monge *1773a*] tries to address that flaw: to show that the surfaces that satisfy the integral of a partial differential equation also satisfy that partial differential equation. For instance, in problem II [Monge *1773a*, 273-275] he constructs the surface-locus of

$$z = M + N\varphi V \tag{6.16}$$

(where M, N and V are given functions of x and y) such that it passes through a curve with projections $y = Fx$ and $z = fx$; in theorem II [Monge *1773a*, 275-280] he proves that for each point of the surface thus constructed the differential equation (independent of the arbitrary function φ)

$$\partial V[N\delta z - N\delta M - z\delta N + M\delta N] = \delta V[N\partial z - N\partial M - z\partial N + M\partial N] \tag{6.17}$$

holds. [Monge *1773b*] is a further exploration of the problem of determining arbitrary functions in integrals of partial differential equations, associating it with finite difference equations. Taton [*1951*, 281] complains about the fastidious and repetitive nature of these memoirs: "ayant mis au point une théorie intéressante, il l'applique à tous les exemples d'équations qu'il sait, sinon intégrer, du moins étudier"[57]. However, the feeling one gets from reading these works (besides lack of patience for all the examples) is that Monge was trying to generalize ever more a theory which had started as a set of very simple examples.

Later, Monge's studies on differential equations in three variables and families of surfaces proceeded in different directions (see section 6.1.3.4). However, and in spite of d'Alembert's objections, Monge always kept his belief in the acceptability of discontinuous functions in the integrals of partial differential equations (see for instance [Monge *Feuilles*, n° 4-iii] for cylindrical surfaces, which have always the differential equation $1 = a\left(\frac{dz'}{dx'}\right) + b\left(\frac{dz'}{dy'}\right)$, even if the curve along which the generating line slides is discontinuous).

It is worth stressing the importance of *construction of differential equations* in Monge's argumentation. True, it was not Monge who brought discussions on constructions to the controversy on arbitrary functions: a great deal of the quarrel between Euler and d'Alembert revolved around the former's construction of the extended curve $n'bAMBaM'$ (page 199 above). But that was a discussion on *one* isolated construction, and only of a curve involved in the solution, not of the equation. Monge treated constructions much more generally: the construction of a certain partial differential equation corresponded to the generation of the surfaces of the family defined either by that construction/generation or by that equation.

---

[57]"having developed an interesting theory, he applies it to all the examples of equations which he can, if not integrate, at least study"

The last famous treatment of the issue of acceptability of discontinuous functions in the 18th century [Arbogast *1791*] was very much influenced by Monge.

[Arbogast *1791*] was the winning entry to the 1787 prize of the St. Petersburg Academy, devoted precisely to the question of whether the arbitrary functions introduced by the integration of differential equations in more than two variables may be discontinuous, or rather correspond only to curves capable of being expressed by algebraic or transcendental equations [Arbogast *1791*, 95].

The only contribution of [Arbogast *1791*] which has received any attention [Grattan-Guinness *1970*, 18; Youschkevitch *1976*, 71] is his introduction of the distinction between *contiguous* and *discontiguous* functions, more or less corresponding to the *modern* idea of continuous and discontinuous. Besides the concern about the change or conservation of the form of a function (that is, its "discontinuity" or "continuity"), we have seen above that some mathematicians of the 18th century noticed the relevance of occurrence or not of "jumps" in the course of a function or of its derivatives. But they lacked words for this distinction; Arbogast [*1791*, 11] proposed "courbes discontigues" and "fonctions discontigues" for those composed of disconnected pieces, while keeping the word "discontinuous" with its old meaning.

But it is interesting to look also at Arbogast's arguments for accepting discontinuous and even discontiguous functions in the integrals of partial differential equations, most of which may be seen as more direct uses of Monge's arguments. For most of his dissertation, Arbogast repeatedly takes a partial differential equation, translates it into a geometrical condition, and then constructs the surfaces that obey this condition. Since these surfaces are so undetermined that they may be subject to pass through a discontinuous or even discontiguous curve (or two such curves, in the case of second-order equations), that is, since the construction can be performed using continuous or even discontiguous curves, these curves must be allowed, and also the corresponding discontinuous or even discontiguous functions must be allowed in the integrals of the original partial differential equations.

The simplest example is that of the equation $\dfrac{dz}{dx} = a$ (where z is supposed to be a function of x and y, so that the equation belongs to a surface) [Arbogast *1791*, 12-14]. This means that any section parallel to the xz plane is a straight line with slope a. Everything else (in particular the sections parallel to the yz plane) is undetermined. Therefore, if AB is the x axis, AC the y axis, and AD the z axis; a straight line KM is drawn on a plane perpendicular to BAC and making an angle with MT whose tangent is a; an arbitrary curve $GIKL$ is drawn on the plane KRN perpendicular to AC; and if finally KM is made to slide along $GIKL$, then it will generate a surface satisfying the equation $\frac{dz}{dx} = a$. Now, the integral of $\frac{dz}{dx} = a$ is $z = ax + \phi.y$, and if we put $AR = b$ the equation of $GIKL$ is $z = ab + \phi.y$, so that the possibility of the curve $GIKL$ being discontinuous and discontinguous[58] is passed on to the function $\phi.y$.

---

[58]In the figure it is possible to notice a point of discontiguity between I and G.

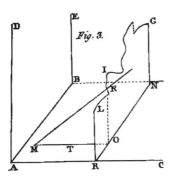

A less convincing example (for a modern reader and probably for some contemporary reader who would agree with d'Alembert) is that of the equation
$\frac{dz}{dx} = \frac{dz}{dy}$ [Arbogast *1791*, 23-25]. The geometrical condition expressed by this
is that if by any point of the surface one takes two sections perpendicular to the
xy plane, one parallel to the x and the other parallel to the y, and if one considers
a tangent to each of these sections, the slopes of these tangents are equal. However, in his construction of the surface Arbogast substitutes an equality between
the two sections for the equality of the tangents, arguing that if the sections are
equal, "leurs élémens seront toujours inclinés de la même quantité au point où
elles se rencontrent"[59]. Thus Arbogast simply considers a straight line on the xy
plane making an angle of 45° with the x axis (clockwise from the x axis), and
imagines it to move freely and irregularly in space, but always keeping the same
direction – he completely bypasses the issue of the existence of the tangents to
the sections (or equivalently whether "their elements" are well defined), admitting
the possibility of these sections being discontinuous and discontiguous (which is
reflected on the possible discontinuity or discontiguity of the function ϕ in the
integral $z = \phi(x + y)$).

But from parts of Arbogast's discussions of objections by Condorcet and
Laplace [Arbogast *1791*, 39; 85-86] it is possible to conjecture why he is not concerned about the existence of tangents: 1 – apparently he regards discontiguous
functions as piecewise contiguous; 2 – if a curve ABC is discontiguous at B, instead of not having a definite value for the differential of the corresponding function
at B, one apparently has two definite values, each applying to one of the branches
AB and BC (so presumably two semi-tangents). In modern terms, Arbogast is
content with left- and right-derivatives. As for difficulties arising from discontinuity, they have to do with "jumps" – that is, discontiguity – not in the function,
but in its differentials, so that similar arguments apply.

Thus we see, in Monge and even more clearly in Arbogast, constructions of

---

[59]"at the point where they meet their elements will always have the same inclination"

equations being used once again in arguments of legitimation – this time, the legitimation of discontinuous ("and even discontiguous") functions.[60] This is likely not a coincidence. The construction of equations (particularly of algebraic equations) was dead as a research subject, but it was still very much alive as a school subject, and was therefore well-known by all mathematicians and available to be used if it were ever appropriate.

6.1.3.3   Lagrange: singular, complete, and general integrals, in geometrical guise

There are more direct connections between geometry and solutions of differential equations than the constructions discussed in the previous section. Some of the most direct ones are related to the problem of singular solutions, through the identification between these and envelopes.

[Lagrange *1774*] not only gives an analytical theory of singular integrals, but it also provides a geometrical interpretation of that theory. In fact, the "third article" of that memoir [Lagrange *1774*, § 21-26] purports to be a deduction through the "consideration of curves" of the theory on singular integrals of first-order ordinary equations that had been set up analytically in the first two "articles". If $V = 0$ is the complete integral of $Z = 0$, where Z is a function of x, y, and $\frac{dy}{dx}$, then V is a function of x, y, and an arbitrary constant a, so that $V = 0$ represents an infinite collection of curves, one for each possible value of a (including plus and minus infinity); naturally, $Z = 0$ also represents these curves; but the key point is that $Z = 0$ also represents the curve that is tangent to all these curves (in modern terms: their envelope), since $Z = 0$ determines $\frac{dy}{dx}$ for each point, and therefore the position of the tangent line, which is shared with the envelope. Considering two infinitely close points of the envelope, corresponding to two infinitely close curves, and making these points coincide, Lagrange characterizes the envelope as formed by "l'intersection mutuelle et successive des courbes données par l'équation $V = 0$, en faisant varier le parametre a"[61] [Lagrange *1774*, § 22]. Since for the same abscissa x the ordinates of two infinitely close curves are y and $y + \frac{dy}{da} da$, the intersection implies $\frac{dy}{da} = 0$; thus the equation of the envelope is obtained by eliminating the parameter a between the two equations that it must satisfy: $Z = 0$ and $\frac{dy}{da} = 0$. Therefore the envelope corresponds perfectly to the singular integral.

Lagrange [*1774*, § 25-26] gives two examples, both of which can be found in [Euler *1756*]. The first is also Euler's first (to find the curves such that all perpendiculars from their tangents to a given point have the same given length), a problem which as seen above has as complete solution a family of straight lines $y - ax - b\sqrt{1 + a^2} = 0$ ((6.6) above) and as singular solution a circle $x^2 + y^2 = b^2$ ((6.5) above). The second (Euler's third) is quite similar, having as complete

---

[60]In the case of Monge there was at least one additional motivation: the determination of the arbitrary functions involved in an integral.

[61]"the mutual and successive intersection of the curves given by the equation $V = 0$ by making the parameter a vary"

solution also a family of straight lines $y - a(x - b) = \sqrt{c^2(1 + a^2) - b^2}$ and as singular solution the ellipse $\frac{(x-b)^2}{c^2} + \frac{y^2}{c^2 - b^2} = 1$. It seems quite likely that Lagrange's process of discovery involved the recognition that the singular solutions found by Euler were envelopes of the complete solutions.

As for partial differential equations, geometrical considerations play a different role. Instead of having a separate "article" for a parallel geometrical deduction of his theory, Lagrange uses geometry twice in the fifth and final "article" (on singular integrals of partial differential equations) to illustrate and explain his concepts of singular, complete and general integral of a first-order differential equation in three variables. In this case, a complete integral $V = 0$ represents a two-parameter family of surfaces, as it contains two arbitrary constants a, b; the singular integral represents the surface that is tangent to all those surfaces (the envelope of the family) [Lagrange *1774*, § 43]. The example given in [Lagrange *1774*, § 44] is not surprising: consider the problem of finding the surfaces such that all perpendiculars from their tangent planes to a given point have the same given length; taking the given point to be the origin of coordinates, the sphere[62]

$$z = \sqrt{h^2 - x^2 - y^2}$$

(where h is the given length) is an obvious solution, but since it does not have any arbitrary constant, it must correspond to the singular integral; a complete integral is represented by the family of planes that are at distance h from the origin

$$z = ax + by + h\sqrt{1 + a^2 + b^2},$$

which of course have the sphere as envelope. The general integral is more complicated; [Lagrange *1774*, § 49] uses the same example: the general integral is the result of eliminating a between

$$z = ax + \phi a \cdot y + h\sqrt{1 + a^2 + (\phi a)^2}$$

and

$$x + \phi' a \cdot y + h\frac{a + \phi a \cdot \phi' a}{\sqrt{1 + a^2 + (\phi a)^2}} = 0.$$

This cannot be done in general, so Lagrange does it for two particular cases of ϕa. $\phi a = m + na$ (for some constant m and n) gives a right cylinder whose axis passes through the origin (and centre of the sphere) and whose radius is h; this is of course tangent to the sphere, although Lagrange does not mention it. $\phi a = \sqrt{k^2 - 1 - a^2}$ (for some constant k) gives a right cone, also tangent to the sphere, although once again this is not mentioned. What Lagrange does mention, is that both the singular solution and each of the surfaces in the general solution are tangent in every point to one of the surfaces in the complete solution; but the singular solution is tangent to *all* the surfaces of the complete solution (it is their envelope), while

---

[62] As in many other occasions, one must read "$\sqrt{}$" as meaning "$\pm\sqrt{}$".

each of the surfaces in the general solution is tangent only to the surfaces in the complete solution that correspond to some particular relation between a and b (if we put $b = \phi(a)$, and then eliminate a between $V = 0$ and $\frac{\partial V}{\partial a} = 0$, we obtain of course the envelope of the one-parameter family $V(x, y, z, a, \phi(a)) = 0$, so that the general integral is the collection of envelopes of one-parameter subfamilies of $V(x, y, z, a, b) = 0$[63]).

  Lagrange returned to geometrical considerations relating to singular integrals in [Lagrange *1779*]. In the first three articles of that memoir he gives examples of problems in plane geometry (on evolutes, "roulettes", and more generally on curves having contact of some order) that are solved by considering singular solutions instead of complete solutions. For instance, the problem of finding the involutes of a given curve is a second-order problem in integral calculus, so that apparently there are two indeterminate elements; nevertheless, there is only one, namely the first point of the involute (in figure 65 in page 114, if BDF is given and AHK is sought, the length, but not the direction, of AB is arbitrary); this is because the involute is the envelope of a family of circles whose centres are on the evolute. Those examples can be seen in coordination with a remark in [Lagrange *1774*, § 56]: the most natural solution of the problem on surfaces above is the sphere, which is not represented in the complete integral, but rather by the singular integral; that shows "la nécessité d'avoir égard à ces sorts d'intégrales pour avoir toutes les solutions possibles"[64]. All this sounds like an answer to Euler's previous objections on singular integrals and especially to [Condorcet *1770-1773*].

  The fourth "article" is quite different, having no direct connection to singular integrals, although it revolves around elimination of constants. Lagrange seeks equations for surfaces composed of lines "of a given nature";[65] for this, he considers the equations of the composing lines and differentiates them relatively to the constants which characterize each line; he then eliminates all the constants, obtaining the desired equation. We will see below that Monge carried this kind of procedure much further.[66]

6.1.3.4   Monge: geometrical integration

We have already mentioned in sections 4.2.2.1 and 6.1.3.2 Monge's association of differential equations in three variables to families of surfaces. In the latter

---

  [63]Although Lagrange does not do it, the example above can also be used to illustrate the diversity of complete integrals: the family of all right cylinders with radius h and axis through the origin is a complete solution (the two arbitrary constants m and n in $\phi a = m + na$ ensure that).

  [64]"the necessity of taking into account this type of integrals to have all the possible solutions"

  [65]Lagrange assumes that these lines must intersect consecutively (or be parallel, which may be interpreted as intersection at infinity). In the case of straight lines this makes him miss the case of skew surfaces [Lacroix *Traité*, I, 501].

  [66]The fifth and final "article" in [Lagrange *1779*] is also very different, but in another sense: it is there that Lagrange presents his method for integrating quasi-linear first-order partial differential equations. The connection with singular integrals is that it is a generalization of a method given in [Lagrange *1774*, § 52]. A geometrical example is given, but it is irrelevant for us here.

section only his early studies, on the determination of arbitrary functions involved in integrals, were addressed. In this section we will look at later developments.

In [Monge *1780*] he put to work several aspects of the association just mentioned, the family in question being that of developable surfaces . We have already seen (page 129) that he obtains in three different ways their differential equation

$$\delta\delta z \cdot \mathrm{d}\mathrm{d}z = (\delta\mathrm{d}z)^2 \qquad (6.18)$$

(where δ still refers to partial differentiation relative to x, while d refers to partial differentiation relative to y). For ruled surfaces he obtains

$$2\delta\left(\frac{-\delta\mathrm{d}z + \sqrt{(\delta\mathrm{d}z)^2 - \delta\delta z\mathrm{d}\mathrm{d}z}}{\mathrm{d}\mathrm{d}z}\right) + \mathrm{d}\left(\frac{-\delta\mathrm{d}z + \sqrt{(\delta\mathrm{d}z)^2 - \delta\delta z\mathrm{d}\mathrm{d}z}}{\mathrm{d}\mathrm{d}z}\right)^2 = 0,$$

of which (6.18) is clearly a particular case [Monge *1780*, 431, 435]. A developable surface is completely determined by its edge of regression: if the latter has projections $y = \psi \cdot x$ and $z = \varphi \cdot x$, then the equation of the surface is

$$z = \varphi \cdot V + (x - V)\,\varphi' \cdot V, \qquad (6.19)$$

where V is such that

$$y = \psi \cdot V + (x - V)\,\psi' \cdot V, \qquad (6.20)$$

and φ' and ψ' are the derivatives of φ and ψ [Monge *1780*, 387, 415]. The first derivation of (6.18) is precisely obtained through differentiation of (6.19) and (6.20) [Monge *1780*, 385-389].

In 1776 Monge received from Condorcet an offprint of [Lagrange *1774*], and he was delighted with it [Taton *1951*, 190-192]. The association between envelopes and singular integrals opened many possibilities for the associated study of surfaces and partial differential equations, as did the elimination of arbitrary elements in finite equations.

Some years later Monge wrote two memoirs on surfaces generated by the movement of space curves.[67] According to Taton [*1951*, 285-286], [Monge *1784-1785*] was written in 1783 and received a favourable report for publication by the Turin Academy in February 1784. We will look only at the first problem studied: that of a surface generated by a circle of constant radius which moves remaining always perpendicular to the space curve described by its centre. If this curve has equations $x = \psi z$ and $y = \phi z$, z' represents the third coordinate of the centre,

---

[67]We cannot exclude the possibility that Monge was inspired in this by the fourth article of [Lagrange *1779*], which appeared in 1781, but apropos of a completely different issue Monge claimed later not to have known [Lagrange *1779*] (in [Monge *1784b*, 118], which according to Taton [*1951*, 289] was submitted only in 1786). The issue there was Lagrange's method for integrating quasi-linear first-order partial differential equations, which appeared in the fifth article of [Lagrange *1779*].

and a is the radius of the circles, then the fact that each point of the surface is on a circle is expressed by

$$(z - z')^2 + (y - \phi z')^2 + (x - \psi z')^2 = a^2 \tag{6.21}$$

and the fact that each point on the surface is on the normal plane to the curve that passes through the centre of the corresponding circle is expressed by

$$z - z' + (y - \phi z')\phi' z' + (x - \psi z')\psi' z' = 0. \tag{6.22}$$

If the curve is given, then all there is to be done is to eliminate z' between these two equations. But if we want the general equation of these surfaces, expressing its generation without regard for a particular curve, then ϕ and ψ are to be considered as arbitrary and be eliminated using differentiation. The clumsy final result is the second-order equation

$$k^4 + ak \left\{ \left(1 + \left(\tfrac{dz}{dy}\right)^2\right)\tfrac{ddz}{dx^2} - 2\tfrac{dz}{dx}\tfrac{dz}{dy}\tfrac{ddz}{dx\,dy} + \left(1 + \left(\tfrac{dz}{dx}\right)^2\right)\tfrac{ddz}{dy^2} \right\} + a^2 \left\{ \tfrac{ddz}{dx^2}\tfrac{ddz}{dy^2} - \left(\tfrac{ddz}{dx\,dy}\right)^2 \right\}$$

$$= 0,$$

where $k^2 = 1 + \left(\tfrac{dz}{dx}\right)^2 + \left(\tfrac{dz}{dy}\right)^2$. Monge [*1784-1785*, 22] does not fail to notice that (6.21) is the equation of the spheres with centre in the curve and radius a, and that (6.22) is the differential of (6.21) relative to z', so that the surface is the envelope of those spheres, or "à la manière de Mr. De la Grange son équation est l'intégrale particulière de l'équation différentielle qui appartient à toutes les sphères"[68].

Notice that elimination of arbitrary elements plays here a double role. The first, which had always been predominant in Monge's studies of classes of surfaces, is in keeping arbitrary the curve used in the generation of the surfaces in a class; the second role can be thought of as obtaining a surface as the envelope of a family of other surfaces, although it is not always explicitly presented that way: the surface may be seen as generated by the movement of a curve (that Monge would later call *characteristic curve*) which is in fact the intersection of two consecutive surfaces in the family. A major difference is that the first role typically involves elimination of arbitrary functions, while the second involves elimination of arbitrary constants. The second role is of Lagrangian inspiration (although of course one can see it in plane geometry since the late 17th-century studies of envelopes, and might see traces of it in space geometry in the elimination of β for obtaining (6.15) in [Monge *1771*, 51-52]); the first role is essentially due to Monge: one can see traces of it in the fourth article of [Lagrange *1779*], but it clearly conforms to Monge's program, and moreover it can be seen applied in [Monge *1773a*, 268], where equation (6.17) is obtained by writing (6.16) as $\tfrac{z-M}{N} = \varphi V$, taking partial differentials relative to x (namely $N\delta z - N\delta M - z\delta N + M\delta N = N^2\varphi' V.\delta V$) and to

---

[68]"in the manner of Mr. De la Grange, its equation is the singular integral of the differential equation which belongs to all the spheres"

y (namely $N\partial z - N\partial M - z\partial N - M\partial N = N^2\varphi'V.\partial V$), and eliminating $N^2V\varphi'V$ between these .

[Monge *1784a*] (submitted to the Paris Academy in July 1785, according to Taton [*1951*, 287]) is an elaboration of the previous memoir. There Monge insists even more on the first role of elimination. An equation for a class of surfaces defined by a form of generation involves arbitrary functions which represent the curve that specifies each member of the class. The fact that a function is arbitrary can be expressed in two ways: either by representing it by a special character; or by eliminating it between the differentials of the finite equation, thus obtaining a partial differential equation for the class of surfaces, where there is no trace of the generating curve [Monge *1784a*, 86]. Monge even develops a new method for the elimination of an arbitrary function: the traditional method was to differentiate relatively to x, then relatively to y, and then eliminate the arbitrary function *and* its derivative (which had been introduced by the differentiations) between the three equations (one finite and two differential); his new method consists in regarding the argument of the arbitrary function as constant (that is, if the finite equation involves $\varphi(\omega)$, where ω is a known function of x, y and z, one puts $\omega = $ const., and then takes the total differential, minding that $\frac{dx}{dy}$ has now a determinate value established by $\omega = $ const.). The main advantage is that no new functions appear.

The second role of elimination gained importance in [Monge *1784c*] (see section 6.1.3.5), and especially later in [Monge *Feuilles*], where many surfaces are studied as envelopes of families of other surfaces. In Lagrangian terms, this does not mean that much attention is paid to singular solutions, but rather to general solutions – we have seen that the geometrical interpretation of a general solution is a collection of envelopes of one-parameter families of surfaces. Nor does Monge dwell much on complete solutions (again, in the Lagrangian sense). He does consider a finite equation $F = 0$ for enveloped surfaces containing two parameters α and β; but immediately (in the same sentence) he takes $\beta = \varphi\alpha$ [Monge *Feuilles*, n° 7-ii]; the two parameters are only useful for him to have a directing plane curve $y = \varphi x$. His first example is a simplified version of the one seen above: a surface enveloping a family of spheres of constant radius a whose centres are on the curve $y = \varphi x, z = 0$ [Monge *Feuilles*, n°s 7-ii – 8-i]; since in this case there is only one arbitrary function, the differential equation is of first order:

$$z^2\left[1 + \left(\frac{dz}{dx}\right)^2 + \left(\frac{dz}{dy}\right)^2\right] = a^2.$$

But one of the most important aspects of [Monge *Feuilles*] for differential equations is its concern with *characteristics*. A characteristic curve of an envelope is the intersection of two consecutive surfaces in the family (in the example above, a vertical circle of radius a). Monge had given in [*1784b*] a method for reducing the integration of a partial differential equation to that of a system of ordinary differential equations. Oddly for Monge, this method did not come then with a geometrical interpretation. This only appeared later, in [*1784c*] and more explicitly in [*Feuilles*]: those ordinary differential equations belonged to the projections onto the coordinate planes of the characteristic curves of the integral surface of the partial differential equation. Among other things, Monge [*Feuilles*, n°s 27-iv – 28-iv] used this method to integrate the equation of minimal surfaces [Taton *1951*, 302-303].

6.1.3.5　Monge: integration of "ordinary" differential equations not satisfying the conditions of integrability

The last example of Monge's geometrical integration we will look at concerns what he called "equations of ordinary differences in three variables" – that is, equations involving *ordinary*, or *total*, differentials of three variables; in the first-order and first-degree case, they correspond, in modern terms, to "Pfaffian equations" (in three variables).[69]

　　We have seen above (section 6.1.3.1) that Clairaut had arrived at a necessary condition (6.13) for a differential equation in three variables to be solvable.[70] For this, Clairaut had assumed that the integral of such a differential equation was composed of one finite equation – or equivalently, that the geometrical construction of such a differential equation resulted in a surface. Euler [*Differentialis*, I, § 307-318; *Integralis*, III, ch. 1] had followed Clairaut's assumption,[71] in a more *functional* manner: using an analogy with finite equations, he had concluded that for a differential equation in three variables to be meaningful, one of those variables had to be a function of the other two: "aequatio differentialis tres variabiles com-

---

[69] Monge also addressed ordinary differential equations in more than three variables, but we will omit them here: he was clearly guided by analogy in that, his reasoning being essentially geometrical.

[70] In fact, a slightly different version of this condition had already been found by Fontaine [Greenberg *1982*, 12, 20-26]. Clairaut, although critical of Fontaine's style, acknowledged his priority [*1740*, 310]. Furthermore, Cousin [*1796*, I, 258] attributed (6.13) to "N. Bernoulli" – presumably Nicolaus (I) Bernoulli, in an extract of a letter published in an article by his cousin Nicolaus (II) Bernoulli [*1720*, 442-443] (see [Engelsman *1984*, 186-187] for the unravelling of this "bibliographical monster", which had been cited by Poggendorff and Fleckenstein as it if were an independent article, with a wrong date, and in the latter case with wrong page numbers – and still Engelsman [*1984*, 231] cites it simply as being § 30 in [Nic. Bernoulli *1720*], apparently not noticing that while it is indeed § 30 in Johann Bernoulli's *Opera Omnia*, it is numbered § 29 in the original publication in the *Actorum Eruditorum Supplementa*, 7 (1721), pp. 310-312, because of a duplication of § 22). Now, a formula somewhat similar to (6.14) does occur in [Nic. Bernoulli *1720*, 443] – namely, $dq = Tq\,dy + R\,dy$, for $dx = p\,dy + q\,da$, where $dp = T\,dx + S\,dy + R\,da$ and dq is the differential of q holding a constant; in modern notation, and noticing that holding a constant makes $dq = \frac{\partial q}{\partial x}dx + \frac{\partial q}{\partial y}dy = \frac{\partial q}{\partial x}p\,dy + \frac{\partial q}{\partial y}dy$, this amounts to $\frac{\partial q}{\partial x}p\,dy + \frac{\partial q}{\partial y}dy = q\frac{\partial p}{\partial x}dy + \frac{\partial p}{\partial a}dy$, whence $p\frac{\partial q}{\partial x} + \frac{\partial q}{\partial y} = q\frac{\partial p}{\partial x} + \frac{\partial p}{\partial a}$, that is, the condition of integrability of $dx = p\,dy + q\,da$. Not only these later developments are not present, but also Bernoulli does not use the formula at all as a criterion for integrability; rather, he uses it to obtain q, given p (i.e., to solve what Engelsman [*1984*] has called the "completion problem"). What really appears in Bernoulli's derivation of that formula for the first time is something else, although essential for (6.13): the equality of mixed second-order differentials – Lacroix noticed this in [Montucla & Lalande *1802*, 344].

[71] In spite of Fontaine's (and to some extent Nicolaus (I) Bernoulli's) priority, it was Clairaut who communicated (6.13) to Euler [Engelsman *1984*, 198].

plectens determinabit, qualis functio una sit reliquarum"[72] [Euler *Differentialis*, I, § 307]. He had also reproduced Clairaut's condition (6.13), but of course with purely analytical proofs [Euler *Differentialis*, I, § 313-316; *Integralis*, III, §,1]; an equation was "real" if it verified this condition, and otherwise it was "imaginariam seu absurdam"[73] [Euler *Differentialis*, I, § 317].

In addition, and for similar reasons, Euler had also declared absurd those equations in which the differentials were raised to powers higher than 1, such as $P dx^2 + Q dy^2 + R dz^2 + 2S dx dy + 2T dx dz + 2V dy dz = 0$, unless they could be reduced to the form $P dx + Q dy + R dz = 0$ [*Differentialis*, § 326; *Integralis*, § 27].

In [*1768*, 15-16], Condorcet challenged this. He accepted that equations "qu'on appelle absurdes" do not have integrals, but not that the related problems are necessarily impossible: given an absurd first-order equation in three variables, the problem is not satisfied by any surface; but if the equation is regarded as representing a curve of double curvature, one projection of which is arbitrary, the problem is not only possible, but even has an infinity of solutions. He might have been thinking of Newton [*Fluxions*, 83]: as Lacroix would point out (see page 253 below), Newton had already used, in order to solve fluxional equations in n variables, the technique of temporarily reducing them to equations in two variables by establishing $n - 2$ relations between the n variables.

But Condorcet did not develop this idea. It was up to Monge to do it, in [*1784c*]. For him, no differential equation in three variables is absurd; those that verify the integrability condition (6.13) belong to curved surfaces, their integrals being single equations, with single arbitrary constants; while those that do not, rather than belonging to no geometrical object, or having no integral, belong to families of curves in space, their integrals being systems of two equations[74]. In more modern (or more Eulerian) terms, an equation relating the differentials of three variables may determine two functions of one independent variable, instead of necessarily one function of two independent variables.

Monge addresses firstly higher-order equations, and his first example [*1784c*, 506-509] is

$$dz^2 = a^2(dx^2 + dy^2), \tag{6.23}$$

which obviously belongs to the curves whose elements make a constant angle with the x, y plane. Therefore, he considers the straight lines that make that angle:

$$x = \alpha z + \beta, \qquad y = z\sqrt{\frac{1}{a^2} - \alpha^2} + \gamma. \tag{6.24}$$

But this is not the "complete" integral of (6.23)[75]: eliminating α in (6.24) gives

---

[72]"a differential equation involving three variables determines which function one of them is of the others"

[73]"imaginary or absurd"

[74]With the sole exception of $M^2 dx^{2m} + N^2 dy^{2m} + P^2 dz^{2m} = 0$, whose integral was the system $x = a, y = b, z = c$: one arbitrary point in space.

[75]That is, it is not the most general one. Monge never followed Lagrange's distinction between "complete" and "general" integrals (see section 6.1.4.2).

$(x - \beta)^2 + (y - \gamma)^2 = \frac{z^2}{a^2}$, that is, the cones with vertices on the x, y plane whose constituent straight lines make that angle;[76] putting $\gamma = \varphi\beta$, i.e., making the vertices follow an arbitrary curve, two consecutive cones will intersect along a straight line, included in (6.24); but the envelope of these straight lines of intersection will also satisfy (6.23); thus the complete integral will be the result of eliminating β between

$$(x-\beta)^2+(y-\varphi\beta)^2 = \frac{z^2}{a^2}, \ x-\beta+(y-\varphi\beta)\varphi'\beta = 0, \text{ and } -1-(\varphi'\beta)^2+(y-\varphi\beta)\varphi''\beta = 0$$

(the reason why there are three equations here instead of two is precisely that β has still to be eliminated; but this cannot be done explicitly, on account of the arbitrariness of φ). A particular case is the thread of a screw with axis perpendicular to the x, y plane.

After a couple more examples, Monge [*1784c*, 518-520] provides a more general picture: given an equation $M = 0$ of a family of surfaces (besides the coordinates x, y and z, M is supposed to involve a parameter α, and an arbitrary function of it $\varphi\alpha$), a partial differential equation $V = 0$ for the envelope of those surfaces may be obtained by eliminating α and $\varphi\alpha$ between $M = 0$, $\left(\frac{dM}{dx}\right) = 0$, and $\left(\frac{dM}{dy}\right) = 0$; but this envelope is composed of its characteristic curves[77], and they in turn intersect two by two, along a curve of double curvature, which Monge calls here the "limit of the envelope", but which he would call in [*Feuilles*] the "edge of regression"[78]; eliminating α and $\varphi\alpha$ between $M = 0$, $\left(\frac{dM}{dx}\right) = 0$, $dM = 0$, and $d.\left(\frac{dM}{d\alpha}\right) = 0$, one obtains an equation $U = 0$ for the edge of regression – an *ordinary* differential equation in three variables, of degree higher than 1, which would be absurd to Euler.

The most important practical consequence of all this is the equivalence, in a sense, between $U = 0$ and $V = 0$. Monge shows how to obtain one from the other without knowing their integrals [*1784c*, 520-5], and that if the integral of $V = 0$ is the result of eliminating α from $M = 0$ and $\left(\frac{dM}{d\alpha}\right) = 0$, then eliminating α from $M = 0, \left(\frac{dM}{d\alpha}\right) = 0$ and $\left(\frac{ddM}{d\alpha^2}\right) = 0$ gives the integral of $U = 0$ [*1784c*, 525-6].

As for linear[79] equations that do not satisfy the integrability condition, Monge [*1784c*, 528-532] applies procedures derived by analogy from the considerations above for higher-degree equations, using auxiliary partial differential equations. In an "addition" at the end of the memoir [*1784c*, 574-576], he remarks that he has not "constructed" any of these linear ordinary equations, and so he gives an example, in order to show "ce que ces sortes d'équations signifient dans l'espace"[80]: the apparent contour of a surface of revolution seen from a point with

---

[76]These cones are made up of straight lines satisfying (6.23), but unlike what Taton [*1951*, 298] says, they do not satisfy (6.23) themselves. The whole point is that these equations belong to families of curves, not to surfaces.

[77]Monge does not use this name here. Instead, he speaks of "curves of intersection".

[78]Here this name is reserved for developable surfaces.

[79]I. e., quasi-linear.

[80]"what is the spatial meaning of this kind of equations"

coordinates a, b, c; this amounts to the curve where that surface is tangent to a conical surface with vertex at that point; it is by combining the partial differential equations of the surface of revolution $py - qx = 0$ and of the conical surface $p(x - a) + q(y - b) = z - c$ that Monge obtains the ordinary differential equation for the apparent contour $[x(x - a) + y(y - b)]dz = (z - c)(x\,dx + y\,dy)$; its integral is given by the system

$$z = \varphi(x^2 + y^2),$$

$$2[x(x - a) + y(y - b)]\varphi' = \varphi - c,$$

where φ is an arbitrary function. After this single example he concludes that any linear first-order ordinary differential equation in three variables not satisfying the integrability condition belongs to the curve of contact of two curved surfaces (each given by a linear partial differential equation).

It is interesting to look at the first attempt to give an analytical version of this, by an author whom Lacroix [*Traité*, II, 629] appreciated particularly: the Italian Pietro Paoli. Given a differential equation in x, y, z that does not satisfy the integrability conditions, Paoli's idea [*1792*, 4-8] is that if one establishes an arbitrary relation $y = \phi.x$, that equation will be transformed into one in two variables x, z – thus necessarily integrable; the integral of the original equation will be the system formed by $y = \phi.x$ and the integral of the secondary equation. Of course this cannot be done in general; but we can obtain a "particular" integral by establishing a particular, rather than arbitrary, relation between x and y; if we include an arbitrary constant α in this relation, that particular integral will have two arbitrary constants (α, and another β originating in the integration of the secondary equation); then, by a Lagrangian procedure of variation of constants, we can obtain the complete integral[81]. He manages to derive from this Monge's procedure for integrating linear equations.

A rather less interesting analytical treatment of these equations was given by a Belgian mathematician, Charles-François de Nieuport [*Mélanges*, I, 211-230], focusing on systems of two or more such equations. It is only worth noting that Nieuport is probably the only author to cite Condorcet (namely [*1768*, 15]) instead of Monge (or even Newton), for the idea of establishing a relation between two of the variables in an equation in three variables that does not satisfy the condition of integrability.

In spite of these reactions by lesser-known mathematicians, this work by Monge was ignored by textbook authors. As late as [*1798*, II, 129-135] Bossut declared equations not satisfying the conditions of integrability to be not real and having no integral. Cousin [*1796*, I, 258-259] was not so radical, but only because he paid much attention to observations by Euler [*Differentialis*, I, § 310, 323-325] and Laplace [*1772a*, 368-370] on the occasional existence of particular integrals of these equations.

---

[81] Of course, Lagrange would call this the "general", rather than "complete", integral.

6.1.4   The formation of differential equations and their complete and general integrals

6.1.4.1   Ordinary differential equations

[Lagrange *1774*] represented somewhat more than a theory of singular solutions. It entailed also a change in the *theory of differential equations*, in an aspect which (at least in theoretical or pedagogical terms) could go beyond the subject of singular integrals: it stressed the formation of differential equations by algebraic elimination of arbitrary elements between a finite equation and its differential(s), as opposed to focusing only on the process of integration (or on that of differentiation, as a simple inverse process). Engelsman [*1980*, 16] put it nicely in the following diagram:

$$\text{Euler}: \qquad Z(x, y, \tfrac{dy}{dx}) = 0 \quad \xrightarrow{\text{integration}} \quad V(x, y, a) = 0,$$

$$\text{Lagrange}: \quad Z(x, y, \tfrac{dy}{dx}) = 0 \quad \xleftarrow{\text{elimination of } a} \quad V(x, y, a) = 0.$$

An early sign of this outside the area of singular integrals is the explanation given in [Lagrange *1774*, § 32] for the fact that a second-order differential equation has two first-order integrals: if instead of eliminating both a and b in (6.11) one simply eliminates a between $V = 0$ and $\frac{dy}{dx} - p = 0$, one will obtain precisely one of those first-order integrals of $Z' = 0$; eliminating b one will obtain a different first-order integral. Lagrange comments that this is "connu des Géometres"[82].

Later, when Lagrange got around to writing his first treatise on the calculus, he introduced a distinction in terminology between the equations that are obtained by immediate derivation of a primitive equation ("prime", "second", etc. equations) and those obtained by combining the primitive equation with its prime equation and/or second equation and/or etc. ("derivative equations"[83]) [Lagrange *Fonctions*, 51].[84] He then explained the occurrence of arbitrary constants in primitive equations obtained from derivative equations (i.e., in solutions of differential equations) by their disappearance through *elimination* between those primitive equations and their prime, second, etc. equations [Lagrange *Fonctions*, 56; *Calcul*, 151]. This should be compared to Euler's explanation, which stressed *differentiation*: to remove the constant a from the equation $x^3 + y^3 = 3axy$, one should divide by xy to obtain $\frac{x^3+y^3}{xy} = 3a$, where the constant a is isolated, so that it disappears by differentiation [Euler *Differentialis*, I, §289]; the arguments about arbitrary constants in the preface to [Euler *Integralis*, I] make no specific reference to differential equations; and arbitrary constants appear casually in the first

---

[82]"known by the Geometers"

[83]The original French being "équations derivées", "derived equations" might be a better translation. But a nicer rendition in English of this distinction would be achieved by calling "derived equations" those obtained by deriving a primitive equation and "derivative equations" those obtained by combining the former.

[84]In [Lagrange *Calcul*, 112] he abandoned the distinction in terminology, calling both kinds of equations either "prime equations" or "first-order derivative equations", etc.

chapters of the second section of [Euler *Integralis*, I] (on differential equations) because the methods used (separation of variables, integrating factors) involved integration of explicit functions. Euler's tendency for analogy between integrating explicit functions and solving differential equations has already been noted twice (sections 5.1.1 and 6.1.2.1).

But did [Lagrange *1774*] really introduce this new conception of formation of differential equations? The idea of conceiving a differential equation as the result of the elimination of arbitrary constants between a finite equation and its differentials can already be found (probably for the first time) in [Fontaine *1764*, 84-85] (in a memoir which according to Fontaine was submitted to the Paris Academy in 1748) – together with the argument we saw above used by Lagrange for the existence of two first-order integrals of a second-order differential equation; a similar one for the existence of three second-order integrals of a third-order differential equation; related ones for the uniqueness of the differential equation derived from a given finite equation and of the finite (complete) integral of a given differential equation of any order [Fontaine *1764*, 86-87]; and finally a claim for priority in these results (included in the table of contents). Fontaine was not concerned with singular solutions; rather, his purpose was the construction of tables of integrals of differential equations, for which he conducted a combinatorial study of possible forms for those equations and their solutions – in fact a study restricted to forms not involving transcendental functions [Gilain *1988*, 93]. Fontaine had a "difficult personality", his work was "of limited scope, often obscure, and willfully ignorant of the contributions of other mathematicians" [Taton *1972*, 54] and "keeping himself aloof, [he] published very little during the bulk of his career, waiting instead until 1764 to bind his unpublished manuscripts together with a few things that had appeared earlier, in the form of complete works"[85] [Greenberg *1981*, 252], so that one might assume that his views on the formation of differential equations were generally overlooked. However, as we will see below, Lacroix (section 6.2.1.1) and Cousin were aware of them, and Condorcet tried to expand on them. One wonders whether Lagrange might have been inspired by Fontaine's work, realizing the potential of that simple idea. And to what did Lagrange refer as "connu des Géometres"? The plain fact of existence of two first-order integrals; or Fontaine's argument? Be as it may, around 1800 there was some public acknowledgement of these ideas to Fontaine [Cousin *1777*, 183; *1796*, I, 196; Montucla & Lalande *1802*, 344].

It does seem very likely that [Lagrange *1774*] brought a much wider acceptance to this conception of formation of differential equations – not in the least because it used it to obtain results that were definitely non-trivial. However, that acceptance varied among the writers of textbooks and treatises in the late 18th

---

[85]Not thoroughly complete, as he published three memoirs in the volumes of the Paris Academy for 1767 and 1768 (thus after [Fontaine *1764*]). We may also notice the contradiction between the inclusion of "a few things that had appeared earlier", namely in the memoirs of the Paris Academy for 1734 and 1747, and the first title of [Fontaine *1764*], which mentions the unpublished character of the works contained within.

century. Bossut used it when presenting singular integrals [*1798*, II, 320-321], but ignored it elsewhere, to the point of arguing for the existence of two first-order integrals of a second-order equation simply by giving examples [*1798*, II, 266-267]. Cousin, on the other hand, not only followed Lagrange in using it for particular solutions [*1777*, 528-549; *1796*, II, 91-105] but also gave Fontaine's argument (crediting it to Fontaine) for the existence of n integrals of order $n - 1$ of any differential equation of order n, and for the uniqueness of its finite integral [*1777*, 181-183; *1796*, I, 194-196].

From 1764 onwards, Condorcet studied the integral calculus in a way much influenced by Fontaine. Like Fontaine, Condorcet tried to have a list of all the possible forms of integrals for each type of differential equation. This led him (as it had led Fontaine) to observe the formation of differential equations from finite equations by differentiation and elimination [Condorcet *1765*, 37-44, 67-69; Gilain *1988*, 91-95] – mainly elimination of transcendental functions, but taking the arbitrary constants with them; [Condorcet *1770*] is more focused on elimination of arbitrary elements (constants and functions).[86] While Condorcet did not share at all Fontaine's lack of social skills, he did share his obscurity of language when writing mathematics, so that his mathematical works are and always were difficult to follow – which was publicly noticed by his friend and admirer S. F. Lacroix in 1813 [Gilain *1988*, 88, 117]. Lacroix also decided not to mention in his *Traité* either of Fontaine's or Condorcet's "general methods of integration", because of their labouriousness [Lacroix *Traité*, II, 251]. However, we will see below that besides the full adoption of Fontaine's conception of the formation of differential equations, Lacroix also made use of some of Condorcet's reflections, namely on partial differential equations.

6.1.4.2 Partial differential equations

What about the formation of partial differential equations: what is the equivalent of Fontaine's conception of ordinary differential equations as the result of elimination of arbitrary constants? Given that in the traditional theory of partial differential equations, as exemplified in [Euler *Integralis*, III], arbitrary functions play a role entirely analogous to that of arbitrary constants for ordinary differential equations, one might expect to see partial differential equations regarded as the result of elimination of arbitrary functions.

But as we have already seen [Lagrange *1774*] takes a clearly different option: a first-order partial differential equation with two independent variables is the result of eliminating two *constants* between a finite equation and its two first-order partial differentials. This has serious consequences for the classification of

---

[86]Condorcet's researches would later evolve into a theory of integration in finite terms [Gilain *1988*], which remained mostly unpublished: his main work on this was a large treatise of integral calculus which was only partly printed (152 of what would be about 1000 pages [Gilain *1988*, 127]; according to Lacroix [*Traité*, 2nd ed, I, xxii-xxiii] those printed pages circulated at the time; but he was only able to study the whole manuscript in 1824 [Gilain *1988*, 110].

types of solutions: the finite equation involving two arbitrary constants is the *complete integral* of the differential equation; the *general integral* is obtained from the complete integral by establishing an arbitrary functional relation between the two constants and then eliminating the one which remains arbitrary – put $b = \phi(a)$ in the complete integral $V(x, y, z, a, b) = 0$, differentiate relative to a alone, and eliminate a (see section 6.1.2.3). The name "general integral" is justified in that it contains the complete integral: we can specify the arbitrary function included in the general integral by giving it a form involving two arbitrary constants and the result is a complete integral[87] (a byproduct of this argument is the conclusion that there are many different complete integrals for the same partial differential equation) [Lagrange *1774*, § 56]. However, since the general integral may be obtained from a complete integral through the process above, it appears that the latter is equally powerful – and it is possible to pass from one complete integral to another through the general integral (in practice, this argument is useless, because it is usually not possible to obtain the general integral *explicitly* from a complete integral – see below).

The historical literature on partial differential equations stresses this scheme as a very important point. For example, [Kline *1972*, II, 532]: "Lagrange's terminology, which is still current, must be noted first to understand his work" (followed by definitions of complete, general and singular integral); [Engelsman *1980*, 19-20]: "Euler's complete solution is characterized by an arbitrary function. [...] Lagrange's complete solution, on the other hand, is characterized by the occurrence of two arbitrary constants. [...] But far from being the final result itself, it is only an intermediary means for arriving at it. [...] Lagrange's new concept of a complete solution and the associated 'variation of constants' method provided a structure for the set of all solutions of a first-order partial differential equation"; [Demidov *1982*, 330]: "The origin of Lagrange's 'theory' [of first-order partial differential equations] is connected with his gradual approach to the new concept of a complete solution".

[Lagrange *Fonctions*, 99-100] is consistent with this: a primitive equation

$$F(x, y, z) = 0,$$

where z is regarded as a function of x and y, has two prime equations:

$$F'(x) + z' F'(z) = 0 \qquad \text{and} \qquad F'(y) + z_, F'(z) = 0$$

(i.e., $\frac{\partial F}{\partial x} + \frac{\partial z}{\partial x}\frac{\partial F}{\partial z} = 0$ and $\frac{\partial F}{\partial y} + \frac{\partial z}{\partial y}\frac{\partial F}{\partial z} = 0$). First-order derivative equations are obtained by combining these three equations in any way; as we have three equations, two constants may be eliminated, so that if we want to determine a function z from an equation in x, y, z, z' and $z_,$, "l'équation primitive entre x, y et z devra contenir deux constantes arbitraires"[88]. In the very next page Lagrange

---

[87]Later, he would deny this inclusion [Lagrange *Calcul*, 372-381]. See section 9.5.3 below.
[88]"the primitive equation between x, y and z must contain two arbitrary constants"

considers the possibility of one of the constants being a function of the other, and concludes that "l'équation primitive qui satisfait *en général* à une équation du premier ordre, doit renfermer une fonction arbitraire"[89] (emphasis added); the primitive equation with two arbitrary constants (i.e., the complete integral) is an intermediate step towards the more general one with an arbitrary function (the general integral), but it is enough to generate it, to generate the singular "primitive equation" (see section 6.1.2.3), and to generate the differential equation. Thus, it seems to occupy the central role in the structure of possible solutions.

However, there are a few problems with giving this scheme such an essential role in Lagrange's theory of partial differential equations, and more generally in the theory of partial differential equations of the late 18th century. The first problem is that other works by Lagrange are *not* consistent with it. After the introduction of this new scheme in [Lagrange *1774*], Lagrange [*1779*] reverted to a more traditional terminology, speaking of a complete integral as containing an arbitrary function. [Lagrange *1779*] is a memoir mainly on geometrical applications of singular integrals (see section 6.1.3.3), but without ever addressing the distinction between complete and general integrals; its fifth "article" has little to do with geometry, apart from some worked examples: it is Lagrange's presentation of his method for integrating (quasi-)linear first-order partial differential equations. Given the partial differential equation

$$\frac{\mathrm{d}z}{\mathrm{d}x} + P\frac{\mathrm{d}z}{\mathrm{d}y} + Q\frac{\mathrm{d}z}{\mathrm{d}t} + \&c. = Z,$$

where $P, Q, ..., Z$ are functions of $x, y, t, ..., z$, Lagrange forms the ordinary differential equations $\mathrm{d}y - P\mathrm{d}x = 0$, $\mathrm{d}t - Q\mathrm{d}x = 0$,..., $\mathrm{d}z - Z\mathrm{d}x = 0$, whose solutions have one arbitrary constant each; from those solutions, these constants can be expressed as functions of $x, y, t, ..., z$; doing this, and calling them $\alpha, \beta, \gamma, ...$, the equation

$$\alpha = \phi(\beta, \gamma, \&c.),$$

where ϕ is an arbitrary function, is an integral of the partial differential equation; "laquelle intégrale sera complette, puisqu'elle contient une fonction arbitraire"[90] [Lagrange *1779*, 153]. When some years later he gave a fuller proof of this fundamental method, he once again used the expression "complete integral" [Lagrange *1785*, §5]. In this particular context, the concept of an integral with arbitrary constants instead of arbitrary functions is in fact irrelevant.

It cannot be said to be entirely irrelevant in a different context: that of Lagrange's method to reduce the integration of a first-order partial differential equation with two independent variables x, y (and one dependent u) to the integration of a (quasi-)linear partial differential equation with an extra variable $p = \frac{\mathrm{d}u}{\mathrm{d}x}$ [Lagrange *1772b*]. Lagrange noticed that it would be enough to find a value for p

---

[89]"the primitive equation which satisfies *in general* a first-order equation must contain an arbitrary function"

[90]"this integral will be complete, as it contains an arbitrary function"

containing one arbitrary constant α; a procedure of variation of this constant α introduces the necessary arbitrary function [Lagrange *1772b*, §6]. At the end of the memoir, in a series of paragraphs unrelated to the method of (quasi-)linearization, Lagrange argues that such a procedure of variation of constants permits us to obtain a value of u with an arbitrary function from one with two arbitrary constants [Lagrange *1772b*, §9-11]. Engelsman [*1980*] correctly points this out as the origin of the new conception of "complete integral" in [Lagrange *1774*] – still, those paragraphs at the end are not related to the main topic of [Lagrange *1772b*]; a solution u with two arbitrary constants α, β does not occur in the (quasi-)linearization method.

It is interesting to look at [Legendre *1787*, 337-348], where first-order non-linear equations are examined. Legendre [*1787*, 337, 340] cites [Lagrange *1772b*] and [Lagrange *1774*] explicitly, and [Lagrange *1779*] implicitly. His version of the complete/general integral – arbitrary constants/function issue may be summarized thus: for an integral to be "complete" it must contain an arbitrary function; a "particular integral" (that is, one without an arbitrary function) which contains as many arbitrary constants as there are independent variables is usually enough to deduce the "complete integral" by variation of constants[91] [Legendre *1787*, 338-340]. It seems reasonable to assume that this was a common scheme (the most common?) by the end of the 18th century: it keeps Euler's terminology, but also acknowledges some importance to integrals with arbitrary constants instead of functions; however, it does not put them in the central place of the theory as [Lagrange *Fonctions*] would do; it also fails to address the issue of the formation of partial differential equations (i.e., should they be studied as the result of elimination of arbitrary constants, or of arbitrary functions?).

Similar schemes may be found in [Bossut *1798*, II, 356-358, 429-434] (integrals of partial differential equations are "completed" by arbitrary functions just like integrals of ordinary differential equations are "completed by arbitrary constants"; integrals of partial differential equations containing arbitrary constants instead of functions only appear very briefly when mentioning singular solutions) and [Cousin *1777*, 283, 702-710; *1796*, I, 253; II, 217-222] (complete integrals of partial differential equations include arbitrary functions; arbitrary constants only appear instead of arbitrary functions when mentioning particular solutions).

But there is a very important difference between these two *traités*. It was seen above that Bossut mostly ignored Fontaine's conception of the formation of ordinary differential equations (except when reporting Lagrange's theory of singular integrals), while Cousin used it in at least one occasion, citing Fontaine. Accordingly, Bossut [*1798*] does not address the formation of partial differential equations – except, insofar as it is necessary, in his very brief account of Lagrange's theory of singular integrals of partial differential equations [Bossut *1798*, II, 429-

---

[91] But not always: Legendre [*1787*, 340] gives two counter-examples in which the integrals thus obtained, although including an arbitrary function, are not as general as the "complete" one (because the functions involved have fewer arguments than the one in the "complete integral").

434].[92] Cousin, on the other hand, often uses the idea that a differential equation
is the result of eliminating an arbitrary function contained in its integral "de
l'ordre immédiatement inférieur"[93] [Cousin *1777*, 667;*1796*, II, 181]. To integrate
$M\frac{dz}{dy} + N\frac{dz}{dx} + Pz + Q = 0$ (where M, N, P and Q are functions of x and y),
he assumes that the "complete integral" has the form $z = \Pi + \Psi\, F{:}(\omega)$ (where
Π, Ψ and ω are unknown functions of x and y and F is the arbitrary function);
he differentiates with respect to x and y separately, eliminates $F{:}(\omega)$ and $F'{:}(\omega)$
between the three equations, and compares the result with the proposed equation
[Cousin *1777*, 295-296; *1796*, I, 260-261]. To integrate $M\frac{dz}{dy} + N\frac{dz}{dx} + V = 0$ (where
M and N are functions of x and y but V is a function of x, y and z), he gives
his own method, apparently submitted to the Paris Academy of Sciences in 1772,
which is similarly based on assuming the form $(B) + F{:}(\omega)$ for the integral ((B)
being a function of x, y and z), differentiating, and eliminating $F{:}(\omega)$ and $F'{:}(\omega)$
[Cousin *1777*, 629-632; *1796*, II, 157-158].

Thus it seems that Cousin extended Fontaine's conception of ordinary differential equations to partial differential equations, not in Lagrange's manner, but
rather according to the natural suggestion at the beginning of this section. This
turns out to be also a new form of Euler's analogy between arbitrary constants
and functions.

This was followed, in a more explicit way, by an important *rival* of Lagrange
in the study of partial differential equations in the late 18th century: Gaspard
Monge. It has been seen in section 6.1.3.4 that Monge gave much importance to
elimination of arbitrary functions. An example given was [Monge *1784a*], a memoir on the determination of equations for classes of surfaces, with an emphasis
on the elimination of the functions that particularize each surface in the class. The
memoir that appears right after this in the volume of memoirs of the Paris Science
Academy for 1784 is also by Monge, but on the integration of partial differential
equations [Monge *1784b*]. There Monge presents Lagrange's method for (quasi-)
linear first-order partial differential equations (which he seems to have developed
independently), and extends it to higher-order and nonlinear equations (this would
later be known as the "method of characteristics", after its geometrical interpretation in [Monge *Feuilles*]). This memoir starts precisely with the elimination of
the arbitrary function φ from

$$U = \varphi V,$$

resulting in

$$\left(\frac{dU}{dx}\right)\left(\frac{dV}{dy}\right) - \left(\frac{dU}{dy}\right)\left(\frac{dV}{dx}\right) = 0.$$

"C'est ce résultat nécessaire, exprimé en quantités différentielles, & délivré de la fonction arbitraire φ, que l'on nomme l'*équation aux dif-*

---

[92]Bossut [*1798*, II, 373-386] reports Lagrange's method for integrating (quasi-)linear first-
order partial differential equations, but not his method of quasi-linearization, which might have
motivated some reference to integrals with arbitrary constants instead of arbitrary functions.

[93]"of immediatly lower order"

férences partielles de la proposée, & dont celle-ci se nomme l'intégrale complète."[94] [Monge *1784b*, 120]

This conception is an important theme in this memoir. An example of its use is Monge's explanation for the nonlinearity of a partial differential equation as a consequence of either an arbitrary function being raised to a power higher than one in the complete integral, or of the arguments of an arbitrary function in the complete integral being given by a nonlinear auxiliary equation; if neither of these situations occur, then the elimination process produces a differential equation that is linear with regard to the highest-order differentials [Monge *1784b*, 164-168].

It was also mentioned above (section 6.1.3.4) that not much attention is paid in [Monge *Feuilles*] to solutions with arbitrary constants (i.e. *complete* solutions in the sense of [Lagrange *1774*]. Accordingly, also there the expression "complete integral" is used for solutions involving arbitrary functions [Monge *Feuilles*, n° 8-iii].

There are enough comparisons made by Monge between the roles of arbitrary constants in integrals of ordinary differential equations and arbitrary functions in integrals of partial differential equations [*1771*, 49; *1770-1773*, 16; *1784a*, 85-86] to assume that, like Cousin, he was extending Fontaine's conception of ordinary differential equations to partial differential equations,[95] in the way most natural to him.

Thus, we can say that Cousin and Monge's scheme is a more elaborate version of the one seen above used by Legendre (and Bossut), with a choice on the formation of partial differential equations: these are the result of elimination of arbitrary functions contained in the complete integral; solutions containing arbitrary constants instead of functions may be useful for particular purposes but are certainly not the central concept.

Condorcet also seems to have had such a scheme in mind. In [*1770*, 151-160], he studied the number of arbitrary constants or functions that may be eliminated between an equation and its differential(s); there he indicated (in a very unclear way) important differences between ordinary and partial differential equations, caused by the fact that partial differentiation of arbitrary functions introduces more unknowns than equations with which to eliminate them. Below we will see Lacroix's much clearer version of this.

---

[94]"It is this necessary result, expressed in differentials and free from the arbitrary function φ, that is called the *partial difference equation of the given [equation]*, and the latter is its *complete integral.*"

[95]We may also notice that Fontaine's conception is very clear in [Monge *1785b*], a memoir on ordinary differential equations.

6.2 Types of solutions of differential equations in Lacroix's *Traité*

6.2.1 Differential equations in two variables and their particular solutions

6.2.1.1 The formation of differential equations in two variables

It has been seen in section 6.1.4.1 that in the late 18th century the adherence to Fontaine's conception of the formation of differential equations varied from referring to it only when dealing with Lagrange's theory of singular solutions to making it a central piece in the presentation of differential equations. Lacroix was definitely a supporter of the latter approach. How relevant he thought it to be can be seen in a footnote signed by him included in [Montucla & Lalande *1802*, 344] (on Fontaine's priorities in the history of differential equations):

> "il ne faut pas oublier que l'on doit à Fontaine la manière d'envisager les équations différentielles comme le résultat de l'élimination des constantes arbitraires entre une équation primitive et ses différentielles immédiates. Cette remarque contient le germe de la théorie de toutes les espèces d'équations différentielles, ou aux différences, et sert de base à l'élégante théorie des *solutions* (ou *intégrales*) particulières, donnée en 1774, par Lagrange, dans les Mémoires de l'académie de Berlin"[96].

Indeed, traces of Fontaine's conception can be seen in Lacroix's *Traité* preceding the sections on particular solutions in volume II. It has been mentioned above (section 3.2.4) that Lacroix included in the first chapter of volume I a section "on differentiation of equations" [Lacroix *Traité*, I, 134-178], corresponding to part of chapter 9 of [Euler *Differentialis*, I]. In that chapter Euler had remarked on the possibility of using differentiation to remove constant, variable, irrational or transcendental quantities. Lacroix [*Traité*, I, 144-147] duly reports this, but with much less emphasis than Euler on the removal of non-constants;[97] and, significantly, he uses algebraic elimination of a constant between a primitive equation and its differential, instead of Euler's procedure of isolating the constant before differentiating (see section 6.1.4.1). Lacroix remarks that although the resulting equation is not the "immediate differential" of the primitive equation, it derives from it in such a way that it expresses the relation that must hold between x, y and $\frac{dy}{dx}$ [Lacroix *Traité*, I, 145].

---

[96]"it should not be forgotten that the manner of viewing differential equations as the result of elimination of arbitrary constants between a primitive equation and its immediate differentials is due to Fontaine. This remark contains the germ of the theory of all the types of differential or [finite] difference equations, and is the basis of the elegant theory of particular *solutions* (or *integrals*) given in 1774 by Lagrange in the Memoirs of the Berlin Academy"

[97]When he later pays more attention to elimination of functions, it is to eliminate arbitrary functions from equations in more than two variables (see section 6.2.2) – something *not* in [Euler *Differentialis*, I, ch. 9].

The chapter on plane geometry in [Lacroix *Traité*, I] is not terribly relevant here, because the theory of plane envelopes is much older than Lagrange's theory of singular solutions. But it is curious to note that just after explaining how to arrive at the equation of the envelope of a family of plane curves, Lacroix remarks that "le procédé par lequel on fait varier les constantes d'une équation, est un des grands moyens de l'Analyse"[98] [Lacroix *Traité*, I, 429-430].

A reference to Fontaine's conception of the formation of differential equations that may seem much more surprising is in volume II, when introducing the method of integrating factors. To explain this method, Lacroix reminds the reader that differential equations are not in general the "immediate result" of the differentiation of a primitive equation, but rather the result of the elimination of an arbitrary constant between such an equation and its "immediate differential" [Lacroix *Traité*, II, 230] – this includes a reference to the passage of the first volume cited above on elimination of constants, which reinforces the impression that in that passage Lacroix had intended to (subtly) prepare the reader for "la théorie de toutes les espèces d'équations différentielles", and especially for Lagrange's "élégante théorie des *solutions* particulières" (see quote above from [Montucla & Lalande *1802*]).

In case the primitive equation is in the form $u = c$, the elimination is immediate: $du = 0$; if in addition it is not divided by any factor, it remains an exact differential. But different situations may occur. Lacroix [*Traité*, II, 234] gives the example of first-degree equations, each of which, according to him, must be the result of the elimination of a constant c between a (primitive) equation of the form $P + cQ = 0$ (where P and Q are functions of x and y) and its differential ($dP + c\,dQ = 0$); this elimination yields

$$QdP - PdQ = 0;$$

however, if we first put $P + cQ = 0$ in the form $u = c$ (i.e., $\frac{P}{Q} = -c$), differentiating we arrive at

$$\frac{QdP - PdQ}{Q^2} = 0;$$

it is the disappearance of the factor $\frac{1}{Q^2}$, along with any possible common factor to QdP and PdQ, that may prevent $QdP - PdQ$ from being an exact differential.

This is quite an unusual explanation: given a differential equation $Pdx + Qdy = 0$, Euler had assumed its complete integral $V(x, y, a) = 0$; considered it put in the form $F(x, y) = a$; then differentiated, resulting in an exact differential equation $Mdx + Ndy = 0$ that must be equivalent to $Pdx + Qdy = 0$; and finally noticed that the equivalence implies that $\frac{P}{Q} = \frac{M}{N}$, i.e., $M = LP$ and $N = LQ$, for some L [Euler *Integralis*, I, §459]. Arguments very similar to Euler's were used by Cousin [*1777*, 198-199; *1796*, I, 204-205] and Bossut [*1798*, II, 124-125]. Bézout [*1796*, IV, 211] simply raised the possibility of making a differential exact

---

[98]"the procedure by which one makes the constants of an equation vary is one of the great methods of analysis"

through multiplication by a convenient factor, without any particular motivation. [Lagrange *Fonctions*] does not address integrating factors.[99]

Second- and higher-order differential equations also receive similar treatments before the study of their particular solutions. For instance, Lacroix reports Fontaine's (and Lagrange's) explanation for the fact that a second-order equation has two "first integrals" (that is, two first-order equations that satisfy it; the primitive equation is its "second integral"): if $U = 0$ is a primitive equation containing two arbitrary constants, c, c_1, and if it is differentiated twice, then a second-order differential equation $W = 0$ results from the elimination of c and c_1 between $U = 0, dU = 0$, and $d^2U = 0$; but there are two possible and distinct intermediate steps, namely either to eliminate c or c_1 between $U = 0$ and $dU = 0$, resulting in different first-order equations – which may be called $V = 0$ and $V_1 = 0$, respectively; both the elimination of c_1 between $V = 0$ and $dV = 0$ and that of c between $V_1 = 0$ and $dV_1 = 0$ will result in $W = 0$; therefore both $V = 0$ and $V_1 = 0$ are *first integrals* of $W = 0$, while $U = 0$ is its *second integral*; similarly a third-order equation has three first integrals and its corresponding primitive equation is its third integral (and an n-th order equation has n first integrals and its corresponding primitive equation is its n-th integral) [Lacroix *Traité*, II, 308-310; Fontaine *1764*, 87; Lagrange *1774*, §32].

Also integrating factors for second-order equations are explained by regarding these as the result of eliminating a constant between a first-order equation and its "immediate differential" – which may cause a factor to disappear [Lacroix *Traité*, II, 335].

6.2.1.2   Particular solutions of first-order differential equations in two variables

Obviously, Lacroix reports not only Fontaine's view on the formation of differential equations but also Lagrange's theory of singular solutions.

However, he adopts Laplace's terminology: "particular integrals" are particular cases of the complete integral; "particular solutions" are solutions not contained in the complete integral, whatever values one might give to the arbitrary constant.[100] In a footnote, Lacroix warns the reader about Lagrange's inverted use of these expressions,[101] and argues for his choice: those solutions which are not contained in the complete integral, "ne s'obtenant point par les procédés de l'intégration, ne doivent pas porter un nom qui rappelle ces procédés"[102] [*Traité*, II, 263]. An

---

[99][Lagrange *Calcul*, 168-177] does, explaining their existence in a way similar to Lacroix's, although more detailed and generalized. But the first edition of [Lagrange *Calcul*] was first printed in 1801 [Grattan-Guinness *1990*, I, 196], three years after [Lacroix *Traité*, II].

[100]There is one detail related to this in which Lacroix's and Laplace's terminologies are different: Lacroix speaks of "complete integrals", while Laplace [*1772a*] spoke of "general integrals".

[101]Which is exaggerated: Lacroix incorrectly says that Lagrange called "particular solutions" the "différens cas de l'intégrale complète" ("several instances of the complete integral") – Lagrange had used the term "incomplete integrals" (see section 6.1.1).

[102]"not being obtained by the procedures of integration, should not bear a name which reminds of these procedures"

argument which Lacroix does *not* invoke, but which might have some weight, is that his choice is consistent with Euler's terminology, unlike Lagrange's.[103]

It is interesting to note that an option which was available at least since the previous year in [Lagrange *Fonctions*, 69], namely "singular primitive equation" (or, adapting to the differential-integral language, "singular integral", or even "singular solution"), is not even mentioned – although material from [Lagrange *Fonctions*] (or at least from Lagrange's lectures at the *École Polytechnique*) is used in this section (see below). The question about why Lacroix ignored this terminology in the first edition of his *Traité* raises once again the issue of whether it was more dependent on [Lagrange *Fonctions*] or on Lagrange's 1795-1796 lectures at the *École Polytechnique*. One could speculate on whether Lagrange did use that terminology in those lectures – he could have introduced it only when writing the book, and Lacroix may have based the passage mentioned below on the lectures, not on the book; the corresponding passage from [Lagrange *Fonctions*] *is* cited in the table of contents [Lacroix *Traité*, II, vi] – but of course the table of contents is the last item to print. Another possibility (which does not exclude this one) is that the bulk of this section of Lacroix's *Traité* (and of its other sections dealing with singular solutions) was already advanced enough when Lacroix knew of this material by Lagrange, so that his choice of terminology was beyond a point of return – the passage inspired by either Lagrange's lectures or [Lagrange *Fonctions*] is quite independent of the rest and could well be a later insertion; not having a good reason to reject "singular solution" Lacroix might have preferred to omit the possibility – but that is not consistent with his *encyclopédiste* approach; besides he did mention Lagrange's new terminology in the second edition, stressing the analogy between "singular primitive equations" and "singular values" (i.e. non-analytic points) of a function [Lacroix *Traité*, 2nd ed, II, 373, 388].[104]

Besides terminology, another small influence from Laplace can be seen in a remark about a distinction to be made between trivial solutions (factors of the given differential equation which do not involve either dx or dy; $\mu = 0$ trivially satisfies $\mu M dx + \mu N dy = 0$) and particular solutions properly speaking [Laplace *1772a*, 344; Lacroix *Traité*, II, 263]. Lacroix does not seem to have noticed Trembley's denial of this distinction (it is possible to transform the equation so that the singular solution appears as a factor) [Trembley *1790-91*, 10] – although he did cite and use that memoir by Trembley (see below).

Apart from these two influences from Laplace and some different examples, Lacroix [*Traité*, II, 263-274] follows closely [Lagrange *1774*, §3-20], that is, the theory of singular integrals of first-order ordinary differential equations: given a primitive equation $U = 0$ in the variables x, y and the constant c, the corresponding

---

[103] And to a minor extent Laplace's, as far as "general integral" goes.

[104] As will be seen below, singular solutions ("singular primitive equations") are introduced in [Lagrange *Fonctions*] in a way that associates them to failures in certain power series. But it should be remarked that the adjective "singular" seems to have been associated with failures in more general power-series expansions (non-analyticity, in modern terms) only in the second edition of [Lagrange *Fonctions*] (dated 1813), and only in the title of chapter 5 – not in its text.

differential equation $V = 0$ is the result of eliminating c between $U = 0$ and $\frac{dU}{dx}dx + \frac{dU}{dy}dy = 0$ (with a reference to the first volume); if this is put in the form $dy = p\,dx$, and if c is regarded no longer as a constant, but rather as a function of x, it will become $dy = p\,dx + q\,dc$; particular solutions are obtained by eliminating c between $q = 0$ and $U = 0$, in case $V = 0$ has particular solutions – otherwise this will result in particular integrals; particular integrals satisfy not only $\frac{dy}{dc} = 0$, but also $\frac{d^2y}{dx\,dc} = 0$, $\frac{d^3y}{dx^2dc} = 0$, etc., while particular solutions satisfy only a limited number of these; particular solutions may be obtained directly from $V = 0$ without access to the complete integral $U = 0$ by putting $\frac{d^2y}{dx^2} = \frac{0}{0}$ or $\frac{d^2x}{dy^2} = \frac{0}{0}$.

The rest of the section on "particular solutions of [ordinary] first-order differential equations" [Lacroix *Traité*, II, 274-284] in fact oscillates between particular solutions and particular integrals. It is broadly dedicated to attempts to find complete integrals from particular solutions (which can be deduced directly from differential equations) and/or from particular integrals (which can sometimes be found from careful examination of differential equations).

Lacroix gives a couple of examples related to the Riccati equation $dy + y^2dx + X\,dx = 0$: if $y = Q$ is a particular integral, then $dQ + Q^2dx + X\,dx = 0$, so that $X\,dx = -dQ - Q^2dx$; $dy + y^2dx - dQ - Q^2dx = 0$ can be solved using an integrating factor.[105] But he remarks that the method involved usually leads to differential equations more difficult to solve than the original.

He then turns his attention to the possibility of using power series for this task. He does that in three articles [Lacroix *Traité*, II, 274-277] two of which are referred to in the subject index as "*Solutions* particulières, procédé de *Laplace*, pour les déterminer par le développement de l'intégrale en série"[106] [Lacroix *Traité*, III, 574]. This is an obvious reference to [Laplace *1772a*], where such series expansions do occur, although not with the purpose of "completing" particular integrals (see page 186 above).

But what Lacroix does here is much closer to (and in fact clearly drawn from) the section in [Lagrange *Fonctions*, 65-69] where singular solutions are introduced, and which is an adaptation of part of [Laplace *1772a*] and of [*Integralis*, I, § 565]. This was a somewhat unusual way of introducing singular solutions, but quite connected to the power-series foundation of the calculus: instead of presenting a few examples of "derivative equations" together with solutions not contained in their complete primitive equations, Lagrange had introduced singular solutions as exceptions to a power-series expansion – an expansion used precisely to "complete" particular primitive equations. In Lacroix's version: let $y = X$ be a particular integral of $dy = p\,dx$, and let $y = V$ represent the complete integral; X is then a function of x and V is a function of x and of an arbitrary constant c, such that $X(x) = V(x, c')$, for some appropriate value c'; thus, the complete integral may

---

[105]This is similar to an example in [Euler *Integralis*, I, §544].

[106]"*Particular* solutions, procedure by *Laplace* for their determination through the series expansion of the integral"

be expanded into

$$y = X + V'\frac{h}{1} + V''\frac{h^2}{1\cdot 2} + V'''\frac{h^3}{1\cdot 2\cdot 3} + \text{etc.}$$

where V', V'', V''', etc. are the values of $\frac{dV}{dc}, \frac{d^2V}{dc^2}, \frac{d^3V}{dc^3}$, etc. for $c = c'$; $h = c-c'$ plays here the role of arbitrary constant. Lagrange [*Fonctions*, 66-67] gives a method (and Lacroix [*Traité*, II, 275-276] reports it) for finding V', V'', V''', etc. using a related expansion for p:

$$P + P'\frac{Wh}{1} + P''\frac{W^2h^2}{1\cdot 2} + P'''\frac{W^3h^3}{1\cdot 2\cdot 3} + \text{etc.}$$

(P, P', P'', etc. are the values of $p, \frac{dp}{dy}, \frac{d^2p}{dy^2}$, etc. for $y = X$). But it had already been shown that there are cases in which series such as these are faulty for particular values of the variable (see section 3.2.5); in those cases the derivatives from some order upwards are infinite and the expansion must involve fractional exponents. An analysis of the more general expansions

$$p = P + Qk^m + Rk^n + \text{etc.} \quad \text{and} \quad y = X + qh + rh^\nu + \text{etc.}$$

leads to the conclusions that if $m < 1$, or equivalently if $y = X$ makes $P'(= \frac{dp}{dy})$ infinite, then the completion is not possible – it is not a particular integral, but rather a particular solution; this means that P' must have the form $\frac{K}{L}$, such that the particular solutions are factors of L [Lacroix *Traité*, II, 276-278]. These results are recognizable as Euler's ($m < 1$) and Laplace's; the characterization of singular solutions as solutions which cannot be completed can also be traced back to Euler [*Integralis*, I, § 565] – Lagrange did [*Calcul*, 237].

To finish the section, Lacroix addresses the relations between particular integrals or solutions and integrating factors, especially a method by Jean Trembley to find the latter from the former [Trembley *1790-91*]. Euler had noticed that, given a differential equation $Mdx + Ndy = 0$, firstly – if z is an integrating factor, then $z = 0$ is a particular integral, as long as it does not make either M or N infinite; and secondly – if $\frac{1}{z}$ is an integrating factor, then again $z = 0$ is a particular integral, as long as it does not make either $M = 0$ or $N = 0$ [*Integralis*, I, §572-574; Lacroix *Traité*, II, 278-279]. Laplace had also noticed that particular solutions make integrating factors infinite (see page 186 above) – they are factors of $z^{-1} = 0$. Trembley's idea was to search for an integrating factor by multiplying the known particular integrals and solutions of a given differential equation[107], each raised to an indeterminate power, and after substituting this product trying to solve for those powers.[108]

---

[107] More correctly, as [Lacroix *Traité*, II, 281-282] puts it: the functions which when equaled to zero yield those integrals/solutions.

[108] [Trembley *1790-91*] is not always very easy to follow: his uses of the expression "particular integrals" are particularly unhelpful (see section 6.1.1).

6.2.1.3   Particular solutions of second- or higher-order differential equations in two variables

Lacroix's explanation for the existence of particular solutions of second- or higher-order differential equations [Lacroix *Traité*, II, 408-409] is, just like Lagrange's, a generalization of the latter's explanation for first-order equations: if $U = 0$ is the complete integral of the second-order equation $V = 0$, then U contains two arbitrary constants, c_1 and c_2, and $V = 0$ is the result of eliminating c_1 and c_2 between

$$U = 0, \quad dU = 0, \quad \text{and} \quad d^2U = 0;$$

now, if c_1 and c_2 are taken as variables, in order to obtain the same results we need to have

$$\frac{dU}{dc_1}dc_1 + \frac{dU}{dc_2}dc_2 = 0$$

and

$$\frac{dU'}{dc_1}dc_1 + \frac{dU'}{dc_2}dc_2 = 0$$

(where, for the sake of abbreviation, $U' = \frac{dU}{dx}dx + \frac{dU}{dy}dy$, that is $U' = dU$ in the cases of c_1, c_2 constant or variable but verifying the first of these conditions); particular solutions are obtained by eliminating c_1, c_2, and $\frac{dc_2}{dc_1}$ between

$$U = 0, \quad U' = 0, \quad \frac{dU}{dc_1}dc_1 + \frac{dU}{dc_2}dc_2 = 0, \quad \text{and} \quad \frac{dU'}{dc_1}dc_1 + \frac{dU'}{dc_2}dc_2 = 0.$$

Nevertheless, the treatment of these particular solutions is mainly inspired by [Legendre *1790*],[109] although with some improvements. Legendre had based his approach on the remark that a singular integral, "reduced to finite form", always contains fewer arbitrary constants than the complete integral (that is, if we have a differential equation $V = 0$ of order n, and a singular solution $W = 0$, say of order $n - i$, the integral of $W = 0$ contains less than n arbitrary constants). Legendre proved this for orders 1 and 2 and claimed that the same reasoning applied to higher orders [Legendre *1790*, 222]. Lacroix, on the other hand, gave a proof for any order: let $V = 0$ be an n-th order differential equation, and let $U = 0$ be its complete integral, containing the arbitrary constants $c_1, c_2, \dots c_n$; $V = 0$ is obtained by eliminating these constants between $U = 0$ and its differentials $dU = 0, d^2U = 0, \dots d^nU = 0$; now suppose $c_1, c_2, \dots c_n$ vary, and let d' represent differentiation relative to them, so that the complete first differential of U is $dU +$

---

[109][Legendre *1790*] was only *published* in 1797, but it was already *printed* in 1794, along with the other memoirs in the Paris *Académie des Sciences'* volume for 1790 – the devaluation of banknotes had prevented its sale in the meanwhile. Given the facts that Lacroix uses this memoir both here and when dealing with particular solutions of partial differential equations (see below), in a volume published in 1798, and that Lacroix had been elected a correspondent of the *Académie* in 1789, it is very likely that he had access to the printed memoir while still unpublished.

$d'U$, where

$$dU = \frac{dU}{dx}dx + \frac{dU}{dy}dy \qquad \text{and}$$

$$d'U = \frac{dU}{dc_1}dc_1 + \frac{dU}{dc_2}dc_2 \ldots \ldots + \frac{dU}{dc_n}dc_n;$$

in order to still satisfy $V = 0$, we need to keep this first differential equal to dU, and in addition the second differential of U (which because of that condition is $d^2U + d'dU$) equal to d^2U, and so on up to the n-th differential of U ($d^nU + d'd^{n-1}U$) equal to d^nU; in other words, we need to have

$$d'U = 0, \quad d'dU = 0, \quad \ldots \quad d'd^{n-1}U = 0,$$

which thanks to the equality of mixed partial differentials can be transformed into

$$d'U = 0, \quad dd'U = 0, \quad \ldots, \quad d^{n-1}d'U = 0;$$

since no differentials of either x or y appear in $d'U$, this set of equations is at most of order $n-1$ relative to x and y, and so will be the result of eliminating the $2n-1$ constants $c_1, c_2, \ldots c_n, \frac{dc_2}{dc_1}, \frac{dc_3}{dc_1}, \ldots \frac{dc_n}{dc_1}$ between the $2n$ equations

$$U = 0, \quad dU = 0, \quad \ldots \quad d^{n-1}U = 0, \quad d'U = 0, \quad dd'U = 0, \quad \ldots \quad d^{n-1}d'U = 0;$$

the integral of this result (which is a particular solution) will therefore contain at most $n - 1$ arbitrary constants.[110]

    This is a smart proof, not only because of its actual generality, but especially because of the casual introduction of the operator d'. To apply this result Legendre had used calculus of variations, something which Lacroix cannot do here, since he is still more than 200 pages away from introducing that method. But this d' is as efficient here as the operator δ in [Legendre 1790].[111] Suppose that Y contains a number of constants (not more than n) and that $y = Y$ satisfies $V = 0$; if those constants are made to vary, $V = 0$ will become $V + d'V = 0$, whence $d'V = 0$, which is of the form

$$M\frac{d^n d'Y}{dx^n} + N\frac{d^{n-1}d'Y}{dx^{n-1}} \ldots \ldots + R\frac{d\,d'Y}{dx} + Sd'Y = 0. \qquad (6.25)$$

Now, if we try to integrate this equation in order to obtain $d'Y = \frac{dY}{dc_1}dc_1 + \frac{dY}{dc_2}dc_2 + \text{etc.}$, we may have two different situations: either $y = Y$ is contained in

---

[110][Houtain 1852, 1181] claims that Legendre's proof (and consequently Lacroix's) rests on a vicious circle. However, I believe that at least in the case of Lacroix the purpose of the proof is not to demonstrate that a singular solution contains less than n arbitrary constants (something which was taken for granted in the 18th century), but rather a simpler consequence: that the finite (or primitive) equation obtained from it (that is, its integral) contains less than n arbitrary constants.

[111][Lagrange 1779, 613-614], addressing singular integrals, introduces this operator using the symbol δ and then suddenly invokes the theory of variations (not the equality of mixed partial differentials) for $\delta dV = d\delta V$.

the complete integral, and thus $d'Y$ contains n arbitrary constants (which are the dc_i's), so that (6.25) is of order n and $M \neq 0$; or $y = Y$ is a particular solution, $d'Y$ must contain less than n arbitrary constants, the equation is of order at most $n - 1$, and therefore $M = 0$ [Legendre *1790*, 222-224; Lacroix *Traité*, 411-412].

Lacroix [*Traité*, II, 417] is less convincing about the correspondence between Legendre's and Lagrange's rules: he simply replaces d' with d to get

$$M \frac{d^{n+1}y}{dx^n} + N \frac{d^n y}{dx^{n-1}} \cdots\cdots + R \frac{d^2 y}{dx} + Sdy + Tdx = 0 \qquad (6.26)$$

whence

$$\frac{d^{n+1}y}{dx^{n+1}} = \frac{-N \frac{d^n y}{dx^n} \cdots\cdots - R \frac{d^2 y}{dx^2} - S \frac{dy}{dx} - T}{M}$$

and since $M = 0$ yields $N \frac{d^n y}{dx^{n-1}} \cdots\cdots + R \frac{d^2 y}{dx} + Sdy + Tdx = 0$, it also yields $\frac{d^{n+1}y}{dx^{n+1}} = \frac{0}{0}$, as Lagrange [*1774*, §35, §37] had indicated. The problem is that d'-differentiation is carried out holding x constant, unlike d-differentiation. So why should the coefficients in (6.25) and (6.26) be the same?

6.2.2   Complete and general integrals and particular solutions of partial differential equations

6.2.2.1   The formation of first-order partial differential equations and their general (and complete) integrals

Given what we saw in section 6.1.4.2, it is natural to ask how does Lacroix present the formation of first-order differential equations in three variables: as the result of elimination of two arbitrary constants between a primitive equation and its two immediate partial differentials like Lagrange [*1774*]; or as the result of elimination of one arbitrary function like Monge [*1784b*]? (In other words, how does he extend Fontaine's formation of ordinary differential equations to partial differential equations?) We will see that although the former possibility is mentioned, Lacroix is much closer to following the latter.

First of all, we may notice that Lacroix had a background of strict adherence to Monge's approach. In the memoir on partial differential equations that Lacroix had submitted to the Paris Academy in 1785 (see appendix A.1) he had expressed this very clearly: starting with the example $z = \varphi : (\alpha x + y)$, Lacroix eliminated $\varphi' : (\alpha x + y)$ between its two differentials $p = \varphi' : (\alpha x + y)\alpha$ and $q = \varphi' : (\alpha x + y)$, arriving at $p - \alpha q = 0$; he then remarked that

> "l'equation différentielle $p - \alpha q = 0$, ou toute autre, peut toujours être envisagée comme produite par l'élimination d'une fonction arbitraire. Cette methode est celle de M. Monge, et s'applique avec elegance aux équations linéaires de tous les ordres: c'est aussi celle dont nous nous

servirons à peu près dans la suite de ces recherches"[112] (see page 355).

The basis of the memoir was in fact an attempt to apply this approach to obtain solutions of non-linear partial differential equations. It was not a very successful attempt, and there seem to be no traces of the specific methods propounded there in his *Traité*; but some basic ideas of Mongean inspiration (formation of partial differential equations, their correspondence to families of surfaces) remain.

Returning to the *Traité*, let us look again at the section "on differentiation of equations" in the first chapter of [Lacroix *Traité*, I]. There Lacroix does allude briefly to the possibility, given an equation $u = 0$ in x, y and z, of eliminating two constants between

$$u = 0, \quad \frac{d(u)}{dx}, \quad \text{and} \quad \frac{d(u)}{dy},$$

the result expressing the relation between the variables x, y, z and the differential coefficients $\frac{dz}{dx}, \frac{dz}{dy}$ [Lacroix *Traité*, I, 176].[113] But he gives much more importance to the possibility of eliminating a function whose form is unknown [Lacroix *Traité*, I, 176-178]. For instance, if we have $z = f(ax+by)$, we can put $t = ax+by$, whence $z = f(t)$, so that

$$\frac{dz}{dx} = f'(t)\frac{dt}{dx} = f'(t) \cdot a \quad \text{and} \quad \frac{dz}{dy} = f'(t)\frac{dt}{dy} = f'(t) \cdot b\,;$$

now $f'(t)$ may be eliminated, yielding

$$b\frac{dz}{dx} - a\frac{dz}{dy} = 0,$$

a differential equation satisfied by $z = ax + by$, $z = \sqrt{ax + by}$, $z = \sin(ax + by)$, or any other equation of the form $z = f(ax + by)$. More generally, if $u = 0$ is an equation in x, y, z and an indeterminate function $f(t)$, where t is a known function of x, y, and z, then $f(t)$ and $f'(t)$ can be eliminated using

$$\frac{d(u)}{dx} = 0 \quad \text{and} \quad \frac{d(u)}{dy} = 0.$$

In the second volume, this latter passage on elimination of a function is referred to as showing that "les équations différentielles du premier ordre se déduis[ent] des équations primitives à trois variables, par l'élimination d'une fonction arbitraire"[114] [Lacroix *Traité*, II, 480], while there seems to be no reference to the former, on elimination of two variables.

---

[112]"the differential equation $p - \alpha q = 0$, or any other, may always be viewed as produced by the elimination of an arbitrary function. This is M. Monge's method, and it applies elegantly to linear equations of all orders: it is also pretty much the one we will use in the course of this research"

[113]For the notation $\frac{d(u)}{dx}$, see page 74.

[114]"first-order differential equation [are] derived from primitive equations in three variables by the elimination of an arbitrary function"

This is *why* arbitrary functions occur in solutions of first-order differential equations with two independent variables, but naturally it is not *how* they appear. Instead, as in [Euler *Integralis*, III, § 7, § 33], an arbitrary function appears when integration is performed holding one of those variables constant: the arbitrary constant thus introduced must be regarded as an arbitrary function of that variable [Lacroix *Traité*, II, 458, 477]. More interestingly, and similarly to [Euler *Integralis*, III, § 73, § 142], an arbitrary function also appears when integrating equations of the form

$$Pp + Qp = 0 \tag{6.27}$$

(where P and Q are functions of x and y, $p = \frac{dz}{dx}$, and $q = \frac{dz}{dy}$): this yields $dz = \frac{q}{P}(Pdy - Qdx)$; if μ is an integrating factor of $Pdy - Qdx$, we can put

$$\mu\, Pdy - \mu\, Qdx = dU;$$

and (since q is indeterminate) $\frac{q}{P\mu} = \varphi'(U)$, so that $dz = \varphi'(U)\, dU$ and therefore

$$z = \int \varphi'(U)\, dU = \varphi(U) \tag{6.28}$$

($\varphi'(U)$ and $\varphi(U)$ being arbitrary functions, subject only to the condition that the former is the derivative of the latter) [Lacroix *Traité*, II, 478-479]. Only after this latter appearance does Lacroix remind the reader of the passage in the first volume on the origin of first-order partial differential equations, establishing a connection between the eliminated function and the one introduced by integration [Lacroix *Traité*, II, 480]. But something similar had happened with ordinary differential equations: arbitrary constants appeared because the methods of solution resort to integration of explicit functions. It was not for *introducing* arbitrary constants that Lacroix invoked the formation of ordinary differential equations by their elimination (see section 6.2.1.1). Nevertheless, those references to the first volume do feel like theoretical explanations for the practical fact of the appearance of arbitrary elements.

Thus Lacroix seems to follow Cousin and Monge in keeping the old Eulerian analogy between arbitrary constants for ordinary differential equations and arbitrary functions for partial differential equations, putting it on the new ground of the formation of the equations by elimination. However, we will see below that Lacroix had very serious reserves about extending this analogy to equations of order higher than one, and that he did not follow Cousin and Monge in their use of the name "complete integrals" for integrals with an arbitrary function.

Lacroix also mentions the possibility of having integrals containing arbitrary constants instead of integrals containing an arbitrary function. He does so several times in the section dedicated to first-order partial differential equations [Lacroix *Traité*, II, 480, 489, 497-499, 516]. But integrals with an arbitrary function are clearly more important, and as we have seen above they seem to be the only ones involved in the formation of partial differential equations; when an integral with arbitrary constants appears it is always a means to obtain another one with an arbitrary function. Just after the reference to the first volume mentioned above, and still addressing equation (6.27), Lacroix notices that if one puts $\frac{q}{P_\mu} = a$, one obtains a result with two arbitrary constants, since this yields $dz = a\,dU$ and therefore

$$z = aU + b; \tag{6.29}$$

he finds "quite remarkable" that although this is obviously less general than the previous result (6.28), it is possible to restore (6.28) from (6.29): varying the constants a and b, we have $dz = a\,dU + U\,da + db$, which is equal to $a\,dU$ provided that $\frac{db}{da} = -U$; thus Lacroix puts $b = \psi(a)$, ψ being an arbitrary function; then $\psi'(a) = U$, whence $a = \psi,(U)$, where $\psi,$ is the inverse function of ψ'; therefore (6.29) becomes $z = U\psi,(U) + \psi[\psi,(U)]$; but $U\psi,(U) + \psi[\psi,(U)]$ is nothing more than an arbitrary function of U, and can be written as $z = \varphi(U)$ [Lacroix *Traité*, II, 480-481].

A similar argument is used for first-order partial differential equations with three independent variables: from the equation $V = aT + bU + c$ it is possible to obtain the more general one $V = \varphi(T, U)$ by varying the arbitrary constants a, b, c [Lacroix *Traité*, II, 489]. This is repeated and generalized when reporting the Lagrange-Charpit method for solving first-order partial differential equations in three variables [Lacroix *Traité*, II, 496-497] (after all, the idea of varying an arbitrary constant to obtain an arbitrary function had first appeared in [Lagrange *1772b*], included in the "first half" of the Lagrange-Charpit method – Lagrange's method for quasi-linearizing first-order partial differential equations). Since the elimination of the arbitrary constants is usually not feasible, general integrals are represented as systems of equations (from which the elimination is supposed to be done, even if only conceptually): if $Z = 0$ is an integral of $dz = p\,dx + q\,dy$ containing the arbitrary constants a and b, and (Z) designates the result of substituting $\varphi(a)$ for b in Z, then the general integral will be represented by

$$(Z) = 0, \qquad \frac{d(Z)}{da} = 0;$$

and analogously, if $Z = 0$ is an integral of a first-order partial differential equation in five variables containing the arbitrary constants a, b, c, e, and (Z) stands for Z with $\varphi(a, b, c)$ substituted for e, the general integral is represented by

$$(Z) = 0, \qquad \frac{d(Z)}{da} = 0, \qquad \frac{d(Z)}{db} = 0, \qquad \frac{d(Z)}{dc} = 0.$$

6.2.2.2 Terminology: "general" and "complete" integrals

Two issues related to our subject are very notably absent from the section on first-order partial differential equations in [Lacroix *Traité*, II]. One is particular solutions: they are addressed only later, together with the particular solutions of higher-order partial differential equations (see section 6.2.2.4).

More importantly, the issue of terminology is not addressed: Lacroix uses occasionally the expression "general integral" for an integral containing an arbitrary function, as opposed to one containing arbitrary constants [Lacroix *Traité*, II, 498, 501, 508, 516], but he never defines explicitly "general integral"; moreover, in this section he does not have any name for integrals containing arbitrary constants instead of arbitrary functions. This is well illustrated by the first occurrence of the expression "general integral":

> "En général, si $Z = 0$ désigne l'intégrale d'une équation différentielle partielle du premier ordre, entre m variables, et que Z renferme $m - 1$ constantes arbitraires, on en pourra tirer l'intégrale générale, qui doit contenir une fonction arbitraire de $m - 1$ quantités différentes."[115]
> [Lacroix *Traité*, II, 498]

Notice the awkward definite article applied to the integral containing arbitrary constants, which is evidently not unique (but Lacroix might claim illustrous antecedents: Lagrange [*1774*, § 57,59-61] repeatedly speaks of "l'intégrale complette" after having argued for the existence of several complete integrals [Lagrange *1774*, § 56]).

But the two most striking points here are on the one hand the use of "general integral" instead of "complete integral" (the expression which Cousin and Monge had used), and on the other hand the lack of conviction in that use.

Later, well into the section on partial differential equations of orders higher than 1, the name "complete integral" is used for integrals with arbitrary *constants*. That is, when Lacroix finally adopts a name distinction between types of integral according to the kind of arbitrary element involved, it is the Lagrangian nomenclature that he adopts (the occasional uses of "general integral" in the section on first-order partial differential equations are certainly only an anticipation of this distinction). A likely reason for this is that it was the *only* nomenclature available: the authors who used "complete integral" for integrals containing arbitrary functions did not have any name for integrals containing arbitrary constants. But even then Lacroix does not seem fully committed to this nomenclature. He introduces it saying that Lagrange uses the name "complete integral" to make a distinction from general integrals [Lacroix *Traité*, II, 555].

For someone who seemed to be so careful about terminology, all this is quite unsatisfactory. It would not have been very difficult to adapt the Laplacian terminology (see footnote 7), using "general integral" also for integrals of explicit functions or of ordinay differential equations containing the appropriate arbitrary constants (as well as for integrals of partial differential equations containing arbitrary functions), and to use the name "complete integral" only for integrals of partial differential equations containing arbitrary constants.

---

[115]"In general, if $Z = 0$ represents the integral of a first-order partial differential equation in m variables, and if $Z = 0$ contains $m - 1$ arbitrary constants, it will be possible to extract from it the general integral, which must contain an arbitrary function of $m - 1$ distinct quantities."

6.2.2.3 Complete and general integrals of second- and higher-order partial differential equations

The issues relating to types of solutions of partial differential equations are more thoroughly addressed in the section on "integration of partial differential equations of orders higher than one" [Lacroix *Traité*, II, 520-608].

It was mentioned above that Lacroix had reserves about the analogies between solutions of ordinary and partial differential equations; that can be seen in this section, where he exposes weaknesses in those analogies. For instance, he gives an example of a second-order equation

$$(x + y)(r - t) + 4p = 0$$

(Lacroix follows the usual conventions $dz = p\,dx + q\,dy$, $dp = r\,dx + s\,dy$, $dq = s\,dx + t\,dy$), which has only one first integral, namely

$$(x + y)(p - q) + 2z = \varphi(y - x), \tag{6.30}$$

instead of two as one would expect by analogy with second-order ordinary equations (see sections 6.1.4.1 and 6.2.1.1); nevertheless it has a second integral (i. e., a primitive equation) [Lacroix *Traité*, II, 534-535]:

$$z = e^{-\frac{2x}{a'}} \left(\int e^{\frac{2x}{a'}} \frac{dx}{a'} \varphi(a' - 2x) + \psi(a') \right) \tag{6.31}$$

(where a' is to be replaced by $x + y$ after the integration). Even stranger seems to be the equation

$$r - t - \frac{2p}{x} = 0 \tag{6.32}$$

[Lacroix *Traité*, II, 547-548]: it does not have any first integral, and yet it has a second integral:

$$z = \varphi(y + x) + \psi(y - x) - x[\varphi'(y + x) - \psi'(y - x)]. \tag{6.33}$$

The reason for the non-existence of first integrals is that it is impossible to eliminate any of the arbitrary functions φ, ψ (each together with its derivatives) between (6.33),

$$p = -x[\varphi''(y + x) + \psi''(y - x)], \tag{6.34}$$

and

$$q = -x[\varphi''(y + x) - \psi''(y - x)] + \varphi'(y + x) + \psi'(y - x); \tag{6.35}$$

while from

$$r = -[\varphi''(y + x) + \psi''(y - x)] - x[\varphi'''(y + x) - \psi'''(y - x)]$$

and

$$t = -x[\varphi'''(y + x) - \psi'''(y - x)] + \varphi''(y + x) + \psi''(y - x)$$

we have

$$r - t = 2[\varphi''(y + x) + \psi''(y - x)],$$

which together with (6.34) gives precisely (6.32) – that is, φ and ψ may be eliminated together, yielding the proposed differential equation, but not separately, which is what would provide the existence of first integrals.[116] Similarly, it is possible to eliminate ψ between (6.31) and its first-order differentials – thus arriving at (6.30) – but to eliminate φ it is necessary to use second-order differentials.

Lacroix's trigger for these reflections was almost certainly [Condorcet *1770*]. That is probably why Lacroix [*Traité*, II, 546] says that this issue, "l'un des plus importans de la théorie des équations différentielles partielles, n'a pas encore été suffisamment éclairci, du moins dans tous les traités qui ont paru jusqu'a ce jour"[117]: a likely allusion to [Condorcet *1770*], which is neither a treatise nor very clear.[118] In his "Compte rendu [...] des progrès que les mathématiques ont faits depuis 1789 [...]" (appendix B) Lacroix would repeat this claim for priority in publication, in a paragraph (page 400) that was not included in [Delambre *1810*].

Lacroix proceeds to clarify the issue, examining the possibility of a second-order partial differential equation being derived from a primitive equation with two arbitrary functions [Lacroix *Traité*, II, 549-553]: if $U = 0$ is a primitive equation in x, y and z, and if it is differentiated to the second order, we have six equations

$$U = 0, \qquad \frac{d(U)}{dx} = 0, \qquad \frac{d(U)}{dy} = 0,$$

$$\frac{d^2(U)}{dx^2} = 0, \qquad \frac{d^2(U)}{dx\,dy} = 0, \qquad \frac{d^2(U)}{dy^2} = 0,$$

(6.36)

so that in general only five quantities may be eliminated; however, if U includes two arbitrary functions $\varphi(t), \psi(u)$, these differentiations introduce four new quantities ($\varphi'(t)$, $\varphi''(t)$, $\psi'(u)$, and $\psi''(u)$), so that we have in total six quantities to eliminate. More generally, if we have a primitive equation with two independent variables and if the differentiations are carried up to order n, we get $\frac{(n+1)(n+2)}{2}$ equations; and if there are m arbitrary functions, each differentiation introduces m quantities, so that there are $m(n + 1)$ quantities to eliminate at order n; the conclusion is that in the worst case scenario it is necessary to have $m(n+1) < \frac{(n+1)(n+2)}{2}$, that is $n > 2m - 2$; in other words, the differentiations must be carried up to order $2m - 1$. In the case of three independent variables, we must have $\frac{(n+1)(n+2)(n+3)}{2} < (n - m + 1)(n - m + 2)(n - m + 3)$.

---

[116]It is possible to eliminate φ and ψ separately using the total differentials dp and dq, but these differentials are of second order, and so are the resulting equations, which are the closest one can have to first integrals [Lacroix *Traité*, 548-549].

[117]"one of the most important in the theory of partial differential equations, has not yet been clarified enough, at least in the treatises published so far"

[118]With a safeguard about the possibility that Condorcet's unpublished treatise might address the subject? In 1798 Lacroix might know its beginning (he knew it in 1810), but he certainly did not know yet the whole manuscript (see footnote 86 above).

Of course, in many situations there are *nice* peculiarities in the equations which allow for simultaneous eliminations, so that some lower order is sufficient. Lacroix [*Traité*, II, 552-555] examines in particular those situations in which the arbitrary functions have the same argument ($u = t$ above): holding that argument constant allows us to treat all the arbitrary functions as constants, which obviously simplifies the elimination procedure.

The fact that from (6.36) it is possible to eliminate five constants motivates the consideration of *complete integrals*, that is, integrals containing arbitrary constants instead of arbitrary functions (see also section 6.2.2.2). More precisely, in the case of two independent variables an n-th partial differential equation may result from the elimination of $\frac{(n+1)(n+2)}{2} - 1$ constants in a primitive equation.

Lacroix [*Traité*, II, 555-556] remarks that this does not solve all the difficulties with elimination, a remark that in fact goes back to [Lagrange *1774*, § 67], and which results from the *conclusion* that a first complete integral must contain two arbitrary constants, a second complete integral must contain five arbitrary constants (i.e., three more than a first complete integral), a third integral must contain nine arbitrary constants (four more than a second complete integral), and so on.[119] The trouble is that it is then necessary to be able to eliminate three constants to go from the second integral to a first integral (and worse, to eliminate four constants to go from the third integral to a second integral), which is generally not possible. Therefore, there are second-order partial differential equations that do not possess complete first integrals.

Naturally, the relationship between complete and general integrals of second-order partial differential equations is an extension of the relationship between primitive equations of first-order equations containing two arbitrary constants and those containing one arbitrary function (page 237 above). A general first integral is obtained from a complete first integral exactly in the same way, since as seen just above a complete first integral contains two arbitrary constants, and a general first integral contains one arbitrary function. As for second integrals, a complete one $U = 0$ contains five arbitrary constants, a, b, a', b', c', which may be regarded

---

[119]This must be because it would be undesirable to sever the ties between *first integral* and *single integration*, *second integral* and *double integration*, etc. Otherwise, the stress on *elimination* instead of *integration* might allow us to consider a first integral of a second-order differential equation in three variables (with primitive equation $U = 0$) as the result of eliminating two of the five constants in $U = 0$ using $\frac{\partial U}{\partial x} = 0$ and $\frac{\partial U}{\partial y} = 0$, so that it would contain *three* arbitrary constants; a second integral of a third-order differential equation in three variables would likewise contain *seven* arbitrary constants, and a first integral of such an equation would contain *four* arbitrary constants (because the six equations (6.36) would permit the elimination of five of the nine arbitrary constants in the primitive equation $U = 0$, or because the three second-order differentials of the primitive equation would permit the elimination of three constants from a second integral); and so on.

as variables as long as

$$\frac{dz}{da}da + \frac{dz}{db}db + \frac{dz}{da'}da' + \frac{dz}{db'}db' + \frac{dz}{dc'}dc' = 0,$$

$$\frac{dp}{da}da + \frac{dp}{db}db + \frac{dp}{da'}da' + \frac{dp}{db'}db' + \frac{dp}{dc'}dc' = 0,$$

$$\frac{dq}{da}da + \frac{dq}{db}db + \frac{dq}{da'}da' + \frac{dq}{db'}db' + \frac{dq}{dc'}dc' = 0;$$

this means that there are three equations to determine five (arbitrary) quantities, so that two of these may be regarded as (arbitrary) functions of the other three:

$$a = \varphi(a', b', c'), \qquad b = \psi(a', b', c');$$

in practice this is even more complicated than its first-order analogue, and too complicated to be useful [Lacroix *Traité*, II, 557-559; Lagrange *1774*, §65-66].

6.2.2.4   Particular solutions of partial differential equations

As has already been mentioned, Lacroix does not address particular solutions in the section on first-order partial differential equations. He only treats the issue (quite briefly) in the section on second- and higher-order equations [Lacroix *Traité*, II, 559-563]. This location seems much more a result of the late introduction of *complete integrals*, rather than some desire for generality: most of these nearly four pages are dedicated to particular solutions of *first-order* partial differential equations.

The order "theory of general/complete integrals" → "particular solutions" reflects (voluntarily?) the historical order: singular solutions of ordinary differential equations had appeared *spontaneously*, as a paradox to be solved (see section 6.1.2.1); while singular solutions of partial differential equations had appeared only in [Lagrange *1774*], not as a problem but rather as a consequence of the very theory which explained them. This is well expressed in Lacroix's introduction of them:

> "La théorie que nous venons d'exposer sur les intégrales des équati-
> ons différentielles partielles [intégrales complètes et intégrales générales],
> montre que ce genre d'équations a aussi ses solutions particulières"[120]
> [Lacroix *Traité*, II, 559].

Naturally, the presentation of these particular solutions is Lagrangian. If $U = 0$ (U containing two arbitrary constants a, b) is the complete integral of a first-order partial differential equation, according to Lacroix all the possibles ways of satisfying that given equation are comprised in the system

$$U = 0, \qquad \frac{dU}{da}da + \frac{dU}{db}db = 0.$$

---

[120]"The theory which we have just set forth on integrals of partial differential equations [complete and general integrals], shows that that kind of equations also have particular solutions"

The general integral is obtained by putting $b = \varphi(a)$ (and eliminating a) – see page 237; but one can also put

$$\frac{dU}{da} = 0, \quad \frac{dU}{db} = 0$$

and eliminate a and b: the result, containing no more arbitrary constants, is "the most particular" solution of the differential equation. Lacroix's single example [*Traité*, II, 560-561] is taken from [Lagrange *1774*, § 42].

But the procedure that Lacroix gives for obtaining particular solutions directly from the differential equations is taken from [Legendre *1790*]. It is of course a development of what he had given for second- and higher-order ordinary differential equations (see section 6.2.1.3). If the given partial differential equation is of first order, its d'-differential (that is, its differential relative to the arbitrary constants appearing in its complete integral[121]) is of the form

$$Pd'\frac{dz}{dx} + Qd'\frac{dz}{dy} + Rd'z = 0,$$

which can be transformed into

$$P\frac{dd'z}{dx} + Q\frac{dd'z}{dy} + Rd'z = 0,$$

a first-order partial differential equation in $d'z$; unless $P = 0$ and $Q = 0$, this equation implies an expression for $d'z$ containing an arbitrary function, which in turn implies a value for z *too general* for a particular integral; thus the particular solution is obtained by combining $P = 0$ and $Q = 0$ with the given partial differential equation [Lacroix *Traité*, II, 561-562].

Lacroix's treatment of particular solutions of second-order partial differential equations amounts to two short paragraphs [*Traité*, II, 561, 563] indicating generalizations of the theory and the procedure above.

6.2.3 Geometrical connections

6.2.3.1 Geometrical interpretation of particular solutions and complete integrals

It may be surprising at first to notice how little space Lacroix devotes in the second volume of his *Traité* to the geometrical interpretation of particular solutions and complete integrals of differential equations in two variables.

For first-order differential equations in two variables there is only § 608 [*Traité*, II, 305-307], half of which is occupied with an example: Euler's problem of the curves whose normals through a given point are all equal (see pages 183-184

---

[121] Legendre (who as already mentioned, called "complete integral" one with an arbitrary function) considered here instead the variation δ relative to the arbitrary function [Legendre *1790*, 235-236].

above), a problem which Lacroix [*Traité*, II, 260-261, 265] had already addressed simply as the equation $ydx - xdy = n\sqrt{dx^2 + dy^2}$, without any geometrical motivation. Now Lacroix notes that the singular solution is a circle tangent to the straight lines which comprise the complete integral, and remarks that this relation is general: particular solutions give envelopes of the curves corresponding to complete integrals. In fact, 1 – a differential equation provides information precisely about the direction of tangents, which are shared with the envelope; and 2 – the procedure for obtaining the equation of the envelope (given in chapter 4 of the first volume) is the same as that for obtaining the particular solution from the complete integral. Geometrical considerations also permit us to arrive at Lagrange's rule for obtaining particular solutions directly ($\frac{d^2y}{dx^2} = \frac{0}{0}$).

The geometrical interpretation of particular solutions of higher-order equations is mentioned even more succinctly [Lacroix *Traité*, II, 418]. The particular integral still belongs to a curve enveloping the curves of the complete integral, but with a higher order of contact (equal to the order of the equation).

Perhaps the reason for this conciseness is that Lacroix had already paid enough attention to envelopes of families of curves in the first volume [*Traité*, I, 427-434] (see the end of section 4.2.1.2). It is enough in the second volume to remark the connection.

Apart from the conciseness, it is interesting to notice the separation between the analytical and geometrical versions of the solutions of differential equations: the geometrical interpretation appears in the chapter on differential geometry, and in the section on geometrical construction of first-order differential equations, both clearly separated from the analytical development of the theory[122]. This separation is quite consistent with Lacroix's ideas about geometrical considerations as depictions of analytical procedures (pages 89 and 105).

Much more surprising than this conciseness is the *absence* of even a remark on the geometrical interpretation of particular solutions, complete integrals and general integrals of partial differential equations. The study of envelopes of families of surfaces in the first volume does not compensate the lack of geometrical versions for these concepts (which are not that simple to understand). The fact that in the second edition Lacroix supplied this interpretation [Lacroix *Traité*, 2nd ed, II, 682-685] supports the verdict that this absence is a *flaw* (a serious one) in the first edition.

6.2.3.2 Construction of differential equations in two variables

One of the sections in the chapter on integration of differential equations in two variables is entitled "De la construction géométrique des équations différentielles

---

[122]The case of higher-order equations is an exception: the analytical study of their particular solutions is accompanied by the very short mention of their geometrical interpretation; perhaps because the section in which this is included is assumedly a miscellany ("General reflections on differential equations and transcendents")

du premier ordre"[123] [Lacroix *Traité*, II, 296-307]. This section is clearly divided into three parts, the first being the only one effectively dedicated to construction of differential equations: the third part [Lacroix *Traité*, II, 305-307], on the geometrical interpretation of particular solutions, was already mentioned above; the second part [Lacroix *Traité*, II, 299-305] is dedicated to the problem of trajectories (given a one-parameter family a curves, to determine a curve that intersects all the given curves in a given angle), which seems to be essentially an example of a geometrical problem solved by integration of a differential equation.

As for the first part, it appears to have a mainly historical interest:

> "Dans les premiers tems on chercha à déterminer par les aires ou même par les arcs de quelques courbes connues, l'ordonnée de la courbe demandée ; depuis on a laissé ces constructions de côté, parce que, quelqu'élégantes qu'elles fussent dans la théorie, elles étoient toujours moins commodes et sur-tout moins exactes dans la pratique, que les formules approximatives qui ont pris leur place."[124] [Lacroix *Traité*, II, 296]

After remarking that usually ("en général") differential equations can only be constructed once their variables are separated,[125] Lacroix gives a construction of $\frac{y\,dx}{dy} = X$ (where X is a function of x) which requires the construction of the logarithmic curve[126] and the quadrature of $\frac{m^2}{X}$ (m is a constant which may be supposed equal to 1) [Lacroix *Traité*, II, 297-298]. This is a generalization (possibly by Lacroix) of Jacob Bernoulli's resolution [*1696*] of an already generalized version of de Beaune's problem: given a curve, to find another where the ratio of the subtangent to the ordinate is equal to the ratio of a constant line m to the sum or difference of the ordinates of the two curves, i. e., $\frac{dx}{dy} = \frac{m}{y \pm Q}$; Lacroix [*Traité*, II, 298-299] duly presents also this application.

In the table of contents Lacroix [*Traité*, II, vi] cites both [Jac. Bernoulli *1696*] and [Joh. Bernoulli *1694*] for this section, but does not use the matter of the latter (a method for constructing non-separable equations; see section 6.1.3.1). Three memoirs of Euler are also cited: two on orthogonal trajectories (the sources for the second part); and one on construction of differential equations using tractorial motion – something that Lacroix [*Traité*, II, 299] quickly dismisses, as being related to mechanics, rather than geometry.

Curiously enough, the most interesting constructions of differential equati-

---

[123]"On the geometrical construction of first-order differential equations"

[124]"In the early period [of the integral calculus] it was sought to determine the ordinate of the required curve by the areas or even by the arc-lengths of some known curves ; later these constructions were abandoned because, however elegant they might be in theory, they were always less convenient and especially less precise in practice than the approximation formulas which took their place."

[125]In a sentence added in the errata, Lacroix [*Traité*, II, 730] explains that this is why in the writings of the early analysts who dealt with integral calculus "to construct a differential equation" is often the same as to integrate it or to separate its variables.

[126]Lacroix suggests a construction by points, or the use of the asymptotic spaces of the hyperbola.

ons in two variables are not in this section. Rather, they occur in the sections on approximation of solutions of first- and second-order differential equations (see section 5.2.4), in awkwardly placed articles on the "possibility" of those equations [Lacroix *Traité*, II, 287, 351-352]. These constructions are geometrical counterparts (depictions) of Euler's "general method" for differential equations. A first-order differential equation gives for each point the value of $\frac{dy}{dx}$, that is, the slope of the tangent to the curve at that point; starting at a point M, one draws the

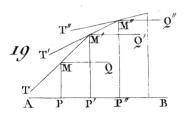

straight line TMM', such that the tangent of the angle $M'MQ$ (where $MQ//AP$) is the value of $\frac{dy}{dx}$ calculated using the abscissa AP and the ordinate PM; next one takes a point P' "infinitely close to P", and draws the straight line $T'M'M''$ in the same way; carrying this on one gets a polygon which will be as closer to the desired curve as the more sides it has. Lacroix concludes from this construction not only that all first-order equations in two variables are "possible" (a conclusion drawn also from the analytical version of the method) but also that each differential equation represents an infinity of curves, since the point M is taken at will.

In the case of a second-order equation, only the second-order coefficient $\frac{d^2y}{dx^2}$ is determined; this means that the terms of the approximating series are of degree at least 2, namely of the form $Y_1 = Y + Y't + Y''\frac{t^2}{2}$. Thus, instead of having tangent straight lines one has osculating parabolas. Also, the first parabola $MM'N$ has two arbitrary elements instead of one, so that in order to draw one needs to fix not only M but also either another point in the parabola or the slope of its tangent at M (i.e., the value of Y); next one takes an "infinitely close" point

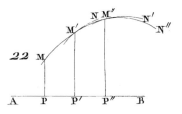

P', which determines the values of $Y_1 = P'M'$ and $Y_1' = Y' + Y''t$ (where $t = PP'$), and therefore the second parabola $M'M''N'$; naturally the process is carried on, and the curve obtained by assemblage of the parabolas will be the closer

to the required curve as the points P, P', P'', etc. are closer to each other.[127] Arguing that it is easy to extend this construction to any order, Lacroix [*Traité*, II, 352] concludes that differential equations in two variables, "qui sont toujours possibles"[128], represent an infinity of curves.

What is the purpose of these constructions? Certainly not historical, like that of Bernoulli's mentioned above. Possibly practical: providing graphical approximations. But the text suggests only theoretical purposes: showing the possibility and infinity of solutions. Their location also suggests purposes similar to those of the geometrical illustration of Euler's "general method" for approximation of explicit functions – only much less developed, as the method is much less developed for differential equations; and the purpose of that was clearly theoretical (sections 5.2.2-5.2.3).[129] Another very likely purpose is that of preparing the reader for the construction of partial differential equations.

6.2.3.3 Construction of differential equations in three variables and arbitrary functions

Chapter 4 of the second volume includes a section "on the geometrical construction of partial differential equations, and on the determination of the arbitrary functions contained in their integrals" [Lacroix *Traité*, II, 608-624]. Naturally, Monge and Arbogast are the main influences (more specifically, according to the table of contents, the memoirs [Monge *1770-1773*; *1773a*] and the dissertation [Arbogast *1791*]); but that of Clairaut [*1740*] is also very clear ([Clairaut *1740*] appears in the table of contents for the first section of the same chapter).

The first construction presented by Lacroix [*Traité*, II, 608-609] is an analogue of the construction of first-order differential equations in two variables based on Euler's general approximation method; we might say it combines that construction with the vertical-section approach present in [Clairaut *1740*] and several of Arbogast's constructions. Given a first-order partial differential equation in three variables $V = 0$, Lacroix considers the value of $\frac{dz}{dy}$ as a function of x, y, z, and $\frac{dz}{dx}$, which are indeterminate; he then takes an arbitrary curve XMm on a plane parallel to the x, z plane BAD, and regards it as a section of the solution surface (along which, of course, y is constant and z and $\frac{dz}{dx}$ are functions of x); for each point M (or m) of that section he draws a straight line MN (resp. mn) on a plane parallel to the y, z plane CAD, having as slope the corresponding value of $\frac{dz}{dy}$; then a plane xNn, parallel to XMm and very close to it, will intersect these straight

---

[127] An alternative construction, unrelated to Euler's "general method" and yielding a polygon instead of an assemblage of parabolas, is relegated to a footnote.

[128] "which are always possible"

[129] In the second edition Lacroix is more direct in dismissing any usefulness of these constructions for approximation, and in explaining that they serve to prove the "reality" of differential equations (see section 9.5.3); in the second edition he also seems less convinced of the practical usefulness of the analytical version of Euler's "general method" for approximating differential equations (see section 9.4.2). The third and later editions of [Lacroix *1802a*] also suggest non-approximative purposes (see section 8.8.2).

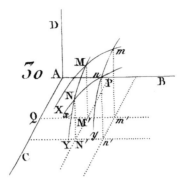

lines in points N, n which may be regarded as belonging to the surface, since the closer the two planes XMm, xNn the less xNn will differ from the section "parallel and consecutive" to XMm; carrying this on, one obtains the desired surface. Thus, concludes Lacroix, the first section XMm is in fact entirely arbitrary, and possibly not even continuous (see below).

Lacroix also uses a similar construction to show the difference in indeterminacy between partial and total differential equations[130]. If we have one of the latter, the differential coefficients $\frac{dz}{dx}$ and $\frac{dz}{dy}$ will both be given, independently of each other, and only a first point M (not a first section XMm) may be taken arbitrarily: the differential equation $\frac{dz}{dx} = p$ allows us to construct the point m, and the equation $\frac{dz}{dy} = q$ to construct the point N; then the point n may be constructed using the former equation, starting at N, or the latter equation, starting at m. For both constructions to give the *same* point n one needs an additional condition (which amounts to the condition of integrability of the original total differential equation): $\frac{dp}{dy} = \frac{dq}{dx}$ (cf. with [Clairaut *1740*], section 6.1.3.1). But this is a parenthesis in the section – the rest of it is entirely dedicated to partial differential equations.

These two constructions are in a certain sense the only constructions of differential equations in this section; true, Lacroix presents a few more constructions, but of *integrals* of differential equations – with some *proofs* that the constructed surface satisfies the respective equation.

The first of these is the construction [Lacroix *Traité*, II, 610-611] of the integral of $Pp + Qq = 0$ (where P and Q are functions of x and y only) – namely $z = \varphi(U)$ (where U is a function of x and y such that $dU = \mu P\,dy - \mu Q\,dx$, and φ is an arbitrary function). This construction had appeared as "Problem I" in [Monge *1773a*, 269-271]. It is a point-wise construction (i.e., for each point M' on the x, y plane BAC, or equivalently for each set of x, y coordinates, we wish to find the z-ordinate $M'M$ of the corresponding point M of the surface). Of course

---

[130]That is, total differential equations *that satisfy the integrability condition*: Lacroix assumes that their construction, like that of partial differential equations, results in a surface

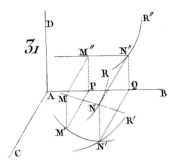

the surface is indeterminate, unless we force it to pass through a given curve NR (with projections $N'R'$ and $N''R''$). The basic idea is that U constant makes z constant: if we draw a curve $M'N'$ of equation $U = a$ on the x, y plane, and if we intersect the cylindrical surface raised on it with the desired surface, we get a curve MN[131] of constant z-ordinate $z = b(= \varphi(a))$; the value of b may be easily obtained by intersecting the curve $M'N'$ with the projection $N'R'$ of NR, and inspecting the z-ordinate QN'' of the intersection N; $M'M$ will be equal to QN''.

The *proof* [Lacroix *Traité*, II, 611-612] that the surface thus constructed effectively satisfies the equation $Pp + Qq = 0$ is also taken from Monge [*1773a*, 271-272]: consider the tangents MX' and MY' to the sections through M parallel to the x, z plane and to the y, z plane, respectively; then $M'X' = \frac{z}{p}$ and $M'Y' = \frac{z}{q}$; consider also $M'N'$ and MN as above; the "element" Mn of MN is in the tangent plane $X'MY'$, and because MN is parallel to the plane BAC, Mn is also parallel to the intersection $X'Y'$ of BAC with $X'MY'$; therefore, $M'n'$ is also parallel to $Y'X'$ and $M'm' : m'n' :: M'Y' : M'X'$; now, if $m'n'$ is dx, then $M'm'$ is $-dy$,[132]

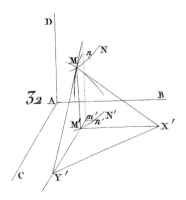

---

[131] The figure is misleading, as MN is not necessarily straight.
[132] Lacroix says that $M'm'$ is dy, keeps all terms of the proportion apparently positive, and only

taken along the curve $M'N'$ of equation $U = a$ (and therefore $P\,dy - Q\,dx = 0$); combining the latter with $dy = -\frac{p\,dx}{q}$ taken from the proportion above gives $Pp + Qq = 0$, as required.

The construction [Lacroix *Traité*, II, 612-613] of the integral $V = \varphi(U)$ of the "general equation" $P\,p + Q\,q = R$ (P, Q, R, U and V are functions of x, y and z) is also taken from Monge [*1773a*, 285-288]. It is also based on the idea that $U = a$ constant makes $V = b$ constant; but it is more complicated, particularly so because the intersection of $U = a$ with $V = b$ must also intersect a given curve NR (through which the constructed surface is supposed to pass). The construction given by Arbogast [*1791*, 30-33] was much simpler, because Arbogast disregarded this condition: the required surface was simply the "continued intersection" of the surfaces of equations $U = a$ and $V = b$. Lacroix does not attempt to report Monge's *proof* [*1773a*, 291-293] of the validity of this construction; instead, he invokes its suitability to the conditions of the integral, and refers the reader to Monge's memoir.

One may wonder why does Lacroix give a construction (probably his own) for "any" first-order partial differential equation – directly from the equation – and then two constructions for less general equations (first-degree) – and which need their integrals? The reason is probably the same as why he prefers Monge's construction to Arbogast's: the *initial condition* should be as general as possible – that is, one should be able to determine the surface that satisfies the equation *and* passes through any given curve of double curvature; the *initial condition* in his construction of first-order equations is a plane curve, and therefore not general enough.

As we would expect, Lacroix uses these constructions also to argue for the admissability of discontinuous functions. When giving the first construction above he remarks that the first section XM is entirely arbitrary, and it is not even necessary for it to be subject to the law of continuity, that is, it does not have to happen "que toutes ses parties puissent être décrites par une même loi, ou dépendent de la même équation"[133] [Lacroix *Traité*, II, 609]. He had mentioned that the differential coefficient $\frac{dz}{dx}$ (which may appear in the equation, and therefore in the expression for $\frac{dz}{dy}$) represents the slope of the tangent to XM; but he seems completely unconcerned about whether $\frac{dz}{dx}$ exists or not in the case of discontinuous XM.

In another article [*Traité*, II, 610] which seems to be about the same issue of discontinuity, Lacroix discusses the equation $p = \mathrm{f}(x, y, z)$. If he were to follow exactly the construction he had just given (taking in account that now he cannot have an expression for $q = \frac{dz}{dy}$), Lacroix would start by fixing an arbitrary constant-x section MNY (see figure on page 248) and then construct the constant-y sections XMm, xNn; instead, Lacroix starts by particularizing a value PM' for y in the

---

argues about signs when reverting to fractional notation – that is, when abandoning geometry and turning to analysis.

[133]"that all its parts may be described by one single law, or depend on the same equation"

equation $\frac{dz}{dx} = f(x, y, z)$ and using it to construct the corresponding constant-y section XM (for which he must fix some arbitrary point); then he does the same with a very similar value PN' for y (for which he must also fix some other arbitrary point); and so on. The only real difference is that with this order it is clearer that the constant-x sections may be completely random, and quite discontinuous. Of course, the fact that $\frac{dz}{dy}$ does not occur in the equation is important for this. But that is not such a particular case as it may seem: Lacroix includes a footnote to say that the equation $Pp + Qq = R$ (that is, any quasi-linear first-order equation) may be reduced to this form, using a change of variables: it is reduced to $\frac{dz}{dx} = \frac{R}{P}$, if y is replaced by a new variable v such that $P\frac{dv}{dx} + Q\frac{dv}{dy} = 0$, which is of course possible[134]. This is a very interesting argument, but once again it overlooks the issue of the *existence* of q (assumed in the equation $Pp + Qq = R$) when the constant-x sections (which correspond to functions of y) are discontinuous.

Lacroix refers more directly to the controversy on discontinuous functions apropos of second-order partial differential equations [*Traité*, II, 618-620]. He gives a construction for the integral $z = \varphi(x) + \psi(y)$ of $\frac{d^2z}{dx\,dy} = 0$, and a *proof* that it satisfies the differential equation (which, predictably, assumes the existence of $\frac{d\varphi(x)}{dx}$[135]). The reason for the choice of this equation is that the vibrating-string equation $r = a^2 t$ is transformed into $\frac{d^2z}{du\,dv} = 0$ by putting $u = x + ay$ and $v = x - ay$ (so that its integral is $z = \varphi(x + ay) + \psi(x - ay)$). Lacroix addresses very quickly the controversy itself, mentioning that it opposed Euler to d'Alembert, but not giving any hint at all of d'Alembert's arguments (nor even of Euler's); he simply expresses his adherence to Arbogast's position, and to his reasonings, "analogues à ceux que je viens de rapporter"[136] – this is not the most *encyclopédiste* passage in Lacroix's *Traité*. Apparently he thought that the issue was settled, and the details were no longer relevant.[137]

6.2.4   Total differential equations not satisfying the conditions of integrability

6.2.4.1   The memoir of 1790

The first sign of interest shown by Lacroix on equations in three variables not satisfying the conditions of integrability appears in the final pages of the memoir that he submitted to the Paris Academy in August 1790 (see appendix A.2, particularly pages 391 ff.).

Naturally, Lacroix follows Monge's approach. He does so much more faithfully (much more geometrically) than he would later do in his *Traité*. However,

---

[134]Provided that P and Q are *well behaved*; but around 1800 they surely were well behaved: discontinuous functions were conceived of only in solutions, expressing initial conditions.

[135]It consists in verifying that $p = \frac{d\varphi(x)}{dx}$ does not vary with y, so that $\frac{d^2z}{dx\,dy} = \frac{d.p\,dx}{dy} = 0$.

[136]"analogous to those I have just reported"

[137]This changed a little in the second edition, because of Laplace (see section 9.5.4).

unlike Monge, he focuses mainly on first-order (quasi-)linear equations, elaborating on their geometrical interpretation: he assumes that any first-order (quasi-)linear ordinary differential equation in three variables is the result of the elimination of p and q between two partial differential equations; in good Mongean fashion, these partial differential equations represent families of surfaces, and in case there are surfaces common to these families, they satisfy the ordinary equation (which in turn satisfies the condition of integrability); in case there are no common surfaces, the ordinary equation represents the curves of contact between the surfaces of the two families. For higher-degree equations, in spite of Monge's results, Lacroix cannot give a full picture of the solutions; nor can he do it for higher-order equations, whose situation is even less clear.

The issue that Lacroix wants to address is the determination of the solutions of these equations that are algebraic. But he does not do much about it: Monge had related the integration of $Mdz + Pdx + Qdy = 0$ to those of $Mp + P = 0$ and $Mq + Q = 0$; Lacroix remarks that other systems of (quasi-)linear partial differential equations will do, as long as they produce $Mdz + Pdx + Qdy = 0$ by combination with $dz = p\,dx + q\,dy$, and that those other equations may be chosen so as to have algebraic integrals.

6.2.4.2   The analytical theory in Lacroix's *Traité*

Lacroix could not fail to treat these equations in his *Traité*. Already in the first volume, in the section on "differentiation of equations", he uses a couple of examples in which variables disappear by differentiation to remark that

> "il n'y a point d'équation différentielle qu'on puisse regarder comme réellement absurde ou insignifiante; il faut seulement entendre qu'une équation différentielle ne se rapporte pas toujours à une seule équation primitive, et que pour y satisfaire il faut en supposer plusieurs, qui quelquefois renfermeront de nouvelles variables."[138] [Lacroix *Traité*, I, 167].

He is more specific in the second volume, when addressing the conditions of integrability for equations in more than two variables [*Traité*, II, 457]: given a differential equation in three variables, we cannot always assume that one of the variables is a function of the others; but Monge has shown that those in which this does not happen are not absurd, rather they belong to an infinity of curves of double curvature, instead of curved surfaces.

Lacroix addresses then only those that do satisfy the conditions, leaving "total differential equations that do not satisfy the integrability conditions" for their own section [*Traité*, II, 624-654], the last one of the large chapter 4 of the second

---

[138]"there are absolutely no differential equations that we may regard as absurd or meaningless; it must simply be understood that a differential equation does not always refer to a single primitive equation, and to satisfy it we must assume several, which sometimes will contain new variables."

volume, on "integration of functions of two or more variables" (which is understandable, even if they do in fact usually refer to *two functions* of *one variable*). This section is roughly divided in two halves, in typical Lacroix fashion: in the first half he gives a purely analytical theory and in the second he gives the geometrical interpretation.

As with Paoli [*1792*], Lacroix's first idea is that if we have a differential equation of the form

$$Pdx + Qdy + Rdz = 0 \qquad (6.37)$$

which does not satisfy the integrability condition

$$P\frac{dR}{dy} - R\frac{dP}{dy} + R\frac{dQ}{dx} - Q\frac{dR}{dx} + Q\frac{dP}{dz} - P\frac{dQ}{dz} = 0, \qquad (6.38)$$

we can change it into a differential equation in two variables only (and thus necessarily integrable) by establishing some relation between x, y and z. For instance, $\frac{dz}{z-c} = \frac{x\,dx+y\,dy}{x(x-a)+y(y-b)}$ does not satisfy (6.38), unless $a = b = 0$; but if we put $y = x$, it becomes $\frac{dz}{z-c} = \frac{2dx}{2x-a-b}$, whose integral is $z - c = C(2x - a - b)$; thus, $\frac{dz}{z-c} = \frac{x\,dx+y\,dy}{x(x-a)+y(y-b)}$ is satisfied by the system

$$y = x, \qquad z - c = C(2x - a - b).$$

An interesting detail here is that Lacroix, unlike Condorcet, Monge, Paoli, or Nieuport, attributes this technique to Newton. In fact, Newton had given it as *the* solution to the "third case" (equations involving fluxions of three or more quantities) of the "second problem" (given an equation containing fluxions, to find the relation between their fluents) of his *Method of Fluxions* [Newton *Fluxions*, 83].

After remarking the serious inconvenience in this technique that one would need to perform a separate integration for each particular relation between x, y and z, Lacroix [*Traité*, II, 625-626] gives Monge's procedure for integrating (6.37), which introduces an arbitrary function in the solution (thus solving the inconvenience). Lacroix's version of this procedure is presented as an adaptation of the method for integrating equations that do satisfy (6.38) – which seems clearer and more natural than the version in [Monge *1784c*].

Naturally Lacroix refers to the problem of determining algebraic solutions. He would even mention this reference in his "Compte rendu [...] des progrès que les mathématiques ont faits depuis 1789 [...]" (appendix B, page 399). Nevertheless, it is only a short reference – one article [*Traité*, II, 626-627]. The most interesting point made is the possibility of choosing the argument of the arbitrary function.

After some remarks on equations in more than three variables, Lacroix proceeds to higher-degree equations. He reports Monge's first example $dz^2 = m^2(dx^2 + dy^2)$ [*1784c*, 506-509] (without the geometrical considerations) and its generalization $F\left(\frac{dx}{dz}, \frac{dy}{dz}\right) = 0$ [*1784c*, 515-516]. Here solutions with three arbitrary constants are obtained easily, and then used to obtain solutions with an arbitrary function by varying the constants[139].

This was certainly the inspiration for what is the core of this section: the analytical theory of the formation of differential equations in three variables that do not satisfy the conditions of integrability [Lacroix *Traité*, II, 634-638]. Lacroix was clearly proud of it: not only did he mention it in his "Compte rendu [...]" (appendix B, page 399), but he even published it in advance as [Lacroix *1798a*].

Of course, as with all other Fontaine-like theories of formation of particular types of differential equations, Lacroix starts with finite equations; since in this case the solutions are composed of two equations, he starts with two equations

$$v = 0 \qquad \text{and} \qquad v' = 0 \tag{6.39}$$

in three variables x, y, z. Now, in (6.39) any two variables are functions of the third one ("and of the constants that may be found" in there); so, Lacroix differentiates (6.39), putting $dy = p\,dx$ and $dz = q\,dx$, resulting in

$$\frac{dv}{dz}q + \frac{dv}{dy}p + \frac{dv}{dx} = 0 \qquad \text{and} \qquad \frac{dv'}{dz}q + \frac{dv'}{dy}p + \frac{dv'}{dx} = 0; \tag{6.40}$$

it is possible to eliminate three constants between (6.39) and (6.40), and the result of this elimination is a first-order differential equation $W = 0$, which does not satisfy the integrability conditions[140]. The equations $v = 0$ and $v' = 0$, containing three constants a, b, c, constitute the *complete integral* of $W = 0$. But, as always, there are other ways to satisfy $W = 0$: the quantities a, b, c may vary instead of being constants, as long as

$$\frac{dv}{da}da + \frac{dv}{db}db + \frac{dv}{dc}dc = 0 \qquad \text{and} \qquad \frac{dv'}{da}da + \frac{dv'}{db}db + \frac{dv'}{dc}dc = 0$$

(so as to keep (6.40)); there are twenty-five ways to satisfy these conditions, from the most particular

$$\frac{dv}{da} = 0, \quad \frac{dv}{db} = 0, \quad \frac{dv}{dc} = 0, \quad \frac{dv'}{da} = 0, \quad \frac{dv'}{db} = 0, \quad \frac{dv'}{dc} = 0 \tag{6.41}$$

to the most general

$$\frac{dv}{da}da + \frac{dv}{db}db + \frac{dv}{dc}dc = 0, \quad \frac{dv'}{da}da + \frac{dv'}{db}db + \frac{dv'}{dc}dc = 0; \tag{6.42}$$

(in [*Traité*, II, 635] – but not in [*1798a*] – Lacroix reports three other possibilities, such as $\frac{dv}{da} = 0$, $\frac{dv}{db} = 0$, $\frac{dv}{dc} = 0$, $\frac{dv'}{da} = 0$, $\frac{dv'}{db}db + \frac{dv'}{dc}dc = 0$).

---

[139]In Monge's version, the process of varying constants was geometrical: one was eliminated in order to obtain conical surfaces out of straight lines, another was put as a function of the last one in order to have the vertices follow a curve, and finally differentiation was performed relative to this last one in order to have the characteristics and the edge of regression.

[140]Well, does not *necessarily* satisfy them. Lacroix concedes later that under certain conditions $v = 0$ and $v' = 0$ may be reduced to a single equation, corresponding to a curved surface.

Presumably all these possibilities, except for (6.42), correspond to *particular solutions* (in different degrees of particularity); however, Lacroix only addresses the case in which the six equations (6.41) are compatible and additionally they reduce $v = 0$ and $v' = 0$ to a single equation – in this case we have a very remarkable "particular solution" belonging to a curved surface.

The *general integral* comes from (6.42): putting $b = \phi(a), c = \psi(a)$, we have instead of $W = 0$ the system

$$v = 0, \quad v' = 0, \quad \frac{dv}{da} + \frac{dv}{db}\phi'(a) + \frac{dv}{dc}\psi'(a) = 0, \quad \frac{dv'}{da} + \frac{dv'}{db}\phi'(a) + \frac{dv'}{dc}\psi'(a) = 0;$$

if one of the functions $\phi(a), \psi(a)$ can be eliminated along with its derivative, then we will have a system of three equations containing one arbitrary function – that is, a general integral.

It is compelling to compare this with Paoli's analytical theory. Not only [Paoli *1792*] appears in the table of contents for this section, Lacroix also cites it in the text [*Traité*, II, 629] – although not in direct relation to the theory of formation of the equations and their types of integrals. Both Paoli's and Lacroix's theories are based on what Lacroix called Lagrange's "general theory of integrals and particular solutions" (see page 399 below). But the similarities end there: in [Paoli *1792*] we see solutions with two arbitrary constants, while Lacroix's complete integrals have three arbitrary constants. Paoli's theory is much more practical, arising from an integration technique, and unconcerned with the *formation* of the equations; Lacroix's theory, with all its similarities to the formations of other types of differential equations, seems to arise from a desire for systematization. It is also clear from what we have seen above that the direct technical source for Lacroix's theory was Monge's work and not Paoli's.

Just after presenting the theory, Lacroix works out another example [*Traité*, II, 636-638]

$$(y\,dx - x\,dy)^2 + (z\,dx - x\,dz)^2 + (y\,dz - z\,dy)^2 = m^2(dx^2 + dy^2 + dz^2)$$

already addressed (geometrically) by Monge [*1784c*, 512-514]; Lacroix's complete integral corresponds to the immediate solution that occurs in Monge's example (the straight lines tangent to a certain sphere); in the end he manages to eliminate one of the arbitrary functions and arrive at a result in the form

$$U = 0, \quad \frac{dU}{da} = 0, \quad \frac{d^2U}{da^2} = 0,$$

where U contains the other arbitrary function.

Lacroix concedes that his theory carries the same practical difficulties as Lagrange's derivation of general integrals from complete integrals for equations of orders higher than 1 (see page 242 above). But for practical purposes there is Monge's "very remarkable correspondence", which Lacroix proceeds to report

[*Traité*, II, 638-643], between the general integral

$$U = 0, \quad \frac{dU}{da} = 0$$

of a first-order partial differential equation $V = 0$ and the general integral

$$U = 0, \quad \frac{dU}{da} = 0, \quad \frac{d^2U}{da^2} = 0$$

of a total differential equation $W = 0$ obtained by eliminating p (or q) between $V = 0$ and $dz = p\,dx + q\,dy$ and then q (or p) between the result $V' = 0$ and $\frac{dV'}{dq} = 0$ (or $\frac{dV'}{dp} = 0$).

6.2.4.3   Geometrical considerations in Lacroix's *Traité*

The articles on geometrical considerations do not bring anything new, being taken up mostly with examples.

The first example leads to the geometrical interpretation in Lacroix's memoir of 1790: if we take two families of surfaces represented by first-order partial differential equations, and combine them with $dz = p\,dx + q,dy$, we obtain a total differential equation; this equation represents the curves along which the surfaces of one family touch those of the other; if the equation satisfies the integrability conditions, then there is a series of surfaces common to the two surfaces, which contain the curves of contact [Lacroix *Traité*, II, 643-645].

The geometrical interpretation of the correspondence between partial and total differential equations is very short – little over half a page [Lacroix *Traité*, II, 649-650]; Lacroix shows succinctly that the procedure to go from $V = 0$ to $W = 0$ (see above) also leads from a "limit surface" (that is, an envelope) to its edge of regression.

Chapter 7

Aspects of differences and series

7.1 Indices

This section is concerned with the indexed (subscript) notation for sequences or series, as in $u_0 + u_1 + \ldots + u_n + \ldots$ This may seem a rather trivial subject, but the dedication of a section to it is justified for three reasons: 1 – the standard reference on the history of notations pays very little attention to it [Cajori *1928-1929*, II, 265-266], making it only a detail in the notations on finite differences, and seemingly giving priority to Lagrange in 1792 (while it had been used nearly twenty years before by Laplace); 2 – the question of "when the subscript notation arose" has been asked very recently, at the end of a paper that illustrates the importance of unifying terminologies and powerful notations [Sandifer *2007*, 299]; and finally 3 – its use by Lacroix has caused some confusion, its creation or its introduction in France being misattributed to him. Thus, Dhombres [*1986*, 156], quoting [Lacroix *Traité*, 2nd ed, I, 33], remarks that "c'est à cette occasion que Lacroix introduit la notation indexée $A_0 + A_1x + A_2x^2 + A_3x^3 + \ldots$"[1]. While Schubring [*2005*, 386] gives a lengthy footnote on the subject, which is worth quoting in full (citations of Lacroix have been adapted):

> "Standard French textbooks up to about 1800 do not give sequences of quantities or variables with a notation identifying the single term of a sequence as part of a generally labeled sequence, for example, a_3 as part of a sequence (a_n) with the general term a_n. Lagrange used letters in alphabetic order to label elements as part of a sequence, for example, the function terms in developing it into a series as P, Q, R, and so forth or coefficients with A, B, C, and so forth. With such an unspecific approach, he was not able to label the last term of a sequence or a general term. It is notable that Crelle shifted to indexed series in the

---

[1]"It is at this point that Lacroix introduces the indexed notation $A_0 + A_1x + A_2x^2 + A_3x^3 + \ldots$"

sections he added to his translation of the *Théorie des fonctions analytiques*, for example: $B_1, B_2 \ldots, B_n$ or P_1, P_2, P_3 with P_n as general term (Lagrange 1823, Vol. 2, 332 ff.). Lacroix had already used general indexed quantities $a_1, a_2 \ldots, a_n$ in both 1798 and 1802, but only in a narrowly restricted field of calculus: within integral calculus to operate with the sequence of approximate values in using approximation to determine integral values [Lacroix *Traité*, II, 135 ff.; *1802a*, 285 ff.]. Lacroix, who had studied the contemporary literature intensively, may have been encouraged to introduce this usage–even though very partial– by the publications of the German school of combinatorics, which used indexed quantities as one of their everyday tools."

We will see that both Dhombres and Schubring were mistaken.[2]

7.1.1 Indices from Leibniz to Laplace

It is a fact that in the 18th century the most common way of naming the coefficients in a power series, or the terms of a (finite or infinite) sequence, was to use the alphabetic order. Thus, for instance, Euler argued for the possibility of expanding any function of z in the form $A + Bz + Cz^2 + Dz^3 + \&c.$, or at least $Az^\alpha + Bz^\beta + Cz^\gamma + Dz^\delta + \&c.$ [*Introductio*, § 59].

Interestingly, a complaint about the insufficiency of letters occurs as early as [Leibniz *1700*, 208]:

> "literas Algebraicas indiscriminatim adhibitas non satis [sunt] utiles, quia ob vagam generalitatem suam non admonent mentem relationis, quam ex prima suppositione sua habent inter se invicem. Hinc ut nonnihil succurramus defectui, solemus interdum (inprimis cum multae adhibendæ sunt) in ordine earum subsidium quærere"[3].

"Their order" might be a reference to the alphabetic order, or it might be a reference to Leibniz's occasional use of numbers in labelling successive points in a construction (for instance, $1C, 2C, 3C$ – see page 174 above[4]). But Leibniz's proposal in [*1700*] was more radical: to use "fictitious" numbers instead of letters –

---

[2]It is not completely clear whether Dhombres, in the sentence quoted above, means that Lacroix *introduces* the indexed notation absolutely (as in, say, "in his first article on differential calculus, Leibniz introduces the d notation"), or only in the context of his book ("in [*1696*] l'Hôpital introduces differentials as infinitely small differences"). If the latter is the case, he was not mistaken. But the former seems much more likely (a few pages earlier he had remarked that Euler had not used that notation [Dhombres *1986*, 153]). In [*1988*, 19] Dhombres and Pensivy were more cautious, speaking only of Lacroix having diffused the modern indexed notation.

[3]"algebraic letters employed indiscriminately are not useful enough, as because of their vague generality they do not remind us of the mutual relation they hold from their introduction. Hence, in order to mitigate somewhat this defect, sometimes (especially when there are many [letters] to be employed) we seek aid in their order"

[4]Notice that these are not subscripts, but rather "old-style numerals". Transcribing to "lined numerals" (more common nowadays), we would have $1C, 2C, 3C$ – but Leibniz did *not* mean $1 \times C, 2 \times C, 3 \times C$.

that is, he put the indices not as subscripts, but in the place of the coefficients themselves, as in $Z = 101Y + 102Y^2 + 103Y^3 + 104Y^4 + 105^5 + \&c$ [*1700*, 207]. This allowed him to use determinant methods, but almost all of his work on this remained unpublished until recently [Knobloch *1994*, 767-769].

Lacroix [*Traité*, 2nd ed, I, xxviii-xxix] saw in [Leibniz *1700*] the inspiration for the German Combinatorial School[5]. But that group of mathematicians only flourished by the end of the 18th century (and only in Germany). Meanwhile, as has already been said, most authors relied on alphabetic order – but often, especially in the "theory of series"[6] and in the calculus of finite differences, they resorted to other notational devices. For instance, Stirling [*1730*, 3] combined the alphabetic order with the special letter T for a general term, and superscript roman numerals (which we now interpret as primes, or accents) for the following ones:

"Terminos seriei initiales designo literis Alphabeti initialibus A, B, C, D, *&c*. A est primus, B secundus, C tertius, & sic porro. Et Terminum quemvis in genere denoto literâ T, atque reliquos ordine succedentes eâdem literâ, adjunctis numeris *Romanis* I, II, III, IV, V, VI, VII, &c. distinctionis gratiâ. Ut si T sit decimus, erit T′ decimus primus, T″ decimus secundus, T‴ decimus tertius, & sic deinceps. Et in genere quicunque Terminus definitur per T, succedentes definientur universaliter per T′, T″, T‴, T^{iv}, &c."[7]

In [*Differentialis*] Euler also used superscript roman numerals, with a slightly different meaning, and printed more clearly as roman numerals: for instance, y being a function of x, he called $y^{\mathrm{I}}, y^{\mathrm{II}}, y^{\mathrm{III}}, y^{\mathrm{IV}}, y^{\mathrm{V}}, \ldots$ the results of substituting $x + \omega, x + 2\omega, x + 3\omega, x + 4\omega, x + 5\omega, \ldots$ for x [Euler *Differentialis*, I, §2].

In Euler's theory of series the notion of index was fundamental, but it did not correspond exactly to these roman numerals; instead of representing the changes in a variable, indices gave the place of a term in a series: "Indices seu exponentes in qualibet serie vocantur numeri, qui indicant quotus quisque terminus sit in ordine: sic, termini primi index erit 1, secundi 2, tertius 3, & ita porro."[8] [*Differentialis*,

---

[5]On the German Combinatorial School, see for example [Jahnke *1993*].

[6]That is, the study of finite or infinite sequences and summation of finite sums or infinite series.

[7]"I denote the initial terms of the series by the initial letters of the alphabet A, B, C, D, etc. A is the first, B the second, C the third, and so on. And I denote an arbitrary term generally by the letter T with the *Roman* numerals I, II, III, IV, V, VI, VII, etc. attached to distinguish them. Thus if T is the tenth term, then T' will be the eleventh , T'' will the twelfth, T''' will be the thirteenth, and so on. And in general, whatever term is defined by T, the succeeding ones will be defined universally by $T', T'', T''', T^{\mathrm{iv}}$, etc." [Stirling *1730*, Eng transl, 21]

[8]"The numbers that indicate the place of each term in order are called indices or exponents. Thus, the index of the first term is one, that of the second is 2, that of the third is 3, and so on."

I, §40]. For notation he often used tables such as

$$\text{INDICES}$$
$$1, \quad 2, \quad 3, \quad 4, \quad 5, \quad 6, \quad 7, \quad \text{\&c.}$$
$$\text{TERMS}$$
$$A, \quad B, \quad C, \quad D, \quad E, \quad F, \quad G, \quad \text{\&c.}$$

and the following example [*Differentialis*, I, §43] is telling of the lack of correspondence between the roman numerals and indices:

$$\text{INDICES}$$
$$1, \quad 2, \quad 3, \quad 4, \quad 5, \quad 6, \quad \text{\&c.}$$
$$\text{TERMS}$$
$$a, \quad a^{\mathrm{I}}, \quad a^{\mathrm{II}}, \quad a^{\mathrm{III}}, \quad a^{\mathrm{IV}}, \quad a^{\mathrm{V}}, \quad \text{\&c.}$$

Naturally, the "general term" of a series or sequence was a fundamental concept for Euler also. He defined it as a function of the index, not only in [*Differentialis*, I, §39], but as early as 1730 [Ferraro *1998*, 293]. And he *was* able to refer
to a general term or to the last term of a sequence, although with cumbersome
notations. For instance: in [*Differentialis*, II, §105] he explains that if we have a
series with general term y

$$
\begin{array}{ccccccccc}
1 & & 2 & & 3 & & 4 & \ldots\ldots & & x-1 & & x \\
a & + & b & + & c & + & d & + \ldots\ldots + & v & + & y
\end{array}
$$

and the term corresponding to the index 0 is A, then v is the general term of the
series

$$
\begin{array}{ccccccccc}
1 & & 2 & & 3 & & 4 & & 5 & \ldots\ldots & x \\
A & + & a & + & b & + & c & + & d & + \ldots\ldots + & v
\end{array}
$$

(and therefore $Sv = Sy - y + A$, S denoting sums); we have also seen on page
152 Euler's use of left-hand accents ($'x$, $'X$) for penultimate values in certain finite
sequences.

Sandifer [*2007*, 288-299] has surveyed Euler's notations for series and indices
– including those we have seen above, $S(1)$, $S(2)$, etc., $S(n)$ (but not $S(n +
1)$) in 1772, and [0], [1], etc., [n] (and [$n + 1$]) in 1773. Sandifer remarks that
Euler developed "several different *ad hoc* indicial notations", but that he "failed to
discover a powerful" one, so that his indicial notations "were not generally adopted
by the mathematical community".

However, I believe that one of Euler's notations is a direct ancestor of our
subscript indices, namely the roman numeral superscript notation. I have only
noticed one occasion in which Euler extends it to refer to a general term: as
$y^{\mathrm{I}}, y^{\mathrm{II}}, y^{\mathrm{III}}, \ldots$ result from substituting $x+\omega, x+2\omega, x+3\omega, \ldots$ for x, he wrote this
one time $y^{(n)}$ for the result of substituting $y+n\omega$ for x [*Differentialis*, I, §23]. But
as far as I know this is an isolated occurrence, and elsewhere Euler's notations for
general terms were entirely separate from the roman numeral superscript notation.

Lagrange [*1759b*], on the other hand, used a (cumbersome) version of Euler's notation to represent general terms; he wrote y^{I} for "le terme qui suit y dans la suite des y"[9], and also y^m for the general term ("the same as y"), m being the "number that denotes the place of the terms", and thus he used indifferently $y^{\mathrm{I}} = Ry + S$ or $y^{m+\mathrm{I}} = Ry^m + S$ [*1759b*, §3-4]. Of course, this invites confusion with exponentiation, which may *partially* explain why the editors of Lagrange's *Œuvres* substituted y_1, y_m, y_{m+1} for $y^{\mathrm{I}}, y^m, y^{m+\mathrm{I}}$ respectively[10] – but this substitution caused Dugac [*1983*, 181] to wrongly ascribe to [Lagrange *1759b*] the first use of the subscript index notation.

The next step in this evolution, resulting in the introduction of the modern subscript notation for indices, appears to be due to Laplace, in 1773. In [*1774*] (submitted in 1772 [Gillispie *1997*, 297]), he still used Lagrange's notation: "si φ exprime une fonction quelconque de x, et que l'on y substitue successivement au lieu de x, 1, 2, 3, &c. on formera une suite de termes dont je designe par y^x, celui qui répond au nombre x"[11]; for double sequences (φ being a function of x and n), of which his "recurro-recurrent" series are a particular case, he used ${}^n y^x$ [*1774*, 353].[12] But in his next memoir on finite difference equations [*1773a*] Laplace adopted a clearer notation, changing the right-hand superscripts into subscripts:

"j'imagine la suite

$$y_1, \ y_2, \ y_3, \ y_4, \ y_5 \ \cdots\cdots\cdots y_x, \ \&c.$$

formée suivant une loi [...] les nombres $1, 2, 3 \ldots x$, placés au bas des y, indiquent le rang qu'occupe l'y dans la suite, ou, ce qui revient au même, l'indice de la série"[13] [Laplace *1773a*, 39].

Laplace really needed a clearer notation in this memoir, not only because of the danger of confusion with exponentiation[14], but also because he wanted to play with indices in different ways: for instance, using ${}^1 H$, ${}^2 H$, ${}^3 H$ for different quantities that

---

[9]"the term that follows y in the sequence of the y's"

[10]They did the same for [Lagrange *1759c*]: compare (my emphases) "si l'*exposant* de y exprime toujours la place qui tient la particule qui parcourt l'espace y, en comptant depuis la premiére F" in [Lagrange *1759c*, 1st ed, 9], with "si l'*indice* de y exprime toujours la place qui tient la particule qui parcourt l'espace y, en comptant depuis la premiére F" in [Lagrange *Œuvres*, I, 55].

[11]"if φ expresses a function whatsoever of x, and if we substitute successively 1, 2, 3, &c. for x, we will form a sequence of terms in which I designate by y^x the one corresponding to the number x"

[12]In this matter, the editors of Laplace's *Œuvres Complètes* (which are *not* complete), were more faithful than those of Lagrange's – they kept these notations.

[13]"I imagine the sequence

$$y_1, \ y_2, \ y_3, \ y_4, \ y_5 \ \cdots\cdots\cdots y_x, \ \&c.$$

formed following a law [...] the numbers $1, 2, 3 \ldots x$, placed in the lower part of the y, indicate the rank occupied by the y in the sequence, or, equivalently, the index of the series"

[14]For instance, in [Laplace *1773a*, 57] we see p^x and ${}^1 p^x$, meaning p and ${}^1 p$ raised to the xth power.

might not have any relation (such as several particular integrals of a given equation [*1773a*, 46]), and $H_1, H_2, H_3 \ldots H_x$ for the terms of a sequence following some law [*1773a*, 41].

Indices were also an essential component of "generating functions", a tool that Laplace developed in [*1779*] and that was to be very important to him (namely being the analytical foundation for his *Théorie analytique des probabilités* [*1812*]).[15] If y_x is a function of x, then

$$ u = y_0 + y_1 \cdot t + y_2 \cdot t^2 + y_3 \cdot t^3 \ldots + y_x \cdot t^x + y_{x+1} \cdot t^{x+1} \ldots + y_\infty \cdot t^\infty $$

is the *generating function* of the variable y_x; and reciprocally, "la variable correspondante d'une fonction génératrice, est le coëfficient de t^x dans le développement de cette fonction suivant les puissances de t"[16] [Laplace *1779*, 211-212]. His first example is that if u is the generating function of y_x, then $u \cdot t^r$ is that of y_{x-r} – which should be enough to show the central role of index manipulation.

It must be remarked that, after Laplace had introduced the subscript notation, it was used by Lagrange for recurrent series / finite difference equations [*1775*; *1792-1793*]. True, he did not use it in [*Fonctions*] nor in [*Calcul*], where power series are fundamental. But the fact is that in these books he did not work with combinatorial properties of the indices of those power series. Therefore he could use what around 1800 was still simpler notations: alphabetical order, or accents similar to the superscript roman numerals used by Stirling – whence our prime or accent notation for derivatives.

It is also true that even in works on finite differences Laplace's notation was not universal. Bossut, in the introduction on finite differences to his treatise on the calculus, used only $_{,}x, x, x', x'', x''', x^{\mathrm{IV}}$ for successive values of the variable [*1798*, I, 7], and a traditional functional notation $\varphi{:}(x)$ when, addressing recurrent sequences, he felt the need for a general term (here indexed by x, of course) [*1798*, I, 76].

But in advanced (or non purely introductory) works a more systematic form of referring to general terms was required, leading to notations more or less equivalent to Laplace's. Prony wrote $z^{-\prime\prime}, z^{-\prime}, z^0$ (or z), $z', z'', z''', z^{\mathrm{IV}}, z^{\mathrm{V}}$ for successive terms, and $z^{(n)}$ (sometimes z^n) for the general term, as well as $z^{(n-1)}, z^{(n-2)}$, with obvious meanings [*1795a*, II, 1-2].[17] The Italian Anton Mario Lorgna, in his memoir [*1786-87*] developing the analogy between differentiation and exponentiation that had been proposed by Lagrange [*1772a*], wrote $y^{0\prime}, y^{1\prime}, y^{2\prime}$ &c. for the successive values of y, and $y^{\lambda\prime}$ for the general term [Lorgna *1786-87*, 412-413]. This notation was meant to keep a distinction, but also an analogy, with the powers $y^0, y^1, y^2, \ldots, y^\lambda$; he also wrote $d^{\lambda\prime}, \Delta^{\lambda\prime}$ for the iterated operators $d^\lambda, \Delta^\lambda$.

---

[15]Euler had already used generating functions, but Laplace "was perhaps the first to exploit fully" this concept [Goldstine *1977*, 127, 185].

[16]"the variable corresponding to a generating function is the coefficient of t^x in the expansion of that function in powers of t"

[17]Towards the final lectures, Prony also wrote $z_0, z_{\prime}, z_{\prime\prime}, \&c \ldots z_{(n)}$ [*1795a*, IV, 544].

7.1.2   Indices in Lacroix's *Traité*

It is clear enough from the previous section that contrary to Schubring's suggestion, Lacroix did not need German encouragement to use subscript indices. But
there is another mistake in the quotation from [Schubring *2005*, 386] given above:
that Lacroix used "general indexed quantities $a_1, a_2 \ldots, a_n$ in both 1798 and 1802,
but only in a narrowly restricted field of calculus: within integral calculus to operate with the sequence of approximate values in using approximation to determine
integral values" – that is, in his version of Euler's "general method" for approximate integration (see sections 5.2.2-5.2.4). It is quite true that Lacroix uses
subscript indices in that context – see for instance equation (5.8), page 160 above.
But this is very far from being the "only field" in which he uses them.

The first use of subscript indices in Lacroix's *Traité* (their introduction, according to Dhombres), is in the first volume, in the Introduction, for the expansion
in power series of a^x :

"Nous supposerons que a^x soit représenté par la série

$$A_0 + A_1 x + A_2 x^2 + A_3 x^3 + \text{etc.}$$

A_0, A_1, A_2 sont des coefficiens indépendans de x, et les chiffres inférieurs
$0, 1, 2$ etc. marquent l'exposant de la puissance de x qui multiplie la
lettre à laquelle ils sont attachés, ainsi A_m sera le coefficient de x^m. Ce
qui m'a déterminé à employer cette notation, quoiqu'elle paraisse un
peu compliquée, c'est que par son moyen il sera facile de découvrir la
loi qui régne entre les valeurs des coefficiens."[18] [Lacroix *Traité*, I, 33]

Lacroix makes effective use of this notation, not only in the expansion of the
exponential function, but also in those of the logarithm, cosine, and sine. For the
exponential function, he uses the functional equation $a^x \times a^u = a^{x+u}$, so that

$$(A_0+A_1 x+A_2 x^2+\text{etc.})\times(A_0+A_1 u+A_2 u^2+\text{etc.}) = A_0+A_1(x+u)+A_2(x+u)^2+\text{etc.}$$

Expanding the product on the right side and the powers on the left, and comparing
the coefficients, Lacroix concludes first that $A_0^2 = A_0$, whence $A_0 = 1$, and thus
the coefficients of x, x^2, x^3, etc. are A_1, A_2, A_3, etc., on both sides; next, analysing
the coefficients of u, ux, ux^2, etc., he sees that

$$A_1 = A_1, \; A_1 A_1 = 2A_2, \; A_1 A_2 = 3A_3, \text{ etc., and in general } A_1 A_{m-1} = mA_m,$$

---

[18]"We will suppose that a^x is represented by the series

$$A_0 + A_1 x + A_2 x^2 + A_3 x^3 + \text{etc.}$$

A_0, A_1, A_2 are coefficients independent of x, and the inferior numerals $0, 1, 2$ etc. mark the
exponent of the power of x that is multiplied by the letter to which they are attached; thus
A_m will be the coefficient of x^m. Although this notation appears a little complicated, I have
decided to employ it because by using it it will be easy to discover the law ruling the values of
the coefficients."

whence

$$A_1 = \frac{A_1}{1}, \; A_2 = \frac{A_1^2}{1 \cdot 2}, \; A_3 = \frac{A_1^3}{1 \cdot 2 \cdot 3}, \text{ etc., and in general } A_m = \frac{A_1^m}{1 \cdot 2 \cdot 3 \cdots m}$$

(A_1, which depends on a, is to be determined later). So far the indexed notation only makes this a little clearer. But Lacroix also needs to confirm that these values for the coefficients satisfy the rest of the equality, and that is where indices really make generalization easier: an arbitrary term from the left side is of the form

$$A_m A_n u^m x^n = \frac{A_1^m}{1 \cdot 2 \cdot 3 \cdots m} \times \frac{A_1^n}{1 \cdot 2 \cdot 3 \cdots n} u^m x^n = \frac{A_1^{m+n}}{1 \cdot 2 \cdots m \times 1 \cdot 2 \cdots n} u^m x^n;$$

now, on the right side, $u^m x^n$ obviously comes from $(x + u)^{m+n}$, and has as coefficient

$$A_{m+n} \frac{(m + n)(m + n - 1) \cdots (m + 1)}{1 \cdot 2 \cdot 3 \cdots n}$$

$$= \frac{A_1^{m+n}}{1 \cdot 2 \cdot 3 \cdots (m + n)} \times \frac{(m + n)(m + n - 1) \cdots (m + 1)}{1 \cdot 2 \cdot 3 \cdots n} = \frac{A_1^{m+n}}{1 \cdot 2 \cdots m \times 1 \cdot 2 \cdots n}$$

as above.[19]

Lacroix regarded this method as important enough to be mentioned in his *Compte rendu [...] des progrès que les mathématiques ont faits depuis 1789* (see appendix B, page 398). In the preface to the second edition of his *Traité*, he also stressed the advantages of his method, namely over those that used infinite or infinitely small quantities (he might have been thinking of [Euler *Introductio*]):

> "La méthode dont j'ai fait usage pour le développement des fonctions, ne s'appuie sur aucune considération de ce genre; aucun terme n'y est négligé; toutes les équations de condition y sont vérifiées en quelque nombre qu'elles soient, par un calcul fondé sur les indices des quantités à déterminer, et très-propre, je crois, à faire sentir les avantages de la symétrie dans les calculs, et la puissance d'une notation quand elle est analogue aux idées qu'elle représente."[20] [Lacroix *Traité*, 2nd ed, I, xix-xx]

---

[19] Around the same time, Fourier, in his lectures at the *École Polytechnique* [1796, 54-55], gave a similar proof for the expansion of a^x, with two differences: 1 – he did not use indices, but rather the alphabetical order $A, B, C, \ldots$; 2 – instead of $a^x \times a^u = a^{x+u}$ he used the property $a^{2x} = (a^x)^2$, which makes calculations much easier, and indices dispensable. In the second edition of his *Traité*, Lacroix mentioned this approach in a footnote, but he preferred $a^x \times a^u = a^{x+u}$ for being more general and expressing the most extensive definition of a^x [*Traité*, 2nd ed, I, 35].

[20] "The method which I used for the expansion of functions does not rely on any consideration of that kind; no term is neglected; all the equations of condition are verified, whatever their number, by a calculation based on the indices of the quantities to be determined, and which I believe to be very proper to make perceive the advantages of symmetry in calculations and how powerful is a notation that is analogous to the ideas for which it stands."

Now, Lacroix does not adopt the subscript index as default notation in the first two volumes of his *Traité*; most often, he keeps the use of alphabetic order for coefficients of series. Still, he does occasionally use subscript indices – probably in those occasions where they do seem useful, even if not terribly so. For instance, in the chapter on the principles of differential calculus (see section 3.2.2), deriving Taylor's theorem, where we find [*Traité*, I, 88]:

X_1, X_2, X_3, etc. for the coefficients in the expansion of the increment of f(x);
X_1', X_1'', X_1''', etc. for the coefficients in the expansion of the increment of X_1;
X_2', X_2'', X_2''', etc. for the coefficients in the expansion of the increment of X_2;
. . .

In precisely the same context, Lagrange had used

p, p', p'', etc. for the coefficients in the expansion of the increment of u;
π, ρ, σ, etc. for the coefficients in the expansion of the increment of p;
π', ρ', σ', etc. for the coefficients in the expansion of the increment of p';
. . .

– somewhat more cumbersome [Lagrange *1772a*, § 4].

The situation in the third volume is a little different. Subscript indices become much more frequent – which is natural, given that it was within the context of series and finite differences that they had appeared, and that this is a more combinatorial subject. In fact, the first numbered paragraph of the third volume starts with a *reintroduction* of indices:

"Supposons qu'on ait une série de la forme

$$A_0 + A_1 x + A_2 x^2 + A_3 x^3 + \text{ etc.}$$

dans laquelle les chiffres inférieurs affectés aux coefficiens des puissances de x, et que je nommerai *indices*, font connoître le rang qu'occupe chaque terme [. . .] si l'on avoit l'expression du terme général $A_n x^n$, qui répond à un indice quelconque, on en déduiroit tous les autres, en donnant à n différens valeurs"[21]. [Lacroix *Traité*, III, 2]

Unlike what this reintroductory example suggests, Lacroix usually abstains from writing 0 as a subscript. This sometimes results in ambiguity (probably intentional) between a variable x and its first value x_0 (in these cases we might see the variable x as distinct from its general value x_n) .

----

[21]"Suppose that we have a series of the form

$$A_0 + A_1 x + A_2 x^2 + A_3 x^3 + \text{ etc.}$$

in which the inferior numerals affected to the coefficients of the powers of x, and which I will call *indices*, display the rank occupied by each term [. . .] if we had the expression for the general term $A_n x^n$, corresponding to an arbitrary index, we would deduce all the others from it, giving different values to n"

Thus, given a sequence $u, u_1, u_2, u_3, \ldots$, the difference Δu is defined as $u_1 - u$; Δu_1 is defined as $u_2 - u_1$; and more generally Δu_{n-1} as $u_n - u_{n-1}$ (and naturally $\Delta^2 u = \Delta u_1 - \Delta u$, and so on). Some calculations follow, giving

$$u_n = u + \frac{n}{1}\Delta u + \frac{n(n-1)}{1\cdot 2}\Delta^2 u + \frac{n(n-1)(n-2)}{1\cdot 2\cdot 3}\Delta^3 u + \text{ etc.} \qquad (7.1)$$

and

$$\Delta^n u = u_n - \frac{n}{1}u_{n-1} + \frac{n(n-1)}{1\cdot 2}u_{n-2} - \frac{n(n-1)(n-2)}{1\cdot 2\cdot 3}u_{n-3} + \text{ etc.} \qquad (7.2)$$

(7.1) is found in [Euler *Differentialis*, § 22] – it is the single occurrence of $y^{(n)}$ for a general term, mentioned in the previous section; (7.2), which requires a systematic notation for general terms, appears in [Euler *Differentialis*, § 10] only as a set of examples, up to $\Delta^5 y = y^{\mathrm{V}} - 5y^{\mathrm{IV}} + 10y^{\mathrm{III}} - 10y^{\mathrm{II}} + 5y^{\mathrm{I}} - y$.

Lacroix [*Traité*, III, 6] also presents (7.1) and (7.2) as the symbolic expressions

$$u_n = (1 + \Delta u)^n \qquad \text{and} \qquad \Delta^n u = (u - 1)^n; \qquad (7.3)$$

in the expansion of $(1 + \Delta u)^n$, one has to remember to change the powers $(\Delta u)^k$ into higher differences $\Delta^k u$; and in the expansion of $\Delta^n u = (u - 1)^n$ one has to remember to change the powers u^n into terms u_n.[22] These symbolic expressions, as such, come from [Lorgna *1786-87*],[23] but Lacroix abstains from expounding Lorgna's "new kind of calculus", which consisted in using the analogy between exponents of powers on one side and indices of iteration on the other to *obtain* formulas. Lacroix limits himself to notice the analogy, both in (7.3) and in Lagrange's

$$\Delta^n u = (e^{\frac{du}{dx}h} - 1)^n \qquad (7.4)$$

(where u is a function of x, $h = \Delta x$, and the powers $\frac{du^k}{dx^k}$ must be changed into the higher derivatives $\frac{d^k u}{dx^k}$).

Naturally, the sections on difference equations are written in the language of indices. Thus, the general first-degree equation is:

$$y_{x+n} + P_x y_{x+n-1} + Q_x y_{x+n-2} \ldots + U_x y_x = V_x$$

(the subscripts in P_x, Q_x, etc. are not indices: they mean that those are functions of x) [*Traité*, III, 188]. It is certainly not necessary to speak here of chapter 2, on generating functions, where Laplace's notations are followed.

---

[22] Notice that 0-powers are included: in the first case, $(\Delta u)^0$ must be changed into $\Delta^0 u = u$, and in the second case, u^0 must be changed into $u_0 = u$.

[23] In [Domingues *2005*, 289] I said that (7.3) come from [Lagrange *1772a*]. I was wrong: [Lagrange *1772a*] gives analogies between powers and higher differences and derivatives (like (7.4)), and it is the inspiration for [Lorgna *1786-87*], but (7.3) are not found there. Incidentally, (7.1) is, but with u_n referred to only verbally [Lagrange *1772a*, § 17].

What seems to be correct in Schubring's and Dhombres' suggestions (and especially in [Dhombres & Pensivy *1988*, 19]) is that Lacroix *diffused* the use of subscript indices. Their use in volume III was obvious enough; but their uses in volumes I and II, limited as they are (although far from being as limited as Schubring has it), probably contributed to their adoption outside the area of "theory of series" and finite differences.

7.2 The "multiplicity of integrals" of difference equations

7.2.1 The peculiar equivalent to singular integrals in finite difference equations

The subject of finite difference equations started with [Lagrange *1759b*]. This memoir consists in applications to linear finite difference equations of existing methods for linear differential equations: separation of variables for first order, and d'Alembert's reduction of higher-order linear equations to systems of first-order ones. According to Wallner [*1908*, 1052] the majority of works on finite difference equations in the 18th century remained dependent on analogies with differential equations. The area in which this analogy was trickier was that of singular integrals.

We will not start by the exact beginning, but by something close enough. On the 30th November 1785 Monge read to the Paris Academy of Sciences a very short memoir [*1785c*] on integration of nonlinear finite difference equations. As usual, this consisted in adapting a method for differential equations (proposed in [Monge *1785b*]). This method involved differentiating the equation enough times as to be able to eliminate all constants, or at least enough times as to obtain a quasi-linear equation. In the case of finite difference equations, there were remarkable consequences. Monge gives the very simple example

$$(\Delta y)^2 = b^2,$$

where b is a constant[24]: the common integration of $\Delta y = \pm b$ gives

$$y = \pm \frac{b}{a} x + A$$

(where the constant $a = \Delta x$ and A is the arbitrary constant); but differentiating $(\Delta y)^2 = b^2$ we obtain

$$2\Delta y \Delta \Delta y + (\Delta \Delta y)^2 = 0,$$

---

[24] *Sic.* Probably what Monge means is that $\Delta b = 0$, that is, that b is constant for values of x that differ by Δx. Euler had already remarked that this is also satisfied when $b = \varphi(\sin \frac{\pi x}{\Delta x}, \cos \frac{\pi x}{\Delta x})$, for constant Δx. This is not terribly important for the subject of multiple integrals, and so I will avoid the issue, using the word "constant" when the author studied uses it, and speaking of "arbitrary quantities" otherwise.

which can be split into the factors

$$\Delta\Delta y = 0 \quad \text{and} \quad 2\Delta y + \Delta\Delta y = 0;$$

the first gives

$$y = \pm\frac{bx}{a} + A$$

as above; the second, however, gives

$$y = C \pm \frac{b}{2}(-1)^{\frac{x}{a}}.$$

The latter is a solution of the given equation which is not contained in $y = \pm\frac{b}{a}x+A$. Thus, Monge had come across a Clairaut-like situation – an extra solution obtained via differentiation. With a surprising difference: the equivalent to the singular integral, namely $y = C \pm \frac{b}{2}(-1)^{\frac{x}{a}}$, also contains an arbitrary constant – C – and is therefore as general as the equivalent to the complete integral.

The reason why this was not the beginning is that precisely one week before, Jacques Charles ("le géomètre") had read to the same academy an even shorter work stating that "there are finite difference equations that have two complete integrals" [Charles *1785b*]. While Monge's observation is similar to Clairaut's and Euler's "paradoxes", Charles's approach is an adaptation of Lagrange's theory of singular integrals. He considers the integral

$$V = 0$$

of a finite difference equation

$$Z = 0,$$

V being a function of x, y and of a constant a not in Z; if V is (finitely) differentiated holding a constant, and if the result is denoted δV, then $Z = 0$ must be the result of eliminating a between

$$V = 0 \quad \text{and} \quad \delta V = 0; ^{25}$$

but if a is also varied, then we get

$$\Delta V = \delta V + R\Delta a; \tag{7.5}$$

the result of eliminating a between $V = 0$ and $\Delta V = 0$ will still be $Z = 0$, provided that $R = 0$; thus a singular integral should be obtained by eliminating a between

$$V = 0 \quad \text{and} \quad R = 0;$$

the problem is that while the equivalent to R in differential equations does not contain da, most often this R does contain Δa; thus, to eliminate a one must

---

[25]Charles [*1785b*, 560] has "$V = 0$, & $V = 0$", which is clearly a typo.

integrate $R = 0$ beforehand, and this introduces an arbitrary constant, which will also appear in the not-so-singular integral. Charles gives two examples, the first of which will suffice. Consider

$$gy = x\Delta y + \frac{\Delta y^2}{4n^2} \tag{7.6}$$

(where the constant $g = \Delta x$), whose complete integral is

$$gy = 2nax + a^2 \tag{7.7}$$

(where a is the arbitrary quantity); the finite difference of this integral, holding a constant, is

$$\Delta y = 2na;$$

and if a is varied, it is

$$\Delta y = 2na + \frac{\Delta a}{g}[2n(x+g) + 2a + \Delta a]; \tag{7.8}$$

(7.8) reduces to $\Delta y = 2na$ by putting

$$2n(x+g) + 2a + \Delta a = 0, \tag{7.9}$$

where, as had been warned, we find Δa; now, the integral of (7.9) is

$$-a(-1)^{\frac{x}{g}} = b + n(-1)^{\frac{x}{g}}\left(\frac{g}{2} + x\right)$$

(where b is an arbitrary quantity), and substituting this value of a in (7.7) we get

$$gy = -n^2x^2 + \left[\frac{gn}{2} + b(-1)^{\frac{x}{g}}\right]^2 \tag{7.10}$$

as a second complete integral of (7.6). Charles also remarks that following this procedure with (7.10) as the first integral, we would arrive at (7.7) – as Wallner [1908, 1053] put it, the singular integral of the singular integral is the original complete integral.

This would have remained as a nice observation, but unfortunately Charles decided to elaborate – in a misguided direction that made him arrive at strange paradoxes. In [1788] he retook (7.6), writing it as

$$y = \frac{x\Delta y}{g} + \frac{\Delta y^2}{4n^2g^2}, \tag{7.11}$$

and writing its two complete integrals as

$$y = 2nax + a^2 \tag{7.12}$$

and

$$y = -n^2x^2 + \left(b\cos.\frac{\pi x}{g} + \frac{gn}{2}\right)^2 ; \qquad (7.13)$$

it must be remarked that $\cos.\frac{\pi x}{g}$ is precisely the same as the $(-1)^{\frac{x}{g}}$ that occurred
in (7.10), as x is a discrete variable with difference $\Delta x = g$, and therefore $\frac{x}{g}$ takes
only integral values. But in order to have a locus for the equation, Charles needs
a continuous x; he divides the abscissa axis into equal segments $TV, VR, RS, \ldots$
of length g, and puts $x = X + g\mu$ – the integer μ indicates the division where x

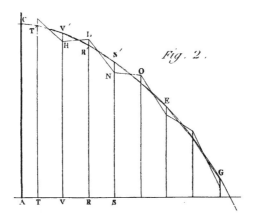

Fig. 2.

lies and, in modern terms, X is x modulo g; (7.13) is thus transformed into

$$y = b^2 + \frac{n^2g^2}{4} - n^2x^2 - nbg\cos.\pi\mu. \qquad (7.14)$$

He then constructs the parabola CEG with equation

$$z = b^2 + \frac{n^2g^2}{4} - n^2x^2$$

(that is, the difference between y and z is $-nbg\cos.\pi\mu$), and since $-nbg\cos.\pi\mu$ is
alternately $-nbg$ and nbg, he alternately adds and subtracts nbg to the division
ordinates $TT', VV', RR', SS', \ldots$, obtaining new points $H, L, N, O, \ldots$ that belong
to (7.14). Then he decides that the polygon $\ldots HLNO\ldots$ obtained by joining
these new points may be regarded as the locus of (7.14); also each side of the
polygon is of the form (7.12) – luckily the first complete integral was a linear
equation; Charles concludes that the polygon must verify (7.11).

As if this were not confused enough, Charles goes on: making the difference
g diminish until it becomes zero, the polygon becomes the parabola CEG, which
therefore must be an integral of "la proposée dans le cas des différences infiniment
petites"[26] – presumably

$$y = \frac{x\,dy}{dx} + \frac{dy^2}{4n^2dx^2}, \qquad (7.15)$$

---

[26]"the given equation in the case of infinitesimal differences"

although he does not write it explicitly. Now, this second integral retains the arbitrary constant b: it is

$$y = b^2 - n^2 x^2 + n\, b\, dx \cos .(\pi\mu). \tag{7.16}$$

Charles's grand conclusion is that singular integrals are in fact only incomplete integrals taken from a second complete integral that no one had noticed before [Charles *1788*, 118].

The last problem to be mentioned is the term $n\, b\, dx \cos .(\pi\mu)$ in (7.16). The real infinitesimal equivalent to (7.13) or (7.14) would have been simply $y = b^2 - n^2 x^2$. But Charles noticed that this would not satisfy (7.15). Thus he decided to keep the "differential term" $n\, b\, dx \cos .(\pi\mu)$, which allowed him to obtain the "true value" of $\frac{dy}{dx}$, namely $-2n(nx + b \cos .\pi\mu) = -2n(nx \pm b)$.[27] The need to keep this "differential term" led him to ramble on about not all (sequences of) polygons that converge to a curve being valid to obtain the tangents to that curve, and about the need, when studying a differential equation, to carefully consider the finite difference equation from which it derives.

Charles goes on with this in what appear to be *two* additions: on page 121 (the seventh of the memoir) there is a sidenote "presented on the 4th March 1790", presumably referring to pages 121 to 132, which suggests that they form one such addition; pages 132 to 139 constitute explicitly a "suite du mémoire". One can imagine the negative reactions at the Academy meetings, and Charles coming up with new examples and arguments. This was probably not easy – he was quite ill by then, suffering from paralysis of his right hand, and was to die the next year [Hahn *1981*, 85-86]. But summing up, we have to conclude that Charles thought too much in terms of finite differences, and was not able to grasp how a limiting process works.

In [*1795a*, IV, 502-509] Prony addressed this subject of "multiplicity of integrals", as an example of the difficulty in dealing with nonlinear finite difference equations (all the examples of multiple integrals were of nonlinear equations, for very good reasons – see below). But he gives only examples taken from [Monge *1785c*], adding geometrical constructions for the two double integrals of $(\Delta z)^2 = a^2$ that Monge had found. He referred the students who would like further details to the memoirs published by Monge and Charles in the volumes of the Paris Academy from 1783 to 1788, and mentioned "paradoxical results", but did not give details.

7.2.2 Biot's work and Lacroix's account

One of Prony's students in the first year of the *École Polytechnique* was Jean-Baptiste Biot. Grattan-Guinness [*1990*, I, 224] says that Biot began his scientific career by taking Prony's advice, mentioned in the previous section, of looking into Monge's and Charles's memoirs on the multiplicity of integrals of finite difference

---

[27] Even then, I cannot understand how this value is supposed to satisfy (7.15).

equations. In fact, Biot's first research work [*1797*] addressed that problem. But the story of Biot's motivation may have been a little more complicated.

Biot completed his studies at the *École Polytechnique* in 1795 [Grattan-Guinness *1990*, I, 188] or 1796 [Frankel *1978*, 37], and he quickly created a relationship of patronage with Lacroix, described in [Frankel *1978*]. Lacroix was preparing a new edition of Clairaut's *Élémens d'Algèbre* to be used in the newly founded *écoles centrales*, and it was Biot who wrote the introduction on arithmetic. In November 1796 Biot applied for a job as teacher of mathematics at the *École centrale* of the Oise department, in Beauvais, with the support of Lacroix, Prony, and Cousin (he was appointed in February 1797).

It is the correspondence dating from Biot's Beauvais period (1797-1800) that best tells us of the relationship between him and Lacroix:

> "Lacroix was the 'master', who suggested problems to his pupil, evaluated his solutions, helped him to become known to other scientists, generated publication and guided his career. Biot was the protégé who worked diligently on the tasks set to him by his 'master', edited and made additions to Lacroix's textbooks, dutifully followed his advice on matters affecting his career and thanked him profusely for his services." [Frankel *1978*, 38]

This relationship probably changed after 1800, when Biot was appointed both as an associate member of the *Institut National* and as a professor of the *Collége de France*. He and Lacroix were now on similar levels. But it is reasonable to assume that Lacroix's patronage had started before Biot's move to Beauvais forced it to be expressed in writing.

Thus Frankel [*1978*, 40] has suggested that [Biot *1797*] may have been written specifically to be included in the third volume of Lacroix's *Traité*. This is probably an exaggeration, as it is somewhat exaggerated to say that [Biot *1797*] "appeared intact" in [Lacroix *Traité*, III]: the changes from [Biot *1797*] to [Lacroix *Traité*, III, 237-247] are not very substantial – there are a few differences in notation, terminology, one less example, occasionally less detail, and the whole is rewritten by Lacroix – but enough not to consider this a section of Lacroix's *Traité* commissioned to Biot. Lacroix's section is rather a close account of Biot's work.[28]

Still, it is very likely that Lacroix suggested the topic to Biot, and maybe even some hints at how to deal with it. It is also very likely that Lacroix had his third volume in mind – that he wished to have a better source on the multiplicity

---

[28]Oddly, Grattan-Guinness [*1990*, I, 224] has said that in the first edition Lacroix "mentioned" Biot's memoir, "and gave a lengthy account of it in the second edition". In fact, the lengthy account of the second edition is virtually identical to that of the first edition [Lacroix *Traité*, 1st ed, III, 237-247; 2nd ed, III, 250-260]. He has also said [*1990*, I, 227] that Lacroix used Biot's paper on mixed difference equations [Biot *1799*] only in the second edition of his *Traité*, but as we will see below that already happened in the first edition. Grattan-Guinness must have underestimated the degree of collaboration between Lacroix and Biot.

of integrals of difference equations than the confused [Charles *1788*] or the laconic [Prony *1795a*].

The similarity between [Biot *1797*] and [Lacroix *Traité*, III, 237-247] and the possibility of Lacroix having suggested the topic are two reasons to address together Biot's work and Lacroix's account. One final reason has to do with dates of publication. Biot submitted his memoir "Considérations sur les intégrales des équations aux différences finies" to the *Institut National* on the 6 Ventose of year 5 (24 February 1797) [Acad. Sc. Paris *PV*, I, 174]. Laplace and Prony were charged with reporting on it, but the report (written by Prony) only appeared over two and a half years later (6 Frimaire year 8 = 27 November 1799) [Acad. Sc. Paris *PV*, II, 45-48]; it recommended either the publication of Biot's memoir in the *Savants Étrangers*, or of the report itself in the *Mémoires*; it was the latter option that was followed, in the volume that appeared in 1801. Biot's memoir was finally published in the *Journal de l'École Polytechnique* in 1802 (it is this version that is cited here as [Biot *1797*]). But as [Lacroix *Traité*, III] had appeared in 1800, Lacroix's account constitutes the first publication of Biot's work.

Still about dates: the publication in the *Journal de l'École Polytechnique* mentions that the memoir had been submitted to the *Institut* on the "6 Ventose year 8"; this is a typo for year 5, corrected in the errata at the end of the volume; moreover, Biot did not submit anything at the meeting held on the 6 Ventose year 8 [Acad. Sc. Paris *PV*, II, 110-114]. He did submit another memoir on the 11 Pluviose year 8 (31 January 1800), but it was on the integration of linear finite difference equations [Acad. Sc. Paris *PV*, II, 87]; this was never published, and the report (of which Laplace and Lacroix were charged) was never made; thus, we do not know what were its contents; but the fact that it was about *linear* finite difference equations indicates that it had little or nothing to do with his memoir on the multiplicity of integrals (which, as has been noted above and will be explained below, occurred only for nonlinear equations); in particular, it was not another version of it, as Grattan-Guinness [*1990*, I, 224] has suggested. Frankel [*1978*, 41] made a similar claim, even quoting a letter from Biot to Lacroix, which he dates of the winter of 1799-1800:

> "I have started again from scratch and I have arrived at the same results but in a much simpler manner... using powers of the second order. You can see that I am profiting from what you tell me, because it was you who engaged me to read your third volume carefully, and the high opinion you have of powers of the second order led me to use them to good advantage."

As I have not seen this letter, which is kept at the David Eugene Smith Collection, in Columbia University, New York, I cannot discuss in detail Frankel's claim that it refers to Biot's memoir on the multiplicity of integrals of finite difference equations; but I find it more likely to refer to Biot's memoir on linear finite difference equations, which may very well have been through two versions. It is noteworthy that the version of the former that we know, published only in 1802, has no second-order powers whatsoever.[29] If there ever was a second version of it, it is not the one published in the *Journal de l'École Polytechnique*. I assume that this published version is the original (or only) one, and that is why I cite it as [Biot *1797*].

After all these introductory considerations, let us examine [Biot *1797*], together with the section in Lacroix's *Traité* "on the multiplicity of integrals of which difference equations are capable" [*Traité*, III, 237-247]. The main differences between the two will be noted. Otherwise, where one reads "Biot" one may also read "Lacroix". As for notation, it is Lacroix's that will be followed.

It could go without saying that Lacroix acknowledges Biot's authorship. As always for manuscripts, he does not include Biot's memoir in the table of contents [Lacroix *Traité*, III, vi], but he cites it at the beginning of the section [*Traité*, III, 237] as the source from where he took what follows.

Biot starts with a similar approach to that of Charles – namely, adapting Lagrange's explanation for singular integrals of differential equations; Lacroix does not fail to highlight the analogy [*Traité*, III, 237]. But instead of using a complete integral and its finite difference, Biot uses a complete integral

$$F\{x, a, y_{x,a}\} = 0 \qquad (7.17)$$

(the notation $y_{x,a}$ is used to exploit the fact that y is a function of x and particularly of a) and the consecutive equation

$$F\{x_1, a, y_{x_1,a}\} = 0; \qquad (7.18)$$

that is, (7.17) is the complete integral of the difference equation $Z = 0$ that results from eliminating a between (7.17) and (7.18).[30] This is of course equivalent to using (7.17) and its difference, since that difference is precisely $F\{x_1, a, y_{x_1,a}\} - F\{x, a, y_{x,a}\} = 0$; but this format is more appropriate than Charles's (7.5), since it allows us to deal better with different values for a. If a is varied along with x, (7.17) becomes

$$F\{x_1, a_1, y_{x_1,a_1}\} = 0; \qquad (7.19)$$

but the same difference equation $Z = 0$ may result using (7.19) instead of (7.18), as long as in these two equations we have $y_{x_1,a_1} = y_{x_1,a}$, that is, as long as we have

$$F\{x_1, a, y_{x_1,a}\} = 0 \quad \text{and} \quad F\{x_1, a_1, y_{x_1,a}\} = 0. \qquad (7.20)$$

Elimination of $y_{x_1,a}$ between these two equations results in an equation in x, x_1, a and a_1 (a difference equation) that gives the law that the values of a must follow

---

[29]"Second-order powers" are generalized factorials. See page 43 above.

[30]This is Lacroix's notation. Biot has x' instead of x_1. The brackets instead of parentheses, as well as the subscript x and a, are in both Biot and Lacroix. In both cases, one must be aware that x and a stand both for the variables (sometimes constant, in the case of a) and for their *first* values, which Lacroix might have noted x_0 and a_0. This ambiguity has been remarked in section 7.1.2.

for $Z = 0$ to be satisfied. Since this a difference equation, it must be integrated in order to get an expression for a, which when substituted in (7.17) will result in a new integral for $Z = 0$; as that expression for a contains an arbitrary quantity, this new integral is "aussi générale que la première"[31] [Biot *1797*, 183], or "encore une intégrale complète [. . .], au lieu d'une intégrale particulière"[32] [Lacroix *Traité*, III, 238].

This "new integral" is not necessarily *new*: if (7.17) is linear in a, then (7.20) gives the trivial equation $a_1 = a$, so that no new integral arises;[33] but if a is raised to some power in (7.17), then there should be a new integral. Incidently, this is why all the examples of multiple integrals were of nonlinear equations: if (7.17) is not linear in a, then the elimination of this quantity between (7.17) and (7.18) should result in a nonlinear difference equation; but neither Lacroix nor Biot make this remark.[34]

At this point a difference in terminology between Lacroix and Biot must be noted: Lacroix [*Traité*, III, 240] calls the new integrals (those truly *new*) "indirect integrals"; while Biot simply calls them "new integrals". Curiously, later in [*1799*, 311] Biot was to refer to this memoir as a "théorie des intégrales indirectes des équations aux différences"[35].

Something that Biot introduces very early in his memoir [*1797*, 183; Lacroix *Traité*, III, 239] is a geometrical interpretation: a difference equation is the locus of a sequence of points corresponding to abscissas that "follow a certain law" (that of $x, x_1, x_2, \ldots$); assigning distinct particular values to a, (7.17) gives us distinct particular integrals

$$F\{x, a, y_{x,a}\} = 0, \quad F\{x, a_1, y_{x,a_1}\} = 0, \quad F\{x, a_2, y_{x,a_2}\} = 0, \quad \ldots,^{36} \quad (7.21)$$

so that (7.20) – or equivalently, the equality $y_{x_1,a_1} = y_{x_1,a}$ – means that the first two particular integrals in (7.21) intersect at a point of abscissa x_1; now, if the successive values of a follow the law mentioned above for $Z = 0$ to be satisfied (that is, the integral of the difference equation obtained by elimination of $y_{x_1,a}$ from (7.20)), then we will have $y_{x_2,a_2} = y_{x_2,a_1}$ as well, so that the second and third particular integrals in (7.21) will intersect at a point of abscissa x_2; and so on; these points of intersection

$$x, \quad x_1, \quad x_2, \quad \ldots \; x_n$$
$$y_{x,a}, \quad y_{x_1,a} = y_{x_1,a_1}, \quad y_{x_2,a_1} = y_{x_2,a_2}, \quad \cdots \quad y_{x_{(n)},a_{(n-1)}} = y_{x_{(n)},a_{(n)}}$$

form a sequence that satisfies $Z = 0$ and is an indirect integral.

---

[31]"as general as the first"

[32]"yet a complete integral [. . .], instead of a particular solution"

[33]This is better explained in [Biot *1797*, 184-185] than in [Lacroix *Traité*, III, 238].

[34]Apparently it was Poisson who first made it [*1800*, 180].

[35]"theory of indirect integrals of difference equations"

[36][Biot *1797*, 183] has $F\{x, a, y_{x\,a}\} = 0$, $F\{x', a', y_{x\,a'}\} = 0$, $F\{x', a'', y_{x\,a'}\} = 0$, which must be a triple typo for $F\{x, a, y_{x\,a}\} = 0$, $F\{x, a', y_{x\,a'}\} = 0$, $F\{x, a'', y_{x\,a''}\} = 0$.

It is when presenting this geometrical interpretation that Biot remarks that the new integral, although coinciding with the original one as far as first-order differences go, deviates from it at second-order differences. Lacroix is a little clearer on why this is so: the sequence above does not necessarily verify $y_{x_2,a} = y_{x_2,a_1}$, and the like. Lacroix also uses a more precise language when explaining that distinct integrals of the same difference equation cannot coincide indefinitely at differences of *all* orders, so that two integrals of one first-order equation should, *in general*, correspond to distinct second-order equations [*Traité*, III, 243]. This property is another analogue between indirect integrals of difference equations and singular integrals of differential equations (see page 189 above). And it is important – Biot uses it to explain Monge's example $\Delta y^2 = c^2$, and similar situations, and to propose a general method for finding new integrals, without recurring to the ordinary integral: differentiating (finitely) the difference equation, if the result is factorizable, then each of the factors corresponds to an integral; but he recognizes the disadvantage of introducing higher-order equations [*1797*, 192].

The refutation of Charles's paradoxes is more detailed in [Biot *1797*, 195-198] than in [Lacroix *Traité*, III, 246-248]. Let us sum it up.

Charles had treated differential equations and their integrals as limits of difference equations and their integrals, respectively; but his handling of limits was *very* naïve. Both Biot and Lacroix agree with Charles that putting $\Delta x = 0, \Delta y = 0$ in a difference equation results in the differential equation that is its limit;[37] but Charles also assumed that putting $\Delta x = 0, \Delta y = 0$ in an integral of the difference equation was enough to obtain an integral of the differential equation, and this was his big mistake. Taking $y_{x_1,a} = \mathrm{f}(x_1, a)$ and $y_{x_1,a_1} = \mathrm{f}(x_1, a_1)$ from (7.20), we get $\mathrm{f}(x_1, a_1) - \mathrm{f}(x_1, a) = 0$; writing a_1 as $a + \Delta a$, Biot argues that this can be written in the form

$$\Delta a \, \mathrm{f}_{\prime}(x_1, a, \Delta a) = 0,$$

whence the two possibilites $\Delta a = 0$ (that is, a is a constant and the original integral results) and

$$\mathrm{f}_{\prime}(x + \Delta x, a, \Delta a) = 0.$$

Now, this difference equation can be integrated, resulting in an indirect integral; but if we put $\Delta x = 0, \Delta a = 0$,[38] it becomes a "primitive equation" [Lacroix *Traité*, III, 247], so that a can be retrieved from it without integration, and therefore without an arbitrary quantity (resulting in a particular solution). Thus, to go from the indirect integral of the difference equation to the particular solution of the differential equation it is necessary to drop the arbitrary quantity.

It was because Charles arrived at false integrals that he needed an extra "differential term". Both Lacroix and Biot remark that this term, destroying the "homogénéité qui fait la base du calcul différentiel"[39] [Biot *1797*, 198; Lacroix

---

[37]Of course this language of "putting $\Delta x = 0, \Delta y = 0$" is also very naïve. But we understand that it means taking the limit as $\Delta x \to 0, \Delta y \to 0$, taking in account the limit $\frac{dy}{dx}$ of $\frac{\Delta y}{\Delta x}$.

[38]That is, if we take the limit as $\Delta x \to 0, \Delta a \to 0$.

[39]"homogeneity that is the basis of the differential calculus"

Traité, III, 247], should have made him realize how wrong he was. The error of concluding that not every inscribed polygon tends to the curve, as the number of sides is assumed infinite, but does not even deserve a counter-argument – it seems to be presented as yet another silly conclusion (not in so many words).

To finish this, we must look at the citations of Charles, where we find one of the few mistakes in Lacroix's references. Biot cites only [Charles *1788*] – which is enough for his purposes. Lacroix has a more historical concern – and gets it wrong: he correctly points out Charles's priority in noticing the multiplicity of integrals of difference equation, but cites [Charles *1785a*] as the place where that happened, instead of [Charles *1785b*]; [Charles *1785a*] is a memoir on difference equations, but not on what Lacroix called indirect integrals, and certainly prior to Charles's discovery of them[40]. To make this mistake worse: 1 - Lacroix [*Traité*, III, vi] cites [Monge *1785c*], which was published in the same volume as [Charles *1785b*]; 2 - he fails to include [Charles *1788*] in the table of contents. This omission (which is not serious, since [Charles *1788*] is mentioned in the main text) was rectified in the second edition, but the confusion between [Charles *1785b*] and [Charles *1785a*] was not [Lacroix *Traité*, 2nd ed, III, xiv, 259]. There is some further evidence that Lacroix did not really know [Charles *1785a*] (see footnote 53 below).

7.3 Mixed difference equations

7.3.1 "Equations in finite and infinitely small differences"

Equations containing both finite differences and differentials appeared for the first time in [Condorcet *1771*]. In this memoir, Condorcet reduced to finite difference equations the determination of the arbitrary functions occurring in integrals of partial differential equations; but in some cases, where those arbitrary functions are originally given by non-algebraic equations, the resulting finite difference equations contain also differentials [*1771*, 51-52]; hence Condorcet dedicating the third "article" of the memoir [*1771*, 56-66] to "équations aux différences finies et infiniment petites"[41].

Condorcet starts by the easy possibilities: if regarding the differentials as new variables in a finite difference equation this finite difference equation is integrable, then we should integrate it – the result will be a differential equation, which we then integrate; and of course if regarding the finite differences as new variables we get an integrable differential equation, then we should integrate it, and then integrate the resulting finite difference equation. But he notices that these two cases do not cover all equations in finite and infinitesimal differences. Therefore Condorcet tries to get a general mode of solution through different means ("more

---

[40][Charles *1785a*] is probably the result of combining several memoirs submitted to the Paris Academy in 1779 and 1780, and possibly one submitted in May 1785 [Hahn *1981*, 84]. [Charles *1785b*], as we have noticed above, was read in November 1785.

[41]"equations in finite and infinitely small differences"

direct principles", according to him). His answer is typically Condorcetian: try to find the form of the solution (how many transcendental functions, and of what types)[42], and then use the method of indeterminate coefficients.

Laplace also occasionally addressed this kind of equation. In [*1779*, 302-305] he applied his calculus of generating functions to "équations aux différences partielles, en partie finies, et en partie infiniment petites"[43]. In [*1782*, 31-53] he addressed approximate integration of linear finite difference equations, also extending it to linear differential equations and linear equations in finite differences and differentials [*1782*, 42-43].

According to Wallner [*1908*, 1065], Lorgna and Paoli also treated these equations (in the latter case using Laplace's generating functions).

Finally, we must mention Jacques Charles – the same of the paradoxical results in finite difference equations. From 1779 to 1785 Charles submitted seven memoirs to the Paris *Académie des Sciences*, in an effort to be elected a member (he was successful in 1785); out of these, two were expressly about "equations containing both finite differences and infinitesimal differences" – one submitted in 1779, and the other submitted in 1785 (the last in the series of seven memoirs) [Hahn *1981*, 84].

Although both these memoirs were recommended for publication in the *Savants étrangers*, none of them was published, at least not in its entirety. [Charles *1785a*] seems to be a combination of some of those seven memoirs, but little survived from these two. The subject of the 1779 memoir was the construction of equations containing both finite differences and differentials, according to the report made by Vandermonde, Bossut and Condorcet [Acad. R. Sc. *PV*, XCVIII, 224r-224v], also quoted by Hahn [*1981*, 84]; while in the one submitted in 1785 Charles reduced the integration of these equations to that of partial equations in finite differences only (according to the reporters[44] [Acad. R. Sc. *PV*, CIV, 80v-81r]). But in the published memoir there are less than two full pages [Charles *1785a*, 584-585] dedicated to "equations containing both differentials and finite differences"; these contain a "problème", indeed solved through a partial finite difference equation, which suggests that it is taken from the memoir submitted in 1785, and a "remarque" on the application of this kind of equation to Lagrange's version of the vibrating string with discrete weights – there is no trace of constructions of equations in both differentials and finite differences (that is, no trace of the 1779 memoir).

Still, this is one of the few publications in the 18th century on equations containing both differentials and finite differences. And its larger part (the "problème") was reprinted, already in 1785, as one of Charles contributions to the

---

[42]Condorcet's underestimation of the variety of transcendental functions is one of the biggest problems with his "general" theory of integration [Gilain *1988*, 93].

[43]"equations in partial differences, partly finite, and partly infinitesimal"

[44]The *procès-verbal* says that the reporters were Lavoisier, Cadet and Darcet, which must be a mistake (these were all chemists). According to Hahn [*1981*, 84] the reporters were Cousin and Condorcet.

Encyclopédie Méthodique – the article "Intégral *(Calcul intégral des équations en différences mêlées)*" [Charles *1785d*].[45] Incidentally, the title of this latter version seems to be the first occurrence of the expression "différences mêlées" ("mixed differences"), which was to become standard with Biot's work and Lacroix's account of it.

7.3.2 Biot's work and Lacroix's account

We saw above that it is *possible* that it was Lacroix who proposed to Biot to study the multiplicity of integrals of finite difference equations. As for the topic of Biot's second submission to the *Institut*, namely mixed difference equations, we *know* that it was suggested by Lacroix, in 1797 [Frankel *1978*, 40].

Biot did not produce a memoir then, but he resumed his research in early 1799, and on the 1st Brumaire of year 8 (23 October 1799) he read to the *Institut* his "Considérations sur les équations aux différences mêlées" [Acad. Sc. Paris *PV*, II, 18]. Laplace, Bonaparte and Lacroix were charged with reporting on it, and the report (written by Lacroix[46]) was read twenty days later, recommending the publication in the *Savans Étrangers* [Acad. Sc. Paris *PV*, II, 30-32]. Unlike what happened to his memoir on integrals of finite difference equations, this recommendation was eventually followed, and the memoir was published in the new series of the *Savans Étrangers*, in 1806 – this is what is cited here as [Biot *1799*]; but of course Biot was not very confident that this would happen (no one would be – the *Savans Étrangers* was not published between 1786 and 1806), and he submitted the memoir also to the *Société Philomatique*, in whose *Bulletin* appeared a summary [Biot *1800*].[47]

The issue with that summary was published in Pluviose year 8 (January-February 1800). That same year appeared the third volume of Lacroix's *Traité*; and most of its final chapter (chapter 4, "On mixed difference equations") is an account of [Biot *1799*] – although it must be said that it does not follow Biot's work as close as the section on the multiplicity of integrals of difference equations. Lacroix starts by mentioning Condorcet and Laplace as the originators of the subject (Biot omits this); then he gives a couple of examples; and only then he picks up the beginning of Biot's memoir. We will also see below that he actually has more to say than Biot on "mixed difference equations in the strict sense". On

---

[45]Another contribution, immediately preceding that one, is the article "Intégral *(Calcul intégral des équations en différences finies)*" [Charles *1785c*], more than half of which is also reproduced from [Charles *1785a*, 574-579]

[46]Both Frankel [*1978*, 41] and Grattan-Guinness [*1990*, I, 227] attribute it to Lacroix, and there is no reason to question this attribution; on the contrary – its terminology ("differences" instead of "finite differences"; "partial differentials" instead of "partial differences"; "indirect integrals"; "differential coefficients") points to Lacroix, and so does a reference to Fontaine's authorship of the "important remark" that a differential equation is the result of elimination of constants between a "primitive equation" and its differentials.

[47]Biot was an *associé-correspondant* of the *Société Philomatique*. Although this summary has an indication "Institut Nat." on the side, the report of the activities of the *Société* states that Biot also read the memoir to its members [Soc. Phil. *Rapp*, IV, 14].

the other hand, it is noticeable that Biot follows Lacroix's notation (x_1 instead of x') and terminology more closely here than in [1797] – to the point of referring to his previous memoir as being about "indirect integrals" [Biot 1799, 311].

[Biot 1799] is divided into two parts, corresponding to the two sections in [Lacroix Traité, III, ch. 4]: the analytical theory and geometrical applications. Although Biot does not cite Condorcet (or anyone else for that matter, except himself on indirect integrals, and Euler as a source of geometrical problems), the analytical theory seems to be a clarification of some parts of that in [Condorcet 1771].[48] Like Condorcet, Biot's starting point is that a mixed difference equation results from combining an equation with its differences and differentials. This is an extension of Fontaine's conception of differential equations (see sections 6.1.4.1 and 6.2.1.1), similar to what Charles and Biot himself had done for difference equations, as Lacroix refers in the report on [Biot 1799] for the Institut [Acad. Sc. Paris PV, II, 30-31]. But Biot is much clearer than Condorcet in how that "combination" happens: elimination of constants.[49]

In the case of first-order equations (the only one considered by Biot), there are four possibilities for this elimination. Writing them as in Lacroix's version[50], the first two consist in eliminating two constants between

$$V = 0, \qquad dV = 0, \quad \text{and} \quad \Delta V = 0$$

or eliminating four constants between

$$V = 0, \qquad dV = 0, \qquad \Delta V = 0 \quad \text{and} \quad d\Delta V = 0;$$

the third possibility consists in eliminating one constant between

$$V' = 0 \quad \text{and} \quad dV' = 0,$$

where $V' = 0$ is already a difference equation (Lacroix notes that $V' = 0$ is obtained by eliminating an "arbitrary function of the type that complete integrals of difference equations" between $V = 0$ and $\Delta V = 0$); while the fourth possibility consists in eliminating an arbitrary quantity between

$$dV' = 0 \quad \text{and} \quad d\Delta V' = 0,$$

where dV' represents "a first-order differential function of two variables" (and presumably is obtained by eliminating a constant between $V = 0$ and $dV = 0$).

---

[48] As has been remarked above, Lacroix does mention Condorcet, but he does not say anything about the contents of [Condorcet 1771], nor establishes any relation between Condorcet's and Biot's theories.

[49] Well, mostly constants. In some cases, more or less obvious, what is intended is elimination of functions of period Δx (constant for values of x that differ by Δx). One reference by Biot [1799, 297] to the possibility of a "more general" characterization may be an allusion to this issue.

[50] The main difference from Biot's version is that the latter uses $V_1 (= V + \Delta V)$ instead of ΔV.

This division into several possibilities suggests another point of contact with [Condorcet *1771*] – Biot's third and fourth cases correspond to Condorcet's easy possibilities: the mixed difference equation obtained in Biot's third case is such that regarding Δy as a new variable we get an integrable differential equation, whose integral is of course $V' = 0$; and the fourth case is such that $dV' = 0$ is the finite difference integral of the mixed difference equation, when dy is regarded as a new variable. Biot [*1799*, 300] calls these two cases "équations aux différences successives" – "successive difference equations", because they result "d'une diffé-rence succédant à une différentiation, ou d'une différentiation effectuée sur une différence"[51] [Lacroix *Traité*, III, 532]. Successive difference equations are easily recognizable because they must satisfy their respective conditions of integrability (an observation that Condorcet would have appreciated); for instance, in the third case, the successive difference equation must satisfy the conditions for integrability of differential equations in three variables – these three variables being x, y and Δy.

When both a finite difference integration and a differential integration can be performed, it is the latter that should be done done first – it only introduces an arbitrary constant, while the former introduces an arbitrary function that is constant for values of x differing by Δx, and this function must be particularized before the differential integration can be performed.

As for Biot's first and second cases, he calls them "*équations aux différences mêlées* proprement dites"[52] [*1799*, 300]. Biot does not give any method for solving them, and he explicitly avoids the complicated topic of the extent ("étendue") of their integrals [Biot *1799*, 303]. Lacroix is a little more helpful: he makes it clearer that this extent problem is similar to that of partial differential equations, and refers the reader to the proper passages in the second volume (see section 6.2.2.3) [*Traité*, III, 534]. And he briefly addresses a method of solution, admittedly very difficult to actually use: to replace Δx with

$$\frac{dy}{dx}\frac{h}{1} + \frac{d^2y}{dx^2}\frac{h^2}{1 \cdot 2} + \frac{d^3y}{dx^3}\frac{h^3}{1 \cdot 2 \cdot 3} + \text{etc.}$$

and $\Delta \frac{dy}{dx}$ with

$$\frac{d^2y}{dx^2}\frac{h}{1} + \frac{d^3y}{dx^3}\frac{h^2}{1 \cdot 2} + \text{etc.,}$$

thus converting the mixed difference equation into an indefinite-order differential equation [*Traité*, III, 533]. He uses this method in one geometrical example (see below).

The only analytical issue about mixed differences in the strict sense that Biot really develops is that of their indirect integrals; he occupies eight pages with this

---

[51]"from a difference succeeding to a differentiation, or from a differentiation effected on a difference"

[52]"*mixed difference equations* in the strict sense"

[*1799*, 303-310]. Lacroix, on the other hand, devotes less than a page [*Traité*, III, 534] to results that are "very analogous" to those on difference equations.

As has already been mentioned, the second sections of both [Biot *1799*] and [Lacroix *Traité*, III, ch. 4] are dedicated to applications to geometrical problems (essentially problems that had been treated by Euler using other means, and namely the problem of reciprocal trajectories). Also both authors present this as the main interest of mixed difference equations [Biot *1799*, 297; Lacroix *Traité*, III, 535]. However, we will not dwell much on this, as they are mostly that – applications.

But there are a couple of issues to point out, constituting two more differences between Biot's memoir and Lacroix's chapter. The first is that all problems treated by Biot lead to successive difference equations, while Lacroix includes one that leads to a mixed difference equation in the strict sense. In this problem he manages to use the method mentioned above of using the series expansion of Δy to reduce the equation to a differential one of indefinite order. And in addition he gives Charles's treatment of this problem – the mixed difference equation in question is the one that Charles had solved in [*1785a*, 584-585] and [*1785d*].[53]

The second issue is that, unlike Biot, Lacroix includes one short paragraph on analytical applications of mixed difference equations [Lacroix *Traité*, III, 543]. He briefly mentions an unpublished work by "Français de Colmar"[54] on the use of mixed difference equations in Laplace's cascade method, and also the original context of mixed difference equations – the determination of arbitrary functions occurring in integrals of partial differential equations.

All things considered, Lacroix's 14-page chapter, although more concise, seems a little more substantial than Biot's 32-page memoir.

---

[53]The fact that in the table of contents Lacroix only mentions [Charles *1785d*] is the final indication that he did not really know, or did not pay attention to, [Charles *1785a*].

[54]François-Joseph Français (1768-1810), who was for some time a teacher in Colmar.

Chapter 8

The *Traité élémentaire*

In 1802 Lacroix published a *Traité élémentaire du calcul différentiel et du calcul intégral* (Elementary treatise of differential and integral calculus) [Lacroix *1802a*]. According to the publisher's list of elementary works by Lacroix, it was "tiré en partie"[1] from the large *Traité* [Lacroix *1802a*, ii]. Indeed it is mostly an abridged version of the latter. It is divided into a "first part: differential calculus", a "second part: integral calculus" and an "appendix: on differences and series". The correspondence between these three parts and the three volumes of the large *Traité* is perfect.

But before we compare [Lacroix *1802a*] with [Lacroix *Traité*] we must see where and how the former fits in the context of Lacroix's pedagogical *œuvre* and in the curriculum of the *École Polytechnique*.

8.1 The *Traité élémentaire de calcul...* and the *Cours élémentaire de mathématiques*

The first edition of the *Traité élémentaire* opens with a *discours préliminaire* entitled "Réflexions sur la manière d'enseigner les Mathématiques"[2]. There Lacroix mentions that he is publishing "la dernière partie du Cours élémentaire [de Mathématiques]"[3] [Lacroix *1802a*, v]. This *Cours* was probably thought of as composed by a set of works advertised in the same volume as being sold at Duprat and collectively referred to as the "collection complète des ouvrages élémentaires, publiés

---

[1]"partly taken"

[2]The full title is "Réflexions sur la manière d'enseigner les Mathématiques, et d'apprécier dans les examens le savoir de ceux qui les ont étudiées" ("Reflexions on the manner of teaching Mathematics, and of evaluating in examinations the knowledge of those who have studied it") [Lacroix *1802a*, v-xxxii]. These "Réfléxions" were afterwards included in [Lacroix *1805*] and therefore omitted from later editions of [Lacroix *1802a*].

[3]"the last part of the elementary course [of mathematics]".

par S. F. Lacroix, membre de l'Institut national"[4]:

1. *Traité élémentaire d'Arithmétique à l'usage de l'École centrale des Quatre-Nations*

2. *Elémens d'Algèbre* [Lacroix *1799*]

3. *Complément des Elémens d'Algèbre* [Lacroix *1800*]

4. *Elémens de Géométrie*

5. *Complément des Elémens de Géométrie, ou Essais de Géométrie sur les plans et les surfaces courbes* [Lacroix *1795*]

6. *Traité élémentaire de Trigonométrie et d'application de l'Algèbre à la Géométrie* [Lacroix *1798b*]

7. *Traité élémentaire du calcul différentiel et du calcul intégral* [Lacroix *1802a*]

This same list of works appears explicitly in two advertisements by the publisher of [Lacroix *1805*] (Courcier, successor of Duprat), as a "Cours de Mathématiques à l'usage de l'Ecole centrale des Quatre-Nations, par S. F. Lacroix, membre de l'Institut national, ouvrages adoptés par le gouvernement pour les Lycées et les Ecoles secondaires, 7 vol. in-8"[5] [Lacroix *1805*, iv, 391]. In 1819, this *cours* (now with the extra adjective "complet") had grown to nine volumes [Lacroix *Traité*, 2nd ed, III, ii], including a *Traité élémentaire de Calcul des Probabilités* and even [Lacroix *1805*, 2nd ed], which was not a textbook, but rather a collection of writings about mathematical education.

However, that same book [Lacroix *1805*] includes an analysis by Lacroix of his "Cours *élémentaire* de Mathématiques pures, à l'usage de l'Ecole Centrale des Quatre-Nations" (our emphasis), where it is made clear that the author thought of it as comprising only items 1-2 and 4-6 above.[6] He does include a few words on the *Traité élémentaire de calcul* [Lacroix *1802a*], probably because it had been written to follow immediately the *cours élémentaire*, but does not dwell on it, since "[il] ne fait point partie du Cours élémentaire"[7] [Lacroix *1805*, 384, 386]. As to item 3, the *Complément des Elémens d'Algèbre* [Lacroix *1800*], it is even more distant from the *cours élémentaire*.

Lacroix does not give a reason for [Lacroix *1802a*] not being part of the *cours élémentaire*, but the fact that it was directed at higher-education students

---

[4]"complete collection of elementary works published by S. F. Lacroix, member of the *Institut national*"

[5]"Course of Mathematics for the use of the *Ecole centrale des Quatre-Nations*, by S. F. Lacroix, member of the *Institut national*, works adopted by the government for the *Lycées* and secondary schools, 7 vols. in-8"

[6]Or even just 1, 2, 4 and 6. Item 5 [Lacroix *1795*] was not "essentiellement partie du cours élémentaire de Géométrie" ("essentially part of the elementary course of geometry") [Lacroix *1805*, 346]. That minimal version of the *cours élémentaire* is the one that appears in the first edition of [Lacroix *Traité*, III] (in the usual advertisement for books by Lacroix), [Lacroix *1795*], [Lacroix *Traité*], and the *Complément des Elémens d'Algèbre* appearing apart. But it is not of much concern here whether [Lacroix *1795*] should be included in Lacroix's *cours élémentaire*.

[7]"[it] is really not part of the *cours élémentaire*"

(although not exclusively – see below) must have been relevant. A much more interesting problem is the status of [Lacroix *1800*]; and although it is not this book that we are studying here, its stronger separation from the *cours élémentaire* had important consequences for [Lacroix *1802a*]. This separation was motivated by Lacroix's views on mathematical education and on what a good curriculum should include:

> "le nombre des matières qui doivent entrer dans l'instruction de la jeunesse est si grand, qu'il faut écarter, quelque intéressant qu'il puisse être en lui-même, tout sujet qui n'est pas d'une application fréquente."[8]
> [Lacroix *1805*, 389]

In other words, the *encyclopédiste* approach that is so clear in [Lacroix *Traité*] was not present in Lacroix's pedagogical works.[9] Instead, he sought to avoid too many metaphysical details, attempts to present all the artifices employed by geometers, and duplications:

> "présenter [les matières aux élèves] sous de points de vue différens, serait les éblouir et non les éclairer"[10] [Lacroix *1805*, 117];

> "ne convient-il pas mieux d'employer le temps des élèves à leur faire connoître des résultats nouveaux, plutôt que des procédés différens pour parvenir au même résultat[?]"[11] [Lacroix *1802a*, x-xiv; *1805*, 177-181]

[Lacroix *1800*] deals with several questions on the theory of equations (symmetric functions of their roots, the fundamental theorem of algebra and complex numbers, etc.) and an algebraic treatment of series: that is, it roughly comprises what was then often referred to as *algebraic analysis* (and also corresponds to the introduction and chapter 3 of [Lacroix *Traité*, 1st ed]) – see the beginning of section 3.2.6. According to Lacroix, these topics were very convenient for those who wished to study pure mathematics, and would even facilitate the study of [Lacroix *1802a*]; but were dispensable for the physico-mathematical applications. Being dispensable, they should be dispensed with in the *cours élémentaire* [Lacroix *1805*, 389-390].

One might ask then, to whom was [Lacroix *1800*] addressed. Its full title does say it is "à l'usage de l'École Centrale des Quatre-Nations"[12], which seems

---

[8]"the number of subjects that must be studied by the youth is so large, that it is necessary to put aside any topic that is not of frequent application, however interesting in itself it may be."

[9]At least it was not present within *each* subject. Lacroix was an ardent supporter of the model of the *écoles centrales*, which offered a much wider range of subjects than either the pre-revolutionary *collèges* or the *lycées* that later replaced them. "[T]he avowed aim of [the *écoles centrales*] was a sound but encyclopedic education, covering all 'positive' knowledge" [Dhombres *1985*, 125]. Dhombres [*1985*, 130] seems to attribute an encyclopedic character also to each of Lacroix's textbooks by extrapolating from the characteristics of [Lacroix *Traité*].

[10]"to present [the subjects to the pupils] under different points of view would be to dazzle, rather than to enlighten them"

[11]"is it not more convenient to employ the pupil's time acquainting them with new results, rather than with different procedures to arrive at the same result[?]"

[12]"for the use of the *École Centrale des Quatre-Nations*"

clear enough: it was a textbook at secondary-school level; but a special, advanced
secondary-school level. According to Dhombres [*1985*, 125, 127], "special classes for higher mathematics ("mathématiques transcendantes") were added to [the
écoles centrales]". They certainly existed in the *lycées* which replaced the *écoles
centrales* in 1802.[13] These special classes seem to solve our riddle, since in the
list [Lycées *1803*] of textbooks adopted in 1803 for the *lycées*, [Lacroix *1800*] and
[Lacroix *1802a*] are chosen for "transcendental mathematics".

The motivation that Dhombres [*1985*] presents for these special classes is the
preparation of pupils for admission to the *École Polytechnique* – this admission was
through a nationwide selection, at first based on information given by more than
22 local examiners, and from 1798 onwards it was carried out by four or five itine-
rant examiners [Belhoste *2003*, 54-56]; the programme for the entrance exams was
published every year. However, this seems to have soon excluded the topics treated
in [Lacroix *1800*] (and to have never included those in [Lacroix *1802a*]): the first
regulation of admission spoke quite vaguely on "connaissance de l'arithmétique et
des élémens d'algèbre et de la géométrie"[14] [Fourcy *1828*, 30; Belhoste *1995*, 73];
after a first year in which the lack of mathematical preparation of the students
caused many difficulties [Langins *1987a*, 76-79], the requirements in algebra were a
little detailed (and probably much enlarged) to include "la résolution des équations
des quatre premiers degrés, et la théorie des suites"[15] [Fourcy *1828*, 82; Belhoste
1995, 73], an expression that might cover a large part of algebraic analysis; but in
1798 they were relaxed back to "l'algèbre jusqu'aux équations du deuxième degré
inclusivement"[16] [Fourcy *1828*, 155; Belhoste *1995*, 73]. A more detailed admission programme, written by Monge, was adopted in 1800. It was sent by the
minister of the interior (Lucien Bonaparte) to the teachers of mathematics of the
écoles centrales throughout the country, together with a letter, containing metho-
dological advices for their teaching, signed by the minister but in fact, according

---

[13]The curriculum at each *école centrale* was decided by a local commission. On mathematics
the law only stipulated that at each *école centrale* there should be one teacher of that subject,
placed at the "second section" (to which only pupils aged 14 and over were admitted). All
subjects being optional for the students, the "special" character of some is doubtful. Moreover,
transcendental mathematics might be taught in some *écoles centrales* but not in others. At
the *École Centrale du Doubs* at Besançon, for instance, the most advanced topic seems to have
been the application of algebra to geometry (no theory of series or calculus) [Troux *1926*, 167-
170]. On the other hand, infinitesimal calculus (which would qualify as transcendental) was
taught at the *école centrale* of Nantes; and yet, very few students from Nantes applied for the
École Polytechnique [Lamandé *1988-1989*, 134-143]. The *lycées*, created by law in 1802, were
on the contrary highly centralized. At each *lycée* there should be six "classes" of mathematics
(two per year, giving a total of three years), taught by three teachers, plus two "classes" of
"transcendental mathematics" (two years, one teacher). Transcendental mathematics included
topics such as "application of differential [and integral] calculus to mechanics and to the theory
of fluids" or "general principles of high physics, especially electricity and optics" [Lycées *1802*,
307].
[14]"knowledge of arithmetic and the elements of algebra and analysis"
[15]"the solution of equations up to the fourth degree and the theory of series"
[16]"algebra up to and including the equations of second degree"

to Belhoste [*1995*, 73], written by none other than Lacroix[17] [Fourcy *1828*, 203-208; Belhoste *1995*, 73-76].[18] This programme remained essentially unchanged until 1854 [Belhoste *1995*, 73].[19] The topics covered in algebra are: the solution of equations of second degree; the proof of Newton's binomial formula for positive integer exponents, using combinations; the composition of equations and their numerical solution, using the method of commensurable factors and approximation; elimination in equations of higher degrees in two unknowns; and finally, the theory of logarithms (apparently as inverse functions of exponentials), explicitly excluding their series expansions from the requirements. All of these required subjects were included in [Lacroix *1799*]. The candidates to the *École Polytechnique* were not compelled to study [Lacroix *1800*] or any similar textbook.

However, the candidates to the *École Polytechnique* were certainly advised to study some matters not required for the entrance exams but taught there in the first year. This was strongly defended by a competitor of Lacroix as textbook writer, Jean-Guillaume Garnier, who was an examiner (and a teacher) of candidates to the *École Polytechnique* and also taught there from 1798 to 1802 (replacing Fourier, away in the Egyptian campaign):

> "pour qu'un candidat soit suffisamment préparé, je pense qu'il faut non-seulement qu'il possède toutes les connoissances énumérées dans le programme d'admission, mais encore qu'il ne soit pas étranger à l'analyse algébrique qui fait partie de l'enseignement mathématique de la première division de l'Ecole"[20]. [Garnier *1801*, vii].

Lacroix might not agree with this (he did not think that teaching algebraic analysis at the *École Polytechnique* was a good idea); but we have seen above that he found some usefulness in his *Complements of algebra* [*1800*] as facilitator of more advanced studies. In 1804 he was appointed teacher of transcendental mathematics at the *Lycée Bonaparte*, where he had to teach algebraic analysis as a secondary-school subject (and he certainly had done the same at the *École Centrale des Quatre-Nations*, possibly only to a few more advanced students).

Summing up, we can picture Lacroix's *cours de mathématiques* as containing several layers:

a) The *cours élémentaire* consisted in items 1, 2, 4 and 6 above (*Traité élémentaire d'Arithmétique, Elémens d'Algèbre* [Lacroix *1799*], *Elémens de Géomé-*

---

[17]Dhombres [*1987*, 95], on the other hand, suspects that the letter had been prepared by the predecessor of Lucien Bonaparte, Laplace.

[18]Thus the *École Polytechnique*, through its entrance exams, would serve as a factor of unification in a highly decentralized educational system. Whether that occurred in the two or three years between this letter and the replacement of the *écoles centrales* by the centralized *lycées*, is a good question.

[19]A very similar programme can be seen in [Éc. Pol. *Concours 1802*] (1802, incidentally, is the year of publication of the first edition of Lacroix's *Traité élémentaire du calcul...*).

[20]"for a candidate to be prepared well enough, I find it necessary not only that he possess all the knowledge detailed in the admission program, but also that he be familiar with the algebraic analysis that is part of the mathematical teaching in the first division of the *École*"

trie and *Traité élémentaire de Trigonométrie et d'application de l'Algèbre à la Géométrie* [Lacroix *1798b*]). This probably corresponded to the usual curriculum in the *écoles centrales* (special classes excepted); it certainly corresponded to the curriculum of "mathematics" *stricto sensu* (that is, excluding transcendental mathematics) in the *lycées*;[21] and also to the required knowledge for admission to the *École Polytechnique*.

b) In addition, item 5 (*Complément des Elémens de Géométrie* [Lacroix *1795*]) was apparently included in Lacroix's teaching at the École Centrale des Quatre-Nations [Lacroix *1805*, 346], at an *elementary* level.

c) The *Traité élémentaire du calcul différentiel et du calcul intégral* [Lacroix *1802a*], in spite of the title, was no longer at an *elementary* level: it was used mainly in higher education; in secondary education it was studied only at special classes. However, it had a close connection with the *cours élémentaire*, as it had been written so as to follow immediately the latter's final part (namely the application of algebra to geometry in [Lacroix *1798b*]), and thus formed a natural continuation [Lacroix *1805*, 384].

d) The *Complément des Elémens d'Algèbre* [Lacroix *1800*] was not more elementary than [Lacroix *1802a*] (being absent from the normal curriculum of mathematics at secondary schools), and was dispensable for the study of applications, so that it stayed outside of the progression from the *cours élémentaire* to [Lacroix *1802a*].

In 1805 these books constituted a *cours de mathématiques* at least in the commercial sense that Courcier would sell them as a set for 28 fr. 50 c. [Lacroix *1805*, iv][22] In 1819 the *cours complet de mathématiques* included two more items, costing in total 38 fr. 50 c. [Lacroix *Traité*, 2nd ed, III, ii][23]:

e) The *Essais sur l'enseignement* [Lacroix *1805*] were a natural complement to the *cours élémentaire*, a useful aid for those teachers who would follow Lacroix's *cours* (especially the *cours élémentaire*).

f) The *Traité élémentaite du Calcul des Probabilités*, first published in 1816, was also included in the 1819 *cours complet*. Unfortunately Lacroix does not seem to have inserted any reference to it in subsequent editions of [Lacroix *1805*].

---

[21]These were precisely the textbooks adopted in 1803 for the six normal "classes" of mathematics [Lycées *1803*].

[22]This apparently meant a modest discount, as bought separately they would cost 29 fr. 50 c. But it may be a misprint, the *Élémens d'algèbre* costing 4 fr., not 5 [Lacroix *1805*, iv, 391].

[23]And very clearly there was no discount.

8.2 Analysis in the early years of the *École Polytechnique*

The history of the teaching of analysis in the early years of the *École Polytechnique* is quite a complicated subject. The first year of the *École* (1794-1795)[24] was chaotic, with frequent changes of staff due to illnesses and political troubles (including imprisonment), and unrealistic syllabi which most students could not follow – resulting in improvised solutions [Langins *1987a*]; in the following years the situation stabilized, but there were only official, fixed programmes of teaching from 1800 onwards. Moreover, the habit of taking down summaries of the lectures only started in 1805, which does not facilitate the understanding of what was going on before that. Still, work has been done on this. [Langins *1987a*] is an excellent study of the first year of the *École*, and [Belhoste *2003*, 235-252] gives a very good survey of mathematics there before and during Lacroix's time.

One very important characteristic of the teaching of analysis at the *École Polytechnique* is its novelty. I believe that Belhoste exaggerates somewhat in his claim that "la méthode analytique n'[a] été enseignée nulle part de manière régulière et complète avant 1794. [...] l'étude des séries et surtout celle du calcul infinitésimal rest[aient] exceptionnelles"[25] [Belhoste *2003*, 234]: Bézout included a section on the calculus in his course for the *Gardes du Pavillon et de la Marine* [Bézout *1796*]; and so did Marie in [La Caille & Marie *1772*], a textbook that he probably followed in his lectures at the *Collège Mazarin*.[26] But the high level of the mathematics taught at the *École Polytechnique* seems really unprecedented – far beyond the level of Bézout's or Marie's textbooks. This means that a lot of experimenting was being done in the early years of the *École*, regarding what could be taught to a large number of students and how.

The first *instituteur* (i.e., professor) of analysis was Lagrange. His lectures are famous because of [Lagrange *Fonctions*], but it is not easy to know what in that book was taught in class. According to Prony [*1795b*] Lagrange's course of analysis in 1795 started with arithmetic (even number systems!), proceeded with the theory of series, and then went on to his power-series version of the calculus (so that [*Fonctions*] corresponds only to this last part). After a few lectures very few students could follow him, and his course was soon regarded as optional, and attended only by the best students.

---

[24]During this first year its name was *École Centrale des Travaux Publics*. But I will ignore this detail here.

[25]"the analytical method was not taught in a regular and complete manner anywhere before 1794. [...] the study of series and especially that of infinitesimal calculus were exceptional"

[26]It may also be relevant that the statutes of the University of Coimbra of 1772 established the regular teaching of differential and integral calculus in the second year of the new Faculty of Mathematics [Univ. Coimbra *Estatutos 1772*, III, pt. 2] – for this teaching the calculus section in Bézout's course was translated into Portuguese; even Belhoste acknowledges that Lagrange taught the calculus in an artillery school in Turin in 1758 and 1759, and that Euler appears to have done the same in St. Petersburg in the late 1720's [*2003*, 477].

Meanwhile, in this first year of the *École Polytechnique* Prony taught a course in "analysis applied to mechanics", with a quite surreal syllabus: his lecture notes [Prony *1795a*] are almost entirely devoted to the calculus of finite differences; they also include a summary of the six lectures where he addressed the fundamental principles of the differential calculus [Prony *1795a*, IV, 543-569]; and near the end he mentions in passing that he also gave lectures on mechanics [Prony *1795a*, 567]. Another surreal aspect of Prony's course is that it was for second- and third-year students, in spite of this being the first year that the *École* functioned (this was a consequence of the "revolutionary courses", and is explained in [Langins *1987a*]).

The first-year students had a course in "analysis applied to geometry". If the programme for this course was similar to that of the corresponding revolutionary course [Langins *1987a*, 130-131], and it probably was, it had three parts: the first part consisted in some advanced algebra (equations up to fourth degree, including approximation methods) culminating in analytic geometry; the second part included the rest of algebraic analysis (series, logarithms and exponentials, elementary probabilities), differential and finite difference calculus, and differential geometry; and the third part was mainly integral calculus (including partial differential equations and the method of variations).

According to Langins [Langins *1987a*, 78] this course was initially given by Monge, but many students could not follow it, and an easier course was given by Hachette (until both Monge and Hachette had to disappear temporarily for political reasons, further confusion ensuing). However, according to Belhoste and Taton [*1992*, 294-299] Monge's course was restricted ("restricted" may not be a good word) to the application of analysis to geometry – i.e., analytic and differential geometry; from this resulted [Monge *Feuilles*]. Presumably, either the students were initially expected to acquire the necessary analysis to be applied in Lagrange's lectures; or the more elementary course by Hachette was meant to cover that. An aspect that resulted from this confusion, and remained for several years, was some lack of correspondence between teaching posts and courses: the teachers of descriptive geometry (Monge and Hachette) would systematically teach part(s) of the course of "analysis applied to geometry" [Langins *1981*, 206]. Of course this makes it harder to understand what was going on. In 1800 the application of analysis to geometry was officially annexed to descriptive geometry [Éc. Pol. *Rapport*, an 9].

In the middle of the confusion, Fourier was recruited in 30 Floreal (19 May) to give a course in (algebraic?) analysis. But he was arrested less than three weeks later for being a Jacobin, and stayed in prison until Vendemiaire (October).

As has been said above, the situation became much more stable afterwards. In years 4 to 6 of the French Republic (1795-1796 to 1797-1798) Fourier gave regular lectures of analysis, "des mathématiques pour tous les élèves"[27] [Belhoste *2003*, 245]. Two manuscripts containing Fourier's own notes survive – one is kept at the *Bibliothèque de l'Institut de France*, and the other at the *École Nationale des*

---

[27]"mathematics for all students"

Ponts et Chaussées; unfortunately none of these has been published. But another one, with notes taken by one of his students (C. L. Donop) has been transcribed and published [Fourier *1796*]; and it gives us a good idea of Fourier's lectures in analysis (except for the integral calculus, which is not included there).

It seems that Fourier gave two courses in analysis: presumably one was for first-year students (but, at least in year 4, open to all students), while the other was for second-year students (and possibly third).[28] The first course was on "algebraic analysis" (the expression occurs, but is not yet predominant). Fourier was helpful in dividing it for us in two parts: "la 1ere considère les équations; la 2^{d} comprend les séries, suites arithmétiques, géométriques et récurrentes, les fractions continues, les logarithmes et le théorème de Côtes"[29] [Fourier *1796*, 19] – although he does not seem to have followed this particular order. The "séries" in the second part included expanding the usual transcendental functions (trigonometrical, exponential, logarithmic). There was some concern with convergence [Fourier *1796*, 89-90, 103]. Fourier sometimes used infinite and infinitesimal quantities, but he also gave alternative, algebraic methods (such as indeterminate coefficients), regarded as more rigorous.

Fourier's other course was on differential and integral calculus. His approach was a mixture of limits with power series. *Foundationally*, it was mainly based on limits: "L'objet du calcul des différences est de trouver le rapport de la différence de la fonction à la différence de la variable. [...] Le calcul différentiel ne considère que la limite de ce rapport"[30] [Fourier *1796*, 114].[31] But the fundamental *technique* used for differentiation was the expansion of the difference of the function into a series of powers of the difference of the variable; then, the limit of

$$\frac{\Delta y}{\Delta x} = A + B\Delta x^2 + C\Delta x^3 + \&\text{c.}$$

is easily obtained as

$$\frac{dy}{dx} = A.$$

But this technique also shared some of the conceptual burden: it is not clear whether he defined the differential dy as the first term in the expansion of Δy

---

[28] Apart or in connection with these he also taught descriptive geometry, Euclidean geometry, statics, hydrostatics and dynamics [Fourier *1796*, xv; Grattan-Guinness *1972*, 6-7]. But these subjects are not our concern here.

[29] "the 1st regards equations; the 2nd comprises series, arithmetic, geometric and recurring sequences, continued fractions, logarithms, and Cotes' theorem"

[30] "The purpose of the calculus of differences is to find the ratio between the difference of the function and the difference of the variable. [...] The differential calculus examines only the limit of that ratio"

[31] Belhoste [*2003*, 245], as well as Lorrain and Pepe [Fourier *1796*, xviii], associate Fourier's use of finite differences in introducing the differential calculus to Prony's lectures of year 3 [Prony *1795a*]. But Fourier uses finite differences in a traditional manner, similar to what Euler [*Differentialis*] and Cousin [*1777*; *1796*] had done, and Bossut [*1798*] was about to do. Prony's use of the calculus of finite differences instead of differential calculus is something quite different, and not necessary to explain Fourier's short references.

(changing Δx into dx), or he gave this only as a means to calculate the differential; but either way, he added that "le calcul différentiel considéré *analytiquement* est le calcul des 1ers termes des différences"[32] (my emphasis) [Fourier *1796*, 118]. Of course the expansions obtained previously in algebraic analysis were applied here. Differentials were used throughout, but derivatives ("fonctions dérivées") also appeared [Fourier *1796*, 172]. Fourier tried to combine an analytical with a geometrical approach: for instance, he introduced the treatment of maxima and minima by studying behaviour of curves [*1796*, 183-190], but followed it with a power-series analysis, using what I call Arbogast's principle [*1796*, 190-192]. Good students would then be able to follow Lagrange and/or Monge.

As for integral calculus, the published manuscript [Fourier *1796*] does not include it. Grattan-Guinness [*1972*, 6-7], based on the Paris manuscripts, mentions "foundations of integral calculus"[33], applications to geometry (probably calculation of areas, and so on), and calculus of variations. Ordinary differential equations were likely to be included, but not partial differential equations. The latter were probably taught by Monge and Hachette, associated with differential geometry.

In May 1798 Fourier was invited to join the scientific expedition that accompanied Napoleon's Egyptian campaign. He accepted and Jean-Guillaume Garnier was recruited to replace him during his absence. Garnier stayed in the *École Polytechnique* until 1802. There are plenty of sources to study Garnier's teaching, but most of them not published (at least in the usual sense) or rare: a manuscript programme of his course of differential and integral calculus, sent to the examiners Laplace and Bossut at the end of year 7 (1798-1799) is kept at [Éc. Pol. *Arch*, III3b]; he published textbooks on algebraic analysis and differential and integral calculus [*1801*; *1800*][34]; and he had printed lecture notes distributed to the students [Garnier *1800-1802*].[35] True, [Garnier *1800-1802*; *1800*; *1801*] are all *contaminated* by the official programmes approved in 1800, when Lacroix was already at the *École*. But the similarities with his personal programme of 1799 and with Fourier's lectures suggest a deep continuity.

But let us start with the time allocation for courses decided by the Council of the *École* on 12 Frimaire year 7 (2 December 1798) [Éc. Pol. *Extraits Conseil*, 62]: first-year students would have a year-long course on "the method of indeterminate coefficients, the theory of higher-degree equations, the application of algebra to geometry, the introduction to differential calculus, and the differential calculus"; while the second-year students would have a 4-month course on integral calculus,

---

[32]"the differential calculus, regarded *analytically*, consists in calculating the first terms in the differences"

[33]Presumably, in this context "foundations" means introductory remarks and integration of explicit functions, not elaborated conceptual work.

[34][Garnier *1801*] is relatively common. But [Garnier *1800*] seems quite rare – no copies at the *École Polytechnique, Bibliothèque Nationale de France*, or British Library; oddly, there are copies in the Faculty of Science of Porto and Science Museum of Lisbon (with some differences between them – see the Bibliography below).

[35]The text of these lecture notes seems very close to that of his published textbooks, although with frequent changes in order.

with applications taken from [Monge *Feuilles*][36]. The analytic and differential geometry implied in the last sentence were certainly taught by Hachette[37]. The rest was taught by Garnier.

It is clear that Garnier gave considerable importance to algebraic analysis. His first-year lecture notes [Garnier *1800-1802*, I-III] contain 16 leaves of algebraic analysis, against 18 of differential calculus, and 9 of integral calculus; and in the preface to that set, he implies that he taught more on algebraic analysis than what was specified in the official programme recently approved[38]. Interestingly, and unlike Fourier's case, his algebraic analysis does not include the expansions in series of transcendental functions – these are only dealt with in the differential calculus. Instead, it mostly addresses the theory of equations.

As for differential calculus, Garnier's approach is very similar to Fourier's: there are introductory sections on finite differences and on limits; then it is *proven* that the increment $f(x + \Delta x)$ of a function may be expanded into a series of powers of Δx; from this follows that the limit of a ratio such as $\frac{\Delta y}{\Delta x}$ is the first term in its expansion; and the differential calculus consists in determining these first terms; the *differential* is "the part of the difference suitable to give the limit, having substituted d for Δ [Garnier *1800*, 380-381; *1800-1802*, II, n° 5]. The main difference from Fourier, as has already been noted, is that Garnier does not have the expansions of transcendental functions beforehand, and so he has to obtain them here;[39] one might think that this is an influence from the official programme approved in 1800, but Garnier's programme of differential calculus of 1799 already used Taylor's theorem for those expansions. We may also notice some greater detail on limits and differences, and less pedagogical use of geometrical considerations; but these may be due to the difference between manuscript notes and printed, more or less published notes – and possibly also to the increase in allocated time to analysis lectures from year 4 to year 9 [Belhoste *2003*, 247].

There is not much to say about the integral calculus, except that Garnier does not address either the calculus of variations or partial differential equations. He explicitly mentions that the teaching on partial differential equations was trusted to Monge and Hachette [Garnier *1800*, 826] or Monge [Garnier *1800-1802*, VI, n° 32].

---

[36]"Cours de Calcul intégral dont on prendra des applications dans la suite des feuilles de l'analyse géometrique de Monge"

[37]Monge was in Egypt. Hachette published that year [Monge & Hachette *1799*] to compensate for the lack of material on space curves in the first edition of [Monge *Feuilles*]

[38]"lorsque le programme nous fut remis, mes leçons d'algèbre [était] préparées et le cours engagé [...] et si le cours d'analyse algébrique que j'ai fait n'est pas textuellement celui qui est exigé, au moins le comprend-il en entier" ("when the programme was sent to us, my lectures in algebra [were] prepared and the course had began [...] and if the course in algebraic analysis that I have given is not word for word the one that is required, at least it comprises it in full")

[39]But, according to an addition ("Note sur les numéros 6, 7, 8 et 9") to [Garnier *1800-1802*, II], in one of his courses Fourier used functional equations ("propriété[s] caractéristique[s]") to obtain the differentials of transcendental functions, and then used these to arrive at their expansions.

8.3   Lacroix in the *École Polytechnique*

In Brumaire year 8 (November 1799) Lagrange resigned from his post of *instituteur* for health reasons. Lacroix was chosen to replace him. Lagrange had suggested that the person who was to replace him should teach obligatory courses. It appears that at first the *Conseil d'Instruction* of the *École* did not wish to follow this suggestion: in the meeting of 28 Brumaire (19 October) Garnier was charged with a first-year course in algebra (45 lectures) and differential calculus (40 lectures), and a second-year course in differential and integral calculus (40 lectures); while Lacroix was invited to give an optional course for the best students (with only one lecture every 10 days) [Éc. Pol. *Arch*, X2c/30, II, 53-54]. But in later meetings there were some discussions on how to improve the course distribution, and at the end of that school year the examinations of first-year students followed Lacroix's programme of algebra and differential calculus, while those of second-year followed Garnier's programme of differential and integral calculus [Éc. Pol. *Arch*, II, 109-110].[40]   For the following year it was decided that Lacroix would teach the second year and Garnier the first year [Éc. Pol. *Arch*, II, 102]. This scheme of alternation, so that each student would have the same teacher for the two years (provided he passed), was kept thenceforth. Lacroix taught first-year courses in 1799-1800, 1801-1802, 1803-1804, 1805-1806, and 1807-1808; and second-year courses the alternating years until 1808-1809 (inclusive). In 1809 he left the post of teacher for the higher-ranking one of permanent examiner,[41] which he kept until 1815.

On 25 Frimaire year 8 (16 December 1799), little over two months after Lacroix's appointment, a new organization for the *École* was decreed. One of the novelties was that a new body, the *Conseil de Perfectionnement*, should fix official syllabi every year. Monge, Garnier and Lacroix prepared the project of syllabus of analysis [Belhoste *2003*, 248]. Lacroix prepared a document that is transcribed in appendix C.2.1 below. It contains the radical proposal of abolishing algebraic analysis. For Lacroix, the subject that was really important for the students of the *École Polytechnique* was the differential and integral calculus; he did not see the point in teaching them the theory of equations – excepting the best students, those attracted by pure mathematics. Thus the binomial formula in the cases of negative or fractionary exponent, and the series expansions of trigonometric and logarithmic functions, would be obtained with differential calculus, using Taylor's theorem. This is fully consistent with what we have seen in section 8.1 on his opinion about [Lacroix *1800*].

---

[40]Considering the summaries of Lacroix's lectures on differential and integral calculus transcribed in appendix C.1 and almost certainly related to this year, it is possible that Garnier gave 25 lectures on algebraic analysis, and Lacroix took over afterwards; or that Lacroix started the course afresh and hence gave only 54 lectures (including algebraic analysis) instead of the 85 assigned.

[41]According to the *Registre de Contrôle des Instituteurs et Agents* [Éc. Pol. *Arch*, X2c26], he had already fulfilled the duties of examiner in 1808 (seemingly in a temporary way), but was only appointed for the post in 1809.

But the programme that was approved by the *Conseil de Perfectionnement* was not the one propounded by Lacroix: it included extensive sections in algebraic analysis, both in the first and second years (see appendix C.2.2). In the following years the programme became less detailed, and the section on algebraic analysis was shortened. But it was never short enough for Lacroix. When we compare the official programmes for 1805-1806 and 1806-1807 with the summaries of Lacroix's lectures in those years (appendices C.3.1 and C.3.2), we see that Lacroix did not teach all of the algebraic analysis that he should in the first year (he missed the expansion of functions using indeterminate coefficients), and he ignored it in the second year (it was his *adjoint* Ampère who gave three lectures on solving 3rd- and 4th-degree equations, after Lacroix had declared the course finished; he had been explicit in 1800 about the limited usefulness of this).

8.4 From the large *Traité* to the *Traité élémentaire*

In the first year(s) that he taught at the *École Polytechnique*, Lacroix used his large *Traité* as a supporting text. We know this from a manuscript syllabus kept at the Wellcome Institute, London (appendix C.1): next to each lecture it indicates the corresponding articles in the *Traité*.

But of course the large *Traité* was not a textbook, and the same manuscript also shows how Lacroix adapted it. The first, obvious, change is the reduction in covered subject matter: the *Traité* addresses much more that what the students at the *Polytechnique* had to (or could) study, and we can see that out of the 403 articles in the first volume of the *Traité*, only about 100 appear in the syllabus.

A second change is in the order in which some topics of differential calculus are treated[42]: the exposition is more driven by pedagogical concerns and less tightly packed into subjects – for instance, Taylor series for functions of one variable appear before the differential calculus of functions of two variables, and maxima and minima appear in the middle of the discussion of special points of curves.

A third change is in foundations: limits instead of power series (more on this in section 8.5 below). Fortunately, Lacroix had addressed limits in the Introduction of the large *Traité*, and so he could support his first lecture on the principles of differential calculus with some articles from the Introduction.

In 1802 Lacroix took the obvious next step: the publication of this adapted version of the large *Traité* (with some further changes) as a book – a textbook, to be followed in his lectures; this was his *Traité élémentaire de Calcul différentiel et de Calcul intégral* [1802a].

Table 8.1 shows the contents of the first edition of the *Traité élémentaire* (succinctly), and how they correspond to the chapters of the large *Traité*.

Here we see a further change in the order of subjects (once again, only for differential calculus): first, functions of one variable, including analytical and geometrical applications; only after that are treated functions of two or more variables.

---

[42]The order in integral calculus is kept.

Part I − differential calculus		Large *Traité*
Topics	Pages	(vol. 1)
Differentiation of functions of one variable and of equations in two variables	1-55	Ch. 1
Maxima and minima of functions of one variable; indeterminacies	55-75	Ch. 2
Application of differential calculus to the theory of curves	75-143	Ch. 4
Change of independent variable; differentiation of functions of two or more variables	143-172	Ch. 1
Maxima and minima of functions of two variables	172-179	Ch. 2
General notions on the application of differential calculus to curves of double curvature and surfaces	179-186	Ch. 5
Part II − integral calculus		Large *Traité*
Topics	Pages	(vol. 2)
Integration of functions of one variable	187-309	Ch. 1
Application of integral calculus to quadrature, rectification, and cubature	309-341	Ch. 2
Integration of differential equations in two variables	341-430	Ch. 3
Integration of functions of two or more variables	430-461	Ch. 4
Method of variations	461-488	Ch. 5
Appendix − Differences and series		Large *Traité*
Topics	Pages	(vol. 3)
Calculus of differences (direct; inverse; equations in two variables)	489-557	Ch. 1
Application of integral calculus to the theory of sequences	557-570	Ch. 3

Table 8.1: Lacroix's *Traité élémentaire de Calcul différentiel et de Calcul intégral*

In the Preface to the second edition of the large *Traité*, Lacroix remarked that in chapter 1 of the first volume he had given the complete exposition of the principles of differential calculus, "at one stroke"; but "dans un livre élémentaire, cette marche retarderait trop les applications, si nécessaires pour soutenir le courage d'un lecteur qui s'engage pour la première fois dans une carrière dont il n'apperçoit pas le but"[43] [*Traité*, 2nd ed, I, xx].[44]

    We can also confirm the difference in size: the three volumes of the first edition of the large *Traité* have 1790 quarto pages in total; the first edition of the *Traité élémentaire* has 574 octavo pages – the latter is about one sixth of the

---

[43]"in an elementary book, that process would delay applications too much, [and they are] so necessary to keep up the heart of a reader who does not notice the goal of a course that he undertakes for the first time"

[44]For some reason, from the third edition of the *Traité élémentaire* onwards Lacroix reverted to an order closer to that of the large *Traité*: differentiation of functions of more than one variable and of equations in two variables before any applications (but change of independent variable and differentiation of equations in more than two variables between applications to planar geometry and space geometry).

former. Naturally, the most advanced subjects are the ones that suffer most in this reduction: differential geometry in space, partial differential equations, and the whole appendix on differences and series (but especially difference equations). For example, in the *Traité élémentaire* partial differential equations have one third of the space dedicated to ordinary differential equations, while in the large *Traité* they have almost the same number of pages; "differences and series" are reduced from about one third of the large *Traité* to one seventh of the *Traité élémentaire*.

With these rearrangements and reduction, and with the loss of such special characteristics as the subject index and the bibliography in the table of contents, Lacroix's *Traité élémentaire* became less encyclopedic than his large *Traité*. But that is only natural in a *textbook*. Its scope was much narrower than that of the large *Traité*. Still, it was a landmark textbook. It was very far from the common textbooks of the 18th century, such as Bézout's, both in content and in mathematical style. Pierre Lamandé [*1988*] compared [Lacroix *1802a*] with Bézout's section on the calculus; the comparison is quite relevant because Bézout's text was popular well into the 19th century. Lamandé remarked the huge gap that existed between Bézout's text (and other pre-revolutionary textbooks) and mathematical research [*1988*, 23].[45] Lacroix's *Traité élémentaire*, on the other hand, pointed in the direction of contemporary mathematics, even if it did not prepare the students for understanding research works (which was the aim of the large *Traité*).

The success and influence of [Lacroix *1802a*] are undeniable. It had five editions in Lacroix's lifetime (1802, 1806, 1820, 1828, 1837), and four posthumous ones (1861-1862, 1867, 1874, 1881)[46]; less than his textbooks on more elementary subjects, but much more than usual for a calculus textbook. Translations were published in Portuguese, English, German (two), Polish, and Italian; in addition, a Greek translation was made but not published (see below). The English translation is famous for its importance in introducing Continental calculus in Britain.

Of course, part of its influence came from being the "reference work" on the calculus in the *École Polytechnique* until about 1815 [Belhoste *2003*, 249].[47] But it must not be reduced to a *Polytechnicien* text. As was mentioned in section 8.1, it was adopted also for the *Lycées*; Lacroix probably used it in the *Faculté des Sciences* and in the *Collège de France*; and only two out of its nine editions appeared during its *Polytechnicien* period. Moreover, it was never a perfect fit for the course of analysis at the *École Polytechnique*: it does not contain algebraic analysis, and it does address partial differential equations. It could and did live a

---

[45]Lamandé has also compared [Lacroix *1802a*] with [l'Hôpital *1696*], in [Lamandé *1998*]. A detail in the title of this paper is quite eloquent: "Une même mathématique?" ("The same mathematics?"). Still, there is a point in common between [Lacroix *1802a*] and [l'Hôpital *1696*]: both were *modern* when they were written; the same cannot be said of Bézout's text.

[46]The posthumous editions, in two volumes, contain extensive endnotes by Joseph Alfred Serret and Charles Hermite, necessary to bring it up to date.

[47]Actually, part of its influence may have been lost before 1815. From 1808, Ampère had introduced some developments of his own on the use of limits; and in 1812, limits were officially replaced by infinitesimals [Fourcy *1828*, 303].

life of its own.

Garnier's texts [*1800*; *1800-1802*] are comparable to [Lacroix *1802a*], if we except their lack of treatment of partial differential equations, calculus of variations, and finite differences. These are important exceptions; but one can imagine that, if Garnier's textbooks had not had such a restricted distribution, they might have been serious competitors. As it happened, [Lacroix *1802a*] was the foremost textbook on the calculus in the early 19th century (when Garnier published enlarged editions [*1811*; *1812*] it was a little too late to make a stand).

Of course, in spite of the differences, much of the quality and modernity of [Lacroix *1802a*] result from the fact that it is a by-product of [Lacroix *Traité*]. Certainly not many textbooks have resulted from such an amount of work.

In the following sections we will look at what happened in the *Traité élémentaire* to the aspects of the large *Traité* that have been studied in chapters 3-7. We will focus mainly on the first edition (1802), but also look at the second (1806) and third (1820) editions, still chronologically close to the large *Traité*; only occasionally will later editions be mentioned.

8.5   The principles of the calculus

The most famous difference between [Lacroix *Traité*] and [Lacroix *1802a*] is foundational: in the latter Lacroix wished "un degré suffisant de rigueur et de clarté"[48], but without the lengths entailed by certain unnecessary details [*1805*, 384], and for this reason he decided to use limits (always calculated naïvely). These "unnecessary details" were almost certainly the whole Introduction of the large *Traité*, and the *proof* that $f(x + k) - f(x)$ may be expanded into a series of powers of k prior to the introduction of differential coefficients.

However, in the first edition of [Lacroix *1802a*] we still find several remnants of the power-series foundation of the large *Traité*. Let us examine the foundations of the calculus in [Lacroix *1802a*, 1st ed].

After defining *function*, *variable* and *constant*, Lacroix explores the relations (and particularly the ratios) between the increments of a variable and of functions of that variable. If $u = ax^2$, putting $x + h$ in the place of x and calling u' the new value of u, we have

$$\frac{u' - u}{h} = 2ax + ah.$$

This ratio is clearly divided into two parts, one independent of and the other dependent on h. As h is supposed to decrease, the ratio keeps approaching $2ax$, not reaching it unless $h = 0$. Thus, $2ax$ is the limit of $\frac{u'-u}{h}$, "c'est-à-dire, la valeur vers laquelle il tend à mesure que la quantité h diminue et dont il peut approcher autant qu'on le voudra"[49] [Lacroix *1802a*, 3].

---

[48]"a sufficient degree of rigour and clarity"

[49]"that is, the value towards it tends as the quantity h diminishes, and which it can approach as much as one might wish".

A similar situation occurs if $u = ax^3$, since in that case

$$\frac{u' - u}{h} = 3ax^2 + 3axh + ah^2,$$

which has $3ax^2$ as limit. Lacroix then remarks that to find such a limit it is enough to consider the first term in the difference

$$u' - u = 3ax^2h + 3axh^2 + ah^3, \tag{8.1}$$

and he extrapolates this for every function. He assumes that the increment of any function can be expanded into a power series of the increment of the variable; but he never states this explicitly – only that "this first term, or this limit" of the ratio between the increments always exists.

Later in the book, when introducing the geometrical applications of the differential calculus, Lacroix emphasizes this point: it is an *analytical fact* that all functions admit a limit in the ratio between their increments and those of the independent variable; consideration of limits allows to express the "law of continuity" in the calculus [*1802a*, 75-76]. The "law of continuity" is not very easy to understand, but refers to the situation in which "les point consécutifs d'une même ligne se succèdent sans aucun intervalle"[50]; in the "calcul" one always presumes an interval between consecutive values, but limits compensate for this. Lacroix *proves* the existence of the limit between the ratios of the increments, by establishing an equivalence between a function and a (graph-)curve and assuming the existence of a tangent at any point of this curve [*1802a*, 76-77].

Back to the beginning of the book: the first term in (8.1) receives the name *differential*, because it is only a portion of the *difference* of the function [Lacroix *1802a*, 4]. It is also given the notation du, so that in this case $du = 3ax^2h$. But in the case of a simple variable, the difference and the differential are the same, that is, $dx = x' - x = h$. Thus, h is replaced by dx "afin de mettre de l'uniformité dans les calculs"[51], and

$$du = 3ax^2dx \qquad \frac{du}{dx} = 3ax^2.$$

$\dfrac{du}{dx}$ is christened *differential coefficient* because it is the multiplier of dx in the expression of the differential. Notice how all these fundamental concepts are introduced by examples, presumed to be generalizable.

The immediate relations between differential and differential coefficient are useful, because in some cases it is easier to find the former and in others to find the latter. It is more direct to substitute $x + dx$ for x, expand the function in powers of dx, and extract the term with the first power; but this requires that one

---

[50]"consecutive points of a line succeed each other without interval"
[51]"to put uniformity in the calculations"

knows how to expand the proposed function, which in some cases demand "secours étrangers"[52] – in those cases, limits often save us that trouble.

Thus, in most cases Lacroix uses power-series arguments: for instance, the differential of $u = a^x$ is obtained by putting $a^{x+dx} - a^x = a^x(a^{dx} - 1)$, and expanding $a^{dx} = (1 + b)^{dx}$ using the binomial theorem [*1802a*, 23-24].[53] But the series expansions of the trigonometric functions are much more involved, and when it comes to $\sin x$ he uses an argument free of power-series: using a few trigonometric identities,

$$\frac{\sin(x + dx) - \sin x}{dx} = (\sin x \frac{\sin dx}{1 + \cos dx} + \cos x)\frac{\sin dx}{dx};$$

and, as dx vanishes, $\sin dx$ becomes 0, $\cos dx$ becomes 1, and $\frac{\sin dx}{dx}$ tends also to 1, so that the right side of this equality tends to $\cos x$ [*1802a*, 33-34].

Lacroix declares that the differential calculus consists in finding "la limite du rapport des accroissemens simultanés d'une fonction et de la variable dont elle dépend"[54] [*1802a*, 5]; in the introduction to the geometrical applications he also expresses the following opinion:

> "Il me paroît maintenant très-évident que la métaphysique précédente renferme l'explication philosophique des propriétés du Calcul différen-tiel et du Calcul intégral, soit par rapport aux recherches sur les cour-bes, soit par rapport à celles qui concernent le mouvement."[55] [Lacroix *1802a*, 76]

Nevertheless, one cannot fail to notice several passages similar to those in [*Traité*, I], where the power-series approach was followed; the definition of the differential as the first term in the development of the difference of the function is striking.

Also striking is the similarity between this approach and those of Fourier and Garnier: limits as the main foundation; power series as the main technique and intervening in the definition of the differential. Where Lacroix departs from his predecessors, he is a little less rigorous: both Fourier and Garnier had tried to prove the general validity of the power-series expansion; Lacroix simply assumes it.

It is not inconceivable that this similarity with Fourier and Garnier is a result of influence from them (or from a tradition in the *École Polytechnique*). But Lacroix's advocacy of limits in [*1802a*] seems quite sincere (and there are several

---

[52]"extraneous assistance"

[53]There is a serious (but common) problem here. Lacroix had obtained the binomial expansion in two ways, but both dependent on the differential coefficient of x^n being nx^{n-1}; and he had only derived this for rational n.

[54]"the limit of the ratio between the simultaneous increments of a function and of the variable on which it depends"

[55]"It now seems to me very clear that the preceding metaphysics comprises the philosophical explanation of the properties of the differential and integral calculus, both in relation to researches on curves and in relation to researches concerning movement."

later texts supporting it). And it is not necessary to invoke such an influence in order to explain his use of power series: it was mentioned in section 3.1.4 that Cousin [*1777*; *1796*], for instance, used power-series expansions in a context of limit-based calculus; moreover, would it not be easier for Lacroix to simply adapt most of the power-series arguments in his large *Traité*, rather than create new ones?

In the second edition there are a few little, but important, changes. Most of the preliminary considerations remain, but Lacroix adds two simple theorems on limits, includes a proof of the chain rule (in the first edition it was simply assumed, in a Leibnizian way), and replaces the assumption of power-series expansion for something a little different, when deriving some differentiation rules involving general functions (such as the product rule, or the chain rule itself). The two theorems on limits are: the limit of a product is the product of the limits; and the limit of the quotient is the quotient of the limits. The former is proved thus: let p and q be the limits of P and Q, respectively; then $P = p + \alpha$, and $Q = q + \beta$, where α and β are "quantités susceptibles de s'évanouir en même tems, après avoir passé par tous les degrés de petitesse"[56]; we have

$$PQ = (p + \alpha)(q + \beta) = pq + p\beta + q\alpha + \alpha\beta,$$

and the limit of the rightmost expression is pq, as we can see by putting $\alpha = 0$ and $\beta = 0$, and noticing that "en donnant aux quantités α et β des valeurs convenables, on peut rendre aussi petite qu'on voudra la différence"[57] [*1802a*, 2nd ed, 8]. As for the limit of the quotient, the argument is similar: the difference turns out to be

$$\frac{P}{Q} - \frac{p}{q} = \frac{q\alpha - p\beta}{q(q + \beta)},$$

and can also be made as small as we wish.[58]

The theorem on the limit of the product is applied to prove the chain rule: let v be a function of u and u be a function of x; let them simultaneously become v', u' and x'; the limits of $\frac{v'-v}{u'-u}$ and $\frac{u'-u}{x'-x}$ will be $\frac{dv}{du}$ and $\frac{du}{dx}$, respectively; therefore the limit $\frac{dv}{dx}$ of $\frac{v'-v}{x'-x} = \frac{v'-v}{u'-u} \times \frac{u'-u}{x'-x}$ will be $pq = \frac{dv}{du} \times \frac{du}{dx}$ [*1802a*, 2nd ed, 9].

Another limit argument is used to derive the differential of the product of two functions u and v. In the first edition, Lacroix had written $u + p\,dx + $ etc. and $v + q\,dx + $ etc., multiplied these series, and extracted the dx term [*1802a*, 1st ed, 9], reproducing [*Traité*, I, 102]. In the second edition, instead of assuming the

---

[56]"quantities capable of vanishing simultaneously, after passing through every degree of littleness"

[57]"assigning appropriate values to α and β, we can make the difference as small as we wish"

[58]Grabiner found these simple arguments "important because they exemplify translations of a verbal limit concept into algebraic language, however simple" [Grabiner *1981*, 84]. That is true, but she appears to speak of them only as examples of a kind of argument that sometimes appeared around 1800; in other words, they are not major breakthroughs. For instance, the Portuguese mathematicians José Anastácio da Cunha and Francisco Garção Stockler had given more sophisticated arguments [Domingues *2004a*], as had l'Huilier.

power-series expansions, he writes the incremented states of u and v as $u + \alpha$ and $v + \beta$; we have

$$\frac{(u + \alpha)(v + \beta) - uv}{\mathrm{d}x} = u\frac{\beta}{\mathrm{d}x} + v\frac{\alpha}{\mathrm{d}x} + \frac{\alpha}{\mathrm{d}x}\beta,$$

and since β vanishes with $\mathrm{d}x$, the limit of this is $u\frac{\mathrm{d}v}{\mathrm{d}x} + v\frac{\mathrm{d}u}{\mathrm{d}x}$ [1802a, 2nd ed, 11-12].

Thus, while a case could be made for a mixture of approaches in the first edition, the second edition has a more clear-cut option for limits.

In the third edition, this option is a little strengthened. There remained in the second edition at least one instance of the assumption of power-series expansion of the difference of any function; now it disappears [Lacroix 1802a, 2nd ed, 6; 3rd ed, 6]. And more importantly, an endnote on "the method of limits" is added [1802a, 3rd ed, 625-631], developing Lacroix's advocacy of limits, mainly based on geometrical arguments. It is interesting to read that the consideration of limits "est aujourd'hui la meilleure base que l'on puisse donner au Calcul différentiel"[59] [1802a, 3rd ed, 628]; this was published in 1820 – his former student Cauchy was then giving these words a meaning that far surpassed Lacroix's.

In fact, in spite of the option for limits, there seems to be no trace in the fourth and fifth editions (1828 and 1837) of Cauchy's new foundations for the differential calculus. Or, to be more precise, the little trace that there is is negative. In 1822 Cauchy published a refutation of Lagrange's power-series foundation (and of similar uncritical uses of power series as representatives of functions): taking $f(x) = e^{-\frac{1}{x}}$, we have $f(0) = f'(0) = f''(0) = \ldots = 0$, so that its Taylor series around zero is null, although $f(x)$ is not. Moreover, this means that for any function $g(x)$, the Taylor series of $g(x) + f(x)$ around zero is indistinguishable from that of $g(x)$, and therefore does not represent any of them in particular. A counter-refutation, attributed by Grattan-Guinness [1990, II, 735-736] to Poisson, appeared quickly, arguing that not all of the differential coefficients of $e^{-\frac{1}{x}}$ in zero are null; and this explanation was included as a footnote in [Lacroix 1802a, 4th ed, 338-339].[60]

8.6 Analytic and differential geometry

First of all, let us note that, contrary to the large Traité, there is practically no analytic geometry in Lacroix's Traité élémentaire de calcul... The place for

---

[59]"is nowadays the best basis we can give to the differential calculus"

[60]Actually, the fact that all the differential coefficients at $x = 0$ of $e^{-\frac{1}{x}}$ are zero had appeared in [Euler Integralis, I, § 327], in [Lacroix Traité, II, 149], and in the earlier editions of [Lacroix 1802a] (in the context of the application of the improved version of Euler's "general method" for approximation of integrals; it is at this point in the fourth edition that Lacroix includes Poisson's explanation). But this did not seem to bother anyone, probably because they were used to the Taylor series to fail in particular points (see section 3.2.5). The detail added by Cauchy to the effect that this destroys what we would call the bijection between functions and Taylor series was certainly more disturbing.

analytic geometry in his *Cours de mathématiques* was the *Traité élémentaire de Trigonométrie [...] et d'Application de l'Algèbre à la Géométrie* [*1798b*] – and when applying the calculus to geometry Lacroix often invokes results from that other textbook, in the form "(*Trig.* 146)" [*1802a*, 80].

The only exception to the absence of analytic geometry is the introduction of polar coordinates and their transformation to and from rectangular coordinates [*1802a*, 134; 136-137]. The context is the study of spirals, which are not treated in [Lacroix *1798b*]. In the large *Traité* polar coordinates also appeared apropos of spirals and separated from the rest of analytic geometry.

On differential geometry of plane curves, or rather "application of differential calculus to the theory of curves" [*1802a*, 75-143], the most important difference relative to the large *Traité* is the exclusive use of limits (recall from above that this section starts with considerations on the metaphysics of the calculus based on limits). We have seen in sections 4.1.2.1 and 4.2.1.2 that in [*Traité*, I, ch. 4] Lacroix had used five approaches to calculate tangents to curves: 1 - using transformation of coordinates (supported by a limit argument); 2 - using the series expansion of the equation of the curve, obtained by algebraic means (supported by Arbogast's principle); 3 - using differential calculus (also supported by Arbogast's principle); 4 - using the method of limits directly; and 5 - using infinitesimals. In [Lacroix *1802a*] there is only one approach (recall from section 8.1 his rejection of duplications in textbooks): consider a given curve, and another having two points in common with the former; if the coordinates of the first point of intersection are x', y', and the general coordinates of the second curve are x, y, then for that first point of intersection we will have $y = y'$; if in addition h is the difference between the abscissas of the points of intersection, then

$$y + \frac{dy}{dx}h + \text{etc.} = y' + \frac{dy'}{dx'}h + \text{etc.},$$

whence

$$\frac{dy}{dx}h + \text{etc.} = \frac{dy'}{dx'}h + \text{etc.};$$

now, dividing the latter equation by h and then taking the limit for $h = 0$, we get $\frac{dy}{dx} = \frac{dy'}{dx'}$; if the second curve is to be a straight line $y = Ax + B$, then $\frac{dy}{dx} = A$, and thus the equation of the tangent of the first curve at x', y' is

$$y - y' = \frac{dy'}{dx'}(x - x').$$

That is, we have a naïve limit argument with a little help from Taylor series.

Similarly, and since three points determine a circle, the osculating circle to a curve is introduced by considering those three points on the given curve, and then examining what happens when the three points coincide [Lacroix *1802a*, 112-114].

There is also some reduction in topics addressed. The most marked absence is that of envelopes – except in the particular case of the evolute, which is seen to be the "limit" of the intersections of the normals [Lacroix *1802a*, 117].

As for differential geometry in space, it is reduced to just some "general notions on the application of differential calculus to the theory of curves of double curvature and of curved surfaces" [Lacroix *1802a*, 179-186]. Only the most simple problems. For space curves: tangent lines, osculating planes and normal planes. For curved surfaces: the "law of continuity" $\frac{\mathrm{d}^2 z}{\mathrm{d}x\mathrm{d}y} = \frac{\mathrm{d}^2 z}{\mathrm{d}y\mathrm{d}x}$, differential equations of sections, tangent planes, and normal planes. No evolutes of space curves, no curvature of surfaces, no families of surfaces. Notice also the order (curves first, then surfaces), reversed from that of the large *Traité*.

I have not noticed any relevant changes in the second edition.

The same cannot be said for the third edition: the space dedicated to differential geometry in space more than triples. There are now three sections. The first is on the "application of differential geometry to the theory of curved surfaces" [Lacroix *1802a*, 3rd ed, 189-205]: apart from what was already in the first and second editions, it includes generation of surfaces (that is, a short introduction to families of surfaces), and curvature of surfaces. The following section is not exclusively on differential geometry, but rather "on singular points of curved surfaces, and on maxima and minima of functions of several variables" [*1802a*, 3rd ed, 205-212]. Finally, the third of these sections is "on the application of differential calculus to curves of double curvature, and on developable surfaces" [*1802a*, 3rd ed, 212-224]: apart from what was already in the first and second editions, and from an introduction to developable surfaces, it contains more details than the large *Traité* on the two "curvatures, or *flexions*" of space curves [*1802a*, 3rd ed, 221-224; *Traité*, 2nd ed, I, 632-633].

8.7 Approximate integration and conceptions of the integral

8.7.1 Conceptions of the integral and approximate integration of explicit functions

Lacroix's conceptual reflections on integrals, treated in sections 5.2.2 and 5.2.3, were naturally appropriate for inclusion in an educational version of his *Traité*.

The syllabus of the first course of analysis given by Lacroix at the *École Polytechnique* effectively includes them, under the heading "de la détermination des Constantes dans les Intégrales"[61] (see page 408). From the articles linked to this entry we can conclude that in the 6th lecture on integral calculus Lacroix spoke about Euler's approximation method (with very few details and no applications), about his interpretation of the integral as a sum or a limit of sums (but did not give either of the two proofs involving limits), about the distinctions between the integral and a given primitive function and between definite and indefinite integrals, and about the geometrical interpretations of all this.

---

[61]"on the determination of the constants in the integrals"

In [Lacroix *1802a*] he was a lot more detailed. In fact, he reproduced practically the entire section on the "general method to obtain approximate values of integrals" from the large *Traité* [Lacroix *1802a*, 284-309]. The extra details were almost certainly not taught at the lectures, but rather left for smart students to read.

Significant alterations were introduced in the second edition of [Lacroix *1802a*]. The articles directly addressing definite and indefinite integrals and arbitrary constants were joined and transferred to the beginning of the section. This made Lacroix's explanations clearer, but also more conventional and less attached to the conception of the integral as sum or limit of sums [*1802a*, 2nd ed, 303-304]:

"Si $\int X \mathrm{d}x = P + C$, P désignant la fonction variable déduite immédiatement du procédé de l'intégration, C la constante arbitraire, et que l'intégrale doive, s'évanouir pour une valeur $x = a$ qui change P en A; on posera l'équation $A + C = 0$, de laquelle on tire

$$C = -A \quad \text{et} \quad \int X \mathrm{d}x = P - A.$$

Sous cette forme l'intégrale $\int X \mathrm{d}x$ n'est plus que la différence entre la valeur que prend la fonction P lorsque $x = a$, et celle qu'elle acquiert pour toute autre valeur de la même variable. Si, par exemple, $x = b$, change P en B, il vient

$$\int X \mathrm{d}x = B - A.\text{"}[62]$$

This, of course, is the "definite integral", taken "from $x = a$ to $x = b$", and so on.

The derivation of the approximation formulas is also different. Lacroix postpones the neglecting of higher-order terms in the Taylor series, but assumes quite early that the subintervals are all equal, so that the formula at which he arrives first is equivalent to (5.11). (5.8) does not occur; instead, we see

$$\int X \mathrm{d}x = A\alpha + A_1\alpha + A_2\alpha \ldots \ldots + A_{n-1}\alpha$$

(the A_i's correspond to the Y_i's in [Lacroix *Traité*]). Notice that because of the change in the order of presentation, the approximation formulas can be introduced as formulas for definite integrals (as is the case for this one).

---

[62]"If $\int X dx = P + C$, P denoting the variable function immediately deduced by the process of integration, C the arbitrary constant, and if the integral ought to vanish for a value of $x = a$, which changes P into A; we shall then have the equation $A + C = 0$, from which we deduce

$$C = -A, \quad \text{and} \quad \int X dx = P - A.$$

Under this form the integral $\int X dx$ is nothing more than the difference between the value of the function P, when $x = a$, and that which it acquires for every other value of the same variable. If, for example, $x = b$, changes P into B, there arises

$$\int X dx = B - A.\text{"}$$

[Lacroix *1816*, 271-272]

Finally, the consideration of limits is very diminished, or even completely gone in this section of [Lacroix *1802a*, 2nd ed].[63] The last formula above is used to explain the conception of the integral as an *infinite sum* – clearly not as a limit of sums – by putting x equal to $a, a + \alpha, a + 2\alpha$, etc., and dx equal to α [Lacroix *1802a*, 2nd ed, 306-307]. Naturally the two proofs that used the property of the integral being the limit of approximating sums are now absent.

Overall, in the second edition this section seems to be much more pedagogically oriented: clearer, more neatly organized. But also, probably for the same reason, less complex and less interesting mathematically.

From the third edition onwards Lacroix returns to limits, reusing material from the second edition of [Lacroix *Traité*] (see section 9.4.1) but keeping the order of [Lacroix *1802a*, 2nd ed]. The introduction to the section is the same, with the explanation of the definite integral quoted above. But after arriving at the formulas equivalent to (5.11) and (5.12), Lacroix sets out to prove their convergence. For this he invokes Arbogast's principle, and concludes that, in case X is increasing and we restrict ourselves to the terms in α,

$$\alpha(A + A_1 + A_2 \ldots + A_{n-1}) < \int X dx < \alpha(A_1 + A_2 + A_3 \ldots + A_n)$$

and that the difference between these two "bounds" for the integral, that is $\alpha(A_n - A)$, can be made always smaller by decreasing α, so that each of them can approach the true value of $\int X dx$ as close as one wishes. As in the original version of the section [Lacroix *Traité*, II, 137; *1802a*, 287], this is given as the justification for the possibility of viewing the integral as a sum of differentials (the difference being that here there is a stronger emphasis on the limit process). Also as in the original version [Lacroix *Traité*, II, 140; *1802a*, 291], the problem of the necessity for the function to be monotonic and non-infinite is addressed by suggesting that the interval of integration be split into several intervals where those conditions hold [Lacroix *1802a*, 3rd ed, 315-317].

After this oscillation in the first three editions, this section did not suffer any more major changes in the two last editions during Lacroix's lifetime. It did however gain a more *modern* look, thanks to a modernization of notation: some use of f(x) for y, and especially the adoption of Fourier's notation $\int_a^b X dx$; hence [Lacroix *1802a*, 4th ed, 324; 5th ed, 341]:

$$\int_a^b X dx = f(b) - f(a)$$

(instead of $\int X dx = B - A$ as above) and the *explicit* conclusion in the introduction

---

[63]It may be said to survive timidly in the passage giving the geometrical interpretation of the approximation method [Lacroix *1802a*, 2nd ed, 310-312], and in the argument that because $A\alpha + A_1\alpha + A_2\alpha \ldots + A_{n-1}\alpha < A_m n\alpha = A_m(b-a)$, where A_m is the largest of $A, A_1, A_2, \ldots A_{n-1}$ and a, b are the limits of integration, then $\int X dx < M(b - a)$, where M is the largest value of X between $x = a$ and $x = b$ (and similarly for $\int X dx > m(b - a)$) [Lacroix *1802a*, 2nd ed, 307]. But this argument might also be interpreted in terms of infinitesimals.

to the section that

$$\int_a^c X \, \mathrm{d}x = \int_a^b X \, \mathrm{d}x + \int_b^c X \, \mathrm{d}x$$

(simply because $f(c) - f(a) = f(b) - f(a) + f(c) - f(b)$) and even, in a footnote, the statement that "$\int_a^b X \, \mathrm{d}x = f(b) - f(a)$ est la limite dont l'expression

$$\alpha \left\{ f'(a) + f'(a + \alpha) \ldots \ldots + f'[a + (n - 1)\alpha] \right\}$$

s'approche de plus en plus, à mesure que le nombre n augmente et que α, qui est $\frac{b-a}{n}$, diminue"[64] [Lacroix *1802a*, 4th ed, 329; 5th ed, 346].

8.7.2 Approximate integration of differential equations

In the first edition of [Lacroix *1802a*] there are two short sections on methods for solving differential equations by approximation: one for first-order [Lacroix *1802a*, 383-387] and another for second-order differential equations [Lacroix *1802a*, 412-415]. Each is a shortened version of the corresponding section in the large *Traité*; there is no trace of the section on successive approximations using integration of "first-degree" differential equations.

The section on first-order equations is a copy of the beginning of the corresponding section in the large *Traité*: undetermined coefficients and Taylor series. The reference to insufficiencies of Taylor series is omitted, as well as Lagrange's method of continued fractions. There is an advantage in this: the use of Taylor series for approximation finishes this section, being immediately followed by the section on the geometrical construction of first-order equations, which opens with the remark on the "possibility" of those equations – because of Taylor series and because of their geometrical construction; this article, which seemed out of place in the large *Traité*, fits nicely here.

Very similar comments can be made about the section on second-order equations. Lacroix says very little about approximation properly speaking, and includes a subsection on "geometrical constructions" [*1802a*, 414-415] with the same text as the corresponding article in [Lacroix *Traité*, II, 351-352].

In the second edition the order is yet improved: there is only one section on approximation methods, for both first- and second-order equations [Lacroix *1802a*, 2nd ed, 420-428] (including a subsection on geometrical constructions of those equations [Lacroix *1802a*, 2nd ed, 426-428]). Lacroix speaks first of first-order equations: undetermined coefficients and Taylor series (including now its insufficiencies – but with simpler techniques to try to overcome them than Lagrange's continued fractions); then second-order equations, similarly to the first edition. As for the subsection on geometrical constructions, see section 8.8.2.

---

[64]"$\int_a^b X \, \mathrm{d}x = f(b) - f(a)$ is the limit which the expression

$$\alpha \left\{ f'(a) + f'(a + \alpha) \ldots \ldots + f'[a + (n - 1)\alpha] \right\}$$

approaches more and more as the number n increases and α, that is $\frac{b-a}{n}$, decreases"

From the third edition onwards Lacroix pays less attention to approximation methods (consistently with what had happened in the second edition of the large *Traité*). This section [Lacroix *1802a*, 3rd ed, 450-454] is shortened (even considering that the geometrical constructions are no longer included here), and most of it is taken up with two examples of use of undetermined coefficients. The use of Taylor series is only alluded to very quickly[65], and Euler's "general method" is not even mentioned (the associated constructions do appear, but without approximative purposes – see section 8.8.2). The section finishes with the remark that these approximation methods are seldom convergent enough, and that in "physico-mathematical" problems one usually just tries to determine small corrections to values that one already knows to be approximate (see pages 156 and 175 ff above, and the end of section 9.4.2 below) – but the methods used for this are too varied to be included in "elements".

8.8   Types of solutions of differential equations

8.8.1   Formation of differential equations and their types of solution

The most significant alterations on this subject from the large *Traité* to the *Traité élémentaire* are a consequence of the radical decrease in attention given to partial differential equations and (a little less so) to ordinary differential equations of degree higher than 1.

The idea that ordinary differential equations are formed by eliminating constants between finite equations in two variables and their differentials is present in the same places as in the large *Traité*: a section on "elimination of constants" [Lacroix *1802a*, 48-50]; the explanation for the method of integrating factors for first-order equations [Lacroix *1802a*, 354, 359-361] (integrating factors for second-order equations are not treated in [Lacroix *1802a*]); the explanation for the existence of n first integrals of an nth-order equation [Lacroix *1802a*, 397-399]; and of course the section on particular solutions of first-order equations [Lacroix *1802a*, 371-385] (particular solutions of second-order equations are also not treated in [Lacroix *1802a*]).

There are no differences in this respect in the second edition. From the third edition onwards, however, we see in two subsections on the number of arbitrary constants and the number of integrals [Lacroix *1802a*, 3rd ed, 402-409] a combination of this idea with a use of Taylor series, inspired by Lagrange [*Fonctions*; *Calcul*]. This is an adaptation of changes introduced in the second edition of the large *Traité* (see section 9.5.1).

As for the section on particular solutions of first-order differential equations in two variables [Lacroix *1802a*, 371-383], it is a close reproduction of [Lacroix *Traité*, II, 262-274], that is, the essential part of the corresponding section in the large

---

[65]In fact, Lacroix refers to a previous article, where Taylor series had been used to argue for the existence of solutions [Lacroix *1802a*, 3rd ed, 402-404].

Traité, with material taken from [Lagrange *1774*]: the explanation for the existence of particular integrals, their characterization as satisfying all the equations $\frac{dy}{dc} = 0, \frac{d^2y}{dx\,dc} = 0, \frac{d^3y}{d^2x\,dc} = 0$, etc. (while particular integrals satisfy only a finite number of these), and the procedure to obtain particular solutions directly from differential equations by putting $\frac{d^2y}{dx^2} = \frac{0}{0}$ or $\frac{d^2x}{dy^2} = \frac{0}{0}$; attempts to obtain complete integrals from particular solutions or particular integrals are entirely omitted. Also omitted are particular solutions of higher-order differential equations in two variables.

There are a couple of significant changes on this in the second edition: first, Lacroix [*1802a*, 2nd ed, 434-436] cites [Poisson *1806*][66] to the effect that the form of a differential equation may be changed so as to have its particular solution as a factor[67]; more importantly, the method given to obtain particular solutions directly from the differential equations [Lacroix *1802a*, 2nd ed, 436-440] is no longer that of [Lagrange *1774*], but rather that of [Lagrange *Fonctions*, 65-69], based on a power-series completion of particular integrals. This had been reported in the first edition of the large *Traité*, but would become much more important in the second one, where we see an expansion of the changes introduced here, with several pages simply reproduced (see section 9.5.1). Lacroix seems to have been quite happy with this new version, so much so that he kept this section practically unchanged in the third edition.

As for reflections on the formation of partial differential equations, the only trace of them in the first and second editions of the *Traité élémentaire* is the reproduction, in the section on differentiation of functions of two or more variables, of the passages from volume I of the large *Traité* on elimination of either two constants or one arbitrary function between a finite equation in three variables and its two first-order partial differentials [Lacroix *1802a*, 168-171]. But unlike in the large *Traité*, neither is later referred to as showing how partial differential equations are formed. As has been noted already, partial differential equations receive much less attention, and particular solutions are not even mentioned.

This changes a little in the third edition. Partial differential equations do not get much more coverage than in previous editions (particular solutions are still entirely absent), but Lacroix includes two new mentions to their formation: a brief reference to the passage on elimination of arbitrary functions when arriving at the solution $N = \varphi(M)$ of $Pp + Qq = R$ [Lacroix *1802a*, 3rd ed, 478]; and a new short article about the limitations of the analogy between arbitrary functions and arbitrary constants [Lacroix *1802a*, 3rd ed, 497-498].

---

[66]Notice the dates: [Lacroix *1802a*, 2nd ed] was also published in 1806.

[67]In spite of this, in the introduction to the section Lacroix keeps a distinction between particular solutions which are simply factors of the differential equation and others "intimately linked" to it [*1802a*, 2nd ed, 429-430].

8.8.2   Connections between differential equations and geometry

Once again, the most relevant modifications are simple consequences of the decrease in importance of partial differential equations. All considerations on their construction are reduced to a short footnote [Lacroix *1802a*, 457], associating the determination of the arbitrary functions involved in their integrals to making the corresponding surfaces pass through given curves, and claiming that those curves and functions may be discontinuous – no details on either the claim or the association.

Still, there are a couple of novelties in organization which are worth mentioning, since they throw light on the geometrical versions of Euler's "general method", showing them openly as *constructions*. It has been mentioned already how the diminution of the section on approximate integration of differential equations allows for that geometrical version to open the section on "geometrical construction of first-order differential equations" [Lacroix *1802a*, 387-396][68]. Moreover, the article giving the geometrical version of Euler's general method for second-order equations, which is also reproduced [Lacroix *1802a*, 414-415], is referred to in the table of contents as "geometrical constructions of [second-order differential] equations" [Lacroix *1802a*, xl].

This neat order is a little affected in the second edition, due to a reorganization of the chapter on differential equations in two variables: the special topics of approximate integration (for both first- and second-order equations), particular solutions, and geometrical problems are treated, in this order, at the end of the chapter (this is an anticipation of the second edition of the large *Traité*, where they have separate chapters). Thus the geometrical version of Euler's "general method", being in the section on approximation, becomes separated again from the geometrical section; still, it is entitled to its own subsection in the table of contents "geometrical constructions of [first- and second-order differential] equations" [Lacroix *1802a*, 2nd ed, x].

This is reverted again from the third edition onwards, due to the decrease in importance given to approximation methods, and especially to the disappearance of the analytical version of Euler's "general method" for differential equations. The corresponding geometrical constructions [Lacroix *1802a*, 3rd ed, 460-462] appear in the middle of the section on "resolution of some geometrical problems", are referred to in the table of contents as "geometrical constructions of differential equations" [Lacroix *1802a*, 3rd ed, ix] and, in case someone might wrongly suspect that they have something to do with approximation, Lacroix had a few pages previously finished the section on approximate integration, saying precisely that he was "terminant [...] ce qui regarde l'intégration approchée des équations différentielles"[69] [*1802a*, 3rd ed, 454].

---

[68]The rest of this section is a shortened version of the one in the large *Traité*, omitting Jacob Bernoulli's construction of $\frac{dx}{dy} = \frac{m}{y \pm Q}$ and some technical details on construction of trajectories, but mostly reproducing it word for word.

[69]"concluding what regards the approximate integration of differential equations"

8.8.3 Total differential equations not satisfying the conditions of integrability

These equations have their own section, albeit a short one [Lacroix *1802a*, 458-461]. It is a plain reproduction of the first two articles of the corresponding section in the large *Traité*: the idea of establishing a relation between x, y and z (and its attribution to Newton), and Monge's procedure for integrating equations $P \, dx + Q \, dy + R \, dz = 0$, presented as an adaptation of the method for integrating them when they do satisfy the condition of integrability.

This section remained unchanged throughout the several editions of the *Traité élémentaire*, except for being moved, from the third edition onwards, from the end of the chapter on "Integration of functions of two, or more, variables" to right after the section addressing the conditions of integrability and the integration of total differential equations that satisfy them (and thus before the integration of partial differential equations).

8.9 Aspects of differences and series

8.9.1 Indices

Subscript indices have a smaller presence in the *Traité élémentaire* than in the large *Traité*. There are two main reasons for this. One, is that their first appearance in the large *Traité* (and one of the most innovative) is in the Introduction, which is absent from the *Traité élémentaire*. The other reason is that the appendix on differences and series has a lesser weight in the *Traité élémentaire* than volume III in the large *Traité*. Moreover, the occurrence of indices in the expansion of arbitrary functions (Taylor's theorem) [Lacroix *Traité*, I, 87-91] disappears with the change of foundations for differential calculus. Still, other occasional occurrences seem to be kept; for instance, in approximate integration.

8.9.2 The "multiplicity of integrals" of difference equations

The subject of the different types of integral of difference equations was clearly too complicated, or at least too finicky, for the *Traité élémentaire*. Lacroix did not include anything on it.

8.9.3 Mixed difference equations

Mixed difference equations seem also too specific for the *Traité élémentaire*. In the first and second edition there is nothing on them. However, from the third edition onwards Lacroix included one short article about them, at the end of the section on difference equations [*1802a*, 3rd ed, 602-603]; but this article only gives two very simple examples, refers the interested reader to the large *Traité*, and cites the authors that had addressed the subject.

8.10 Translations of the *Traité élémentaire*

Around 1800 French mathematical (and generally scientific) books seem to have circulated widely in Europe. Among them, Lacroix's textbooks were very popular. It is not easy to give a quantitative perspective on this, but at least in good British and Portuguese libraries it is certainly easy to find copies of them (although not always of the earlier editions).

Translating French textbooks into other languages was also a common activity [Grattan-Guinness *2002*, 20-24]. Once again, Lacroix's textbooks were popular targets. Below we will see translations of his *Traité élémentaire* into Portuguese (made in Brazil), English, German, Polish (made in what is nowadays Lithuania), Italian, and Greek. Notice that translations of some of his other textbooks into Portuguese, Italian, and Greek will also be mentioned; several more German and Spanish translations are mentioned in [Grattan-Guinness *2002*, 39-40]; we have mentioned in section 4.1.1.2 an English translation, made in America, of his trigonometry text, and this was part of a series of translations of European textbooks that also included Lacroix's Arithmetic and Algebra [Ackerberg-Hastings *2004*, 7]; Danny Beckers [*2000*] has discussed the unfaithful Dutch translation of Lacroix's algebra textbook; farther away, an English translation of his algebra textbook was made and printed in Calcutta to be used at the local Hindu College [Aggarwal *2006*, 111]; and it would be surprising if this list were exhaustive.

8.10.1 The Portuguese translation (Rio de Janeiro, 1812-1814)

Lacroix's *Traité élémentaire* was translated into Portuguese during a very peculiar period in Portuguese history. In November 1807 the royal family fled to Brazil from a French invasion, only to return in June 1821. During those nearly 14 years, Rio de Janeiro was the capital of Portugal. Naturally, this situation had far-reaching consequences for Brazil, including the foundation of the first printing press and of the first higher-education institutions.

Among these institutions was the *Royal Military Academy* (*Academia Real Militar do Rio de Janeiro*), created in 1810 by the Prince Regent, John (later king John VI). It had a 7-year course, of which four years were devoted to mathematics [C.P.Silva *1992*, 51-57]. Several French textbooks were translated to be used by the students of this Academy, and among them several by Lacroix, including the *Tratado Elementar de Calculo Diferencial e Calculo Integral* [Lacroix *1812-1814*].[70]

---

[70] And also the *Tratado elementar de Arithmetica* (1810, translated by Francisco Cordeiro da Silva Torres e Alvim; I have not seen this book, but it is mentioned by Inocêncio [*DBP*, II, 367] and Circe M.S. Silva [*1996*, 82]), the *Elementos d'Algebra* (1811, translation of [Lacroix *1799*], also by Francisco Cordeiro da Silva Torres), the *Tratado Elementar de Applicação de Algebra á Geometria* (1812, partial translation of [Lacroix *1798b*], with an appendix on geometry in space, by José Victorino dos Santos e Souza), and the *Complemento dos Elementos d'Algebra* (1813; I have not seen this book, which is mentioned in [*NUC*, CCCX, 654]).

Like all those textbooks, the [Lacroix *1812-1814*] was published in Rio de Janeiro by the Royal Press (Impressão Regia). It appeared in two volumes: the first in 1812, dedicated to differential calculus, and corresponding to the first part of [Lacroix *1802a*]; the second in 1814, dedicated to integral calculus, and corresponding to [Lacroix *1802a*, 187-461] – that is, the second part minus the method of variations. Thus, it is an incomplete translation, as the method of variations and the appendix on finite differences and series are missing.

The translator was Francisco Cordeiro da Silva Torres (often called Francisco Cordeiro da Silva Torres e Alvim). Silva Torres was born in Ourém (European Portugal) in 1775 and died in Rio de Janeiro in 1856. He was at the time of this translation sergeant-major in the Royal Corps of Engineers and a lecturer at the Royal Military Academy (according to Clóvis P. Silva [*1992*, 56] he taught higher algebra, analytic geometry, and differential and integral calculus). He stayed in Brazil after the independence (1822) and became viscount of Jerumarim, state councillor, etc. Apart from translating Lacroix, Silva Torres also published a few works on weights and measures and on finance [Inocêncio *DBP*, II, 367; IX, 281-282; C.M.S. Silva *1996*, 82].

As to the translation itself, there is nothing to say, except that it was clearly made from the first edition of [Lacroix *1802a*] – although the second had already been published in 1806; presumably it was not easily available in Rio de Janeiro.

In spite of this, and of the incompleteness of the translation, the students of the Royal Military Academy of Rio de Janeiro were undoubtedly well served with this textbook; at least much better than their colleagues at the University of Coimbra: the adopted textbook there was still Bézout's, and would be until the late 1830's, when it was replaced by Francoeur's.[71]

In Brazil, this translation remained as the adopted textbook for a long time, and was probably still used in 1871. It is also remarkable that what seems to have been the first textbook on the calculus written by a Brazilian, José Saturnino da Costa Pereira, in 1842, was entitled "Elementos de Calculo Differencial e de Calculo Integral, segundo o systema de Lacroix" – i. e., "Elements of differential and integral calculus, following Lacroix's system" [C.M.S. Silva *1996*, 84].

8.10.2 The English translation (Cambridge, 1816)

The most famous, and probably most interesting, translation of Lacroix's *Traité élémentaire* was the English one, published in Cambridge in 1816 by George Peacock (1791-1858), Charles Babbage (1791-1871) and John Herschel (1792-1871).

During the 18th century the British method of fluxions had grown apart from the Continental differential and integral calculus. In the beginning of that century the difference was mainly one of notation and a few distinct conceptions. But

---

[71]It seems that not many copies of the Brazilian editions of Lacroix crossed the Atlantic to Portugal. At least, they are not very common in Portuguese libraries nowadays.

from the 1740's onwards the British were not able to follow Continental developments such as partial differential equations [Guicciardini *2003*, parts 2 and 3]. At the University of Cambridge mathematics had a prominent role in education and particularly in examinations; but it was seen as a mere exercise in reasoning, and there were no incentives for doing research nor simply for keeping up to date with external research.

By the late 18th century and the early years of the 19th, a number of mathematicians tried to change this state of affairs. It is only fair to mention the Scot William Wallace (1768-1843), who held teaching posts at Perth Academy (1794-1803), the Royal Military College (1803-1819), and the University of Edinburgh (1819-1838). Wallace published in 1815 an 86-page article on "Fluxions" in the *Edinburgh Encyclopædia* [Wallace *1815*], using the differential notation and including "partial fluxions" (i.e., partial differentials) and "fluxional coefficients" (i.e., partial derivatives) [*1815*, 433].[72] In spite of the words "fluxion" and "fluent", this was in fact a complete account of the (Continental) differential and integral calculus – the first one in Britain [Guicciardini *2003*, 120]. However, partly because this was an encyclopedia article instead of a book, and partly because he used limits instead of Lagrangian power series, his contribution was disregarded by more influential British mathematicians, and soon forgotten [Panteki *1987*; Craik *1999*, 253].

A more influential figure was Robert Woodhouse (1773-1827), a fellow of Gonville and Caius College, Cambridge (from 1795), Lucasian Professor of Mathematics (1820-1822), Plumian Professor of Astronomy and Experimental Philosophy (1822-1827), and director of the Cambridge Observatory (from 1824). Starting in 1790's, Woodhouse was also a reviewer of mathematics for the London *Monthly Review*. This made him read the works of French mathematicians, and soon he was a Lagrangian. He published in 1803 a book entitled *Principles of Analytical Calculation*, where he adopted the power-series approach (although criticizing some details of Lagrange), and used the differential notation, as well as Arbogast's *D* operator [Guicciardini, *2003*, 126-131; Philips *2006*, 70-71].[73] Later, he published books on trigonometry and the calculus of variations, that according to Philips [*2006*] had much greater influence in Cambridge education than his 1803 book.

Woodhouse was certainly also an inspirational figure for the famous *Analytical Society*. This society was formed in 1812 by a group of undergraduate students, among whom were Babbage, Peacock, and Herschel, later to be active researchers in mathematics. The Analytical Society started as a joke on societies devoted to distributing Bibles – instead, it would distribute Lacroix's *Traité élémentaire*, as

---

[72]"Fluxional coefficient" is of course evocative of Lacroix's "differential coefficient". Wallace gave a long list or works on the calculus, both British and Continental, [*1815*, 388-389], but lamenting the absence of up-to-date books in English. Anyone wishing to study it "beyond its mere elements" should recur to Euler's books, or French treatises – among the latter, he stressed Lacroix's large *Traité*.

[73]But apparently this was not a *complete* account of the calculus, rather just a reflection on its principles.

a way of propagating the "pure *d*-ism against the Dot-age of the University" (that is, the Continental *dx* against the Newtonian *ẋ*) [Guicciardini *2003*, 135; Enros *1983*, 26-27]. Setting the joke aside, the society was formed and met regularly, discussing "analytics" and putting out a volume of memoirs in 1813.

The society dissolved in early 1814, but nearly three years later its three more prominent members published [Lacroix *1816*] – a partial English translation of [Lacroix *1802a*, 2nd ed], with additions.

The division of labour between the three of them, according to the "advertisement" [Lacroix *1816*, iii-iv] was thus: Babbage translated part 1 (differential calculus); Peacock and Herschel translated part 2 (integral calculus); Peacock alone wrote twelve endnotes (A-M)[74] on the differential and integral calculus; Herschel alone wrote four more endnotes (N-Q) on differential equations and the calculus of variations, and an appendix on differences and series to replace that of Lacroix.

Some of the endnotes are quite extensive. In particular, those written by Peacock are one of the main points of interest in this translation: the advertisement tells us that they "were principally designed to enable the Student to make use of the principle of Lagrange" – that is, to compensate for the fact that in the *Traité élémentaire* Lacroix had "substituted the method of limits of D'Alembert, in the place of the more correct and natural method of Lagrange". Let us see some of the most important examples.

Note (A) [Lacroix *1816*, 581-596] is in fact about limits. Peacock gives a historical account of them (and also of infinitesimals and indivisibles), starting with the "Method of Exhaustions", and he establishes some results, probably taken from the Introduction of Lacroix's large *Traité*. In particular, he gives Arbogast's principle (and Lacroix's counter-example), and uses it for the pinching theorem for power series, useful for geometrical applications (see section 3.2.6 and page 122 above).

Note (B) [Lacroix *1816*, 596-620] is the most substantial one, and the one that most closely corresponds to the design announced in the "advertisement" – that is, it is an attempt to establish the differential calculus on a power-series basis (still using the differential notation). Peacock acknowledges that he used [Lagrange *Calcul*] and [Lacroix *Traité*] to write the note, but probably he used the latter more than the former. The last pages of this note are dedicated to comparisons between the power-series approach, the method of limits, infinitesimals, and the method of fluxions (including a criticism of the fluxional notation).

Note (D) [Lacroix *1816*, 622-633] is dedicated to finding the differentials of exponential and trigonometric functions, by other means than those used in [Lacroix *1802a*] – using power series, of course.

Note (G) [Lacroix *1816*, 654-660] is on "the application of differential calculus to the theory of curves, without the introduction of limits" – power series again, and the pinching theorem proved in note (A).

---

[74]There is no note (J).

Some other notes give details that Lacroix had omitted (or much reduced) in
the *Traité élémentaire*. For instance, note (F) [Lacroix *1816*, 647-654] addresses
the particular values for which Taylor's series was seen to fail.

According to the "advertisement", Herschel's appendix on differences and series purported to include "many important subjects [...] which had been either
entirely omitted, or very imperfectly considered" in Lacroix's. Clearly, one such
important subject was the "determination of functions from given conditions"
[Lacroix *1816*, 544-550] – that is, functional equations, a favorite topic for Herschel
and Babbage [Grattan-Guinness *1994*, 559-560]. But there is an overall increase
in size: Herschel's appendix occupies about 20% of the book (endnotes excluded)
against about 15% for Lacroix's.

One gets the distinct feeling that this translation aimed at a kind of compromise between the large *Traité* and the *Traité élémentaire*.

As for the influence of [Lacroix *1816*], the traditional view was quite enthusiastic: "the year 1816, in which Lacroix's shorter work was translated into English
[...] witnessed the triumph in England of the methods used in the Continent"
[Boyer *1939*, 265-266]. This opinion is no longer held by historians of the period
[Enros *1983*; Guicciardini *2003*; Philips *2006*]. There had been precursors, like
Wallace and Woodhouse; and the actual reform in Cambridge teaching was a slow
process, in which the role of [Lacroix *1816*] is not clear. But eventually it was
seen as a landmark, at least by research mathematicians. When De Morgan finished his book on *The Differential and Integral Calculus*, he expressed its extent
by saying that it was "more than double in matter of the Cambridge translation
of Lacroix, and full half as much as the great work of the same author in three
volumes quarto" [De Morgan *1836-1842*, iii].

8.10.3   The German translations (Berlin, 1817; 1830-1831)

There were two (or possibly three) German translations of Lacroix's *Traité élémentaire du calcul*....

I have not been able to consult the first of these. The following information
is taken from the catalogues [*NUC*, CCCX, 657] and [*GV*, LXXXIII, 198]: it
had the title *Handbuch der Differential- und Integral-Rechnung*; it was made by
C. F. Bethke, from the second French edition[75], and it was published in 1817 by
G. Reimer in Berlin. [*NUC*, CCCX, 657] indicates the publisher as Realschul-
buchhandlung, while [*GV*, LXXXIII, 198] indicates Reimer; but this is not so
strange – Georg Andreas Reimer (1776-1842) had taken over the *Buchhandlung
der Königlichen Realschule* (Bookstore of the Royal Secondary School) in 1801
[Gruyter *History*].

I have not been able to locate any biographical information on C. F. Bethke.

Another translation, with the same title, but by a different translator ("Dr.
Fr. Baumann") was published also by Reimer (and naturally also in Berlin) in 1830-

---

[75]"Nach der zweiten durchgesehen und verbesserten Original-Ausgabe."

1831, in three volumes. The first volume (1830) contains the differential calculus; the second volume (1831) contains the integral calculus, minus the method of variations; and the third volume (1831) contains the calculi of variations and of differences. This second translation was made from the fourth French edition[76].

This translator was certainly Franz Baumann, professor of mathematics at the University of Münster from 1826 to 1832;[77] he took his doctorate at Göttingen, and his habilitation in Bonn [Nastold & Forster *1980*, 429]. In the preface Baumann explains his preference for Lacroix's textbook with the facts that it follows the method of limits[78], and that it opens the way to his large *Traité*; and he justifies this new translation by the poor quality of the previous one (at least according to his students)[79] [Lacroix *1830-1831*, I, vi].

Baumann included several footnotes throughout the three volumes (usually short comments or alternative calculations). They are signed "B", while Lacroix's original footnotes appear signed "L".

As for the possible third translation, we have seen in section 2.6.1.2 that it is unlikely that Reimer published yet another translation of the *Traité élémentaire* in 1833.

8.10.4   The Polish translation (Vilnius, 1824)

[Lacroix *1802a*] was translated into Polish in Vilnius, nowadays the capital of Lithuania, but at the time recently incorporated in the Russian Empire, as a result of the 1795 partition of the Polish-Lithuanian Commonwealth. Polish was then the main language of the higher classes in Lithuania, and had been replacing Latin as a teaching language [Venclova *1981*; Yla *1981*].

I have not consulted this translation. The online library catalog of Vilnius University[80] gives its title as *Traktat początkowy rachunku różniczkowego i całkowego*, place of publication Wilno (the Polish name for Vilnius), publisher A. Marcinowski, and date of publication 1824. As subtitle there are also the indications "przełożony na język polski z drugiego wydania przez Zacharyasza Niemczewskiego; poprawiony i wydany przez Michała Pełkę Polińskiego" – that is, "translated into Polish from the second edition by Zacharyasz Niemczewski; corrected and edited by Michał Pełkę Poliński".

The translator Zacharyasz Niemczewski[81] (1766-1820) was of peasant origin.

---

[76]"Nach der vierten verbesserten und vermehrten Original-Ausgabe (1828)"

[77]The translator's preface is dated at Münster, Westphalia [Lacroix *1830-1831*, I, vi].

[78]Which, besides being much easier to grasp than other approaches, was beginning to be generally acknowledged as the true foundation of the differential calculus ("die Grenzen-Methode, die man jetzt allgemein als die wahre Grundlage der Differential-Rechnung anzuerkennen anfängt").

[79]"...über deren Werth ich mich nur dahin erklären mag, dass ich, nach dem, was ich von meinen Schülern darüber vernommen, daran zweifeln muss, dass jemand, dem beide Sprachen gleich bekannt sind, lieber das Deutsche als das Französische zum Führer wählen werde."

[80]<http://lanka.vu.lt>, accessed on 26 December 2006.

[81]In Lithuanian: Zakarijas Niemčevskis [Venclova *1981*; Yla *1981*] or Zacharijus Nemčevskis [Banionis *2001*, 43].

Apart from mathematics he also contributed to Lithuanian studies, for instance writing a short French-Lithuanian dictionary [Venclova *1981*; Yla *1981*]. In 1799 he started teaching applied mathematics at Vilnius University, from where he had graduated. From 1802 to 1808 he stayed in Paris to pursue further studies in mathematics, namely at the *École Polytechnique*[82] (where he attended Poisson's lectures on analysis). Returning to Vilnius, he lectured from 1810 to his death in 1820 on differential and integral calculus, following Lacroix's *Traité élémentaire*, and on mechanics, following Francoeur's *Traité de mécanique élémentaire* [Gyachyauskas *1979*, 169; Banionis *2001*, 56-57]. It was certainly for his lectures that Niemczewski translated these two textbooks, as well as Biot's *Essai de géométrie analytique*. But he did not publish any of these translations.[83]

As we have seen, it was Michał Pełkę Poliński (1784-1848) who accomplished the publication of Niemczewski's translation of Lacroix's *Traité élémentaire*, in 1824 – four years after Niemczewski's death. Poliński had also studied for some time in Paris, but in the *Faculté des Sciences* (where he was a student of Lacroix), not in the *École Polytechnique*[84]. From 1819 until the closure of Vilnius University in 1832 he taught several mathematical subjects, from algebra to analytical mechanics. He also published textbooks on trigonometry and geodesy [Gyachyauskas *1979*, 169].

8.10.5   The Italian translation (Florence, 1829)

An Italian translation was printed in Florence, at the press of Francesco Cardinali, in 1829, with the title *Trattato Elementare del Calcolo Differenziale e del Calcolo Integrale*. Unfortunately, there is not much that can be said about this translation. It does not indicate who the translator was, nor where it was meant to be used.

Italian translations of several other textbooks by Lacroix had been published or were later published, and Florence appears to have been the main (or sole) place of publication: the *Elementi d'algebra* appeared in 1809, and the *Elementi di geometria* in 1813, both in Florence, both with Piatti as publisher [Pepe *2006*, 3]; at the time, Florence (and a great part of northern and middle Italy) was part of the French Empire. But the influence of French textbooks remained after the fall of the French Empire in 1815, at least in Tuscany (where Florence is located) and in Naples. In 1834 a *new* edition of Lacroix's *Trattato elementare di applicazione dell'algebra alla geometria* was prepared by the professor of the University of Pisa, Filippo Corridi (1806-1877) – once again in Florence, and once again published by Piatti [Pepe *2006*, 16]. Although the publisher of the translation of [Lacroix *1802a*] was a different one, one may conjecture that Corridi is a good candidate for having been the translator.

---

[82]He appears in Fourcy's list of foreign students for 1804, as "Niemezewski" [Fourcy *1828*, 387].

[83]Another translation of Biot's *géométrie analytique*, by Antoni Wyrwicz, was published in 1819 (by the same publisher, Marcinowski). Wyrwicz (or Virvichyus [Gyachyauskas *1979*, 170]) also taught at Vilnius University.

[84]He does not appear in Fourcy's list of foreign students [Fourcy *1828*, 387-389].

As for the translation itself: it was based on the French fourth edition (published in 1828 – just one year previously), and it appears to be faithful.

8.10.6   The Greek translation (unpublished; Corfu, 1820's)

We have seen in section 2.6.2 Ioannis Carandinos' activity in the 1820's as a translator of contemporary mathematical works into Greek. We have also seen that many of these translations were not published, and are now lost. Among these was not only Lacroix's large *Traité* (partially), but also several of Lacroix's textbooks, including the *Traité élémentaire* [Phili *1996*, 318].

Chapter 9

The second edition of Lacroix's *Traité*

9.1 Overview of the second edition

We do not know the print-run of the first edition of Lacroix's *Traité*, but it must have sold well. [Lacroix *1805*] includes, just before the table of contents, a list of other works by Lacroix "that can be found in the same bookstore" (Courcier). We find the several textbooks in his *Cours de Mathématiques*, with their respective prices, and the large *Traité*. But the latter does not have a price; instead, it carries the indication "rare et épuisé"[1].

The three volumes of the second edition came out in 1810, 1814, and 1819.

The first issue that comes to mind is the nine-year interval between the first and the third volumes. Recall that in the first edition the corresponding interval was only of three years (or at most five, if we account for the fact that part of volume I was printed and distributed to some people already in 1795 – see section 3.2.1). I cannot explain this difference. But it is noticeable that the coherence of the second edition suffered from this: the third volume finishes with a 132-page set of "corrections and additions", nearly all dedicated to volumes I and II, and nearly all consisting in "additions" – material that had come to Lacroix's knowledge or mind after the printing of volumes I and II.

Tables 9.1-9.3 show the chapters of the second edition, comparing them with those in the first edition. We can see that some of the larger chapters were sub-divided – namely, the chapter on differential equations in two variables in vol. II, and the chapters on the calculus of differences and on several mixtures of integral calculus with series in vol. III.

Something that cannot be seen in these tables but is also present is a similar subdivision of many sections. For instance, the section on the "application of

---

[1]"rare and out-of-print"

1st edition		2nd edition		
chapter	pages	chapter	pages	(some) topics covered
Preface	iv-xxix	Preface	i-xlviii	History of the calculus
Table of contents	xxx-xxxii	Table of contents	xlix-lvi	Contents and bibliography
Introduction	1-80	Introduction	1-138	Functions, series and limits; series expansion of functions; "imaginary" (i.e., complex) numbers
Ch. 1: Principles of differential calculus	81-194	Ch. 1: Principles of differential calculus	139-248	Differentiation of functions; differentiation of equations
Ch. 2: Main analytical uses of differential calculus	195-276	Ch. 2: Use of differential calculus to expand functions	249-326	Differential methods for expansion of functions in series
		Ch. 3: Particular values of diff. coefficients	327-388	Indeterminacies; maxima and minima
Ch. 3: Digression on equations	277-326	(suppressed; "imaginary" numbers transferred to Introduction)		
Ch. 4: Theory of curves	327-434	Ch. 4: Theory of curves	389-500	Analytic and differential geometry of plane curves
Ch. 5: Curved surfaces and curves of double curvature	435-519	Ch. 5: Curved surfaces and curves of double curvature	501-652	Analytic and differential geometry of surfaces and space curves

Table 9.1: Volume I of the second edition (1810), compared with the first edition

differential calculus to the theory of curved surfaces" [*Traité*, 1st ed, I, 465-504] is divided into the sections on the "application of differential calculus to the theory of contact of surfaces", "theory of curvature of surfaces", and "generation of surfaces" [*Traité*, 2nd ed, I, 563-572, 572-588, 588-615].

These new divisions constitute clear improvements in the structure of the *Traité*, making it even easier to use as a reference text.

As for content, the second edition is a little larger than the first, in spite of a couple of passages having been removed[2]. The bibliography grew considerably (although a new graphical arrangement for the table of contents exaggerated this in terms of number of pages; beware this in tables 9.1-9.3). Many new de-

---

[2]The principal passages removed were from the first volume: the sections on symmetric functions, and on the elementary aspects of analytic geometry on the plane.

1st edition		2nd edition		
chapter	pages	chapter	pages	(some) topics covered
Table of contents	iii-viii	Table of contents	vii-xxi	Contents and bibliography
Ch. 1: Integration of functions of one variable	1-160	Ch. 1: Integration of functions of one variable	1-155	Antiderivatives; approximate integration
Ch. 2: Quadratures, cubatures and rectifications	161-220	Ch. 2: Quadratures, cubatures and rectifications	156-224	Areas, volumes and arc-lengths
(partly in chapters 3 and 4)		Ch. 3: Integration of differential functions of several variables	225-249	Conditions of integrability
Ch. 3: Integration of differential equations in two variables	221-452	Ch. 4: Integration of diff. eqs. in two variables	250-372	Solutions of ordinary differential equations
		Ch. 5: Particular solutions of diff. eqs.	373-408	"Particular" (i.e, singular) solutions
		Ch. 6: Approximate integration of diff. eqs.	409-446	Approximation methods
		Ch. 7: Geometrical applications of diff. eqs. in two variables	447-470	Geometrical problems
		Ch. 8: Comparison of transcendental functions	471-502	Logarithmic, trigonometric and elliptic functions
Ch. 4: Integration of functions of two or more variables	453-654	Ch. 9: Integration of equations in three or more variables	503-720	Partial differential equations
Ch. 5: Method of variations	655-724	Ch. 10: Method of variations	721-816	Calculus of variations

Table 9.2: Volume II of the second edition (1814), compared with the first edition

velopments were included; for instance, a short account of Cauchy's early work on definite integrals appears in [*Traité*, 2nd ed, III, 497-500]. But there are no major modifications. As Grattan-Guinness has said, "the general impression is still that the main streams and directions of the calculus had been amplified and enriched, rather than changed in any substantial way" [*1990*, I, 267]. Moreover, it seems that Lacroix missed some signs of what were to be substantial novelties: for instance, although he included some references to Gauss in the bi-

| 1st edition | | 2nd edition | | |
chapter	pages	chapter	pages	(some) topics covered
Table of contents	iii-viii	Table of contents	vii-xxiv	Contents and bibliography
Ch. 1: Calculus of differences	1-300	Ch. 1: Direct calc. of differences	1-74	Finite differences; interpolation
		Ch. 2: Inverse calc. of differences of explicit functs.	75-194	Σ-integration; summation of series; interpolation
		Ch. 3: Integration of difference equations	195-321	Difference equations
Ch. 2: Theory of sequences derived from generating functions	301-355	Ch. 4: Theory of sequences derived from generating functions	322-373	Generating functions
Ch. 3: Application of integral calculus to the theory of sequences	356-529	Ch. 5: Application of integral calc. to the theory of sequences	374-411	Summation of series; interpolation
		Ch. 6: Evaluation of definite integrals	412-528	Use of series and infinite products
		Ch. 7: Definite integrals applied to the solution of differential and difference equations	529-574	Transcendental functions
Ch. 4: Mixed difference equations	530-544	Ch. 8: Mixed difference equations	575-600	Difference-differential equations
Subject table	545-578	Subject table	733-771	Subject index
Corrections and additions	579-582	Corrections and additions	601-732	Some corrections; mainly additions

Table 9.3: Volume III of the second edition (1819), compared with the first edition

bliography [*Traité*, 2nd ed, III, xi-xii], he omitted both Gauss' 1813 paper on the hypergeometric series [Grattan-Guinness *1970*, 145-146] and Gauss' proofs of the Fundamental Theorem of Algebra;[3] neither did he mention Argand's geometrical representation of complex numbers [Grattan-Guinness *1990*, I, 256-259]. He did

---

[3]As is well known, Gauss gave four proofs of the Fundamental Theorem of Algebra [Kline *1972*, 598-599]. The fourth proof appeared only in 1850, and the first proof was given in his doctoral dissertation, which Lacroix probably did not know. But the second and third proofs [*1814-1815b*; *1814-1815c*] appeared in the same volume of the Royal Society of Göttingen as a paper on approximation of integrals [Gauss *1814-1815a*] which was cited by Lacroix [*Traité*, 2nd ed, III, xii].

mention Fourier's (unpublished) works on heat theory [Lacroix *Traité*, 2nd ed, III, 501, 562-564] and Cauchy's early works on definite integrals [*Traité*, 2nd ed, III, 497-500], but a more detailed analysis would be needed to determine how well he understood them.

Still, there were some modernizations. In the chapter on the calculus of variations Lacroix introduced Lagrange's power-series approach. And we will see in the following sections that he updated the particular aspects that have been studied in chapters 3-7.

9.2 The principles of the calculus

Between the publication of [Lacroix *Traité*, 1st ed, I] and that of [Lacroix *Traité*, 2nd ed, I], that is, between 1797 and 1810, there were a considerable number of publications on the foundations of the calculus [Grattan-Guinness *1990*, 195-223]. To start with, Lagrange published [*Fonctions*] (still in 1797) and [*Calcul*] (in different forms, between 1801 and 1806). Still in or around the *École Polytechnique*, there appeared papers by Poisson [*1805*], Ampère [*1806*], and Paul René Binet [*1809*] on Taylor series, the derivative, and their relations (Poisson staying in a power-series framework, Ampère and Binet moving towards limits). Outside the *École Polytechnique*, there was also interest in series and algebraic views on the calculus; the most important outcome was Arbogast's 1800 book on the "calculus of derivations". Meanwhile, Lacroix himself published the first two editions of the *Traité élémentaire de Calcul...*, following a limit-based foundation for the calculus (see section 8.5). So, how does all this reflect in the second edition of the large *Traité*? The first point to make is that Lacroix keeps the power-series foundation, instead of changing to the method of limits. Why? We saw in section 8.5 that he now saw limits as embodying the proper metaphysics of the calculus, at least as far as applications went. And he is very explicit in the Preface about his preference for limits in teaching [*Traité*, 2nd ed, I, xxiv-xxvi]. Perhaps he was not willing to do a major reform of the large *Traité* (but would it be that major?). Perhaps he still thought of power series as more appropriate for an *analytical* treatise, rather than a textbook (but in that case, it would have been helpful if he had said so explicitly). Perhaps he simply did not want to lose face – writing a whole treatise based on power series was one of the main motivations he had presented in the first edition. I do not have an answer for this. But this issue was probably not so important as it might seem; the important point was still the "rapprochement des Méthodes" (section 3.2.8). Even in his preference for limits in teaching, Lacroix remarks that each foundation offers some facilities, and one could not foresee important discoveries that might be provided by them [*Traité*, 2nd ed, I, xxiv].

Keeping the power-series foundation does not mean that Lacroix does not introduce modifications. He does – and they might be interpreted as facilitating the "rapprochement des Méthodes", by making the power series less fundamental. Recall that in the presentation of the principles of the calculus in the first edition, one of the main points was the process of *derivation* for "any" function f – that is, that the coefficients in the series expansion of f($x + k$) can be obtained by a recursive process; the "derived functions" were then introduced in one lot. In other words, Taylor's theorem was embedded in this presentation. In the second edition, the derivation process appears only in an example (x^n) [*Traité*, 2nd ed, I, 144-145]; the first-order differential and first-order differential coefficient are introduced together (as the first term in the expansion of the increment of the function, and as the coefficient of the increment of the variable in that term) [*Traité*, 2nd ed, I, 146-147], and separately from higher-order differentials and differential coefficients. The latter only appear after the differentiation of common functions, in a section dedicated to Taylor series [*Traité*, 2nd ed, I, 160-169]. In this new structure, the differential coefficient could be defined as the limit of the ratio between the increments, as Lacroix himself acknowledges [*Traité*, 2nd ed, I, 146] – hence the possible interpretation that it is more appropriate to accommodate different foundations.

The proof of Taylor's theorem is an adaptation of [Poisson *1805*], assuming the existence of a development of the form $N + Ph^\alpha + Qh^\beta + Rh^\gamma + Sh^\delta +$ etc. (that is, the exponents of h are not assumed, but rather *proved* to follow the sequence 1, 2, 3, 4, etc.). In the argument it is necessary to use the fact that the second term in the expansion of $(h + k)^m$ is of the form $Mh^{m-1}k$, but not the whole binomial expansion.

Notationally, Lacroix abandons the Lagrangian $f'(x), f''(x)$, etc. – which is natural, given the loss of importance of the derivation process. Instead, he uses the Eulerian p, q, etc. – as he had done in most of the first edition (together with the Leibnizian $\frac{dy}{dx}, \frac{d^2y}{dx^2}$, etc.), apart from the presentation of the principles of the calculus. We have seen in section 3.2.8 that already in the third volume of the first edition Lacroix had criticized [Lagrange *Fonctions*] for its exclusive use of the accent notation.

The section on alternative foundations [*Traité*, 2nd ed, I, 237-248] is greatly changed. Even the name is different: "Reflexions on the metaphysics of differential calculus and on its notations", instead of "method of limits" (which did not quite cover its contents, in either edition). Lacroix's explanation of the Leibnizian calculus is mostly unchanged. But the passage on limits is completely different. The calculations of some differential coefficients using limits simply disappear (possibly because most of them were in the *Traité élémentaire*). Instead, Lacroix gives a glimpse of Landen's "residual analysis" [Guicciardini *2003*, 85-88], and interprets it in terms of limits: the differential coefficient is the limit of $\frac{f(x')-f(x)}{x'-x}$, corresponding to Landen's "special value" of this quotient (for $x' = x$).

An interesting point is the discussion on the existence of this value, whatever the function; this is acknowledged by Lacroix as a "difficulté" [*Traité*, 2nd ed, I, 240]. Being a "difficulté" does not mean that Lacroix actually doubts its general existence: we have seen that in the *Traité élémentaire* he had claimed this existence to be an *analytical fact* (section 8.5); and in the Preface to the second edition of the large *Traité* he repeats the claim [*Traité*, 2nd ed, I, xxv]. But it is something that needs to be proved. Lacroix gives two ways to prove it. The first, only sketched, consists in using the possibility of expanding $f(x + h)$ into a series $f(x) + Ph + Qh^2$ + etc. (which had been *proved* following [Poisson *1805*]) – P is the special value of $\frac{f(x+h)-f(x)}{h}$ for $h = 0$. The second, given in a long footnote, is Binet's *proof* [*1809*] that $\frac{f(x+h)-f(x)}{h}$ cannot become infinite or zero when h tends to zero (whence it has an assignable limit) except for particular values of x. Ampère had also given a *proof* of this, in [*1806*]; but apparently even he recognized that Binet's was simpler[4]; and Lacroix does not cite [Ampère *1806*] here.[5]

The second half of the section on alternative foundations consists in an enlarged version of the footnote in [*Traité*, 1st ed, III, 10-12] criticizing novel but unnecessary notations. In 1810 Lacroix was aware of a few more targets, namely notations for the differential coefficient employed by Pasquich, Grüson, and Kramp [*Traité*, 2nd ed, I, 247]. He does not address Arbogast's "numerous notations" because he thinks that they do not really refer to differential coefficients.

Arbogast's calculus of derivations is relegated to the last section of chapter 2, "investigations on the development of functions of polynomials" [Lacroix *Traité*, 2nd ed, I, 315-326]. It is associated to the German Combinatorial School, and especially to Kramp; but it is presented without Arbogast's and Kramp's notations, and mainly through an adaptation by Paoli, connecting it to the usual differentiation. In the Preface [*Traité*, 2nd ed, I, xxviii-xxxii] Lacroix is quite critical of both the German Combinatorial School and of Arbogast's calculus of derivations; in a draft letter written about this time to François-Joseph Français,[6] he expanded this criticism: briefly, they did not really offer anything that the usual calculus did not offer, and they were not practical for applications (namely, physical applications).

9.3 Analytic and differential geometry

9.3.1 Analytic geometry

Between 1797 and 1810 analytic geometry became a standard subject in mathematics education, and several textbooks on it were published [Taton *1951*, 132-133] – including [Lacroix *1798b*]. We could expect it to disappear from the second edition of [Lacroix *Traité*]. But that is not what happens.

In the case of analytic geometry on the plane, the preliminary paragraphs do disappear – they were quite elementary, and had been transferred to [Lacroix *1798b*]. But the same could not be said about the investigation of singular points;

---

[4]It was Ampère who reported Binet's proof to the *Société Philomatique*, saying that Binet proposed to demonstrate this theorem "d'une manière plus simple qu'on ne l'a fait jusqu'à présent" ("more simply than what has been done until now") [Binet *1809*, 275].

[5]He does cite it later apropos of its other subject: the remainder of Taylor's series [*Traité*, 2nd ed, I, 388, III, 399-400].

[6]This draft is kept at [Lacroix *IF*, ms 2400], and is transcribed in [Grattan-Guinness *1990*, III, 1325-1329]. The letter was not sent, because meanwhile Français died (in October 1810).

nor about the applications of coordinate transformation;[7] nor about the applications of series expansions;[8] nor, finally, about polar coordinates[9] – none of these topics is to be found in [*1798b*], and so they are kept in [*Traité*, 2nd ed, I].

As for analytic geometry in space, the situation is different: Lacroix keeps even the most elementary results. At the start of chapter 5, he justifies this option by saying that that this way he offers a "more complete whole" ("ensemble plus complet"). But he is probably more sincere in the Preface [*Traité*, 2nd ed, I, xxxvii], explaining that he tried to give a version of analytic geometry in space even more independent of geometrical considerations than that in the first edition. In the previous 15 years a lot of work had been done on the systematization of three-dimensional analytic geometry, associated to the teaching in the *École Polytechnique* and elsewhere;[10] and Lacroix was clearly motivated by that to improve this chapter (he could not really do this in [*1798b*], which was too elementary, and only had an appendix on three dimensions).[11]

The first section, "on the point, the plane, and the straight line" [*Traité*, 2nd ed, I, 501-527], is for a great part rewritten, and actually doubled in size. The distance formula is much more prominent than in the first edition. Lacroix uses a definition of plane that had been proposed by Fourier in a debate at the *École Normale (de l'an 3)* [Monge *1795*, 318-319]: a plane is a set of points equidistant from two given points. He says that this definition does not have the simplicity required to be used in the "elements of geometry", but that it provides an elegant means to arrive at the equations of the plane and of the straight line in space. Transformation of coordinates gains a section [*Traité*, 2nd ed, I, 528-542], which has almost three times the space that had been dedicated to that topic in the first edition (this increase is mainly due to the addition of several trigonometrical relations). The study of second-order surfaces [*Traité*, 2nd ed, I, 542-563] is about double in size compared to the first edition (partly because he gives more attention to Euler's classification). It is noticeably influenced by the "first part" of [Monge *Feuilles*, 3rd ed], by Monge and Hachette (and with a little participation by Poisson).

In "additions" at the end of the third volume Lacroix continues keeping track of works on analytic geometry, particularly in space [*Traité*, 2nd ed, III, 646-654]. He cites Gabriel Lamé, Aléxis Petit, (Joseph-Baltazar?) Bérard, and even "M.

---

[7]There is even a new application of this: the determination of infinite branches of curves [Lacroix *Traité*, 2nd ed, I, 408-413]. New, that is, in the sense that it was not in the first edition – Lacroix attributes the procedure he reports to du Séjour and Goudin.

[8]Lacroix adds four pages on series with decreasing exponents [*Traité*, 2nd ed, I, 417-421].

[9]We have seen in section 8.6 that polar coordinates is the only topic of analytic geometry appearing in [Lacroix *1802a*] instead of [Lacroix *1798b*].

[10]Throughout the three sections on analytic geometry in space we can see references to works by l'Huilier, Monge, Carnot, Puissant, Biot, and Hachette and Poisson, published after the first edition of the *Traité*; and this is in the text of the sections, not just in the bibliography in the table of contents.

[11]In spite of the improvement, there is a clear editorial flaw: the equation of the sphere is derived twice [Lacroix *Traité*, 2nd ed, I, 508, 519] – and this is not a sign of *encyclopédisme*, since the derivation is precisely the same.

Yvory" (i.e., James Ivory).[12]

9.3.2 Differential geometry

The application of differential calculus to the study of curves on the plane does not suffer many changes in the second edition. There is some rearrangement of topics and sections, and a couple of passages are rewritten to achieve a clearer systematization. For instance, just after the determination of tangents and asymptotes there is a new section on the differentials of arc length and of area under a curve [*Traité*, 2nd ed, I, 431-436]; in the first edition this was included in the section on transcendental curves. The section on the theory of osculation is divided into two: one on the general theory of contact, and the other on the osculating circle; the latter also includes evolutes, and the limit-based approach to osculation . The most thoroughly rewritten section is the new one on "determination of singular points" [*Traité*, 2nd ed, I, 456-470], where Lacroix tries to systematize the methods for characterizing the several kinds of such points.

There are more changes in the sections on differential geometry in space – although maybe not as much, or not as deep, as one might expect, given the popularity of the subject in the early 18th century, in and around the *École Polytechnique*. But the situation was different from that of analytic geometry in space: the fundamental outlook of the subject had been established in [Monge *Feuilles*], and the work done on it (mainly by Monge and Lancret) was research work, not experiments on its systematization.

The first change to note, as usual, is the multiplication of sections[13]. The section on the "theory of contact of surfaces" is very similar to the corresponding articles in the first edition. The section on the "theory of curvature of surfaces" does not have many changes either, but one of them is worthy of note: the inclusion of a version of Monge's [*Feuilles*, n°s 19-20; 3rd ed, 122-132] determination of the lines of curvature of the ellipsoid, very shortened and adapted to the fact that it is presented before integral calculus [Lacroix *Traité*, 2nd ed, I, 584-586].[14] The section "on generation of surfaces" again has only changes in detail, except for the inclusion of some remarks by Monge for simplifying the elimination of arbitrary functions [Lacroix *Traité*, 2nd ed, I, 612-615].

The main change to the section on "curves of double curvature", although not much more extensive, is more substantial. In its final pages [*Traité*, 2nd ed, I, 632-636] Lacroix reports the work of Michel-Ange Lancret [Grattan-Guinness *1990*,

---

[12]Ivory is cited because of his 1809 paper on attractions of spheroids, a paper that caused a sensation among Parisian mathematicians, although for much more than analytic geometry [Grattan-Guinness *1990*, I, 418-422].

[13]Recall that in the first edition there were only two: one on surfaces and another on curves of double curvature.

[14]This finishes with a reference to a couple of papers (or a couple of versions of a paper) on optics by Étienne Louis Malus [Grattan-Guinness *1990*, I, 473; Struik *1933*, 115], which do not appear in the table of contents (the version submitted to the *Institut* had received a favourable report by Lacroix).

I, 261-263; Struik *1933*, 115-116]: the notions of "first and second flexion" (more or less infinitesimal equivalents of modern first curvature and second curvature, or curvature and torsion), but especially an introduction to his work on "développoïdes" (a generalization of evolutes, arising not from normals to the curve, but rather from straight lines at a fixed angle).

The last section, "on the development of curves traced on curved surfaces" [*Traité*, 2nd ed, I, 636-652], is in a certain sense one of the most interesting, because its content is due to Lacroix himself. It consists in a revised and somewhat shortened version of the first part of the memoir submitted to the Academy of Sciences of Paris by Lacroix in 1790 (see appendix A.2). It revolves around two problems: given a curve on a developable surface, what does it become when the surface is developed into a plane; and reciprocally, given a curve on a plane, what does it become when that plane is enveloped onto a surface.

The most important work on differential geometry published in the 1810's was Charles Dupin's *Développements de Géométrie* (1813) [Struik *1933*, 117-118]. We should expect to see traces of it in the third-volume "additions". Indeed, it is added to the bibliography [Lacroix *Traité*, 2nd ed, III, xxii]. But in the "additions" properly speaking, even though Lacroix includes many new details on differential geometry in space [*Traité*, 2nd ed, III, 654-677], I could not find any that would seem to be drawn from Dupin's book.[15]

9.4 Approximate integration and conceptions of the integral

9.4.1 Approximate integration of explicit functions and conceptions of the integral

The section on Euler's "general method" of approximation suffered several modifications in the second edition [Lacroix *Traité*, 2nd ed, II, 130-150]. Similarly to what he had done in the second edition of [*1802a*] (see section 8.7.1), Lacroix assumes almost from the start that the differences $a_1 - a$, $a_2 - a_1$, $a_3 - a_2$, etc. are all equal, and postpones the neglect of their higher powers, so that the first (and main) formulas of approximation are those of Euler's "improved" method: (5.11) and (5.12). The equality of the subintervals raises the issue of what to do when the function to be integrated has marked differences in its rate of change; but this is solved quite simply by first splitting the interval of integration appropriately and then applying the method to each of the resulting intervals [Lacroix *Traité*, 2nd ed, II, 141-143].

However, the articles on the "nature of integrals", arbitrary constants and definite integrals are kept essentially unchanged from the first edition; the main

---

[15]These new details are not necessarily *new* – that is, not necessarily posterior to 1810; even d'Alembert is cited [*Traité*, 2nd ed, III, 671, 672-673].

difference is that now, more sensibly, they are fused into one. Naturally they appear after the derivation of (5.11) and (5.12).

The main modification from the first edition is in Lacroix's examination of the convergence of (5.11) and (5.12) [Lacroix *Traité*, 2nd ed, II, 135-136] (in the first edition this convergence had simply been assumed). The first part of it may be inspired by [Lagrange *Fonctions*, 45-46]: assuming for simplicity's sake that all of $Y', Y_1'', \ldots Y'', Y_1''$, etc. are positive, Arbogast's principle guarantees that, for values of α small enough, (5.11) always takes values smaller than $\int X\,dx$ and (5.12) takes values alternately smaller and larger than $\int X\,dx$ ("alternately" in the sequence: neglecting α^2 and higher powers of α; neglecting α^3 and higher powers of α; etc.). In case we neglect α^2 and higher powers of α, this means that (for values of α small enough)

$$\alpha\{Y' + Y_1' + Y_2' \ldots + Y_{n-1}'\} < \int X\,dx < \alpha\{Y_1' + Y_2' + Y_3' \ldots + Y_n'\}.$$

Now, the difference between these two approximations is $\alpha(Y_n' - Y')$, which can be made as small as wished by increasing n (which does not affect Y_n'), that is, by decreasing α; and of course this difference is larger than the error associated to any of these two approximations. Lacroix concludes that, "même en se bornant à la première ligne des formules"[16] (5.11) and (5.12), it is possible to obtain values for $\int X\,dx$ as approximate as one may wish.[17] The phrase "même en se bornant à la première ligne" seems to imply that if those two approximations converge to the true value, then the same must happen to any truncation of formulas (5.11) and (5.12).

Naturally, Lacroix presents this as the explanation for the possibility of viewing the integral as the sum of "an infinite number of elements" – and he adds in a footnote that the word "infinite" is being used only as an abbreviation for "the larger the number of elements", the closer the approximation.

The two proofs that used the property of the integral being the limit of approximating sums in the first edition are absent from the second edition: one, that $\alpha(Y' + Y_1' + Y_2' \ldots + Y_{n-1}')$ and $\alpha(Y_1' + Y_2' + Y_3' \ldots + Y_n')$ are bounds for the integral, is substituted by the argument invoking Arbogast's principle in the beginning of the proof above; the other, that $\int X\,dx$, taken between $x = a$ and $x = b$, is positive if X is always positive in the same interval, is simply dispensed with in this section[18].

But the reason why the latter had been included in the first edition – namely the proposition that $m(b - a) < \int X\,dx < M(b - a)$, where the integral is taken

---

[16]"even if we restrict ourselves to the first line of the formulas"

[17]This contradicts Grabiner's assertion that "Lacroix did not try to prove that the true value of the integral of an arbitrary function differs from the approximating sums by less than any given quantity for sufficiently small subintervals" [Grabiner *1981*, 152]. She seems to have read the section on approximation only in the first edition of Lacroix's *Traité*, and to have assumed that it was unchanged in the second.

[18]A similar result had been proved by other means in [Lacroix *Traité*, 2nd ed, I, 382].

between $x = a$ and $x = b$ and m and M are the smallest and largest values of X in that interval – is given as a result of both sums $\alpha(Y' + Y_1' + Y_2' \ldots + Y_{n-1}')$ and $\alpha(Y_1' + Y_2' + Y_3' \ldots + Y_n')$ being contained between $n\alpha Y_m = (b-a)Y_m$ and $n\alpha Y_M = (b-a)Y_M$, where Y_m and Y_M are the smallest and largest of $Y', Y_1', Y_2' \ldots Y_n'$. This is precisely the same argument that had been used in [Lacroix *1802a*, 2nd ed, 307], with the significant difference that there it could be seen as invoking the idea of integral as infinite sum, while here it certainly invokes the idea of integral as limit of sums.

Thus we see that the idea of integral as limit of sums is addressed in the second edition of Lacroix's *Traité* in a different way from the first edition. But it is certainly kept, and even reinforced, insofar as the convergence of the approximating sums receives a proof.

9.4.2   Approximate integration of differential equations

The methods for approximate integration of differential equations are the subject of one of the four new chapters (more precisely chapter 6 [Lacroix *Traité*, 2nd ed, II, 409-446]) which corresponds to sections from chapter 3 in the second volume of the first edition, but with several modifications. Having their own chapter does not mean that they play a larger role than in the first edition. Quite the contrary: Lacroix explains in the *avertissement* [*Traité*, 2nd ed, II, v-vi] that he has suppressed more than added. He had done so because there were too many methods, proper for specific applications, and it would be useless to expound them all, separated from the applications and therefore deprived of interest.

This chapter is divided into three sections, the first of which is dedicated to power series. Lacroix starts by referring to Taylor series (influenced by Lagrange [*Fonctions*; *Calcul*], he had already given greater importance than in the first edition to Taylor series in establishing fundamental properties of integrals of differential equations, namely number of arbitrary constants, number of "first integrals", and even the existence of solutions [Lacroix *Traité*, 2nd ed, II, 294-298] – see section 9.5.1). After this he quickly mentions Euler's "general method", but also quickly dismisses it because it demands too many calculations and because each step is affected by the error of the previous one – something which does not happen in the case of explicit functions. The larger part of the section [Lacroix *Traité*, 2nd ed, II, 411-426] is occupied with the method of undetermined coefficients – mostly by examples, of both first- and second-order. Finally Lacroix quickly refers to the method he had extracted from Lagrange's memoir on continued fractions (pages 108 and 157), as he had done in the first edition, but adding that it is hardly useful because the series obtained by using it are usually not very convergent, and their general terms not easy to understand [Lacroix *Traité*, 2nd ed, II, 427].

The second section [Lacroix *Traité*, 2nd ed, II, 427-434] is dedicated precisely to Lagrange's method of continued fractions, and it has almost no difference from the corresponding articles in the first edition [Lacroix *Traité*, II, 288-296].

The third and final section [Lacroix *Traité*, 2nd ed, II, 435-446] is dedicated to the "use of first-degree differential equations to integrate by approximation", that is to the methods used in obtaining approximations of planetary orbits, mentioned above (pages 156 and 175 ff.). Meanwhile new work had been done by Laplace, Lagrange and Poisson on subjects close to this, namely on the stability of the planetary system, and more general variational mechanics leading to Poisson and Lagrange brackets [Grattan-Guinness *1990*, I, 371-385]. Lacroix was well aware of this new work, having reviewed and praised one of Poisson's papers [Grattan-Guinness *1990*, I, 380]. He includes the most relevant new papers (by Lagrange and Poisson) in the table of contents for this section, but only mentions them briefly in the text [Lacroix *Traité*, 2nd ed, II, 443]. Nevertheless, a great deal of the section is written anew, with clear improvements – Lacroix seems more comfortable with the subject, and the series of mistakes in the first edition is gone. Also, the astronomical motivation is acknowledged [Lacroix *Traité*, 2nd ed, II, 443-446].

9.5 Types of solutions of differential equations

9.5.1 Differential equations in two variables and their particular solutions

There was no reason for Lacroix to abandon, from the first to the second edition, his point of view on the formation of differential equations in two variables by elimination of constants between finite equations and their differentials. And he did not. Most traces of it remain: for instance, the elimination of constants leading to differential equations [Lacroix *Traité*, 2nd ed, I, 197-198], the explanation for the method of integrating factors for first-order equations [Lacroix *Traité*, 2nd ed, II, 260-261], and of course the new chapter on particular solutions [Lacroix *Traité*, 2nd ed, II, 373-408].

Nevertheless, the importance of that point of view decreases, as is clear in a few changes inspired by Lagrange [*Fonctions*, 54-58; *Calcul*, 151-167] (resp. a section and a chapter, both with the title "Théorie générale des équations dérivées, et des constantes arbitraires"[19]). Although the "nature" of differential equations corresponded to their formation by algebraic elimination of constants between primitive and derivative equations [*Fonctions*, 56], Lagrange had also used Taylor series to explore arbitrary constants [*Fonctions*, 55; *Calcul*, 160-165]. In the second edition of his *Traité*, Lacroix includes a new section "on the successive integrals of higher-order differential equations" [Lacroix *Traité*, 2nd ed, II, 292-298], whose references in the table of contents are precisely those two passages by Lagrange, and which opens mentioning "la théorie générale de la liaison qui existe entre les équations différentielles et leur intégrales successives"[20]. This "general

---

[19]"General theory of derivative equations, and of the arbitrary constants"

[20]"general theory of the connection between the differential equations and their successive

theory" includes the formation of differential equations by algebraic elimination of constants, with its consequences on the number of "first", "second", etc. integrals; but naturally Lacroix also follows Lagrange in using Taylor series in this – he uses them to reinforce the conclusions on the number of arbitrary constants, and to conclude that every differential equation in two variables is possible (i.e., has a solution, even if we cannot find it), provided that the highest-order differential coefficient is a real function of the others and of the variables [Lacroix *Traité*, 2nd ed, II, 296]. We thus see, as in [Lagrange *Fonctions*; *Calcul*], a shared *foundation* of differential equations.

As for particular solutions of differential equations in two variables, they gain a separate chapter (chapter 5 [Lacroix *Traité*, 2nd ed, II, 373-408]). This new chapter is divided into one small introduction and three sections: "liaison des solutions particulières avec les intégrales"[21], "comment les solutions particulières se tirent des équations différentielles"[22], and "application de ce qui précède, à l'intégration"[23].

The first section has little novelty: Lacroix combines the explanations for the existence of particular solutions of first-order and higher-order differential equations, which had been separate in the first edition. The only change to be noticed, related to higher-order equations, is that now Lacroix pays a little more attention to particular solutions of non-finite particular solutions (which are themselves differential equations) – he mentions Lagrange, and his name of "double" and "triple" particular solutions for these, introduced in [Lagrange *Calcul*, 199-202].

The second section is quite a different matter. Lacroix must have been impressed by [Poisson *1806*] – he had already cited it in the second edition of the *Traité Élémentaire* – and its influence here is clear. Poisson had adopted as a point of departure the characterization of particular solutions as solutions which cannot be completed by an arbitrary constant [*1806*, 61]. In the first edition Lacroix had reported this characterization [*Traité*, II, 274-277] (adapting a passage from [Lagrange *Fonctions*]) but in a subsidiary manner. In the second edition, it is *the* way to study particular solutions directly from the differential equations. The first few pages of this section [Lacroix *Traité*, 2nd ed, II, 383-387] reproduce [Lacroix *1802a*, 2nd ed, 436-442]: an adaptation of [Lacroix *Traité*, 1st ed, II, 274-277]. Next, in an article with some historical remarks [*Traité*, 2nd ed, II, 388], Lacroix mentions Lagrange's terminology of "singular primitive equations", and how it refers to an analogy with "singular values" (for which Taylor series fails) – something which he had failed to notice in the first edition. Another clear influence from [Poisson *1806*] is the important issue of the possibility of transforming a differential equation possessing a particular solution so that the latter appears as a factor. In the first edition, Lacroix had distinguished the particular solutions which are factors (and thus apparently trivial) from the *real* particular solutions;

---

integrals"
[21]"connection between particular solutions and integrals"
[22]"how to obtain particular solutions from the differential equations"
[23]"application of this to integration"

this in spite of Trembley [*1790-91*] having already stated that that transformation is always possible. It seems that Lacroix was more convinced by Poisson [*1806*, 70-71], so that he reports his proof [Lacroix *Traité*, 2nd ed, II, 389].[24] Finally, the study of particular solutions of differential equations of order higher than 1 [Lacroix *Traité*, 2nd ed, II, 392-399] is also admittedly based on [Poisson *1806*].

In the third section Lacroix mostly retakes, from the first edition, Trembley's method to find integrating factors from particular integrals and solutions. A few novelties result from Lagrange's factorization of the derivative of a "derivative equation" into a singular primitive equation times something which corresponds to the complete primitive equation [Lagrange *Calcul*, ch. 15] – which explains why certain differential equations are easier to integrate after being differentiated.

9.5.2 Partial differential equations and their particular solutions

From Lacroix's memoir of 1785 (appendix A.1), through the first edition of his *Traité*, to the second, one can observe a decrease in Lacroix's confidence in the point of view that partial differential equations result from the elimination of arbitrary functions. It has already been seen how in the first edition he expresses his reservations about the analogy between arbitrary constants and arbitrary functions in those eliminations.

Not that this point of view is abandoned. But the reserves on the analogy gain relevance. The first encounter with them is now in the first volume, in a new section "on the elimination of indeterminate or arbitrary functions" [Lacroix *Traité*, 2nd ed, I, 230-237] where they follow immediately the introduction of those eliminations.

In the second volume, the first reference to the formation of partial differential equations occurs somewhat later than in the first edition, near the end of the section on "integration of first-order partial differential equations" (which in fact addresses only those of "first degree"): Lacroix remarks that the method he has been using (Lagrange's method for quasi-linear equations) assumes the solution to be $V = \varphi(U)$ in the case of three variables, $V = \varphi(T, U)$ in the case of four variables, and so on; and since no limitation appears, those forms are general and the origin of first-order partial differential equations indicated in the first volume "ne souffre aucune exception, lorsque les coefficiens différentiels ne passent pas le premier degré"[25] [Lacroix *Traité*, 2nd ed, II, 545]. As for partial differential equations of degree higher than 1, he does not seem so certain here that those forms are indeed the most general [Lacroix *Traité*, 2nd ed, II, 564-565], although in a later addition he seems to have been convinced by Poisson [Lacroix *Traité*, 2nd ed, III, 705-708]. Also the section on "integration of partial differential equations

---

[24]And modifies the introductory article to the chapter [Lacroix *Traité*, 2nd ed, II, 373]: now particular solutions "paraissent d'abord de deux sortes" ("appear at first to be of two kinds"), instead of "sont de deux sortes" ("are of two kinds") [Lacroix *Traité*, 1st ed, I, 389].

[25]"does not admit any exception, as long as the differential coefficients do not go above first degree"

of order higher than 1" opens with a complaint about the ignorance of the general forms of the integrals [Lacroix *Traité*, 2nd ed, II, 575-576].

The relationships between complete and general integrals are addressed in three (or four) places: 1 – just after the remark quoted above on the assumption of the forms $V = \varphi(U)$ or $V = \varphi(T, U)$, Lacroix [*Traité*, 2nd ed, II, 545-546] repeats from the first edition their derivation from $V = aU + b$ and $V = aT + bU + c$; 2 – a new section "on the various forms of the integrals of partial differential equations" [Lacroix *Traité*, 2nd ed, II, 658-667] repeats those reflections on this in the first edition which had not been moved to the first volume, including the introduction of Lagrange's terms "complete" and "general" integrals (in an addition, Lacroix [*Traité*, 2nd ed, III, 710-711] reports later work by Ampère [*1815*] on the number of arbitrary elements, based on the differentiation-elimination process); 3 – the section on construction of partial differential equations now includes an article on the geometrical interpretation of types of solutions (see below).

As for particular solutions of partial differential equations, the new (short) section on them [Lacroix *Traité*, 2nd ed, II, 667-672] repeats the Lagrangian explanation from the first edition, but replaces the procedure for obtaining them by Legendre [*1790*] with the one by Poisson [*1806, 114-116*].

9.5.3   Geometrical connections

The second edition includes a new chapter on "geometrical applications of differential equations in two variables" [Lacroix *Traité*, 2nd ed, II, 447-470], which corresponds for the most part to the section on "geometrical construction of first-order [ordinary] differential equations" in the first edition. There are a couple of new problems (namely a construction for curves of the form $s = f(p)$, where s is the arc length, and the determination of curves whose radius of curvature is equal to the normal). But the most interesting modifications are the inclusion of the construction analogous to Euler's "general method", and a significant expansion of the space dedicated to particular solutions.

It has been remarked that in the first edition the most interesting constructions of ordinary differential equations occur in connection to approximation, more specifically to Euler's "general method" (see sections 5.2.4 and 6.2.3.2). In the second edition Lacroix is quite dismissive of the application of this method to differential equations (see section 9.4.2), and moves its geometrical version to the chapter on geometrical applications [Lacroix *Traité*, 2nd ed, II, 451-452] – a more natural location. Their purpose is now much clearer: these constructions "ne saurai[en]t guères être utile[s] dans la pratique; mais [ils] prouve[nt] que ces équations expriment toujours quelque chose de réel"[26], confirming or reinforcing a conclusion which Lacroix had already obtained using power series (see section 9.5.1).[27]

---

[26]"are hardly ever useful in practice; but they prove that these equations always express something real"

[27]In the case of second-order equations, Lacroix gives also a construction using osculating circles, simpler than the one resulting from Euler's "general method", which involves osculating

The five-and-a-half-page new section on the (geometrical) "meaning of particular solutions" [Lacroix *Traité*, 2nd ed, II, 465-470] has over twice the space that had been dedicated to that issue in the first edition. Still, most of it is yet dedicated to examples: a great deal of this expansion comes from an adaptation of [Lagrange *Calcul*, 263-268] – a discussion of a problem solved by Leibniz, and of why Leibniz had arrived only at the particular (singular) solution. However, Lacroix now includes also the remark that any given curve corresponds to the particular solutions of several differential equations (that of its tangents, that of its osculating circles, and so on); and alludes briefly to the geometrical problems posed by the possibility of removing the particular solution from the differential equation (after having transformed the latter so that the former becomes a factor – see section 9.5.1), directing the reader to Poisson's work [*1806*, 75, 117-123].

The only significant modification to the section on geometrical construction of partial differential equations is the correction of an important flaw in the first edition: the addition of an article [Lacroix *Traité*, 2nd ed, II, 682-685] on the geometrical meaning of complete and general integrals and particular solutions. General integrals are associated to the two Mongean types of generation of surfaces: by movement of a line in space, and by continued intersection of a family of surfaces (that is, as an envelope). Integrals directly of the form $V = \varphi(U)$ correspond to surfaces generated by the movement of a line in space – a line with equations $U = a$ and $V = \varphi(a) = b$. General integrals obtained from complete integrals by variation of constants, being expressed by systems of the form

$$F[x, y, z, a, \varphi(a)] = 0 \quad \text{and} \quad \frac{dF}{da} = 0$$

(from which a is to be eliminated), correspond to envelopes of families of surfaces parametrized by a; these latter surfaces correspond to instances of the complete integral, each family corresponding to a particular function φ, that is to a particular relation between the two constants in the complete integral. Lacroix also mentions here the correspondence between Monge's *characteristics* (given by the above system, without eliminating a) and the ordinary differential equations appearing in the Lagrange-Charpit method. And naturally he also refers to the tangency between the surfaces given by particular solutions and those given by complete and general integrals. But about half of the article is taken up with the issue of whether the complete integral is contained or not in the general integral. Lagrange had originally assumed that the general integral contains the complete integrals [*1774*, §56] (see also section 6.1.4.2 above), but he had later [*Calcul*, 372-381] changed his mind, based on two arguments. Firstly, the example (written here with the notation used by Lacroix) of the equation $px + qy = z$; its complete integral is $z = ax + by$ and therefore one may regard as its general integral the result of eliminating a between $z = ax + y\varphi(a)$ and $x + y\varphi'(a) = 0$; but there is no

---

parabolas. Tournès [*2003*, 469] remarks that although the determination of centres of curvature and osculating circles had long been an important problem, he has not found any instance of this kind of inverse problem prior to the second edition of Lacroix's *Traité*.

function φ such that this elimination yields $z = ax + by$. Secondly, the geometrical interpretation of the complete and general integrals: the latter is formed by the successive intersections of the former (for a particular φ), suggesting that they are essentially distinct. Lacroix reports Lagrange's view, especially the example above, but he does not seem to adhere to it: $z = x\varphi\left(\frac{y}{x}\right)$ is also a general integral of $px + qy = z$, and putting $\varphi\left(\frac{y}{x}\right) = A + \frac{By}{x}$ one obtains $z = Ax + By$, and so he concludes that that exception does not affect general integrals represented by one single equation (Lagrange himself had given this apparent counter-example [*Calcul*, 374-377]; it seems that this muddle was a matter of definition – in [*Calcul*, 371] Lagrange had *defined* "general primitive equation" not as one containing an arbitrary function, but rather as one obtained from the complete primitive equation $F(x, y, z, a, b) = 0$ by putting $b = \varphi a$ and eliminating a using $F'(a, \varphi a) = 0$, so that he did not refer to $\frac{z}{x} = \varphi\left(\frac{y}{x}\right)$ as a "general" primitive equation, only as a "simpler and more general form" of the primitive equation [*Calcul*, 374]).

9.5.4   Continuity of arbitrary functions

As is implicit in the previous section, the articles where Lacroix addresses the possible discontinuity of arbitrary functions remain practically unchanged in the second volume of the second edition. But he adds an announcement: he will treat Laplace's opinion on the subject in the third volume [Lacroix *Traité*, 2nd ed, II, 686].

Laplace had stated his opinion in [*1779*, 298-302]. He regarded partial differential equations as particular cases of partial finite difference equations; the solutions of the latter might be constructed as polygons, and "lorsqu'on passe du fini à l'infiniment petit, ces polygones se changent dans des courbes qui par conséquent peuvent être discontinues"[28] [*1779*, 300]. Of course, "discontinuity" is to be understood with its 18th-century meaning of absence of a general expression. In fact, Laplace proceeded by remarking that in order for an n-th order partial differential equation to "subsist", there can be no jumps between consecutive values of the dependent variable, nor of its derivatives up to order $n - 1$ – so as to ensure that the n-th derivatives (or rather, the n-th "differences, divided by the respective powers" of the independent variables) are finite quantities.

Arbogast had argued against Laplace's opinion in his dissertation on arbitrary functions [*1791*, 79-86], and Lacroix seems to have thought that this settled the issue, so that in the first edition he did not even mention Laplace apropos of this controversy.

But Laplace repeated his old stand in [*1812*, 72-80]. As after all the issue was not consensual, Lacroix included a new short section "sur la nature des fonctions arbitraires des intégrales aux différentielles partielles "[29] [*Traité*, 2nd ed, III, 307-311]. Most of the section is taken up with reporting Laplace's argumentation

---

[28]"when we pass from the finite to the infinitely small, these polygons change into curves, which may thus be discontinuous"

[29]"on the nature of the arbitrary functions of the integrals of partial differentials"

(which must be the reason for putting this section in the third volume, as it starts with a difference equation). But in the last article [*Traité*, 2nd ed, III, 310-311] Lacroix briefly refers to Arbogast's counter-arguments, and remarks that Lagrange's final opinion was a complete agreement with Monge (that is, the full acceptance of discontinuous functions), that Laplace's opinion was first put forward by Condorcet (with a different argument), and finally that Poisson had recently expressed an opinion similar to that of Laplace. Clearly, Lacroix had not changed his mind about the admissability of discontinuous functions; but the presentation of the problem is more balanced in the second edition than in the first – even though d'Alembert's case is never described.

9.5.5 Total differential equations not satisfying the conditions of integrability

The section on "total differential equations that do not satisfy the integrability conditions" remains mostly unchanged from the first edition in [Lacroix *Traité*, 2nd ed, II, 690-720]. The most relevant of the small changes are a different proof of part of Monge's correspondence between partial and total differential equations [*Traité*, 2nd ed, II, 707-709] and an increase in caution on statements about generality of solutions: Monge's procedure for integrating first-degree equations, which in in the first edition led to "la solution la plus générale que l'on puisse obtenir"[30] [*Traité*, II, 625], now simply provides "une solution remarquable par sa forme et son étendue"[31] [*Traité*, 2nd ed, II, 692]; then, when presenting his theory of the formation of these equations, Lacroix doubts the generality of what in the first edition was the "general integral" – an argument involving a Taylor series for z as a function of x makes him think that there should be an arbitrary constant independent of the arbitrary function [*Traité*, 2nd ed, II, 703-704].

But the most relevant new material occurs in an "addition" in the third volume, rather than in the second volume. Between the publication of the first edition and that of the second volume of the second edition (1814), no important new work had appeared on total differential equations in three variables. [Pfaff *1815*], on the other hand, was important enough for the name *Pfaffian equation* to be still nowadays used for first-order linear total differential equations in more than two variables. To overcome difficulties in the Lagrange-Charpit method, which was not practical with more than two independent variables, Pfaff reduced the integration of a partial differential equation in n variables to that of a total differential equation in $2n - 1$ variables, and gave a method to solve the latter [Demidov *1982*, 333-334]. In [*Traité*, 2nd ed, III, 711-712] Lacroix gives a short, but appreciative, account of [Pfaff *1815*]. However, he remarks Pfaff's acknowledgement that Monge had suggested total differential equations as the "key" for integrating partial differential equations. He also remarks Paul Binet's priority, of which Pfaff

---

[30]"the most general solution that might be obtained"
[31]"a solution remarkable for its form and extension"

was certainly unaware, in a fundamental result on the number of equations in the solution of a total differential equation.[32]

9.6 Aspects of differences and series

9.6.1 Indices

The only noteworthy difference from the first edition that I have noticed in the use of subscript indices, is the disappearance of their occurrence in the derivation of Taylor's theorem. This is a consequence of the loss of importance of the derivation process, and of the adoption of Poisson's proof [*1805*] (see section 9.2). Indices might still make it easier to read (at least for modern eyes) but not that much.

9.6.2 The "multiplicity of integrals" of difference equations

After the publication of the first edition, Poisson took over Biot's mathematical subjects – the multiple integrals of finite difference equations, and mixed difference equations [Grattan-Guinness *1990*, I, 189-190, 223-231]. In the second edition, Lacroix reported Poisson's new results.

The section on "multiplicity of integrals" of difference equations [*Traité*, 2nd ed, III, 250-267] has a very clear organization. First, we find what is practically a reprint of the same section in the first edition (that is, Lacroix's account of Biot's work) [*Traité*, 2nd ed, III, 250-260]. The shorter remainder is dedicated to [Poisson *1800*] and [Poisson *1806*]. In [*Traité*, 2nd ed, III, 260-264] Lacroix reports Poisson's conclusion [*1800*] that there are even more integrals for difference equations than those studied by Charles, Monge and Biot; Poisson's new integrals contain arbitrary functions subject only to take integer values when the argument assumes integer values. In [*Traité*, 2nd ed, III, 264-267] Lacroix gives very short accounts of Poisson's arguments in [*1806*] for the existence of those new integrals, and also for the existence of particular solutions of difference equations. No one had noticed the existence of particular solutions, because they are not obtained by variation of constants (which instead leads to indirect integrals); Poisson arrived at them via his characterization of particular solutions as solutions that cannot be completed by arbitrary constants. We have already remarked on the good impression of [Poisson *1806*] that Lacroix had.

9.6.3 Mixed difference equations

As expected, the main modifications to the chapter on mixed difference equations are additions drawn from Poisson's work. In 1806 Poisson published a memoir

---

[32]Binet had submitted a memoir with this result to the *Institut* in August 1814 [Acad. Sc. Paris *PV*, V, 385]. Lacroix and Poisson had been charged with reporting on it, but Binet had withdrawn it "for perfecting". It appears to have never been published.

on mixed difference equations [Grattan-Guinness *1990*, 230-231], where he rejected Lacroix's suggestion for solving mixed difference equations in the strict sense – namely transforming them into indefinite-order differential equations through power-series expansions for Δy and $\Delta \frac{dy}{dx}$; Lacroix himself had not been enthusiastic about this method, recognizing that it was often difficult to apply. Instead, Poisson applied Laplace's cascade method to linear first-order mixed difference equations. And Lacroix reports this [*Traité*, 2nd ed, III, 579-584]. He also reports Poisson's more thorough treatment of a geometrical problem already addressed by Biot [*Traité*, 2nd ed, III, 591-595]. However, he maintains unchanged his sole example of a geometrical problem leading to a mixed difference equation in the strict sense – thus not using the cascade method, but rather the series expansion of Δy and, alternatively, Charles's solution.

The only other noteworthy modification is that the article mentioning analytical applications is transformed into a short section [*Traité*, 2nd ed, III, 598-599] on "mixed and partial difference equations" – that is, equations involving both partial differentials and partial differences. In fact, one of the examples that he had given in the first edition of analytical applications of mixed difference equations – the one studied by François-Joseph Français – led to mixed *partial* difference equations. He also gives an example from [Laplace *1779*]. But most of the section is dedicated to one example by Lacroix's favorite Italian author Pietro Paoli.

Chapter 10

Final remarks

10.1 Originalities, both real and misattributed

10.1.1 Originalities in Lacroix's *Traité*

In the Preface of the first edition of the *Traité* Lacroix made a declaration of modesty:

> "Parmi beaucoup de choses extraites des ouvrages des grands Géomètres de nos jours, il se trouvera peut-être quelques détails qui m'appartiendront ; mais je ne disputerai pas là-dessus, et je me contenterai de ce qu'on voudra bien me laisser."[1] [*Traité*, I, xxviii]

We have seen that he did not keep this promise entirely. In the Preface of the second edition, and in his "Compte rendu [...] des progrès que les mathématiques ont faits depuis 1789 [...]" (appendix B) he claimed priority for some details: his use of indices in proving the power-series expansions of transcendental functions (section 7.1.2); the change of independent variable without consideration of constant differentials (section 3.2.4); a proof of Newton's theorem on the sums of powers of the roots of an equation [*Traité*, I, 283-286]; remarks on limitations in the number of arbitrary functions in integrals of higher-order partial differential equations (section 6.2.2.3); and the analytical theory of the different kinds of integral of total differential equations in three variables that do not satisfy the conditions of integrability (section 6.2.4.2). To this, we can also add the section on the "development of curves traced on surfaces" in the second edition, adapted from Lacroix's 1790 memoir (appendix A.2).

A different kind of original contribution is in terminology. There are a few expressions that appear to have been introduced in Lacroix's *Traité*: "differential

---

[1]"Among many things extracted from the works of the great Geometers of our time, one may find perhaps a few details belonging to me; but I will not dispute over them, and I will be content with what one is willing to leave me."

coefficient" (see section 3.2.2); "partial differential", instead of "partial difference" (see section 3.2.3); "indirect integral" (see section 7.2.2); possibly "symmetric function" and "total differential equation" (see page 37); and least successfully, "first-degree differential equations", instead of "linear differential equations" (see page 35).

Apart from these, there are a couple of issues in which modern readers see innovations: the systematization of analytic geometry, particularly on the plane (section 4.1.2); and the exploration of the conception of the integral as a limit (section 5.2.3). It is interesting that these two issues appear now much more relevant than his claimed originalities. Perhaps this is so because they are related to what Lacroix did best: to expound mathematics, rather than to achieve new results or techniques.

10.1.2   Misattributions of originality

Some innovations have been misattributed to Lacroix's *Traité*.

We have seen one of these in section 7.1: the introduction (or introduction in France) of subscript indices. Lacroix may have contributed to their diffusion; but Laplace had already used them extensively.

Another situation that might be regarded as a misattribution relates to the so-called "Faà di Bruno's formula" for the nth derivative of a composite function. The Italian Francesco Faà di Bruno (1825-1888) gave that formula in 1855, but he was not the first one. In [*2002*] Warren Johnson unearthed several precursors of Faà di Bruno, among which is Lacroix, in the "corrections and additions" at the end of the second edition [*Traité*, 2nd ed, III, 629]. Johnson recognized that Arbogast had given several particular cases and a "prose rule for writing the general case", and that Lacroix had drawn on Arbogast's work for writing the general formula, but he stated that Arbogast seemed never to have "written down Faà di Bruno's formula as such", thus apparently giving priority to Lacroix [*2002*, 230]. Alex Craik has argued convincingly that "'Faà di Bruno's formula' was first stated by Arbogast in 1800" [*2005*, 128]. Of course, the issue here is semantical: can a "prose rule" qualify as a formula? I believe that nearly every historian of mathematics would agree with Craik.

Our final case is a more clear-cut misattribution. It deals with fractional calculus: calculus with derivatives of non-integer order. According to [Ross *1977*, 76-77], Leibniz toyed with the idea of a differential of order $\frac{1}{2}$; in [*1730-1731*] Euler suggested using interpolation to obtain such differentials; but it was in [Lacroix *Traité*, 2nd ed, III, 409-410] that appeared "the first mention of a derivative of arbitrary order in a text" – for $y = x^m$, Lacroix writes[2]

$$\frac{\mathrm{d}^n y}{\mathrm{d}x^n} = \frac{m!}{(m-1)!}x^{m-n} = \frac{\Gamma(m+1)}{\Gamma(m-n+1)}x^{m-n},$$

---

[2]I am copying Ross's use of factorials and Legendre's Γ symbol; Lacroix actually wrote $[m]^m$ instead of $m!$, and $\int \mathrm{d}x \left(1\frac{1}{x}\right)^m$ instead of $\Gamma(m+1)$.

and putting $y = x$ and $n = \frac{1}{2}$ he gets

$$\frac{\mathrm{d}^{\frac{1}{2}}y}{\mathrm{d}x^{\frac{1}{2}}} = \frac{2\sqrt{x}}{\sqrt{\pi}}.$$

Two similar but shorter accounts (omitting Euler altogether) have appeared more recently, in educational journals [Doyle *1996*, 16; Debnath *2004*, 487-488]; [Doyle *1996*] even has a section named "Lacroix's formula for D^{m}". The first point to make is that although both Ross and Debnath only cite the second edition of Lacroix's *Traité* (Doyle does not cite Lacroix directly), and indeed all of them stress the year 1819, this passage is already present, precisely in the same form, in [Lacroix *Traité*, 1st ed, III, 390-391]. But more seriously, it is also present (modulo notations for factorials and π) in [Euler *1730-1731*, 56-57] – a work that Ross knew and cited, and from which Lacroix acknowledged to have taken this.[3]

10.2 Impact

How to assess the impact of a book that did not intend to introduce any major innovation? Let us examine some leads.

First of all, it is undeniable that Lacroix's *Traité élémentaire* was hugely successful – not only in France (being adopted for some time in the *École Polytechnique* and in the *Lycées*, and having several editions even after that), but also in several other European (and American) countries (see section 8.10). If we take the *Traité élémentaire* as a by-product of the large *Traité*, then the latter must partake of the obvious educational influence of the former.

We can also examine what happened to the terminological innovations mentioned above. "First-degree differential equations" never had any success – the word "linear" proved too appealing. "Indirect integral" was used for some time (namely by Biot). "Partial differentials" quickly gained ground: compare in appendices C.2.2 and C.3.1 the programmes of the *École Polytechnique* for 1800-1801 and 1805-1806 – the former has "notion of partial differences" and the latter "notions on partial differentials" (the change actually occurred in 1802-1803). True, older mathematicians stuck with "partial differences" (for instance, Monge in [*Feuilles*, 3rd ed], published in 1807); but they eventually lost – this was such an obviously sensible suggestion... "Total differential equation" and especially "symmetric function" were also very successful (although it is not clear whether they were introduced by Lacroix). Finally, "differential coefficient": this expression has disappeared in the meantime, but throughout the 19th century it was an extremely popular name for the derivative [Anonymous *1900*]. Of course, the popularity of "differential coefficient" and of "partial differential" (and possibly of "total diffe-

---

[3]Lützen [*1990*, 779] had already noticed this misattribution, while classifying [Ross *1977*] as an "excellent paper" [Lützen *1990*, 306].

rential equation") result at least as much from the *Traité élémentaire* as from the large *Traité*.

Focusing strictly on the large *Traité*, we can invoke some pieces of evidence that add up to form the picture of a treatise fundamental in the formation of a generation or two of mathematicians.

In [*1843*, 3] Libri reminded his listeners that for 45 years (that is, since its publication), Lacroix's *Traité* had been "le compagnon inséparable de tous les géomètres, [...] le guide sûr et fidèle de tous ceux qui aspirent à se faire un nom dans les mathématiques"[4]. This could be disregarded for being said at a funeral eulogy. But Libri added an anecdote, about a young would-be scientist in the 1820's, to whom Laplace had said:

> "Vous êtes fort heureux actuellement d'avoir le grand ouvrage de M. Lacroix; quand j'ai commencé à étudier, il m'a fallu dix ans de travaux pour y suppléer."[5]

We do not have any assurance that this is a true story, but it certainly is believable.

In fact, the greatest merit in Lacroix's *Traité* is in accomplishing the purpose of making the 18th-century calculus, in all its details, much more easily accessible and fruitful to the 19th-century mathematicians. We have evidence that he succeeded in that. Let us give three examples.

The first example relates to a Portuguese mathematician, Francisco Garção Stockler (1759-1829). A staunch supporter of d'Alembert, Stockler gave in [*1805*] his opinion on the vibrating-string controversy (naturally, he held that the arbitrary functions involved had to be "continuous"). The point that interests us here is that he did not fail to express his disagreement with Arbogast's arguments, although he had not had access to [Arbogast *1791*]. How? He relied on the short account given by Lacroix [Stockler *1805*, 183]. Thus, through Lacroix a mathematician in 1805 Lisbon had easy access, even if second-hand, to an argument published in St. Petersburg in 1791.

The second example has already been mentioned in section 5.2.3: Cauchy's definition of definite integral comes from 18th-century techniques for approximation of integrals, particularly Euler's "general method"; although Cauchy had some direct knowledge of Euler's work, Grabiner has argued that Lacroix's account was "Cauchy's most likely immediate source" [*1981*, 151].

The final example, possibly the strongest evidence for direct influence of Lacroix's *Traité*, was mentioned in page 38: the work of Paul Charpit connecting two methods by Lagrange was known for a long time only through Lacroix's account; if it were not for Lacroix, there would not be a "Lagrange-Charpit method".

These leads and the considerations above on originalities seem to confirm that Lacroix's *Traité* did have a significant impact; but that this impact had very

---

[4]"the inseparable companion of all geometers, [...] the reliable and faithful guide of all those who aspire to acquire a reputation in mathematics"

[5]"You are very fortunate to have nowadays the great work of M. Lacroix; when I started my studies, it took me ten years of labour to make up for it."

little to do with its original contributions; rather, what made it so relevant was the non-trivial fact that it was a well-organized, truly comprehensive, up-to-date, and advanced-level survey of the calculus.

10.3 Issues of affiliation, style, and method

10.3.1 Lagrange vs Monge; algebra vs geometry

It has been noticed that in the late 18th century the dominant approach to the calculus was *algebraic* [Fraser *1989*]. This statement does not apply simply to Lagrange's power-series foundation: algebraic views (usually called "analytical") had been gaining ground since the beginning of the century, and were already quite strong in Euler. The typical example is the change in the object of the calculus: from curves to functions – that is, "analytical expressions".

The only major mathematician in late 18th century France who took a different stand was Monge. We have seen in sections 6.1.3.2, 6.1.3.4, and 6.1.3.5 his application of *geometrical* reasonings to differential equations.

This distinction between a Lagrangian-algebraic style and a Mongean-geometric style poses us a question: what happens in Lacroix's *Traité*, given that Lacroix was a disciple of Monge and that he chose Lagrange's power-series foundation? Was he a Lagrangian, or was he a Mongean? An easy answer is to say that Lacroix was both: in an eclectic style, he was Lagrangian in the chapter on the principles of the calculus, and Mongean in the chapters on geometrical applications.

But I believe that the situation was not so simple, and that it needs to be desimplified in order to understand Lacroix's standpoint. The description above passes over the fact that the differences between Monge and Lagrange lie not only in style, but also in subjects. Both studied partial differential equations, but other than that, they usually addressed different topics. There is no book by Monge on the calculus, and Lagrange's contributions to differential geometry are limited (in [*Fonctions*, 168, 184, 187] he recommended Monge's works). There is, of course, the creation of "analytic geometry" (instead of "application of algebra to geometry"), for which they are jointly credited. But this is precisely a case in which they concurred, rather than competed with different points of view. We might also recall Monge's appreciation of Lagrange's geometrical interpretation of singular integrals. Thus, the Lagrangian and Mongean styles were not so incompatible – and in fact they had to be conciliated if one was to address from an advanced standpoint both the calculus and its applications to geometry.

In Lacroix's *Traité*, for most particular topics, we see a typical late 18th-century algebraic-analytical approach – nothing else should be expected from a book intended to pave the way for future researchers. And there is ample evidence that Lacroix was sympathetic to "analysis". His defense of analytic geometry (and its comparison to Lagrange's *Méchanique analitique*) is one example [*Traité*, I, xxv]; his description of Lagrange's suggestion of foundation for the calculus in

[*1772a*] as "idées lumineuses"[6] is another [*Traité*, I, xxiv]; his clear sympathy for Fontaine's conception of formation of differential equations by elimination of arbitrary constants (see chapter 6) is another; his defense that textbooks should be written so as to ultimately lead to Lagrange's *Mécanique analitique* and Laplace's *Mécanique celeste* are yet another [*1802a*, xviii; *1805*, 205-206]; finally: "j'ai apporté le plus grand soin à donner aux formules cette symétrie qui les fait presque deviner, et dont les écrits de Lagrange offrent tant d'examples"[7] [*Traité*, I, xxviii]. This predominance of the Lagrangian style is also true for the only topic in which the two mathematicians competed, namely partial differential equations – even when presenting work by Monge (his treatment of second-order quasi-linear equations) he sticks to analytical considerations [*Traité*, II, 524-535].

Of course we see Mongean influence in the geometrical sections. But above all, we see it in an aspect of the overall structure: whenever possible, Lacroix tries to give geometrical depictions of analytical situations, separate from the "analytical course" (see page 89); this is the case for chapters 4 and 5 of the first volume, and for several sections and passages on geometrical applications of differential, difference, and mixed equations, in the second and third volumes.[8] This attempt at separation went as far as Lacroix constructing an analytical theory for a topic that had been given a geometrical treatment by Monge: total differential equations in three variables not satisfying the conditions of integrability (see section 6.2.4).

I believe that the result is not an eclectic compilation (as Lacroix accused Cousin of having done), but rather an effective "rapprochement des méthodes"[9] – an *encyclopédiste* approach.

10.3.2   Encyclopedism and *encyclopédisme*

My use of two versions of the same word in the title just above is deliberate. What I intend to argue in this section is that Lacroix's *Traité* is both encyclopedic in scope and *encyclopédiste* in methodology.

The encyclopedic scope should not need much arguing for. It is clear from chapter 2 that it covers every topic in the calculus and differential geometry relevant around 1800, with a level of detail that could not be matched by smaller works (two examples of very different worth: [Bossut *1798*] and [Lagrange *Fonctions*]). The only important omission one might point is that of applications other than geometrical (namely, applications to mechanics) – but Lacroix chose to remain within the confines of pure mathematics.[10]

---

[6]"brilliant ideas"

[7]"I have taken great pains to give to formulas that symmetry which makes one almost guess them, and of which so many examples can be found in the writings of Lagrange"

[8]With the exception of the construction argument for the admissability of discontinuous functions in integrals of partial differential equations. But once again, a situation in which Lagrange himself referred to Monge [Truesdell *1960*, 295].

[9]"conciliation of methods"

[10]Perhaps some other important omissions might be pointed out in the second edition (see page 323).

As for the *encyclopédiste* methodology, it can be seen in the attempts at systematization, and in the reporting of all relevant points of view.

Systematization is reflected on the existence of a subject index and of a magnificent bibliography, but also in the division into chapters with clearly defined subjects. That this latter characteristic is not so trivial might be seen by comparison with [Cousin *1796*], and to a lesser extent even [Lacroix *1802a*]. In [Cousin *1796*, I], much of elementary differential and integral calculus is effectively presented in a chapter entitled "on the method of the ancient geometers known under the name of method of limits"; later, after applications of the method of limits to mechanics, there is a three-page introduction of the differential notation and of the infinitesimal and fluxional approaches, followed by a short chapter "on differential calculus", dealing mainly with conditions for exact differentials, and with changes of independent variables; still in the first volume, there is a chapter on "integral calculus in general", which gives a survey of the whole subject, including partial differential equations and finite difference equations; but [Cousin *1796*, II] is entirely dedicated to more advanced results in integral calculus – starting with integration of rational functions. Trying to find some particular result in [Cousin *1796*] is not an easy task! Lacroix's *Traité élémentaire* is of course not so confused, but as we have seen in section 8.4, pedagogical reasons lead to having differentiation of functions of more than one variable appear only after the applications of differentiation of functions of one variable.

That Lacroix tries to report all relevant points of view should be clear from several passages above, particularly sections 3.2.7 and 4.2.1.2: Lacroix chose the power-series foundation, but he did not exclude limits, nor even infinitesimals (even if they were "less rigorous"), and he used all of them, especially in geometrical applications. We may also recall that he defined the integral as an antiderivative, but treated it also as a limit of sums, which in addition explained the Leibnizian conception of sum of infinitesimals (see section 5.2.3); and let us not forget chapter 2 of the third volume, where he used Laplace's generating functions to address subjects already treated in chapter 1 using the usual calculus of differences.[11]

One should not assume that all the relevant points of view are presented with the same weight, or the same level of detail. This is certainly not the case: power series are more important than limits and limits more important than infinitesimals; the integral is essentially the "primitive function", and can also be seen as the limit of a sum; generating functions have one sixth of the space dedicated to the calculus of differences. Lacroix actually made choices about the best (or more relevant) approaches. But he did not exclude the others.

This attempt to report several approaches is one of the most famous aspects of Lacroix's *Traité* [Grabiner *1981*, 79-80; Grattan-Guinness *1990*, I, 141-142]. However, it has been challenged by Gert Schubring [*2005*, 374-379]: "the total structure of the work does not take the 'encyclopedic' form suggested". Schu-

---

[11]There is the occasional flaw: in the first edition, the chapter on the calculus of variations omits the power-series approach; in [*Fonctions*] Lagrange had already tried to use it for variations [Fraser *1985*, 181-182].

bring is concerned only with the foundations of the calculus (thus overlooking for
instance the case of chapter 2 of the third volume), and he argues that Lacroix
bases his whole presentation on the method of limits. If I understand it correctly,
Schubring's case has three points: 1 - Lacroix only used infinitesimals in the applications to curves, and even there only occasionally; 2 - he explained the use of
infinitesimals as an abbreviation for the method of limits; and more importantly 3
- the Introduction in the first volume "develops the limit method as a precondition
and basis for applying the development into series" (that is, Lacroix's discussion of
convergence of series was a precondition for the use of the power-series foundation).
Let us analyze them in turn.

1 - Schubring seems not to recognize the presence of infinitesimal considerations unless one of the expressions "infinitesimal" or "infinitely small" appears
explicitly. Thus, he overlooks considerations of "consecutive normals" (to a surface)
[Lacroix *Traité*, I, 478] and similar situations. He also overlooks the explanation
for the integral as an infinite sum (see the quotation in page 166 above), and the
use of "consecutive values" to obtain the basic rules of the calculus of variations
[Lacroix *Traité*, II, 657].[12] It is true that Lacroix did not develop a full version of
the calculus based on infinitesimals, parallel to one based on limits and one based
on power series; but he did give its basic elements, and used infinitesimals when
it seemed appropriate.

2 - Yes, Lacroix regarded the method of infinitesimals as inferior in principle,
and suggested that its proper understanding was as an abbreviation of the method
of limits (quoting Leibniz) [*Traité*, I, 423-424]. But he did not develop this suggestion (unlike, say, Cauchy [*1821*, 26-34]). So, where Schubring sees a metaphysical
dismissal of infinitesimals, I see simply a "rapprochement des méthodes".

3 - The most critical point (the one where I think Schubring is most mistaken) is the supposed grounding of the power-series foundation on the convergence
of series. Lacroix regarded a series expansion of a function as representing the
function even if it was not convergent; it had to be convergent only if one wanted to use its value (see the quotation in page 83 above). The differential, being
simply the first term in the series expansion of the increment of the function, appears quite unrelated to particular values, and therefore to matters of convergence;
convergence of the series is relevant only for applications (calculations). Moreover, Schubring does not explain why Lacroix claimed to be following different
foundational approaches in the large *Traité* and in the *Traité élémentaire*.

Schubring classifies Lacroix as "propagator of the méthode des limites" [*2005*,

---

[12]It is not only in reference to Lacroix that Schubring misses uses of infinitesimals: he claims
that the discussion on curves of double curvature in [Monge *Feuilles*] has "absolutely no reference
to the *infiniment petits*" [Schubring *2005*, 379]. Compare with [Monge *Feuilles*, n° 32; 3rd
ed, 343-344]: "Par un point A de cette courbe, soit mené un plan $MNOP$ perpendiculaire à
la tangente en A; par le point a infiniment proche, soit pareillement mené un plan $mnOP$
perpendiculaire à la tangente en a [...] tous les points de l'arc infiniment petit Aa [...]" ("Let a
plane $MNOP$ be drawn through a point A of that curve, and perpendicular to the tangent in
A; let a plane $mnOP$ be similarly drawn through the infinitely close point a, and perpendicular
to the tangent in a [...] all the points in the infinitely small arc Aa [...]").

372]. This is correct, as far as the *Traité élémentaire* goes; less so in the second edition of the large *Traité*; and even less in the first edition.

The best way to understand Lacroix's *encyclopédisme* is probably to compare his approach to Lagrange's and Cauchy's. While Lagrange and Cauchy each picked a principle and tried to construct the calculus on it, Lacroix clearly agreed with Laplace that "the reconciliation of methods [...] serves to clarify them mutually" (see section 3.2.8). He also had a more conjunctural reason for adopting this approach, a reason best explained in the Preface of the second edition: when discussing which foundation to adopt in teaching, he says that the answer was difficult

> "dans l'état actuel de la science, puisqu'une route dont on ne fait qu'apercevoir l'entrée, peut conduire à des découvertes importantes, et que chacun des points de vue sous lequel on a envisagé le passage de l'Algèbre au Calcul différentiel, donne à ce calcul des formes qui, pour le moins, offrent des facilités particulières dans la solution de certains problèmes"[13] [Lacroix *Traité*, 2nd ed, I, xxiv];

and when justifying the duplication in volume 3 caused by considering calculus of differences and generating functions:

> "dans l'état actuel de la science, où elle est circonscrite de tous côtés par des limites qu'on cherche à franchir, on ne sait sur quoi doivent s'appuyer les considérations qui leveront les difficultés où l'on est maintenant arrêté"[14]. [Lacroix *Traité*, 2nd ed, I, xlvi]

One might argue that this is the normal state of science. But it is certainly true that in 1810 one could not foresee the road that would be taken by men such as Cauchy.

10.4 Some further questions

There are several questions that remain unanswered.

In this study I have focused mainly on the composition of the first edition of the *Traité*. The reflections above on impact do not intend to have the same strength that any of the conclusions about the book itself. A particular question that would be interesting to pursue more thoroughly is the influence of the *Traité* in research in the period 1800-1820. I suspect that much of the research carried out by people like Poisson and Ampère referred to Lacroix's *Traité* as background,

---

[13]"in the present state of science, since a road whose entry is only glimpsed may lead to important discoveries, and each point of view that has been used for the passage from algebra to the differential calculus gives this calculus forms that, at the least, offer particular facilities in solving some problems"

[14]"in the present state of science, where it is surrounded from all sides by obstacles that one tries to overcome, we do not know on what should lean the considerations that will remove the difficulties where one is halted"

and in some cases may have been triggered by passages in it (an example of this was mentioned in page 124 above – Ampère and Carnot on intrinsic coordinates). But so far this is not much more than a suspicion.

Another issue that has not been fully explored here is the second edition. Was it as up to date in 1819 as the first edition in 1800? I suspect not. But once again, this is not more than a suspicion. In order to answer this question, one would probably have to choose other aspects than those I picked for studying the first edition.

Apart from these, there are of course many questions about Lacroix to be explored. A good biography is still to be written. His teaching at the *Faculté des Sciences* and at the *Collège de France* has not been studied, as far as I know. His textbooks still offer many opportunities for research.

Some studies of his textbooks and of [Lacroix *1805*] have stressed the philosophical context [Lamandé *2004*; Panteki *2003*, 284-290]: influences from d'Alembert and Condillac, seen in his "moderate sensualism" and in the importance given to algebra as a language. These studies might profit from considering also the large *Traité*: Lacroix's care with notation and terminology is very marked; and his *encyclopédisme* may echo not only the [*Encyclopédie*], but more specifically d'Alembert's rejection of the *esprit de système* [Lamandé *2004*, 58] – that is, the use of one single principle to explain everything (precisely what Lagrange had done in [*Fonctions*]).

Appendix A

Two memoirs by Lacroix

A.1 "Memoire sur le Calcul intégral aux différences parti-elles", 1785

This memoir was sent to Monge from Rochefort in July 1785, and presented to the *Académie Royale des Sciences* of Paris in the meeting of 14 December 1785. The manuscript is in the Archive of the *Académie*, in the *pochette* for that meeting.

Bound with the manuscript there is an alternative version for the two first pages. This was ignored in this transcription.

There is also a complete (but quite shorter) alternative version of the memoir, kept in the same *pochette*, but clearly dating from a few years later. In the first page, we can read the following footnote: "*Note* ce memoire a ete presenté en 1785, et est anterieur de plusieurs années a ceux que M. Monge a donné sur le même sujet dans les memoires de l'academie des sciences pour 1786" (the memoirs for 1786 were published in 1788). This alternative version was not transcribed.

Condorcet and Monge were charged with reading the memoir and reporting on it. Their report was presented and approved in the meeting of 11 February 1786 (it can be found both in the *pochette* for that meeting and in the *procès-verbal* [Acad. R. Sc. *PV*, CV, 28r-30v]; a transcription may be found here, just after the memoir). According to that report the memoir should have been published in the *Mémoires de Mathématiques et de Physique, présentés à l'Académie Royale des Sciences*, commonly known as *Savants Étrangers*, dedicated to memoirs submitted by non-members. But this never happened: the publication of this collection slowed down and halted precisely in 1786; many other works shared the fate of this one [Acad. Sc. Paris *Guide*, 121].

Memoire sur le Calcul intégral aux différences partielles.

par M. Lacroix
Professeur à Rochefort

1785.

Mémoire sur le calcul integral aux différences partielles

Je me propose dans ce memoire de ramener l'integration des équations aux differences partielles qui ne sont pas lineaires à l'integration des equations lineaires de ce genre; et de trouver une forme générale qui puisse representer l'intégrale de ces sortes d'equation. [Crossed out: ensuite je tacherai d'apliquer ces formules à la Géométrie dans l'espace pour obtenir des constructions pour chacune de mes resultats; disciple de Mr Monge c'est à lui que je dois les connoissances que j'ay pû acquerir dans ces parties de la Géométrie transcendante dont il est un des inventeurs; heureux si en m'occupant d'objets pareils je ne reste pas[?] audessous[?] de mon modèle.]

Pour proceder avec ordre je reprend les déffinitions qui doivent servir de base à ces methodes:

Le calcul intégral aux differences partielles est l'art de trouver la composition des fonctions de variables quelconques par la relation donnée entre leurs coefficiens differentiels.[1]

(1)   Si l'on à une fonction z de deux variables x, y; de cette forme $z = \varphi : (\alpha x + y)$; il est évident qu'il doit exister entre les coefficiens differentiels de cette fonction une relation telle qu'ils ne puissent appartenir qu'à des fonctions composées de cette maniere; or voici comment on peut trouver cette relation: en differentient une fois par rapport à x et l'autre par rapport à y on aura, $\left\{ \begin{array}{l} p = \varphi' : (\alpha x + y)\alpha \\ q = \varphi' : (\alpha x + y) \end{array} \right\} : \varphi'$ exprime ce que devient la fonction φ après la différentiation. Si on elimine entre ces deux équations $\varphi' : (\alpha x + y)$ on aura $p - \alpha q = 0$, equation qui ne renferme plus qu'une relation entre les coefficiens differentiels p et q; et qu'on peut regarder comme un caractere auquel reconnoitra se telle ou telle quantité peut être fonction de $(\alpha x + y)$.

Remonter de la relation différentielle que nous venons d'obtenir a la fonction $z = \varphi : (\alpha x + y)$ voila le calcul intégral aux differences partielles.

(2)   Il est évident que toute fonction de deux variables de quelque maniere qu'elle soit composée aura toujours pour différentielle $dz = p\,dx + q\,dy$; c'est donc dans cette formule qu'il faudra substituer une valeur de p en q ou de q en p tirée de l'equation donnée, et alors on intégrera la proposée comme une équation différentielle ordinaire: mais il se présente une méthode plus naturelle, l'equation différentielle $p - \alpha q = 0$, ou toute autre, peut toujours être envisagée comme produite par l'élimination d'une fonction arbitraire. Cette methode est celle de M. Monge, et s'applique avec elegance aux équations linéaires de tous les ordres: c'est

---

[1] J'apelle coefficiens differentiels les termes $\frac{dz}{dy}, \frac{dz}{dx}$; de maniere que la differentielle 1.re d'une fonction z composée de deux variables sera $dz = \frac{dz}{dx}dx + \frac{dz}{dy}dy$; nous ferons pour abreger dans le courant de ce memoire $\frac{dz}{dx} = p$ et $\frac{dz}{dx} = q$, on aura donc $dz = p\,dx + q\,dy$.

aussi celle dont nous nous servirons à peu près dans la suite de ces recherches; je
ne sache pas que l'auteur ait rien publié sur les equations non lineaires.

Probleme.

(3)   Trouver la relation des coefficiens differentiels d'une fonction d'une quantité
inconnue donnée par deux equations.

Soient $z = M + \varphi(\omega)$ et $\frac{\mathrm{d}(M)}{\mathrm{d}\omega} = -\varphi'(\omega)$; ω est la quantité inconnue; M
est une fonction connue de x, y, ω. Si on différentie par rapport a x et a y, M
etant composée de x, y et ω, sa differentielle contiendra trois termes, savoir la
differentielle par rapport à x, que nous representerons ainsi, δM; sa différentielle
par rapport a y, sera ∂M; et sa differentielle par rapport à ω, dM; on aura donc:

$$p = \frac{\delta M}{\mathrm{d}x} + \frac{dM}{\mathrm{d}x} + \varphi'.(\omega)\frac{\mathrm{d}\omega}{\mathrm{d}x}$$

$$q = \frac{\partial M}{\mathrm{d}y} + \frac{dM}{\mathrm{d}y} + \varphi'.(\omega)\frac{\mathrm{d}\omega}{\mathrm{d}y}$$

en employant la seconde des equations primitives on reduira les 2. equations précédentes à celles-ci:

$$\left.\begin{array}{c} p = \dfrac{\delta M}{\mathrm{d}x} \\[2mm] q = \dfrac{\partial M}{\mathrm{d}y} \end{array}\right\}$$

equations dans lesqu'elles il restera x, y, ω, combinés avec des quantités constantes;
c'est [sic] equations peuvent donner par l'elimination de ω des equations de tous
les degrés; c'est ainsi que je suposerai produites toutes les equations que j'aurai à
traiter.

[Margin note: n^{te} a mettre au bas. Il existe encore des equations elevées,
celles où la quantite qui est sous la fonction est connue et où cette derniere se
trouve a differentes puissances]

Cette maniere d'envisager les equations aux différences partielles peut se
rendre par la géométrie d'une façon très claire; si on pose dans $z = M + \varphi(\omega)$, $\omega =$
const., cette equation sera celle d'une surface courbe dont ω serait le parametre et
la 2e, $\varphi.(\omega) = \frac{dM}{\mathrm{d}\omega}$, exprime ce que deviendrait la surface courbe posée ci-dessus si
le parametre ω variait, [crossed out: unreadable] il s'en suit que le systeme de ces
deux equations représente la surface engendrée par les intersections consécutives
d'une surface courbe donnée avec elle même, changeante par la variation d'un
paramettre.

[Crossed out: On voit aisément qu'il n'y à aucune surface courbe qui ne
puisse être engendrée de cette maniere ce qui confirme l'assertion que jai fait plus
haut.]

Cette forme contient l'intégrale de l'equation linéaire du premier ordre. Car
lorsque les equations $\left.\begin{array}{c} p = \frac{\delta M}{\mathrm{d}x} \\[1mm] q = \frac{\partial M}{\mathrm{d}y} \end{array}\right\}$ sont lineaires par rapport a ω, l'equation qui en

resulte sera lineaire par rapport aux differentielles p et q; dans ce cas ω n'entrera pas dans $\frac{dM}{d\omega}$, on aura donc en appellant L ce que devient cette quantité $\varphi'.(\omega) = L$ ou $\omega = \Psi.(L)$, et par consequent $z = \omega L + \varphi(\omega)$, ou $z = L\Psi.(L) + \varphi.\{\Psi.(L)\}$; on remarquera que φ' depend de φ, Ψ en dependra aussi et parconséquent cette expression se reduira a $z = K + \varphi(\omega)$: K est fonction connue de x, y seulement. On sait d'ailleurs que cette forme est celle de l'intégrale des equations linéaires du premier ordre.

(4.) On pourra toujours par le moyen des formules posées précédemment réduire l'intégration des equations aux différences partielles de quelque degré qu'elles soient à celle des équations différentielles ordinaires ainsi qu'on va le voir.

En effet si on décompose l'equation aux differences partielles proposée en deux autres par l'introduction d'une nouvelle indeterminée qu'on supose avoir été éliminée; alors on tirera les valeurs de p et de q, qu'on substituera dans les équations $\left.\begin{array}{c} p = \frac{\delta M}{\mathrm{d}x} \\ q = \frac{\partial M}{\mathrm{d}y} \end{array}\right\}$ il s'agit donc d'intégrer $\frac{\delta M}{\mathrm{d}x}\,\mathrm{d}x + \frac{\partial M}{\mathrm{d}y}\,\mathrm{d}y$ pour avoir M; ayant substitué son exprséssion dans la formule générale on aura l'integrale demandée exprimée par deux équations si la proposée n'est pas linéaire.

Je remarquerai ici que l'intégrale demandée pourra se présenter sous differentes formes ce qu'il est aisé d'expliquer, car la décomposition de l'équation proposée pourra toujours se faire de plusieurs manieres, ainsi chacune d'elles donnera une intégrale différente; mais il est toujours possible de ramener ces résultats les uns aux autres [crossed out: unreadable].

[Crossed out: (5) Je ne m'arrêtterai gueres aux equations de 1.er ordre que M.r Euler à traité dans le 3.e volume de son calcul intégral avec beaucoup d'étendue; les constructions geometriques etant le seul motif pour le quel j'ai parlé de cet ordre.]

(5) Je choisirai pour premier exemple de cette méthode l'equation $p^2 + apq + bp + cq + hq^2 = m$, tous les coefficiens sont des quantités constantes. Il faut décomposer cette équation en deux autres, telles qu'eliminant de ces nouvelles équations une indeterminé ω il en résulte la proposée; il y a plusieurs manieres de remplir cette condition; et c'est dans le choix de ces moyens que consiste l'addresse du calcul.

Nous suposerons que la proposée a été produite par ces deux équations:

$$\left.\begin{array}{c} p + Bq + C = \omega \\ p + B'q + C' = \frac{A}{\omega} \end{array}\right\}$$

on aura par l'elimination

$$p^2 + (B + B')pq + (C' + C)p + (B'C + C'B)q + B'Bq^2 + CC' = A,$$

comparant cette equation terme à terme avec la proposée, on en deduira les suivantes $a = B + B'$, $\quad b = C + C'$, $\quad c = B'C + C'B$, $\quad h = B'B$, $\quad m = A - CC'$;

suposant qu'on ait tiré de ces équations les valeurs des indéterminées qu'elles renferment: on aura, $p = \dfrac{B'C - BC' + B\omega - \frac{B'A}{\omega}}{B - B'}$, et $q = \dfrac{C' - C + \omega - \frac{A}{\omega}}{B - B'}$ par conséquent la question sera reduite à intégrer

$$\frac{\delta M}{\mathrm{d}x}\mathrm{d}x + \frac{\partial M}{\mathrm{d}y}\mathrm{d}y = \frac{\left\{B'C - BC' + B\omega - \frac{B'A}{\omega}\right\}}{B - B'}\mathrm{d}x + \frac{\left\{C' - C + \omega - \frac{A}{\omega}\right\}}{B - B'}\mathrm{d}y,$$

en regardant ω comme constant; substituant l'intégrale de cette quantité dans la formule générale $\left\{ \begin{array}{l} z = M + \varphi : (\omega) \\ \frac{dM}{d\omega} = -\varphi'(\omega) \end{array} \right\}$ on aura

$$z = \frac{\left\{B'C - BC' + B\omega - \frac{B'A}{\omega}\right\}}{B - B'}x + \frac{\left\{C' - C + \omega - \frac{A}{\omega}\right\}}{B - B'}y + \varphi : (\omega)$$

$$\frac{1}{B - B'}\left\{Bx + \frac{AB'x}{\omega^2} + y + \frac{Ay}{\omega^2}\right\} = -\varphi' : (\omega)$$

on voit aisement pourquoi je n'ai pas ajouté de constante.

Les equations $p\,q = 1$, $p^2 + q^2 = 1$, et d'autres semblables traitées par M. Euler ont leurs intégrales comprises dans les équations précédentes.

(6)   Auparavant d'aller plus loin dans cette matiere je remarquerai ici que toutes les equations aux différences partielles qu'on pourra donner entre p q et des quantités constantes, auront pour intégrales deux équations de cette forme:

$$\left. \begin{array}{l} z = x\,F : (\omega) + y\,f : (\omega) + \varphi : (\omega) \\ x\,F' : (\omega) + y\,f' : (\omega) + \varphi' : (\omega) = 0 \end{array} \right\},$$

dans lesquelles F et f sont des fonctions connues de ω, et F', f' leurs differentielles par rapport à cette variable. Cela est évident d'après le procédé développé dans l'article precedent; on peut encore s'en assurer de la maniere suivante: toutes les equations que renferme la classe dont nous venons de parler peuvent être réduites à cette forme $Q = 0$, Q exprimant une fonction connue de p, q et quantitités constantes; soient $p = F : (\omega)$, $q = f : (\omega)$, les deux equations desquelles éliminant ω il resulte la proposée; j'aurai en substituant dans les formules générales:

$$z = x\,F.(\omega) + y\,f : (\omega) + \varphi : (\omega)$$

$$x\,F' : (\omega) + y\,f' : (\omega) + \varphi' : (\omega) = 0$$

(7)   C'est là l'equation des surfaces developables; ainsi nous pouvons conclure que la classe d'equations dont nous venons de parler appartient aux surfaces de ce genre. Je puis ecrire les equations précédentes sous cette forme

$$\left\{ \begin{array}{l} z = x\,\pi : (z') + y\,\psi : (z') + z' \\ x\,\pi' : (z') + y\,\psi' : (z') + 1 = 0 \end{array} \right\}.$$

Je fais $\varphi : (\omega) = z'$, et j'entends par cette derniere indéterminée la coordonnée d'un certain point de l'espace. La 1ere de ces deux equations en regardant z' comme constant ... est l'equation d'un plan qui passerait par le point dont z' est la coordonnée et l'origine des coordonnées se trouve au point du plan de x, y qui est le pied de la coordonnée z'

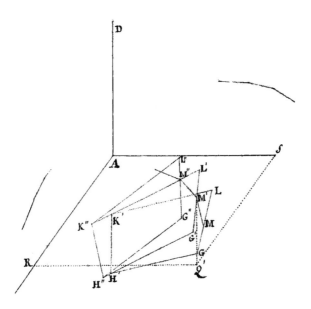

Figure A.1: First figure in the manuscript of the memoir submitted by Lacroix in 1785 [Académie des Sciences de l'Institut de France]

Soit Q le point d'ou partent les coordonnées $QM = z'$; pour rapporter les coordonnées au point A par exemple, il suffira de mettre dans les équations ci dessus pour x, $x - x'$; et pour y, $y - y'$; x' et y' etant les cordonnées $Q'R$ et $Q'S$ du point Q: on aura par cette substitution

$$z - z' = (x - x')\,\pi : (z') + (y - y')\,\psi : (z')$$

$$(x - x')\,\pi' : (z') + (y - y')\,\psi' : (z') + 1 = 0$$

Supposons que $M'M''$ soit l'element d'une courbe a double courbure, la 1ere des equations précéndente [sic] appartiendra au plan normal $L''K''H''G''$ et la 2e sera celle du plan consécutif $L'K''H''G'$ et les projections de cette courbe à double courbure seront $\frac{dx'}{dz'} = \pi : (z')$, $\frac{dy'}{dz'} = \psi : (z')$; l'assemblage de ces deux equations appartiendra donc à la surface formée par les intersections $K'H', K''H'', \&c \ldots$ des plans normaux consécutifs de la courbe à double courbure $MM'M''$ on sçait que cette surface est toujours developpable.

La classe particuliere d'équations dont nous nous occupons dans ce moment renferme toutes les surfaces formées par les intersections des plans normaux des courbes à double courbure dont la relation des deux projections est donnée; car dans la transformation des équations $\left\{ \begin{array}{l} z = x\,F : (\omega) + y\,f : (\omega) + \varphi(\omega) \\ x\,F' : (\omega) + y\,f' : (\omega) + \varphi' : (\omega) = [0] \end{array} \right\}$, j'ai fait $\varphi : (\omega) = z'$ on en peut conclure $\omega = \Delta : (z')$ et parconséquent les equations ci-dessus deviennent

$$\left\{ \begin{array}{l} (z - z') = (x - x')F : \{\Delta : (z')\} + (y - y')f' : \{\Delta : (z')\} \\ 0 = 1 + (x - x')\,F' : \{\Delta : (z')\} + (y - y')\,f' : \{\Delta : (z')\} \end{array} \right\}$$

On remarquera que F et f sont des fonctions connues, Δ reste arbitraire par ce qu'il depend de φ.

(8)   Il suit de là que déterminer la fonction arbitraire de l'intégrale précédente c'est se proposer ce Probleme de Geométrie dans l'espace: connaissant la relation des projections d'une courbe à double courbure sachant de plus que la surface formée par les intersections consécutives des plans normaux de cette courbe doit passer par une autre courbe a double courbure donnée, trouver les equations de la 1.ere courbe.   Soit RO la courbe cherchée et MM' la courbe à double cour-

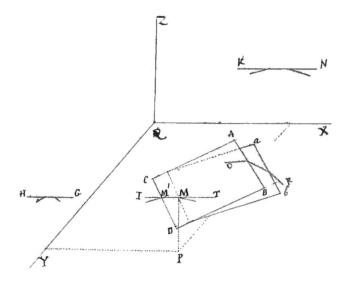

Figure A.2: Second figure in the manuscript of the memoir submitted by Lacroix in 1785 [Académie des Sciences de l'Institut de France]

bure donnée; la solution de cette question se réduit à déterminer les coefficiens de l'equation $z - z' = c\,(x - x') + c'(y - y')$; du plan $ADBC$ qui est normal à la courbe RO, de maniere que ce plan passe par la tangente de la courbe MM': pour cela soient $\left\{ \begin{array}{l} x = K : (z) \\ y = L : (z) \end{array} \right\}$ les equations des projections de la courbe donnée MM'; il faudra chercher la coordonnée PM, z'', du point de rencontre de la courbe MM' avec le plan $ABCD$; alors on formera pour ce point les equations des deux projections GH et NK de la tangente TI; représentons les par $\left\{ \begin{array}{ll} z - z'' = \frac{1}{K' : (z'')}(x - x'') & (1) \\ z - z'' = \frac{1}{L' : (z'')}(y - y'') & (2) \end{array} \right\}$, le plan $ABCD$ passant nécéssairement par le point M, je puis ecrire ainsi son équation $z - z'' = c(x - x'') + c'(y - y'')$ (3) il est evident que mettant pour $(x - x'')$, sa valeur prise de l'equation (1) dans (3) on doit retrouver pour $z - z''$, l'equation (2), ce qui donne $\frac{c'}{1 - c\,K' : (z'')} = \frac{1}{L' : (z'')}$, mais $c = F : (\Delta : (z'))$, $c' = f : (\Delta : (z'))$ ainsi mettant dans cette derniere equation la valeur de c' et de c, celle de z'' qui a été trouvée précédemment, on aura l'expression de la fonction arbitraire $\Delta : (z')$. D'après ce procédé il est aisé de voir que la surface représentée par l'intégrale passera par la courbe MM', car les deux plans normaux consecutifs a la courbe OR passant par les deux tangentes consecutives de la courbe MM', les intersections de ces plans passeront toutes par cette courbe.

Quoi que je n'aye pas employé la 2e équation comme elle est différentielle de l'autre par rapport à z' seulement, il n'en est pas moin clair que le plan normal suivant qu'elle represente doit passer par la tangente consécutive, les fonctions $F : (\Delta : (z'))$ et $f : (\Delta : (z'))$ ayant été déterminées pour cela: on pourra moyennant les equations précédentes combinées avec $\left. \begin{array}{l} \dfrac{\mathrm{d}x'}{\mathrm{d}z'} = \pi : (z') \\[2mm] \dfrac{\mathrm{d}y'}{\mathrm{d}z'} = \psi : (z') \end{array} \right\}$ qui représentent les projections de la courbe à double courbure dont les intersections des plans normaux consecutifs forment la surface demandée, on pourra dis-je resoudre à la fois ces deux questions trouver la surface demandée avec cette condition qu'elle passe par une courbe donnée; et en même tems déterminer la nature de la courbe generatrice.

L'intégrale proposée sous sa premiere forme $\left\{ \begin{array}{l} z = x\,F : (\omega) + y\,f : (\omega) + \varphi : (\omega) \\ x\,F' : (\omega) + y\,f' : (\omega) + \varphi' : (\omega) = 0 \end{array} \right.$ appartient à la surface formé par les intersections continuelles d'un plan mobile par la variation d'une quantité (ω); en la considérant sous ce dernier point de vue on pourra en déterminer la fonction arbitraire de même que nous l'avons fait ci devant en la transformant ainsi $\left. \begin{array}{l} z - z' = x\,F : \{\Delta : (z)\} + y\,f : \{\Delta : (z')\} \\ x\,F' : \{\Delta : (z')\} + y\,f' : \{\Delta : (z')\} + 1 = 0 \end{array} \right\}$ on aura toujours l'equation $\frac{c'}{1 - c K' : (z'')} = \frac{1}{L' : (z'')}$, seulement elle ne contiendra plus que z'.

(9)   Soit l'equation plus générale $XP + YQ = 0$, dans la quelle X, Y sont des fonctions de x et de y; P et Q representent des fonctions de p et de q; on séparera l'equation en deux autres en faisant $XP = \omega$; parconséquent $p = F : (\frac{\omega}{X})$ [crossed out: unreadable] et $q = f : (\frac{\omega}{Y})$, substituant dans les formules du N.º (3) on aura pour intégrale $\begin{cases} z = \int \mathrm{d}x\, F : (\frac{\omega}{X}) + \int \mathrm{d}y\, f : (\frac{\omega}{Y}) + \varphi(\omega) \\ \mathrm{d}\left\{ \frac{\int \mathrm{d}x\, F : (\frac{\omega}{X}) + \int \mathrm{d}y\, f : (\frac{\omega}{Y})}{\mathrm{d}\omega} \right\} + \varphi'(\omega) = 0 \end{cases}$ $\int \mathrm{d}x\, F : (\frac{\omega}{X})$ et $\int \mathrm{d}y : (\frac{\omega}{Y})$ [sic], se rapporteront toujours aux quadratures; si l'on s'etait proposé l'equation $YP + XQ = 0$ on la rameneraie à la précédente en la mettant sous cette forme $\frac{P}{X} + \frac{Q}{Y} = 0$.

L'equation $P + Q = 0$, P etant fonction de p et x, Q de q et y ne presente aucune difficulté, on fera dans ce cas $P = \omega$ et $Q = \omega$, donc $p = F : (\omega, x)$, $q = f : (\omega, y)$, l'intégrale sera de la même forme que la précédente et se rapportera de même aux quadratures.

Si P etait fonction de p et y, et Q de q et x alors il faudrait chercher les moyens de rendre $\frac{\delta M}{\mathrm{d}x} \mathrm{d}x + \frac{\partial M}{\mathrm{d}y} \mathrm{d}y = \mathrm{d}x\, F : (\omega, y) + \mathrm{d}y\, f : (\omega, x)$ differentielle complete.

Soit enfin l'equation $N = 0$, N etant composée de $p\ q\ x,\ y$, on voit qu'en y appliquant la méthode employée précédemment son intégration se rapportera à celle de la formule $\frac{\delta M}{\mathrm{d}x} \mathrm{d}x + \frac{\partial M}{\mathrm{d}y} \mathrm{d}y$; $\frac{\delta M}{\mathrm{d}x}$ et $\frac{\partial M}{\mathrm{d}y}$ contenant tous les deux x, y et ω [crossed out: nous reviendrons par la suite sur cet objet; en attendant] Nous remarquerons que si l'equation $N = 0$ ne contenait que p, q et x ou p, q et y elle s'integrerait très facilement en résolvant l'equation dans le 1.er cas par rapport a p et dans le 2.e rapport q, [alors] p ou q seront les indeterminées[2].

(10)   La forme posée N.º (3) renferme toutes les intégrales des equations aux différences partielles entre p, q, x et y mais lorsque la 3.e variable z, y entre ou qu'elles s'y trouvent toutes il est évident qu'il faut avoir recours a une forme plus générale, et on apperçoit aisément que cette forme se saurait être autre que celle-ci

---

[2]Si on multiplie l'equation $N = 0$ par un facteur μ on aura $\mu N = 0$ et décomposant cette equation en 2 autres $p = F(\omega, \mu)$, $q = f(\omega, \mu)$, on aura donc à intégrer $\mathrm{d}x\, F : (\omega, \mu) + \mathrm{d}y\, f : (\omega, \mu) = \mathrm{d}z$, mais pour que le prem.er membre de cette equation soit [crossed out: immédiatement] intégrable il faut qu'on ait

$$\frac{\mathrm{d}\{F(\omega, \mu)\}}{\mathrm{d}y} = \frac{\mathrm{d}\{f(\omega, \mu)\}}{\mathrm{d}x}$$

les différences partielles de μ dans cette equation n'y monteront qu'au 1.er degré; j'aurai donc réduit l'integration de l'equation $M = 0$ à celle d'une equation linéaire aux différences partielles du 1.er ordre. En appliquant à l'equation $V p^n + T q^h = 0$ on aura $p = \sqrt[n]{\frac{\omega}{\mu V}}$, $q = \sqrt[h]{-\frac{\omega}{\mu T}}$; $\mathrm{d}z = \mathrm{d}x \sqrt[n]{\frac{\omega}{\mu V}} + \mathrm{d}y \sqrt[h]{-\frac{\omega}{\mu T}}$, μ sera donné par l'equation $\frac{\mathrm{d}\{\sqrt[n]{\frac{\omega}{\mu V}}\}}{\mathrm{d}y} = \frac{\mathrm{d}\{\sqrt[h]{-\frac{\omega}{\mu T}}\}}{\mathrm{d}x}$. On ne peut dissimuler que la difficulté ne soit quelquefois aussi grande que [crossed out: dans] pour l'equation proposée.

$$\left\{ \begin{array}{l} F\!:\!(z) = M + \varphi\!:\!(\omega) \\ \dfrac{dM}{d\omega} + \varphi'\!:\!(\omega) = 0 \end{array} \right\} \text{ pour le cas ou } z \text{ se trouve séparé des autres variables;}$$

car alors la fonction arbitraire ne saurait contenir que x, y et des quantites constantes. Enfin si l'equation proposée contenait x, y, z d'une maniere à ne pouvoir

etre separées alors la forme générale serait $\left\{ \begin{array}{l} S + \varphi\!:\!(\omega) = 0 \\ \dfrac{dS}{d\omega} + \varphi'(\omega) = 0 \end{array} \right\}$: elle contient

les deux autres, il est a remarquer que dans ce cas la fonction arbitraire pourra contenir les 3 variables x, y et z.

Cherchant par la différentiation la relation des coefficiens differentiels p et q

de z dans ces deux formules on aura pour la 1.re $\left\{ \begin{array}{l} F'\!:\!(z) \cdot p = \dfrac{\delta M}{dx} \\ F'\!:\!(z) \cdot q = \dfrac{\partial M}{dy} \end{array} \right\}$ et pour la

seconde $\left\{ \begin{array}{l} \dfrac{\delta S}{dx} = 0 \\ \dfrac{\partial S}{dy} = 0 \end{array} \right\}$; dans l'un de ces resultats $\dfrac{\delta M}{dx}$ et $\dfrac{\partial M}{dy}$ contiennent x, y, ω;

dans l'autre $\dfrac{\delta S}{dx}$, et $\dfrac{\partial S}{dy}$ renferment $x\ y\ z$ et ω.

(11)   Nous allons nous occuper d'abord de l'usage des equations:

$\left\{ \begin{array}{l} F'\!:\!(z) \cdot p = \dfrac{\delta M}{dx} \\ F'\!:\!(z) \cdot q = \dfrac{\partial M}{dy} \end{array} \right\}$. Il est évident qu'on obtiendra par l'elimination de ω entre

celles-ci une équation d'un degré quelconque par rapport a p et q [crossed out: unreadable] Si on se propose l'equation $N = 0$, qui contiene p, q, x, y et z les deux equations précedentes donnent

$$F'\!:\!(z) \cdot p\,dx + F'\!:\!(z)\,q\,dx = \frac{\delta M}{dx}\,dx + \frac{\partial M}{dy}\,dy = F'\!:\!(z)\,dz.$$

Ainsi l'intégration de la proposée sera réduite comme ci devant à l'intégration de $\dfrac{\delta M}{dx}dx + \dfrac{\partial M}{dy}dy$, dans laquelle ω est regarde comme constant; ce qui donnera M; et substituant cette valeur dans la 1re formule N.^{o} (10) on aura l'intégrale demandée; il ne sagit donc que de composer l'equation aux différences partielles proposée en deux autres et déterminer les valeurs de p, et q: il faut faire en sorte que les valeurs puissent etre de cette forme $\left\{ \begin{array}{l} p = F'\!:\!(z)K\!:\!(\omega) \\ q = F'\!:\!(z)k\!:\!(\omega) \end{array} \right\}$ nous allons en donner quelques exemples.

Soit $PQ = Z$, je supposerai la fonction Z formee du produit $Z' \times Z''$. Alors je ferai $P = \dfrac{Z'}{\omega}$ et $Q = \omega Z''$, il faudra decomposer ces equations en facteurs de la forme ci dessus; on remarquera qu'on peut determiner Z' ou Z'' a cet effet puisqu'il n'y a qu'une equation donnée entr'elles. Substituant dans $\frac{\delta M}{dx}\,dx + \frac{\partial M}{dy}\,dy$ on aura a integrer $K : (\omega)\,dx + k : (\omega)\,dy$ et l'integrale de la proposée sera

$$\left\{ \begin{array}{l} \int F' : (z)\,dz = \int \{ K : (\omega)\,dx + k : (\omega)\,dy \} + \varphi : (\omega) \\[2mm] \dfrac{d\{ \int (K : (\omega)dx + k : (\omega)dy) \}}{d\omega} + \varphi' : (\omega) = 0 \end{array} \right\}$$

Les integrales des equations $Z = p^n q^h$ et $p^n q^h X Y Z = A$ sont comprises dans celle ci; pour la 1re on a

$$\left\{ \begin{array}{l} \displaystyle\int \dfrac{dz}{Z'} = x\,\sqrt[n]{\omega} + \dfrac{y}{\sqrt[n]{\omega}} + \varphi : (\omega) \\[3mm] \dfrac{d\{ x\sqrt[n]{\omega} + \frac{y}{\sqrt[n]{\omega}} \}}{d\omega} + \varphi' : (\omega) = 0 \end{array} \right\}$$

on determine Z' par cette condition: $Z = Z'^{n+b}$.

L'integrale de la 2^{e} est

$$\left\{ \begin{array}{l} \displaystyle\int Z'dz = \int \left\{ dx\,\sqrt[n]{\dfrac{\omega}{X}} + dy\,\sqrt[h]{\dfrac{A}{\omega Y}} \right\} + \varphi : (\omega) \\[4mm] \dfrac{d\left\{ \int (dx\,\sqrt[n]{\frac{\omega}{X}} + dy\,\sqrt[h]{\frac{A}{\omega Y}}) \right\}}{d\omega} + \varphi' : (\omega) = 0 \end{array} \right\}$$

on a pour Z' la meme equation que precedemment.

Quand on aurait $Z = $ fonc. transcendante on pourrait se servir de cette methode: soit pour exemple $Z = \mathrm{Log}.R$ et $Z = \sin.R$. En passant aux nombres, on aura pour la 1re equation $e^Z = R$ ou $e^Z = p^n q^h X Y$, parconsequent son in-

$$\left\{ \begin{array}{l} \displaystyle\int \dfrac{dz}{e^{Z'}} = \int \left\{ dx\,\sqrt[n]{\dfrac{\omega}{X}} + dy\,\sqrt[h]{\dfrac{1}{Y\omega}} \right\} + \varphi : (\omega) \\[4mm] \dfrac{d\left\{ \int (dx\,\sqrt[n]{\frac{\omega}{X}} + dy\,\sqrt[h]{\frac{1}{Y\omega}}) \right\}}{d\omega} + \varphi' : (\omega) = 0 \end{array} \right\}$$

et Z' est donne par

l'equation suivante $hZ' + nZ' = Z$.

La 1e equation proposée etant mise sous cette forme $\mathrm{Log}.(\sqrt{1 - Z^2} + Z\sqrt{-1})\sqrt{-1} = p^n q^h X Y$, je ferai $\mathrm{L}.(\sqrt{1 - Z^2} + Z\sqrt{-1})\sqrt{-1} = Z'^{n+h}$ alors l'integrale generale deviendra

$$\int \left\{ \mathrm{L}.(\sqrt{1 - Z^2} + Z\sqrt{-1})\sqrt{-1} \right\} dz = \int \{ dx\,\sqrt[n]{\tfrac{\omega}{X}} + dy\,\sqrt[h]{\tfrac{1}{Y\omega}} \} + \varphi : (\omega)$$

$$\frac{d\{ \int (dx\,\sqrt[n]{\frac{\omega}{X}} + dy\,\sqrt[h]{\frac{1}{\omega Y}}) \}}{d\omega} + \varphi' : (\omega) = 0$$

Toutes ces integrales se rapporteront aux quadratures ainsi on pourra y parvenir au moins par approximation.

Nous n'insisterons pas d'avantage sur les équations qui peuvent se rapporter a cette forme; nous en citerons seulement quelques unes; $ap^3 + bpq^2 + cpq^2 + eq^3 = z$ s'integre aisément en la supposant produite par l'elimination de ω entre ces deux equations $\left\{ \begin{array}{l} Az'p + Bz'q + \omega + \frac{1}{\omega^2} = 0 \\ A'z'p + B'z'q + \omega = 0 \end{array} \right\}$.

Toutes les equations différentielles qu'on pourra comparer avec cel- les qui résulteraient de l'elimination de ω entre les equations suivantes

$$\left\{ \begin{array}{l} Az'p + Bz'q + K:(\omega) + C = 0 \\ A'z'p + B'z'q + k:(\omega) + C' = 0 \end{array} \right\}$$ leur integrale pourra se rapporter a l'intégrale générale posée precédemment. Si les coefficiens de la proposée sont des
constantes et des fonctions de z elle s'intégrera sans aucune restriction, ou bien si
les variables x et y sont separées dans les equations linéaires qu'on obtiendra par
la décomposition, hormis ce cas il faudra satisfaire à la condition d'intégrabilité
requise pour la fonction $\frac{\delta M}{\mathrm{d}x}\mathrm{d}x + \frac{\partial M}{\partial y}\mathrm{d}y$.

(12) Si l'équation $N = 0$ ne pouvait pas etre decomposée ainsi que nous l'avons
dit ci-dessus alors son intégrale se rapporterait a la forme $\left\{ \begin{array}{l} s + \varphi :(\omega) = 0 \\ \frac{ds}{d\omega} + \varphi' :(\omega) = 0 \end{array} \right\}$
la quantité s qui entre dans cette formule se détermine par l'intégration de $\frac{\delta s}{\mathrm{d}x}\mathrm{d}x +$
$\frac{\partial s}{\partial y}\mathrm{d}y = 0$ ce qui se voit par les deux equations $\frac{\delta s}{\mathrm{d}x} = 0, \frac{\partial s}{\partial y} = 0$ auxquelles on arrive
après avoir fait disparaitre la fonction arbitraire.

Pour faire usage de ce résultat il faut obtenir par l'introduction de l'indéter-
minée ω, deux equations où p et q soient au 1er degré ou en tirer les valeurs de ces
quantités que nous designerons ainsi [crossed out: et on aura $\left\{ \begin{array}{l} p - \frac{\delta z}{\mathrm{d}x} = 0 \\ q - \frac{\partial z}{\mathrm{d}y} = 0 \end{array} \right\}$]

$$p = M, \quad q = N$$

parconséquent on intégrera [crossed out: $p\,\mathrm{d}x - \frac{\delta z}{\mathrm{d}x}\mathrm{d}x + q\,\mathrm{d}y - \frac{\partial z}{\mathrm{d}y}\mathrm{d}y = 0^3$]
$\mathrm{d}z - M\mathrm{d}x - N\mathrm{d}y = 0$ comme une équation à trois variables, en observant que
[crossed out: $\frac{\delta z}{\mathrm{d}x}\mathrm{d}x + \frac{\partial z}{\mathrm{d}y}\mathrm{d}y = \mathrm{d}z$] $p\,\mathrm{d}x + q\,\mathrm{d}y = \mathrm{d}z$, alors on aura s; substituant
cette valeur dans l'intégrale generale posée plus haut, on aura celle de l'équation
differentielle proposée.

L'intégrabilité de $\frac{\delta s}{\mathrm{d}x}\mathrm{d}x + \frac{\partial s}{\partial y}\mathrm{d}y = 0$ est assujettie aux mêmes conditions que
celle des équations à trois variables dont elle depend. [sidenote: après qu'on a
mis pour $p\,\mathrm{d}x + q\,\mathrm{d}y$ sa valeur $\mathrm{d}z$] Aussi se trouve-t'il très peu de cas desquels on
puisse obtenir la solution.

(13) Pour déterminer par le calcul la fonction arbitraire généralement dans toutes les equations aux différences partielles du 1er ordre il faut avoir recours aux
considérations geométriques; or nous savons que le sistême des deux équations qui
representent l'intégrale [crossed out: signifie] appartient à une surface courbe formée par les intersections consécutives d une autre surface courbe, changeante par
la variation d'un parametre, avec elle meme; la condition qu'on se propose pour
déterminer cette surface est quelle doit passer, par une courbe à double courbure
donnée; et par un procedé analogue à celui du N.º (8) nous allons en déterminer
la fonction arbitraire: soient $\left\{ \begin{array}{l} x = f:(z) \\ y = F:(z) \end{array} \right\}$ les projections de cette courbe je fais

---

^{3}[Crossed-out footnote: On pourrait faire sur ces equations une operation analogue a celle
qui est indiquée dans la note du n° 9; mais le resultat m'a paru [crossed out: unreadable] fort
compliqué et fort loin de faire esperer quelque succès.]

$\omega = c$ par conséquent $\varphi\,\omega = c'$ et l'équation de la surface [crossed out: vient]

devient $\left\{ \begin{array}{l} s + c' = 0 \\ \frac{ds}{dc} + c'' = 0 \end{array} \right\}$, je chercherai [crossed out: l'] son intersection [crossed out: de la surface] [crossed out: de la courbe?] [crossed out: 1ere equation] avec la courbe à double courbure [added: n'employant la 1.?^{ere} equation] et j'aurai z'', je chercherai pour ce point les équations des projections de la tangente de la courbe

a double courbure qui seront $\left\{ \begin{array}{l} z - z'' = \dfrac{1}{f'z''}(x - x'') \\ z' - z'' = \dfrac{1}{F'z''}(y - y'') \end{array} \right\}$ $\begin{array}{l}(1)\\[6pt](2)\end{array}$

Posant l'equation du plan tangent de [crossed out: ma] la surface courbe j'aurai alors $z - z'' = P(x - x'') + Q(y - y'')$ (3) mais P et Q sont des fonctions de $\omega.x''.y''$ qui sont elles même des fonctions de c, c', on aura donc en mettant dans l'equation (3) pour $x - x'$ sa valeur $(z - z'')f'(z'')$ prise dans l'equation (1), $(z - z'')(1 - Pf'(z'')) = Q(y - y'')$, equation qui doit être dentique avec (2), ce qui donne $\dfrac{Q}{1 - Pf':(z'')} = \dfrac{1}{F':(z'')}$; d'ou on peut tirer $c' = H(c)$ ou $\varphi(\omega) = H:(\omega)$;

substituant dans la proposée on aura $\left\{ \begin{array}{l} s + H : (\omega) = 0 \\ \frac{ds}{d\omega} + H'(\omega) = 0 \end{array} \right\}$ equation qui ne renferme plus que des fonctions connues de ω, et eliminant cette arbitraire on aura l'intégrale particuliere cherchée.

Il est évident que la surface ainsi trouvée remplira les conditions requises car elle est formée par la suite des intersections consecutives d'une surface sur laquelle la courbe à un de ses elemens.

Je n'ai point parlé dans cette determination de la 2e equation, car comme elle n'est que la differentielle de la 1.ere en regardant ω comme variable, elle passera néccéssairement par la courbe à double courbure donnée au point infiniment voisin.

Il peut arriver que la valeur de z'' ne soit pas toujours réelle; mais on déterminera pour la rendre telle la constante arbitraire ajoutée pour l'intégration et dont je n'ai point palé [sic] jusqu'à présent parce que je l'ai toujours regardée comme comprise dans la fonction arbitraire.

Essai sur les equations aux differences
partielles du 2e ordre et des ordres superieures.

(14)   Si nous faisons $\dfrac{\delta p}{\mathrm{d}x} = r, \dfrac{\partial p}{\mathrm{d}x} = s = \dfrac{\delta q}{\mathrm{d}y}$ [sic; should be $\dfrac{\partial p}{\mathrm{d}y} = s = \dfrac{\delta q}{\mathrm{d}x}$], $\dfrac{\partial q}{\mathrm{d}y} = t$; nous aurons pour la differentielle generale du $2.^e$ ordre d'une fonction z de deux variables, $\mathrm{d}^2 z = r\,\mathrm{d}x^2 + 2s\,\mathrm{d}x\,\mathrm{d}y + t\,\mathrm{d}y^2$: $\mathrm{d}y$ et $\mathrm{d}x$ sont regardés comme constans. Cela posé nous avons regardé les équations du $1.^{er}$ ordre comme provenues par l'elimination d'une fonction arbitraire; cette maniere d'envisager les equations aux différences partielles peut s'appliquer à tous les ordres. Si la variable z était exprimée par l'assemblage de deux fonctions differentes il faudrait pour éliminer ces fonctions differentier deux fois et l'on obtiendrait une équation entre les coefficiens différentiels r, s, t du 2e ordre et ceux du $1.^{er}$ ordre; cette relation peut etre regardée comme un caractere auquel on reconnaitra quelles sont les quantités qui peuvent se rapporter a la fonction z.

La forme des intégrales premieres de cet ordre se decouvre aisement d'après ce qu'on vient de dire: $M + \varphi(V) = 0$ est l'intégrale premiere de toutes les equations du $2.^e$ ordre ou r, s, t ne passent pas le 2 degré; car cette equation ne donnera par l'elimination de $\varphi(V)$ que des equations lineaires entre r, s, et t, si V ne contient que x y ou z; et si V renferme p ou q, il en resultera des equations du $2.^e$ degré entre r, s et t; mais jamais cette forme n'en produira de plus elevées. Il est encore aisé d'appercevoir que si p et q sont linéaires dans M et qu'ils n'entrent point dans V, non plus que z, ils seront aussi au $1.^{er}$ degré dans l équation différentielle resultante: quant aux equations ou r, s et t passent le $2.^e$ degré leur intégrale peut-être representée par le sisteme de deux equations ainsi qu'on l'a vu pour le $1.^{er}$ ordre.

J'elimine $\varphi(V)$ en differentiant l'equation $M = \varphi(V)$ (a); et j'ai

$$\left.\begin{array}{l} \dfrac{\delta M}{\mathrm{d}x} = \varphi'.(V)\dfrac{\delta V}{\mathrm{d}x} \quad (1) \\[3mm] \dfrac{\partial M}{\mathrm{d}y} = \varphi'.(V)\dfrac{\partial V}{\mathrm{d}y} \quad (2) \end{array}\right\}$$

et enfin $\dfrac{\delta M}{\mathrm{d}x}\dfrac{\partial V}{\mathrm{d}y} - \dfrac{\partial M}{\mathrm{d}y}\dfrac{\delta V}{\mathrm{d}x} = 0$ (b); équation qui peut representer généralement les equations aux différe[nces] partielles de $2.^e$ ordre, pourvu que les coefficiens differentiels de cet ordre ne passent pas le $2.^e$ degré ainsi que nous l'avons remarqué.

(15)   Nous tirerons des formules précédentes le moyen d'intégrer les equations du $2.^e$ ordre qui s'y rapportent. Si on sépare l'equation (b) en deux autres au moyen de la nouvelle indeterminée ω, on fera $\dfrac{\delta M}{\mathrm{d}x} : \dfrac{\delta V}{\mathrm{d}x} = \omega$ et $\dfrac{\partial M}{\mathrm{d}y} : \dfrac{\partial V}{\mathrm{d}y} = \omega$: equations de la forme (1) et (2), on aura donc $\dfrac{\delta M}{\mathrm{d}x}\mathrm{d}x + \dfrac{\partial M}{\mathrm{d}y}\mathrm{d}y = \omega(\dfrac{\delta V}{\mathrm{d}x}\mathrm{d}x + \dfrac{\partial V}{\mathrm{d}y}\mathrm{d}y)$; l'intégrale

de la proposée dependra par conséquent de celle de deux formules differentielles ordinaires. Soit pour exemple l'équation $Ar + Bs + Ct = 0$, pour la ramener à la forme précédente il faut observer que le terme Bs provenant de $\dfrac{\partial p}{\mathrm{d}y}$ et $\dfrac{\delta q}{\mathrm{d}x}$ doit renfermer deux parties; je ferai donc $\frac{B}{A} = \alpha + \alpha'$, alors la proposée sera changée en celle ci, $r + \alpha s + \alpha's + \frac{Ct}{A} = 0$, et se decomposera de la maniere suivante $r + \alpha s = \omega$ et $\alpha'\left(s + \frac{Ct}{A\alpha'}\right) = -\omega$ parconséquent $\dfrac{\delta M}{\mathrm{d}x}\mathrm{d}x + \dfrac{\partial M}{\partial y} = \mathrm{d}p + \alpha\mathrm{d}q$, et $\dfrac{\delta V}{\mathrm{d}x}\mathrm{d}x + \dfrac{\partial V}{\partial y}\mathrm{d}y = \omega(\alpha'\mathrm{d}x - \mathrm{d}y)$; l'intégrale de la proposée sera donc reduite a celle de $\mathrm{d}p + \alpha\mathrm{d}q = 0$ et $\alpha'\mathrm{d}x - \mathrm{d}y = 0$, α et α' seront données par les equations $B = \alpha + \alpha'$ et $\frac{C}{A} = \alpha\alpha'$; ou ce qui revient au même ils seront les racines de l'equation $A\alpha^2 - B\alpha + C = 0$: Nous remarquerons a ce sujet qu'on sera maitre d'echanger α et α' dans les formules $\mathrm{d}p + \alpha\mathrm{d}q = 0$ et $\alpha'\mathrm{d}x - \mathrm{d}y$; il peut en resulter quelque fois des simplifications.

D'après ce qui précède on voit clairement que $Ar + Bs + Ct = 0$ sera intégrable toutes les fois que $\mathrm{d}p + \alpha\mathrm{d}q = 0$ et $\alpha'\mathrm{d}x - \mathrm{d}y$ ne renferment que les variables dont-ils contiennent les différentielles; ce qui aura lieu lorsque l'equation $A\alpha^2 - B\alpha + C = 0$ sera décomposable en deux facteurs $\alpha' - F:(x,y) = 0$ et $\alpha - f:(p,q) = 0$ et dans ce cas, si $\mathrm{d}p + \alpha\mathrm{d}q = 0$ et $\alpha'\mathrm{d}x - \mathrm{d}y$ ne sont pas intégrables par eux-mêmes, on pourra toujours trouver deux facteurs μ et μ', l'un fonction de p et q et l'autre de x et y; qui les rendront intégrables; nommant $\mu\mathrm{d}p + \alpha\mu\mathrm{d}q = \mathrm{d}P$, et $\mu'\alpha'\mathrm{d}x - \mu'\mathrm{d}y = \mathrm{d}T$ l'intégrale de la proposée sera $P = \varphi(T)$.

L'equation $vP'r + \{vQ' - up\}s - uQ't = 0$ contient celles qui satisfont à la condition posée ci-dessus: v et u sont deux fonctions de x et y; Q', P' contiennent p et q; l'equation $A\alpha^2 - B\alpha + C = 0$ devient $vP'\alpha^2 + \{uP' - vQ'\}\alpha - uQ' = 0$, les facteurs sont $\alpha + \frac{u}{v} = 0$ et $\alpha - \frac{Q'}{P'} = 0$, prenant le premier pour la valeur de α', on aura $P'\mathrm{d}p + Q'\mathrm{d}q = 0$ et $u\,\mathrm{d}x + v\,\mathrm{d}y = 0$; d[onc?] l'intégrale sera $\int\{P'\mathrm{d}p + Q'\mathrm{d}q\} = \varphi : \int\{u\mathrm{d}x + v\mathrm{d}[y]\}.$

[Crossed out: Si on prenait $\alpha' = \frac{Q'}{P'}$ et $\alpha = \frac{u}{v}$ on aurait $v\,\mathrm{d}p - u\,\mathrm{d}y = 0$ et $Q'\mathrm{d}y - P'\mathrm{d}x = 0$ équation qu'on ne peut pas intégrer généralement; l'analyse dans ce cas ne saurait fournir d'autre resultat car sil'on intégrait ces deux dernieres formules on trouverait pour la proposée deux intégrales premieres; on pourrait donc parvenir par l'elimination a une intégrale finie représentée par une seule equation ce qui est contraire à ce qu'on a vu ci-dessus.]

Il y a quelques équations qui ne sont pas comprises dans celle que nous venons de traiter et qu'on intégre par une reduction particuliere; telles sont $q^2r - 2pqs + p^2t = 0$ et $x'r + 2xys + y^2t$. On a pour la 1.ere α et $\alpha' = -\frac{p}{q}$, parconsequent $q\,\mathrm{d}p - p\,\mathrm{d}q = 0$ et $p\,\mathrm{d}x + q\,\mathrm{d}y = 0[?]$; si on observe que $p\,\mathrm{d}x + q\,\mathrm{d}y = \mathrm{d}z$, on aura pour intégrale prémière $\frac{p}{q} = \varphi(z)$. La 2.e traitée de la même maniere donne $x\,\mathrm{d}q + y\,\mathrm{d}q = 0$ et $x\,\mathrm{d}y - y\,\mathrm{d}x = 0$; et a pour intégrale prémière $px + qy - z = \varphi:\left(\frac{x}{y}\right)$: dans ces deux cas les valeurs de α etant egales on ne peut pas arriver à l'intégrale finie par l'elimination.

(16) Soit l'equation $Ar + Bs + Ct + W = 0$ j'aurai en operant comme précédemment $r + \alpha s + \frac{W}{A} = \omega$, et $\alpha' \left\{ s + \frac{Ct}{A\alpha'} \right\} = -\omega$; l'intégration de la proposée sera donc reduite à celle de $\mathrm{d}p + \alpha\,\mathrm{d}q + \frac{W\mathrm{d}x}{A} = 0$ et $\alpha'\mathrm{d}x - \mathrm{d}y = 0$. Si on suppose A, B, C constantes et W fonction de x et y, la proposée s'intégrera complettement en faisant $s\{\alpha'\mathrm{d}x - \mathrm{d}y\} = T$; on en tirera une valeur de y en T, x et constantes, on substituera cette valeur dans W, le terme $\frac{W\mathrm{d}x}{A}$ s'intégrera alors par les quadratures, T, devant y etre regardé comme constant: si W etait composé seulement de p et q on metrait l'equation $\mathrm{d}p + \alpha\,\mathrm{d}q + \frac{W\mathrm{d}x}{A} = 0$ sous cette forme $\frac{\mathrm{d}p + \alpha\,\mathrm{d}q}{W} + \frac{\mathrm{d}x}{A} = 0$, la difficulté serait alors d'intégrer la quantité $\frac{\mathrm{d}p + \alpha\,\mathrm{d}q}{W}$; et toutes les fois que cela sera possible l'intégrale de la proposée sera $P + \frac{x}{A} = \varphi(\alpha x - y)$. Si A, B, C sont des fonctions de x, y, p et q, et que W ne contienne que x et y il se présente alors une classe d'equations qui peut se ramener au cas précédent; c'est celle qui rend $A\alpha^2 - B\alpha + C = 0$ décomposable en deux facteurs de la forme $\alpha - F(x,y) = 0$ et $\alpha - f(p,q) = 0$; elle donne $\mathrm{d}p + f(p,q)\mathrm{d}q + \frac{W\mathrm{d}x}{A} = 0$ et $F(x,y)\mathrm{d}x - \mathrm{d}y = 0$; il est aisé de voir que si le terme $\frac{W\mathrm{d}x}{A}$ renferme seulement x, y, la difficulté sera seulement d'intégrer la quantité $p + f : (p,q)\mathrm{d}q$; W etant composée de p et q on aura $\frac{\mathrm{d}p + f(p,q)\mathrm{d}q}{W} + \frac{\mathrm{d}x}{A} = 0$ et $F : (x,y)\mathrm{d}x - \mathrm{d}y = 0$

La formule $Xr + (Y + XN)s + YNt = W$ dans laquelle X, Y sont des fonctions de x, y; N et W contiennent p et q, est une de plus générales de cette classe; on a $\alpha = \frac{Y}{X}$ et $\alpha = N$, les formules à intégrer sont $\frac{\mathrm{d}p + N\mathrm{d}q}{W} + \frac{\mathrm{d}x}{X} = 0$, $Y\mathrm{d}x - X\mathrm{d}y = 0$, si l'intégrale de la quantité $\frac{\mathrm{d}p + N\mathrm{d}q}{W}$ est P, celle [sic] $\mu Y\mathrm{d}x - \mu X\mathrm{d}y$, T; l'intégrale de la proposée sera $P + \int \frac{\mathrm{d}x}{X} = \varphi(T)$; $\int \frac{\mathrm{d}x}{X}$ se reduira toujours aux quadratures par le procédé employé ci-dessus: on trouverait de même l'intégrale de $Nr + \{NX + M\}s + MXt = W$, M, N etant fonction de p q et X de x, y, W de x, y, ou de p et q; et celle de $MYr + \{NY + MX\}s + NXt = W$.

Enfin soit l'equation $Ar + Bs + Ct + W = 0$ dans laquelle A, B, C, W, soient des fonctions de p, q, x, y, et que W contienne en outre z; je ferai $W = W' + W''$, je comprendrai dans W' tous les termes dont la forme indiquera qu'ils proviennent d'une différentiation par rapport a x, et W'' renfermera ceux produits en differentiant par rapport a y; l'equation etant separée en deux autres ainsi qu'on à toujours fait dans le courant de ce mémoire; on aura à integrer $\mathrm{d}p + \alpha\,\mathrm{d}q + \frac{W'\mathrm{d}x}{A} + \frac{W''\mathrm{d}y}{A\alpha'} = 0$ et $\alpha'\mathrm{d}x - \mathrm{d}y = 0$; on aura entre α et α' les équations suivantes $\frac{B}{A} = \alpha + \alpha'$ et $\frac{C'}{A\alpha'} = \alpha$: on est bien éloigné de pouvoir resoudre ce cas généralement.

17. Je ne m'arreterai pas davantage sur cette classe d'equations qui à été traitée par plusieurs geometres, je passe a celle dont l'intégrale peut être comprise dans la formule $M = \varphi(V)$; V renfermant p ou q. Si on differentie cette equation pour

eliminer la fonction on aura[4]

$$\left.\begin{array}{l} \dfrac{\mathrm{d}(M)}{\mathrm{d}p}r + \dfrac{\mathrm{d}M}{\mathrm{d}z}p = \varphi'(V)\left\{\dfrac{\mathrm{d}V}{\mathrm{d}q}s + \dfrac{\mathrm{d}V}{\mathrm{d}x}\right\} \\[3mm] \dfrac{\mathrm{d}(M)}{\mathrm{d}p}s + \dfrac{\mathrm{d}M}{\mathrm{d}z}q = \varphi'(V)\left\{\dfrac{\mathrm{d}V}{\mathrm{d}q}t + \dfrac{\mathrm{d}V}{\mathrm{d}y}\right\} \end{array}\right\} \quad \text{et}$$

enfin
$$\left\{\begin{array}{l} \dfrac{\mathrm{d}M}{\mathrm{d}p}\cdot\dfrac{\mathrm{d}V}{\mathrm{d}q}rt - \dfrac{\mathrm{d}M}{\mathrm{d}p}\cdot\dfrac{\mathrm{d}V}{\mathrm{d}q}s^2 + \dfrac{\mathrm{d}M}{\mathrm{d}z}\dfrac{\mathrm{d}V}{\mathrm{d}q}pt - \dfrac{\mathrm{d}M}{\mathrm{d}z}\dfrac{\mathrm{d}V}{\mathrm{d}q}qs \\[3mm] +\dfrac{\mathrm{d}M}{\mathrm{d}p}\cdot\dfrac{\mathrm{d}V}{\mathrm{d}y}r - \dfrac{\mathrm{d}M}{\mathrm{d}p}\cdot\dfrac{\mathrm{d}V}{\mathrm{d}x}s + \dfrac{\mathrm{d}M}{\mathrm{d}z}\dfrac{\mathrm{d}V}{\mathrm{d}y}p - \dfrac{\mathrm{d}M}{\mathrm{d}z}\dfrac{\mathrm{d}V}{\mathrm{d}x}q = 0.(g) \end{array}\right.$$

Pour intégrer cette equation il faut faire

$$\left\{\dfrac{\mathrm{d}M}{\mathrm{d}p}r + \dfrac{\mathrm{d}M}{\mathrm{d}z}p\right\} : \left\{\dfrac{\mathrm{d}V}{\mathrm{d}q}s + \dfrac{\mathrm{d}V}{\mathrm{d}x}\right\} = \omega \quad \text{et} \quad \left\{\dfrac{\mathrm{d}M}{\mathrm{d}p}s + \dfrac{\mathrm{d}M}{\mathrm{d}z}q\right\} : \left\{\dfrac{\mathrm{d}V}{\mathrm{d}q}t + \dfrac{\mathrm{d}V}{\mathrm{d}y}\right\} = \omega$$

alors on aura $\dfrac{\mathrm{d}M}{\mathrm{d}p}\mathrm{d}p + \dfrac{\mathrm{d}M}{\mathrm{d}z}dz = \omega\left\{\dfrac{\mathrm{d}V}{\mathrm{d}q}\mathrm{d}q + \dfrac{\mathrm{d}V}{\mathrm{d}x}\mathrm{d}x + \dfrac{\mathrm{d}V}{\mathrm{d}y}\mathrm{d}y\right\}$ et parconséquent
$M = \varphi(V)$. [crossed out: unreadable]

    Avant d'embrasser les equations qui se rapportent à ces formules générales supposons que la fonction arbitraire ne doive contenir que la variable q et que M renferme p, x, y, z, alors on aura $\dfrac{\mathrm{d}M}{\mathrm{d}p}\{rt - s^2\} + \dfrac{\mathrm{d}M}{\mathrm{d}z}\{pt - qs\} + \dfrac{\mathrm{d}M}{\mathrm{d}x}t - \dfrac{\mathrm{d}M}{\mathrm{d}y}s = 0$. Et il faudra pour intégrer les equations de ce genre les décomposer en deux autres de la maniere suivante $\frac{1}{s}\{\frac{\mathrm{d}M}{\mathrm{d}p}r + \frac{\mathrm{d}M}{\mathrm{d}z}p + \frac{\mathrm{d}M}{\mathrm{d}x}\} = 0$ et $\{\frac{\mathrm{d}M}{\mathrm{d}p}s + \frac{\mathrm{d}M}{\mathrm{d}z}q + \frac{\mathrm{d}M}{\mathrm{d}y}\}\frac{1}{t} = 0$ ce qui donnera $M = \varphi(q)$.

    Si on se propose d'intégrer l'equation $A\{rt - s^2\} + B\{pt - qs\} + Ct - C's = 0$, on voit aisément qu'il faudra faire $\dfrac{Ar + Bp + C}{s} = \omega$ et $\dfrac{-As - Bq - C'}{t} = -\omega$; on en tirera parconsequent $A\mathrm{d}p + B.\mathrm{d}z + C\mathrm{d}x + C'\mathrm{d}y = \omega.\mathrm{d}q$ et si le premier membre est intégrable immediatement on aura $M = \varphi(q)$ pour l'intégrale 1.re complette de la proposée. L'equation differentio-differentielle des surfaces developpables, $rt - s^2 = 0$, est un cas particulier de la proposée. Son intégrale premiere sera parconséquent $p = \varphi(q)$; on a vu (5) comment on arrivait à l'intégrale finie de cette équation. Je ne m'etendrai pas sur les différens cas d'intégrabilité de la quantité $A\mathrm{d}p + B\mathrm{d}z + C\mathrm{d}x + C'\mathrm{d}y$.

    Si on se fut proposé l'equation $A\{s^2 - rt\} + B\{ps - qr\} + Cs - C'r = 0$, on l'intégrerait comme la précédente en faisant $\dfrac{As + Bp + C}{r} = \omega,$ $\dfrac{-At - Bq - C'}{s} = \omega$; et l'intégrale sera $\int\{A\mathrm{d}q + B.\mathrm{d}z + C\mathrm{d}x + C'\mathrm{d}y\} = \varphi(p)$.

---

[4][Sidenote:] J'ai supposé que M contienne p et z $V, x, y,$ et q

Soit proposée l'equation $A\{rt - s^2\} + B\{pt - qs\} + Cr - C's + Np - N'q = 0$ qui se rapporte a la forme générale (g) posée plus haut; pour la traiter nous la supposerons produite par l'elimination de ω dans ces deux equations $\frac{ar+bq}{a's+e'} = \omega$ et $\frac{as+bq}{a't+e} = \omega$, et comparant l'equation resultante avec la proposée nous aurons:

$$A = aa' \qquad B = a'b \qquad C = ea \qquad C' = e'a \qquad N = be \qquad N' = be'$$

ces equations etant au nombre de six et ne renfermant que cinq inconnues il en resultera des equations de condition pour que la proposée puisse etre decomposée ainsi et si elles sont satisfaites on aura à intégrer $adp + bdz = \omega\{a'dq + e'dx + edy\}$: on tire des equations que nous a donné lidentification de l'equation resultante avec la proposée, $\frac{A}{B} = \frac{a}{b}$, $\frac{A}{C} = \frac{a'}{e}$, $\frac{A}{C'} = \frac{a'}{e'}$, $\frac{B}{N} = \frac{a'}{e}$, $\frac{B}{N'} = \frac{a'}{e'}$, $\frac{C}{N} = \frac{a}{b'} = \frac{C'}{N'}$; d'où il resulte entre les coefficiens A, B, C, les equations de condition:

$$AN - BC = 0 \qquad AN' - BC = 0 \qquad CN - C'N = 0$$

les deux premieres etant verifiées la 3.e s'ensuit, on pourra alors negliger les deux dernieres equations de la 1.ere suite $\left.\begin{array}{l} N = be \\ N' = be' \end{array}\right\}$ et se donnant à volonté une des cinq inconnues a, a', b, e, e', on déterminera les quatre autres au moyen de $\left.\begin{array}{l} A = aa' \\ B = a'b \\ C = ea \\ C' = e'a \end{array}\right\}$; on remarquera que a' doit toujouts etre fonction de q, x, y et a, b, ne doivent contenir que p et z pour qu'on puisse intégrer $adp + bdz = \omega\{a'dq + e'dx + edy\}$; a et b etant des fonctions quelconques de p et z, on pourra rendre le 1.er membre $adp + bdz$ intégrable en le multipliant par un facteur. Si on a[?] e' ou $e = 0$, et que a' et e ou e' renferment seulement q et x ou q et y, alors la proposée serait réduite a $\left.\begin{array}{l} A\{rt - s^2\} + B\{pt - qs\} + Cr = 0 \\ \text{ou } A\{rt - s^2\} + B\{pt - qs\} + C's = 0 \end{array}\right\}$; on aurait à intégrer les deux equations differentielles ordinaires a deux varia- bles $\left\{\begin{array}{l} adp + bdz = 0 \\ a'dq + e'dx = 0 \text{ ou } a'dq + edy = 0 \end{array}\right.$, M etant l'intégrale de la 1.ere et V celle de la 2.e, on aura pour intégrale de la proposée $M = \varphi(V)$. On trou- vera d'ailleurs beaucoup d'autres cas où on obtiendra la solution complette de l'equation $A\{rt - s^2\} + B\{pt - qs\} + Cr - C's + Np - N'q = 0$. Si on proposait $A\{s^2 - rt\} + B\{ps - qr\} + Cs - C't + Np - N'q = 0$, en la traitant de la meme maniere on obtiendrait $M = \varphi(V)$ pour intégrale, en faisant attention que dans ce cas M contiendrait q et z; V serait composé de p, x, y.

Reprenons la formule $M = \varphi(V)$ et supposons que M et V contiennent $x\ y\ z\ p$ et q, ce qui est le cas le plus général; en différentiant et eliminant la fonction arbitraire on obtiendra une equation qu'on pourra envisager comme produite par l'elimination de ω entre deux equations de la forme suivante $\dfrac{ar + bs + cp + e}{a'r + b's + c'p + f} =$

ω et $\dfrac{as + bt + cq + e'}{a's + b't + c'q + f'} = \omega$ alors on aura à intégrer

$$a\,\mathrm{d}p + b\,\mathrm{d}q + c\,\mathrm{d}z + e\,\mathrm{d}x + e'\mathrm{d}y = \omega\{a'\mathrm{d}p + b'\mathrm{d}q + c'\mathrm{d}z + f\,\mathrm{d}x + f'\mathrm{d}y\};$$

les equations qui pourront se rapporter à ce cas seront comprises dans celle qui suit $A\{rt - s^2\} + B\{pt - qs\} + C\{ps - qr\} + Dt - D'r + Es + Np - N'q = 0$; identifiant la resultante avec cette equation proposée on obtiendra des equations pour déterminer les coefficiens. [Crossed out: Je crois etre fondé a dire que toute equation qui ne pourra pas etre ramenée aux formes précédentes soit en la multipliant par un facteur ou autrement, n'aura pas son intégrale première représentée par une seule equation.]

    Quant a l'intégrale complette des equations que nous venons de traiter la difficulté pour l'obtenir se réduit à intégrer généralement l'equation du 1.er ordre $M = \varphi(V)$, qui représente l'intégrale première de ce genre d'equations; en y appliquant les procédés qu'on à donné précédemment pour le 1.er ordre on reduira la question à l'intégration de formules differentielles ordinaires; il y aura sans doute beaucoup de cas dans les quels on ne pourra pas avoir l'intégrale complette.

(18)   Les equations aux différences partielles du 2.e ordre qui ne sont pas comprises dans les précédentes, auront pour intégrale 1.re un systême de deux equations entre lesquelles il reste à eliminer une indéterminée, ainsi qu'on l'a dit (14);

elle pourra etre généralement représentée par cette forme $\left\{\begin{array}{l} M = \varphi(\omega) \\[4pt] \dfrac{dM}{d\omega} = \varphi'(\omega) \end{array}\right\}$: on

obtiendra   en   différentiant   et   en   eliminant   la   fonction   arbitraire

$\left\{\begin{array}{l} \dfrac{\mathrm{d}M}{\mathrm{d}p}r + \dfrac{\mathrm{d}M}{\mathrm{d}q}s + \dfrac{\mathrm{d}M}{\mathrm{d}z}p + \dfrac{\mathrm{d}M}{\mathrm{d}x} = 0 \\[8pt] \dfrac{\mathrm{d}M}{\mathrm{d}p}s + \dfrac{\mathrm{d}M}{\mathrm{d}q}t + \dfrac{\mathrm{d}M}{\mathrm{d}z}q + \dfrac{\mathrm{d}M}{\mathrm{d}y} = 0 \end{array}\right\}$ ces deux équations des quelles ω etant éli-

miné il resultera une equation aux différences partielles du 2.e ordre dont le degré dépendra de la maniere dont ω entrera dans les equations précédentes.

    Cela posé on voit aisément que pour traiter une equation dans laquelle r, s et t passent le 2.e degré, ou qui ne peut pas se rapporter aux précédentes il faut la décomposer en deux autres par l'introduction d'une nouvelle indéterminée ω; ou bien déterminer les coefficiens de deux equations où r, s, t soient au premier degré telles qu'éliminant d'entrelles l'indéterminé ω il resulte la proposée: pour faire usage de ces equations il faut qu'on en puisse tirer deux autres de

cette forme $\left\{\begin{array}{l} ar + bs + cp + e = 0 \\[4pt] a's + b't + c'q + e' = 0 \end{array}\right\}$ et qu'on ait entre les coefficiens les rela-

tions suivantes $\dfrac{b}{a} = \dfrac{b'}{a'}, \dfrac{c}{a} = \dfrac{c'}{a'}$; les conditions etant remplies on aura a intégrer $\mathrm{d}p + B\mathrm{d}q + C\mathrm{d}z + E'\mathrm{d}y = 0$ equation différentielle ordinaire dans laquelle ω est regardé comme constant.

    En suivant ce procédé l'intégration des équations aux différences partielle [sic] du 2.e ordre sera ramenée à celle des equations différentielles ordinaires. Nous ne nous etendrons pas davantage sur l'application de cette méthode qui se voit assez d'après ce qu'on à dit pour le 1.er ordre.

19 Nous terminerons en disant un mot sur la détermination des fonctions arbitraires dans le cas ou l'intégrale complette sera representée par deux equations.

Si elle est de cette forme
$$\begin{cases} z = M + N\varphi(\omega) + [\psi(\omega)?] \\ \dfrac{dM}{d\omega} + \varphi(\omega)\dfrac{dN}{d\omega} + N\varphi'(\omega) + [\psi'(\omega) = 0?] \end{cases}$$
la condition sera qu'elle doit passer par deux courbes a double courbure données. Par un procédé analogue a celui du numero (13.) et qu'on déduira aisément de ce dernier, on obtiendra deux equations telles que $\dfrac{Q}{1 - P\,f' : (z''')} = \dfrac{1}{F' : (z'')}$ et $\dfrac{Q'}{1 - P'k : (z''')} = \dfrac{1}{K' : (z''')}$; je suppose les equations de projections de la courbe a double courbure $\begin{cases} x = k : (z) \\ y = K : (z) \end{cases}$; z''' sera la coordonnée du point d'intersection de la surface representée par $z = M + Nc' + c''$, avec la deuxieme courbe a double courbure, et cette quantité sera determinée comme z'' dans l'article cité; du reste les dénominations y seront les mêmes: au moyen des deux equations precédentes on trouvera la valeur de $\varphi(\omega) = c'$, $\psi(\omega) = c''$ en $\omega = c$ et autres quantités, ce qui donnera la composition de ces fonctions.

20 En traitant les equations du 3e ordre et celles des ordres superieures ainsi que nous venons de faire pour le 1er et le 2e, on obtiendrait des resultats analogues a ceux qui se trouvent dans ce memoire: nous observerons cependant que le nombre des cas qui echappent a la Methode augmente a mesure qu'on s'occupe des ordres plus elevés. Lorsqu'on passe le premier ordre, en traitant les cas generaux on tombe dans des equations differentielles ordinaires dont le nombre de variables devient plus grand d'une unité a chaque ordre oú l'on s'eleve; il peut arriver que ces equations soient immediatement integrables; ou en les multipliant par des facteurs; ce dernier cas conduit a des equations aux differences partielles, le plus souvent aussi compliqués que la proposee; pour ecarter ces difficultés on fait des restrictions qui ne menent qu'a des cas tres particuliers: enfin les equations peuvent etre absurdes. Les integrales successives presentent encore des difficultés insurmontables dans beaucoup de cas. Tel est a peu près l'etat du calcul aux differences partielles dont on ne s'est gueres occupé jusqu'a present que parraport aux applications qu'on avait en vue. Il y a plusieurs points sur la fin de ce memoire que leur etendue ne m'a pas permis de developper, tels que l'application aux ordres superieurs; si ces recherches peuvent ne pas deplaire j'y reviendrai par la suite.

A.1.1   Report by Condorcet and Monge

M. M. De Condorcet et Monge
11 fevrier 1786

Nous Commissaires nommés par L'académie avons examiné un mémoire, *sur le Calcul intégral des Equations aux differences partielles qui ne sont pas linéaires,* présenté par M.^r La Croix.

Lorsque les fonctions arbitraires, qui se trouvent dans une équation intégrale sont toutes linéaires, et que les quantités qui entrent sous ces fonctions sont toutes données immédiatement, l'equation aux differences partielles, à la qu'elle on est conduit en fesant evanouir les fonctions arbitraires, est elle même toujours linéaires; Mais 1.° si les fonctions arbitraires sont élevées à differentes puissances dans l'integrale, 2.° si les quantités sous les fonctions ne sont données que par d'autres equations et que dans ce cas elles ne soient pas linéaires partout, l'équation aux differentes partielles à la qu'elle on arrive, est toujours elevée.

On peut donc dire qu'il y a deux especes d'équations aux differences partielles qui ne sont pas linéaires. Les unes ont pour integrale finie une equation unique, l'integrale des autres ne peut être exprimée en quantités finies, que par le système de deux equations entre les qu'elles il faut eliminer une indeterminée qui se trouve sous les fonctions arbitraires. Les equations de la premiere espèce sont en général plus faciles à traiter que celles de la seconde, & c'est de celles ci que M.^r La Croix s'occupe dans le mémoire dont il sagit.

Le procédé qu'il employe pour le 1er ordre, consiste en général à regarder la proposée comme le résultat de l'elimination d'une certaine indeterminée entre deux equations; il cherche ces deux equations, qui lui fournissent les valeurs des deux differences partielles, et les valeurs substituées dans la forme générale $\mathrm{d}z = p\,\mathrm{d}x + q\,\mathrm{d}y$, donnent une equation aux differences ordinaires; il integre cette equation en regardant comme constante l'indéterminée dont les differences ne sont pas employées, et il complete l'integrale en ajoutant une fonction arbitraire de cette constante. En suite pour exprimer que cette quantité a été regardée comme constante dans l'intégration, il differencie l'integrale en ne fesant varrier que l'indeterminée, et le resultat qui est toujours en quantités finies, est la seconde équation qui sert à eliminer l'indeterminée.

On voit par lá que le résultat au quel on est conduit dans ce cas n'est pas d'une forme nécéssaire, car il y a plusieurs systhêmes d'équations dont le resultat de lelimination est le même; aussi nous connaissons deja plusieurs familles de surfaces courbes, telles que les surfaces dévelopables qui peuvent être exprimées en quantités finies de plusieurs manieres tres differentes; ce qui est un avantage. Chacune de ces expressions enonce en effet un caractère particulier de ces sortes de surfaces et par conséquent une manière distincte de les engendrer.

L'auteur suppose ensuite à l'integrale certaines formes particulieres, et donne pour chacun de ces cas des equations qui par l'elimination de l'indeterminé reproduisent la proposée, ce qui le conduit a l'integration de certaines equations non lineaires assez générales.

Pour les equations du second ordre, M.^r La Croix s'occupe d'abord de celles dont l'integrale premiere peut être exprimée par une équation unique; il donne un caractere auquel on peut les reconnaître dans un très grand nombre de cas: et d'après certaines formes qu'il suppose à cette integrale, il trouve quelles sont les equations à coëfficiens variables qu'il peut integrer par un moyen analogue a celui que nous venons de rapporter.

Enfin il passe aux equations dont l'integrale premiere ne peut être exprimée que par le systhême de deux équations. Entre lesqu'elles il faut eliminer une indeterminée. Il regarde pareillement la proposée comme le resultat de l'elimination de cette indeterminée entre deux autre equations quil trouve et lorsque ces deux équations sont les differentielles d'une même equation prises par rapport à chacune des variables principales, ou peuvent être ramenée à cet état, il integre la proposée et la reduit aux differences premieres.

Nous pensons que ce memoire mérite l'approbation de l'academie et d'être imprimé dans le recueil de ceux des savants etrangers.

_____

Fait au Louvre le 11 fevrier 1786.

Signé le Mis DE CONDORCET.   et   MONGE

A.2 "Memoire sur les surfaces developpables et les equations aux differences ordinaires a trois variables", 1790

According to the *procès-verbaux* (minutes) of the meetings of the *Académie Royale des Sciences* (of Paris), this memoir was read by Lacroix himself on the 1st September 1790 (the indication in the title page that it was read in August 1790 must therefore be a mistake). Lacroix had been a correspondent member of the *Académie* for a year.

Lagrange, Condorcet and Monge were charged with reporting on the memoir, but apparently never did.

The manuscript is in the Archive of the *Académie*, in the *pochette* of the session of 1 September 1790. There are some references to figures in the text, but unfortunately none is found in the manuscript.

The introductory paragraph is a second version, glued over the original one.

This manuscript is much rougher than that of the 1785 memoir. In this transcription, words and sentences between angle brackets < > stand for additions written on the margin of the manuscript. Most of the crossed-out passages have been left out.

A revised and somewhat shortened version of the first part of the memoir (up to article XI) was eventually published as a section "on the development of curves traced on surfaces" in [Lacroix *Traité*, 2nd ed, I, 636-652].

Memoire sur les surfaces developpables et les equations aux differences ordinaires a trois variables*

par M. De La Croix, Correspondant de l'academie,
Professeur de Mathematiques de l'Ecole d'artillerie
a Besançon

---

*Lu au mois d'août 1790

Memoire sur les surfaces developpables
et sur les equations differentielles a trois variables

Plusieurs geometres se sont occupes des surfaces developpables; M. Monge a donné le premier leur equation aux differences partielles et son integrale, il a fait des applications tres interessantes de ces recherches a la theorie des ombres et des penombres, et a montré comment on pouvait determiner celles de ces surfaces qui doivent passer par des courbes a double courbure données. J'ai cru que les questions suivantes completeraient cette theorie, et leur solution fait l'objet de ce memoire que j'ai terminée par quelques remarques sur les equations differentielles a trois variables. Voici ces questions:

Etant donnée une courbe quelleconque sur une surface developpable, trouver ce qu'elle devient dans le developpement de cette surface et reciproquement *une courbe etant donnée sur un plan trouver ce quelle devient lorsqu'on l'enveloppe sur la surface donnée.* On peut toujours reconnaitre par l'equation aux differences partielles si une surface proposée est developpable ou non, et la solution des questions precedentes donne les moyens d'en developper une portion quelleconque terminée de toutes parts par des courbes connues.

art I.

On sait que toute surface developpable doit etre considerée comme l'assemblage d'une infinité de plans infiniment longs <infiniment etroits>, et que si chacun de ces plans tourne autour de sa commune intersection avec son consecutif, comme sur une charniere on pourra etendre cette surface sur un plan sans qu'il y ait aucun pli ou aucune solution de continuité. J'appellerai dans le cours de ce memoire arrêtes de la surface proposée, les lignes qui sont les intersections de deux des plans consecutifs qui la forment. Ces lignes sont tangentes a la surface dans toutes leur etendue et comme elles sont deux a deux dans le même plan elles se coupent reciproquement; leurs points d'intersection forment une courbe a double courbure appellée arrête de rebroussement par M. Monge. Elle est remarquable en ce qu'elle peut seule determiner la surface proposée. Nous diviserons toutes les surfaces developpables en trois classes savoir

1.° les surfaces cilindriques ou celle dont les arrêtes sont paralleles

2.° les surfaces coniques, dont les arrêtes concourrent toutes a un même point.

3.° les surfaces developpables dont les arretes se coupent deux a deux suivant une courbe a double courbure.

Et nous nous occuperons de chacune de ces classes en particuliere, la derniere donnera lieu a la solution generale qui renfermera toutes les precendentes.

II Lemme.

Soient[5] $\left.\begin{array}{l} z - z' = a(x - x') \\ z - z' = b(y - y') \end{array}\right\}$ $\left.\begin{array}{l} z - z' = a'(x - x') \\ z - z' = b'(y - y') \end{array}\right\}$ les equations de deux droites qui se coupent dans un point dont les coordonnées sont x', y' et z'. Si on imagine que ces droites se meuvent parallelement a elles memes jusqu'a ce que le point d'intersection soit a l'origine des coordonnées leurs equations se reduiront en $\left.\begin{array}{ll} z = ax & z = a'x \\ z = by & z = b'y \end{array}\right\}$. En décrivant de ce point comme centre et d'un rayon $= r$, une sphere <dont l'equation sera $x^2 + y^2 + z^2 = r^2$> elle coupera ces droites en deux points pour lesquels on aura $z = \dfrac{r}{\sqrt{\frac{1}{a^2} + \frac{1}{b^2} + 1}}$, $z'' = \dfrac{r}{\sqrt{\frac{1}{a'^2} + \frac{1}{b'^2} + 1}}$. Leur distance sera la corde de l'angle formé par les deux droites, et elle aura pour expression $\sqrt{(x - x'')^2 + (y - y'')^2 + (z - z'')^2} = \sqrt{(z - z'')^2 + (\frac{z}{a} - \frac{z''}{a'})^2 + (\frac{z}{b} - \frac{z''}{b'})^2}$: en mettant pour z et z' leurs valeurs il viendra pour le quarré de cette expression

$$r^2 \left\{ 2 - \frac{2 \left(\frac{1}{aa'} + \frac{1}{bb'} + 1\right)}{\sqrt{\left(\frac{1}{a^2} + \frac{1}{b^2} + 1\right)\left(\frac{1}{a'^2} + \frac{1}{b'^2} + 1\right)}} \right\}$$ et par les formules de trigonometrie

on a $4 \sin^2 \frac{1}{2} A = 2 - 2 \cos A$ en prenant le rayon $= 1$ parconsequent

$$\frac{\frac{1}{aa'} + \frac{1}{bb'} + 1}{\sqrt{\left(\frac{1}{a^2} + \frac{1}{b^2} + 1\right)\left(\frac{1}{a'^2} + \frac{1}{b'^2} + 1\right)}}$$ sera le cosinus de l'angle formé par les droites don-

nées et $\dfrac{\sqrt{\left(\frac{1}{a'b} - \frac{1}{ab'}\right)^2 + \left(\frac{1}{a'} - \frac{1}{a}\right)^2 + \left(\frac{1}{b'} - \frac{1}{b}\right)^2}}{\sqrt{\left(\frac{1}{a^2} + \frac{1}{b^2} + 1\right)\left(\frac{1}{a'^2} + \frac{1}{b'^2} + 1\right)}}$ en sera le sinus.

III

Cela posé, toute courbe a double courbure tracée sur une surface cilindrique quelleconque, aura pour developpement une courbe plane faisant avec des ordonnées paralleles dans chacun de ces points des angles égaux <a ceux> que font ses elemens sur la surface cilindrique avec les arrêtes de cette surface. Cela est evident.

Les surfaces cilindriques etant formées de lignes droites paralleles entr'elles les equations de l'une quelleconque de ces droites seront $\left.\begin{array}{l} z - z' = a(x - x') \\ z - z' = b(y - y') \end{array}\right\}$ celles de la tangente de la courbe a double courbure proposée seront $\left.\begin{array}{l} z - z' = \frac{dz'}{dx'}(x - x') \\ z - z' = \frac{dz'}{dy'}(y - y') \end{array}\right\}$ on aura donc pour le cosinus de l'angle forme par ces deux lignes

$$\frac{\frac{1}{a\frac{dz'}{dx'}} + \frac{1}{b \cdot \frac{dz'}{dy'}} + 1}{\lambda \sqrt{\frac{1}{\left(\frac{dz'}{dx'}\right)^2} + \frac{1}{\left(\frac{dz'}{dy'}\right)^2} + 1}}, \text{ en faisant pour abreger } \lambda = \sqrt{\frac{1}{a^2} + \frac{1}{b^2} + 1}, \text{ ou bien}$$

---

[5]Il faudra prendre $\left.\begin{array}{l} x - x' = a(z - z') \\ y - y' = b(z - z') \end{array}\right\}$ par ce moyen on évitera les fractions.

(W) $\dfrac{\frac{1}{a}dx' + \frac{1}{b}dy' + dz'}{\lambda\sqrt{dx'^2 + dy'^2 + dz'^2}}$. Il n'est pas besoin d'avertir que $\frac{dz'}{dx'}, \frac{dz'}{dy'}$ ne sont point des differences partielles de z' mais seulement les rapports des differentielles des coordonnées prises dans les equations de projection de la courbe a double courbure proposée. D'ailleurs lorsque nous aurons a parler de differences partielles nous ferons $dz = p\,dx + q\,dy$, p et q exprimeront alors les coefficiens differentiels du premier ordre.

Si on designe par u et v des coordonées planes et rectangulaires le cosinus de l'angle formé par une courbe et ses ordonnées a pour expression $\frac{du}{\sqrt{du^2+dv^2}}$, et l arc de cette courbe est represente par $\sqrt{dv^2 + du^2}$ on aura donc les deux equations

$$\left.\begin{array}{c}\sqrt{dx'^2 + dy'^{[2]} + dz'^2} = \sqrt{dv^2 + du^2} \\[2mm] \dfrac{\frac{1}{a}dx' + \frac{1}{b}dy' + dz'}{\sqrt{dx'^2 + dy'^2 + dz'^2}} = \dfrac{dv}{\sqrt{dv^2 + du^2}}\end{array}\right\}$$ le premier membre de la premiere se re-

duira toujours a une fonction de x' et dx' en employant les equations de projection de la courbe proposée et celui de la seconde a une fonction de x' seulement: on parviendra par l'elimination de cette variable a l'equation du developpement cherché.

Si les arrêtes du cilindre étaient perpendiculaires au plan des $x', y[']$ on aurait alors $\frac{1}{a}, \frac{1}{b} = 0$ et la seconde de nos equations se reduirait à $\frac{dz'}{\sqrt{dx'^2+dy'^2+dz'^2}} = \frac{dv}{\sqrt{du^2+dv^2}}$.

Lorsque le developpement est une ligne droite on a $\frac{dv}{\sqrt{dv^2+du^2}} = $ Const. d'ou il suit $\frac{1}{a}dx' + \frac{1}{b}dy' + dz' = $ Const. $\times \sqrt{dx'^2 + dy'^2 + dz'^2}$ equation elevée a trois variables, qui ne satisfait pas aux conditions d'integrabilité et qui appartient a toutes les courbes a double courbure tracées sur les surfaces cilindriques, dont le developpement est une ligne droite. Lorsque $\frac{1}{a}, \frac{1}{b} = 0$ le resultat precedent se change dans cet autre $\left.\begin{array}{c} dz' = \text{Const.} \times \sqrt{dx'^2 + dy'^2 + dz'^2} \\[2mm] \text{ou}\quad dx'^2 + dy[']^2 = \text{Const.} \times dz'^2 \end{array}\right\}$ qui appartient a toutes les helices tracées sur des surfaces cilindriques quelleconques.

IV

On peut resoudre la même question de la maniere suivante.

Si l'on mene un plan perpendiculaire aux arrêtes de la surface cilindrique proposée il la coupera dans une courbe dont le developpement sera une ligne droite perpendiculaire a ces arrêtes: cela est evident par soi meme. Si l'on rapporte la courbe proposée a celle-ci en prenant pour coordonnées la portion des arrêtes de la surface cilindrique comprise entre les deux courbes, et les arcs de la seconde, on aura l'équation du developpement de la courbe proposée en coordonnées rectangulaires.

Nous allons exprimer cette solution analytiquement: pour cela nous designerons par u l arc de la section perpendiculaire et par v la coordonnée prise sur le[s] arrêtes: l'equation du plan dans le quel se trouve cette section sera $z +$

$\frac{1}{dx} + \frac{1}{dy} = $ Const. $\left.\begin{array}{l} y - y' = \beta(x - x') \\ z - z' = \alpha(x - x') \end{array}\right\}$ etant les equations d'une arrête en

eliminant y' entre l'equation de la surface cilindrique et celle du plan posé plus haut on aura $z'' = f(x'')$, pour l'equation de lune des projections de la section perpendiculaire. Mais la distance d'un point quelconque de cette courbe au point de la proposée qui lui correspond dans la direction de l'arrête a pour expression $v = \sqrt{(x'' - x')^2 + (y'' - y')^2 + (z - z'')^2}$ qui devient $(x'' - x')\sqrt{1 + \alpha^2 + \beta^2}$ en mettant pour $y'' - y'$ et $z'' - z'$ leurs valeurs. Enfin si nous representons par $z' = F(x')$ l'une des equations de projection de la courbe a double [sic] proposée nous aurons entre x' et x'' considerées comme cordonnées de deux points pris sur la même arrête <l'equation suivante> $f(x'') - F(x') = \alpha(x'' - x')$: parconsequent on arrivera a celle du developpement cherché en u et v en eliminant x' et x'' entre

$\left.\begin{array}{l} v = (x'' - x')\sqrt{1 + \alpha^2 + \beta^2} \\ f(x'') - F(x') = \alpha(x'' - x') \\ du = k'dx'' \qquad\qquad \text{ou } u = k : (x''). \text{ Si la section perpendiculaire} \\ \qquad\qquad\qquad\qquad \text{aux arrêtes du cilindre est rectifiable} \end{array}\right\}$

V

La methode de larticle III sapplique egalement aux courbes tracées sur une surface conique, mais alors le resultat est presenté en coordonnées polaires et l'on peut se dispenser d'employer larc de la courbe proposée comme on va le voir. L equation generale des surfaces coniques est $\frac{z' - \gamma}{x' - \alpha} = \varphi(\frac{y' - \beta}{x' - \alpha})$ d'ou il suit qu'on peut avoir a la fois $\frac{z' - \gamma}{x' - \alpha} = $ Const et $\frac{y' - \beta}{x' - \alpha} = $ Const: on en tirera pour les equations des projections de l'arrête qui passe par le point dont les coordonnées sont $x'\ y'$ et z'

$\frac{z' - \gamma}{x' - \alpha} = \frac{z - z'}{x - x'} \quad \frac{y' - \beta}{x' - \alpha} = \frac{y - y'}{x - x'}$, ou $\left.\begin{array}{l} z - z' = \left(\frac{z' - \gamma}{x' - \alpha}\right)(x - x') \\ z - z' = \left(\frac{z' - \gamma}{y' - \beta}\right)(y - y') \end{array}\right\}$: parconse-

quent le cosinus de l'angle formé par les elemens de la courbe a double courbure proposée et l'arrête de la surface conique sera

$$\frac{\frac{x' - \alpha}{z' - \gamma} \cdot \frac{dx'}{dz'} + \frac{y' - \beta}{z' - \gamma} \cdot \frac{dy'}{dz'} + 1}{\left\{\sqrt{\left(\frac{x' - \alpha}{z' - \gamma}\right)^2 + \left(\frac{y' - \beta}{z' - \gamma}\right)^2 + 1}\right\} \cdot \sqrt{\left(\frac{dx'}{dz'}\right)^2 + \left(\frac{dy'}{dz'}\right)^2 + 1}} =$$

$$\frac{(x' - \alpha)dx' + (y' - \beta)dy' + (z' - \gamma)dz'}{\sqrt{\{(x' - \alpha)^2 + (y' - \beta)^2 + (z' - \gamma)^2\}(dx'^2 + dy'^2 + dz'^2)}} =$$

$$\frac{d.\sqrt{(x' - \alpha)^2 + (y' - \beta)^2 + (z' - \gamma)^2}}{\sqrt{dx'^2 + dy'^2 + dz'^2}}.$$

On aura en nommant v la partie de l'arrête interceptée entre le sommet du cone et la courbe proposée, u larc de cercle decris de ce point comme centre avec un rayon $= 1$, on aura dis-je les deux equations suivantes

$$\left.\begin{array}{c} v = \sqrt{(x'-\alpha)^2 + (y['] - \beta)^2 + (z'-\gamma)^2} \\[4pt] \dfrac{d.\sqrt{(x'-\alpha)^2+(y'-\beta)^2+(z'-\gamma)^2}}{\sqrt{dx'^2+dy'^2+dz'^2}} = \dfrac{dv}{\sqrt{dv^2+v^2du^2}} \end{array}\right\}$$

la premiere equation sera entierement algebrique entre v et x', le premier membre de la seconde le sera paraport a x et l'elimination conduira a une equation differentielle du premier ordre entre v et u qui sera celle du developpement cherché. On aurait pu arriver directement a l'expression du cosinus de l'angle formé par un element de la courbe cherchee et une arrête quelleconque de la surface conique en remarquant que le cosinus de langle $DNI.$(fig2) $= \frac{d(D.N)}{d(arc NI.)} =$ $\frac{d\sqrt{(x'-\alpha)^2+(y['] -\beta)^2+(z['] -\gamma)^2}}{\sqrt{dx'^2+dy'^2+dz'^2}}$. Dans le cas ou cet angle serait constant on aurait $\frac{d.\sqrt{(x'-\alpha)^2+(y'-\beta)^2+(z'-\gamma)^2}}{\sqrt{dx'^2+dy'^2+dz'^2}} = $ Const: équation elevée a trois variables qui appartient a toutes les courbes a double courbure tracées sur une surface conique [crossed out: quelleconque et dont le developpement est une ligne droite] qui font le meme [angle?] avec toutes les arrêtes. <Toutes les courbes contenues dans cette equation auraient pour developpement une spirale logarithmique.>

S il etait droit on aurait $d.\sqrt{(x'-\alpha)^2 + (y'-\beta)^2 + (z'-\gamma)^2} = 0$ ce qui fait voir que la courbe proposée serait l'intersection de la surface conique avec la sphere decrite de son sommet comme centre, et qu'elle aurait pour developpement un arc de cercle.

[crossed out: VI

Nous tirerons de ce qui vient d'etre dit une maniere d'arriver a la solution du probleme analogue a celle de l'article IV. Pour cela nous rapporterons la courbe proposée a celle que fournirait l'intersection du cône avec la sphere decrite de son sommet comme centre et dont nous venons de voir que le developpement est un cercle en prenant pour coordonnées l'arc de cette derniere et la ligne de l'article precedent.]

On peut presenter les deux equations qui contiennent la solution du Probleme sous cette forme qui peut être commode dans quelques cas

$$\left.\begin{array}{c} v = \sqrt{(x'-\alpha)^2 + (y'-\beta)^2 + (z'-\gamma)^2} \\[6pt] du = \dfrac{\sqrt{dx'^2 + dy'^2 + dz'^2 - dv^2}}{v} \end{array}\right\}$$

l'elimination de x' se fera en arc ici avec la plus grande facilité en partant de la premiere equation.

VI

<Nte au lieu des calculs preliminaires pour cette formule il suffira d'observer qu'on y peut arriver par les expressions du memoire et la [citer ?] tout de suite d'après M. Monge.>

Nous allons passer au cas des surfaces developpables en general, et nous commencerons par chercher l'equation du developpement de l'arrête de rebroussement de ces surfaces.

Soit $NN'N''N'''$&c (fig 3) cette courbe, puisqu'elle est formée par les intersections des arrêtes consecutives $PN, P'N', P''N''$&c l'angle $<(NN'N'')>$ deux quelconques de ses elemens sera supplement de celui qui font entr'elles les deux arrêtes qui leur repondent. On voie de plus que lorsqu'on developpe la surface donnée cet angle ne change pas, mais seulement les angles consecutifs qui etaient dans differens plans sont ramenés au même. Il suit de la que le rayon de courbure absolu de la courbe proposée [ne ch]ange point.

En nommant x', y' et z' les coordonnées d'un point quelconque de l arrête de rebroussement, $\left.\begin{array}{l} z - z' = \frac{dy'}{dx'}(x - x') \\ y - y' = \frac{dy'}{dx'}(y - y') \text{[sic]} \end{array}\right\}$ seront les equations des projections de sa tangente ou ce qui revient $<$au même$>$ celles des arrêtes de la surface developpable donnée; si on imagine, comme dans l'article Ier [sic; should be IIème], une sphere décrite du point dont les coordonnées sont z' x' et y' comme centre et d'un rayon $= 1$ on aura pour le point d'intersection de la tangente de l'arrête de rebroussement et de cette sphere

$$\left.\begin{array}{l} z - z' = \dfrac{[?]dz'}{\sqrt{dx'^2+dy'^2+dz'^2}} \ \ldots \ dN \\[2mm] x - x' = \dfrac{dx'}{\sqrt{dx'^2+dy'^2+dz'^2}} \ \ldots \ dM \\[2mm] y - y' = \dfrac{dy'}{\sqrt{dx'^2+dy'^2+dz'^2}} \ \ldots \ dh \end{array}\right\}$$

et pour celui de la tangente consecutive $N+dN, M+dM, h+dh$; parconsequent $\sqrt{dN^2 + dM^2 + dh^2}$ sera l'expression de la corde ou de l'arc infiniment petit compris entre les deux tangentes consecutives de l'arrête de rebroussement. En effectuant les calculs on aura $\sqrt{dN^2 + dM^2 + dh^2} =$

$$\frac{\sqrt{\{dx'd^2y' - dy'd^2x'\}^2 + \{dx'd^2z' - dz'd^2x'\}^2 + \{dy'd^2z' - dz'd^2y'\}^2}}{dx'^2 + dy'^2 + dz'^2}$$

$<n^{te}$ Je n'ai pas employé la formule de l'article I [sic; should be II] parce [que] en differenciant parapport aux quantites a', b' seulement elle se reduit a zero en y supposant en suite $a' = a, b' = b$. On sent[?] que cela doit être, puisque l'angle etant infiniment petit du premier ordre, le cosinus $= 1$, ou est à son maximum; sa differentielle premiere est nulle. Il faut alors pousser jusqu'aux secondes puissances des differentielles et il m'a paru plus simple de chercher le resultat a priori.$>$ Cette expression nous conduira aisement a celle du rayon de courbure. En effet considerons deux elemens consecutifs MM'' et MM' qui sont toujours dans un meme plan, et soient menes les rayons de courbures absolues MC et $M'C$ et decris le cercle osculateur qui se confondra avec les deux elemens de la courbe; on voit que l'angle $M'CM$ est egal a $LN'M$ formé par le cote MM' et le prolongement de $M'M''$ si l'on prend $Cm = 1$ on aura les deux secteurs semblables MCM' et mCm' qui donneront $1 : mm' :: MC : MM'$ d'ou il suit MC ou le rayon de courbure $= \frac{mm'}{MM'}$ [sic] et a cause que l'angle MCM' est infiniment petit l'expression du quarré du rayon de courbure absolu sera

$$\frac{(dx'^2 + dy'^2 + dz'^2)^3}{(dx'd^2y' - dy'd^2x')^2 + (dx'd^2z' - dz'd^2x')^2 + (dy'd^2z' - dz'd^2y')^2}$$

Nous representerons comme dans les articles précedens les coordonnées rectangulaires sur le plan du developpement par u et v et en ne prennant aucune differentielle pour constante, nous aurions pour arriver au developpement cherché les equations suivantes:

$$\left. \begin{array}{l} \dfrac{\{du\,d^2v - dv\,d^2u\}^2}{(du^2 + dv^2)^3} = \dfrac{\{dx'd^2y' - dy'd^2x'\}^2 + \{dx'd^2z' - dz'd^2x'\}^2 + \{dy'd^2z' - dz'd^2y'\}^2}{\{dx'^2 + dy'^2 + dz'^2\}^3} \\[2mm] du^2 + dv^2 = dx'^2 + dy'^2 + dz'^2 \end{array} \right\}$$

Les secondes membres de ces equations pourront toujours etre reduits a des fonctions de x' et de ses differences et par l'elimination on obtiendra un resultat en u, v et leurs differences. Il suit de ce qui precede que toutes les courbes a double courbure dont le rayon est constant, considerées comme arrête de rebroussement de surfaces developpables ont un cercle pour developpement et que leur equation est

$$\frac{\{dx'd^2y['] - dy'd^2x'\}^2 + \{dx'd^2z' - dz'd^2x'\}^2 + \{dy'd^2z' - dz'd^2y'\}^2}{(dx'^2 + dy'^2 + dz'^2)^3} = C.$$

VII

L'equation generale des surfaces developpables peut etre mise sous cette forme
$$\left. \begin{array}{l} z - \psi(q) = x\varphi(q) + y(q) \\ -\psi'(q) = x\varphi'(q) + y \end{array} \right\}.$$
C'est ainsi qu'elle resulte de l integration de l equation aux differences partielles $p = \varphi(q)$: $\psi'(q)$ et $\varphi'(q)$ representent $\frac{d\psi(q)}{dq}, \frac{d\varphi(q)}{dq}$. Si on fait dans ce systeme d'equations $q = $ const, il appartiendra a une ligne droite et si l'on suppose qu'elle passe par un point de la surface dont les coordonnees soient x' y' et z' les equations de ses projections seront
$$\left. \begin{array}{l} z - z' = \frac{p\,dq - q\,dp}{dq}(x - x') \\ y - y' = \frac{-dp}{dq}(x - x') \end{array} \right\}$$
en mettant p au lieu de $\varphi(q)$ et $\frac{dp}{dq}$ au lieu de $\varphi'(q)$.

Si le point dont les coordonnées sont accentuées est pris sur l'arrête de rebroussement l'arrête qui passera par ce point sera tangente a cette courbe les equations de ces projections seront
$$\left. \begin{array}{l} z - z' = \frac{dz'}{dx'}(x - x') \\ y - y' = \frac{dy'}{dx'}(x - x') \end{array} \right\}$$
d'ou il suit
$$\left. \begin{array}{l} \frac{p\,dq - q\,dp}{dq} = \frac{dz'}{dx'} \\ \frac{-dp}{dq} = \frac{dy'}{dx'} \end{array} \right\}$$
<bien entendu qu'il faudra accentuer les variables dans p, q, puisque ces formules supposent tacitement que le point est sur l'arrête de rebroussement.> L'une quelleconque de ces deux equations jointe a celle de la surface developpable donnée fera connaitre l arrête de rebroussement. On les deduirait l'une de l'autre en employant $dz = p\,dx + q\,dy$.

On peut encore arriver a ce resultat d'une autre maniere. Si l'on prend l'equation du plan tangent a la surface proposée $z - z' = p(x - x') + q(y - y')$

en la differentiant deux fois de suite parapport a x' y' et z' en observant que

$-dz' = -p\,dx' - q\,dy'$ il viendra $\begin{cases} 0 = y - y' + \frac{dp}{dq}(x - x') \\ 0 = -dy' - \frac{dp}{dq}dx' + d\left(\frac{dp}{dq}\right)(x - x') \end{cases}$ d'ou on

tire $\begin{cases} x - x' = \frac{dy'dq + dx'dp}{dq\,d\left(\frac{dp}{dq}\right)} \\ y - y' = -\frac{dp}{dq}\frac{\{dy'dq + dx'dp\}}{dq.d\left(\frac{dp}{dq}\right)} \\ z - z' = \left(\frac{p\,dq - q\,dp}{dq}\right)\left\{\frac{dy'dq + dx'dp}{dq.d\left(\frac{dp}{dq}\right)}\right\} \end{cases}$ <n^{tes} on observera qu'en regardant

les arretes de la surface comme couchées[?] dans toute leur etendue[?] sur le plan tangent et sur la surface on aura $d^2z = 0$ ou $dp\,dx + dq\,dy = 0$ en prenant dy et dx constantes.>

Mais si le point d intersection des trois plans tangents consecutifs est pris sur la surface courbe, ou ce qui revient au meme si les coordonnées z, x, y sont les memes que z' x' y' on aura alors $dy'dq + dx'dp = 0$. Cette supposition ne peut avoir lieu que pour l'arrête de rebroussement de la surface developpable proposée; on a <donc> pour cette courbe $\frac{dy'}{dx'} = -\frac{dp}{dq}$ comme on l'a vu plus haut.

VIII

Quand on a l'arrête de rebroussement d'une surface developpable, et le developpement de cette courbe, il est facile d'obtenir l'equation du developpement d'une courbe quelleconque tracée sur cette surface. Il ne faut pour cela que rapporter l'une de ces courbes a l'autre, en coordonnées qui ne soient pas susceptibles de changer de valeur dans le passage de la surface courbe au plan. Soient donc x', y', z' les coordonnées de la courbe proposée, x'', y'', z'' celle de l'arrête de rebroussement

$\left.\begin{array}{l} z - z'' = \frac{dz''}{dx''}(x - x'') \\ y - y'' = \frac{dy''}{dx''}(x - x'') \end{array}\right\}$ seront les equations des projections de sa tangente ou

de l arrête de la surface qui passe par le point dont les coordonnées sont x'', y'' et z'': representons encore par $y'' = f(x'')$ et $y = F(x)$ les equations des projections sur le plan des x, y de l'arrête de rebroussement et de la courbe proposée. On aura pour le point de cette derniere qui se trouve sur le prolongement de l'arrête $y' - y'' = \frac{dy''}{dx'}(x' - x'')$ ou

$$F(x') - f(x'') = \frac{dy''}{dx''}(x' - x'') \qquad (1):$$

de plus v etant la partie de l arrête interceptée entre la courbe proposée et l'arrête de rebroussement, on aura $v = \sqrt{(x' - x'')^2 + (y' - y'')^2 + (z' - z'')^2}$ ou en chassant $(y - y')$ et $(z - z'')$

$$v = (x' - x'')\sqrt{1 + \frac{dy''^2}{dx''^2} + \frac{dz''^2}{dx''^2}} \qquad (2)$$

enfin designant par du l'arc de l'arrête de rebroussement on aura

$$du = \sqrt{dx''^2 + dy''^2 + dz''^2} \qquad (3).$$

En employant les projections de l'arrête de rebroussement on reduira les equations (1) (2) et (3) a ne renfermer que les variables x'', x', u et v et en eliminant les deux premieres on aura un resultat exprimé par les deux dernieres, qui sera l'equation du developpement cherché. L equation qu'on obtiendra se construira en prenant sur la tangente MM' <(fig 5)> de l'arrête de rebroussement developpée MX une partie $MM' = 0$ et le point M' appartiendra au développement cherche: il serait d'ailleurs très aisé de changer les coordonnées u et v en coordonnées rectangulaires, et nous aurons occasion de le faire dans la suite.

IX

On pourrait demander d'arriver a l'equation du developpement d'une courbe a double courbure tracée sur une surface developpable sans employer l'arrête de rebroussement de cette surface. On y parviendra en cherchant l'expression de l'angle $MM'A$ (fig 6) formé par une [sic] element de la courbe a double courbure et par l'arrete correspondante AB. En faisant varier les quantites relatives a la courbe a double courbure seulement on aura l'angle $AM'I$. formé par le prolongement de l'element consecutif de la courbe et la même arrête; la difference $MM'I$. de ces deux angles, qui ne changent point lorsqu'on les etend sur un meme plan, se trouve <etre> alors l'angle des deux tangentes consecutives du developpement cherché.

Pour mettre cette solution en calcul nous rappellerons ici les formules de l'article II et de l'article VII. Le cosinus de l'angle $MM'A$ a pour expression $\dfrac{\frac{1}{aa'} + \frac{1}{bb'} + 1}{\sqrt{(\frac{1}{a'^2} + \frac{1}{b'^2} + 1)(\frac{1}{a^2} + \frac{1}{b^2} + 1)}}$ et les equations des droites MM' et AB sont

$$\left. \begin{array}{l} z - z' = \frac{dz'}{dx'}(x - x') \\ y - y' = \frac{dy'}{dx'}(x - x') \end{array} \right\} \quad \left. \begin{array}{l} z - z' = (\frac{p\,dq - q\,dp}{dq})(x - x') \\ y - y' = (\frac{q\,dp - p\,dq}{dp})(x - x') \end{array} \right\} \text{ v art VII.}$$

Nous ferons pour abreger $p\,dq - q\,dp = dn$ et la formule du cosinus se changera dans la suivante

$$\frac{\frac{dq}{dn}\frac{dx'}{dz['] } - \frac{dp}{dn}\frac{dy'}{dz'} + 1}{\sqrt{(\frac{dq^2}{dn^2} + \frac{dp^2}{dn^2} + 1)(\frac{dx'^2}{dz'^2} + \frac{dy'^2}{dz'^2} + 1)}} = \frac{dq\,dx' - dp\,dy' + dz'\,dn}{\sqrt{(dq^2 + dp^2 + dn^2)} \times \sqrt{(dx'^2 + dy'^2 + dz'^2)}}$$

Cette formule etant differentie en regardant dp, dq, dn comme constans ainsi que le comporte l'etat de la question, et faisant $\sqrt{dx'^2 + dy'^2 + dz'^2} = ds$ on aura

$$\frac{1}{\sqrt{dq^2 + dp^2 + dn^2}} = \left\{ \frac{ds[dq\,d^2x' - dp\,d^2y' + dn\,d^2z] - d^2s[dq\,dx' - dp\,dy' + dn\,dz]}{ds^2} \right\} :$$

nous passerons ensuite de la differentielle du cosinus a celle de l'arc en prenant la premiere avec un signe contraire et divisant par le sinus, dont l expression donnée a la fin de l article II se change par les substitutions convenables en

$$\frac{\sqrt{(dx'\,dp + dy'\,dq)^2 + (dx'\,dn - dz'\,dq)^2 + (dy'\,dn + dz'\,dp)^2}}{\sqrt{dq^2 + dp^2 + dn^2} \times \sqrt{dx'^2 + dy'^2 + dz'^2}},$$

et il viendra pour la differentielle de l'arc

$$\frac{\{dq\,dx' - dp\,dy' + dn\,dz'\}d^2s - \{dq\,d^2x' - dp\,d^2y' + dn\,d^2z'\}ds}{ds\sqrt{(dx'dp + dy'dq)^2 + (dx'dn - dz'dq)^2 + (dy'dn + dz'dp)^2}} \qquad (dW)$$

En employant les equations de la courbe proposée cette formule se reduira a une fonction de x' seulement car on voit qu il faudra mettre dans dp, dq et dn au lieu de y' et z' leur valeur en x' tiree de ces equations, pour que l'arrête que l on considere soit celle qui passe par le point pris sur la courbe proposée.

Si l'on met au lieu de ds et d^2s leurs valeurs $\sqrt{dx'^2 + dy'^2 + dz'^2}$, $\frac{dx'd^2x['] + dy'd^2y' + dz'd^2z'}{\sqrt{dx'^2 + dy'^2 + dz'^2}}$ on aura après les reductions

$$\frac{(dx'dp + dy'dq)\{dy'd^2x' - dx'd^2y'\} + (dx'dn - dz'dq)\{dx'd^2z' - dz'd^2x'\} + (dy'dn + dz'dp)\{dy'd^2z' - dz'd^2y'\}}{(dx'^2 + dy'^2 + dz'^2)\sqrt{(dx'dp + dy'dq)^2 + (dx'dn - dz'dq)^2 + (dy'dn + dz'dp)^2}}$$
$$(dW')$$

Dans le cas ou la courbe proposée serait elle même l arrête de rebroussement de la surface developpable a cause de $\frac{dz'}{dx'} = \frac{dn}{dq}$, $\frac{dz'}{dy'} = -\frac{dn}{dp}$, $\frac{dx'}{dy'} = -\frac{dq}{dp}$ (art VII) la formule precedente se reduit a $\frac{0}{0}$: et cela doit avoir lieu necessairement comme dans la remarque de l'article VI, puisqu alors la ligne MM' tombe sur la ligne AB et qu'on a le $\cos MM'A = 1$ sa differentielle premiere $= 0$ et le sinus du meme angle $= 0$. Nous reprendrons l'expression <de la differentielle> du cosinus de l angle $MM'A$ trouvée plus haut et après y avoir mis pour ds et d^2s leurs valeurs nous la differencions en regardant dp, dq, dn et les differences secondes comme constantes a fin d'arriver au second terme de la difference generale des cosinus lequel sera en faisant abstraction des formes qui s'evanouissent

$$\frac{1}{1\cdot2}\left\{ \frac{(dp\,d^2x' + dq\,d^2y')(dy'd^2x' - dx'd^2y') + (dn\,d^2x' - dq\,d^2z')(dx'd^2z' - dz'd^2x')}{\left(\sqrt{dp^2 + dq^2 + dn^2}\right)\cdot(dx'^2 + dy'^2 + dz'^2)^{\frac{3}{2}}} + \right.$$
$$\left. + \frac{(dn\,d^2y' - dp\,d^2z')(dy'd^2z' - dz'd^2y')}{\left(\sqrt{dp^2 + dq^2 + dn^2}\right)\cdot(dx'^2 + dy'^2 + dz'^2)^{\frac{3}{2}}} \right\}6$$

et en faisant les substitutions relatives au cas proposé on a

$$\frac{(dy'd^2x' - dx'd^2y')^2 + (dx'd^2z' - dz'd^2x')^2 + (dy'd^2z' - dz'd^2y')^2}{1\cdot2(dx'^2 + dy'^2 + dz'^2)^2}: \quad \text{mais cette}$$

quantite est le sinus verse du petit angle cherche, sa corde ou l'arc qui le mesure etant moyenne proportionnelle entre le diametre et la quantite precedente, on sera conduit par ces considerations au resultat de l'article (VI).

Pour achever la solution de la question generale qui nous occupe dans ce moment nous [crossed out: prendrons] egalerons $\left\{ \frac{du\,d^2v - dv\,d^2u}{dv^2 + du^2} \right\}$, expression de l'angle formé par deux tangentes consecutives de la courbe plane dont les coordonnées sont u et v, et nous l'egalerons a (dW). Il ne faudra plus qu'eliminer

---

[6]The expression within brackets appears as a single fraction in the manuscript.

x' entre les deux equations suivantes

$$\frac{du\,d^2v - dv\,d^2u}{du^2 + dv^2} = (dW)$$

$$\sqrt{du^2 + dv^2} = \sqrt{dx'^2 + dy'^2 + dz'^2}$$

pour arriver a l'equation du developpement cherché. On pourrait mettre la premiere des equations ci dessus sous cette forme

$$\frac{du\,d^2v - dv\,d^2u}{(du^2 + dv^2)^{\frac{3}{2}}} = \frac{(dW)}{\sqrt{dx'^2 + dy'^2 + dz'^2}}$$

alors son premier membre pourrait toujours etre ramene a une fonction de x' seul et sans differentielles.

Si la courbe proposée avait pour developpement une ligne droite, on aurait alors $du\,d^2v - dv\,d^2u = 0$ et par consequent le numerateur de la quantite (dW) doit etre nul ce qui donnera toujours une equation <elevée> du second ordre a trois variables qui appartiendra a toutes les courbes a doubles courbures tracées sur une <famille de> surfaces developpables et qui deviennent une ligne droite lorsque cette surface est etendue sur un plan. <Il faudra eliminer p et q ainsi que leurs differentielles au moyen de l'equation differentielle partielle de la surface proposée et de $dz = p\,dx + q\,dy$.>

<n^{te} Si l'on chasse p dans le numerateur de l'expression qui est au bas de la page 9 [the expression marked (dW)] a l'aide de $dz = p\,dx + q\,dy$ et que l'on fasse $\sqrt{dx^2 + dy^2} = ds' = const$ l'equation $dW = 0$ se change en cette autre $(ds'^2 + dz^2)d^2y = (dy\,dz - q\,ds^2)d^2z$ qui appartient a la courbe que forme un fil plie librement[?] sur une surface. Elle est la plus courte de toutes celles qu'on peut mener entre ses extremites. Elle a ete donnée sous ce dernier point de vue par J. Bernoulli dans le tome IV de ses oeuvres et ensuite sous l autre par M. Monge dans le tome X des Savans etrangers.>

X

Nous pouvons a l'aide des formules precedentes resoudre les differentes questions relatives au developpement des surfaces courbes et de leurs parties. En effet le cas le plus general est celui du developpement d'une portion de surface developpable terminée de toutes parts par des courbes a double courbure données. On arrivera a la solution en rapportant ces courbes a l'arrête de rebroussement de la surface proposée, cette derniere etant developpée, il sera facile de construire le developpement des autres et l'espace compris entre les nouvelles courbes qui en resulteront sera lui même le developpement cherché.

Nous allons parcourir succinctement quelques cas particuliers qui offrent des facilités.

1.° Les surfaces cilindriques se developperont ainsi que les courbes tracées sur elles de la maniere la plus facile en employant la methode de l'article IV, c est <a

dire> en rapportant les courbes proposées a la section perpendiculaire aux arrêtes dont le developpement est une ligne droite.

On pourrait encore employer dans la question qui nous occupe, la courbe qui sert de base a la surface proposée sur l'un quelconque des plans coordonnées, celui des x, y par exemple, les equations qui terminent l'article III deviennent

$$\left. \begin{array}{c} \dfrac{\frac{1}{a}dx' + \frac{1}{b}dy'}{\lambda\sqrt{dx'^2 + dy'^2}} = \dfrac{dv}{\sqrt{du^2 + dv^2}} \\ \sqrt{dx^2 + dy^2} = \sqrt{du^2 + dv^2} \end{array} \right\}$$

et lorsqu'on aura le developpement de cette courbe, il sera très aise d'y rapporter toutes celles qu'on pourra proposer sur les surfaces cilindriques, en prenant pour coordonnées ses arcs, et les arrêtes de ces surfaces.

2.º Pour les surfaces coniques l'arrête de rebroussement se reduit a un point; mais toutes les courbes tracées sur ces surfaces, pourront etre rapportées aux mêmes coordonnées polaires comme on l'a indique dans l'article V. À l'egard de leurs bases on aura les equations necessaires pour arriver a son developpement en effaçant tous les termes affectes de z' dans les equations qui terminent l'article V.

3.º Nous n'ajouterons rien à ce qui a eté dit au commencement de cet article, sur les surfaces developpables en general relativement a l'arrête de rebroussement, nous nous bornerons a observer que dans le cas où l'on voudrait developper directement la courbe qui sert de base a ces surfaces sur l'un quelconque des plans coordonnées, celui des x, y par exemple il faudra faire z' et $dz' = 0$ dans les equations qu'on a trouvées dans l'article precedent. En operant la meme reduction dans l'expression du cosinus de l'angle formé par une arrête et par l'element de la courbe proposée, elle conviendra a l'angle formé par la tangente a chaque point du developpement, et l'arrête de la surface qui passe par ce point – cette formule pouvant toujours etre ramenée a une fonction de x' et parconsequent a ne renfermer que les coordonnées u et v du developpement, il sera donc aisé de mener pour chaque point de la base developpée l'arrête correspondante: si l'on rapporte ensuite aux arcs de cette base et aux arrêtes de la surface proposée toutes les courbes qui seront tracées sur elle on en aura les developpemens avec facilite.

XI

Nous avons donné dans les articles precedens les moyens de trouver pour tous les cas ce que devient une courbe tracée sur une surface developpable. En partant des mêmes formules on pourra resoudre la question inverse: celle de trouver ce que devient une courbe tracée sur un plan qu'on enveloppe autour d'une surface developpable.

1.º Pour les surfaces cilindriques il ne faudra qu'eliminer u et v entre entre [sic] l'equation de la courbe plane et les equations de l'article III, on aura pour resultat une equation en x', y', z' et leurs differentielles, qui sera celle de la courbe cherchée. Ce que je viens de dire suppose que la courbe plane soit rapportee a des coordonnées rectangulaires, dont les arrêtes de la surface cilindrique fassent partie, et cela est toujours possible.

En employant la section perpendiculaire aux arrêtes du cilindre on aura a eliminer x'', u et v entre les équations de l article IV et l'equation de la proposee.

2.º Pour les surfaces coniques, en joignant aux equations de l'article V celle de la courbe plane rapportee aux coordonnées polaires u et v, et eliminant entre elles trois ces quantites, on aura pour resultat une equation qui exprimera conjointement avec celle de la surface conique, la nature de la courbe cherchee.

3.º La question se reduit aux mêmes termes ainsi que la solution pour les surfaces developpables en general, en employant les equations de l'article IX. Si l'on voulait faire usage de l'arrête de rebroussement il faudrait alors avoir recours aux equations de l article VIII; mais elles supposent que la courbe plane soit rapportée aux arcs et aux tangentes du developpement de l'arrête de rebroussement. Cette transformation, quoique sans difficulte, n'etant par très ordinaire, nous allons en donner ice les formules.

Supposons, comme cela est toujours possible, qu'on ait les equations de la courbe plane proposée et celle du developpement de l'arrête de rebroussement en coordonnées rectangulaires qui leur soient communes, soient $\left.\begin{array}{l} Y' = F(X') \\ Y'' = f(X'') \end{array}\right\}$ ce [sic] deux expressions. C'elle [sic] de la tangente a la premiere courbe sera $Y - Y' = \frac{dY'}{dX'}(X - X')$, et la partie de cette droite interceptee entre les deux courbes aura pour expression

$$v = (X'' - X')\sqrt{1 + \frac{dY'^2}{dx'^2}} \qquad (1)$$

mais parce que le point de la seconde courbe qu'on considere doit se trouver sur la tangente de la premiere on aura necessairement

$$f(X'') - F(X') = \frac{dY'}{dX'}(X - X') \quad (2)$$

enfin l'arc de la premiere est

$$du = \sqrt{dY'^2 + dX'^2} \qquad (3).$$

Ces trois equations pourront etre reduites a ne renfermer que u, v, X' et X'', on en pourra déduire par l'elimination un resultat en u et v qui sera l'equation cherchée.

[crossed out: XII.

Une courbe plane etant regardée comme le developpement de l'arrête de rebroussement d'une surface developpable, on peut demander l'equation de cette surface. Le probleme est indeterminé et la même courbe appartient a une infinité de surfaces mais il est toujours possible d'arriver a l equation aux differences partielles du premier ordre que les represente.]

XII
Remarques sur les equations aux differences ordinaires a trois variables

La differentielle de toute fonction a trois [sic] variables etant representee par $dz = p\,dx + q\,dy$, il peut arriver que les coefficiens p et q soient donnés a priori en fonction de x, y et z ou qu'on [crossed out: ait entre eux] des relations entr'eux et ces variables. Le premier cas appartient aux differences ordinaires et le second aux differences partielles.

Si l'on a deux equations entre p, q et les variables x, y, z on en pourra tirer une equation aux differences ordinaires a trois variables, la quelle appartiendra a une surface courbe lorsqu'elle satisfera a l'equation de condition et a une infinite de courbes a double courbure si elle n'y satisfait pas. Les considerations geometriques rendent bien evident ces faits deja connues par l'analyse.

On sait que les equations aux differences partielles peuvent etre rapportees a la generation des surfaces courbes et expriment des proprietes qui appartiennent a toutes celles d'une meme famille. Lors donc qu'on regardera p, q, x, y et z comme des quantités communes entre deux equations aux differences partielles du premier ordre, ou ce qui revient au même lorsqu'on supposera que les surfaces courbes auxquelles elles appartiennent ont le même plan tangent, s'il existe une surface courbe qui jouisse a la fois des proprietes caracteristiques des deux familles, elle sera le lieu de l'equation resultante puis qu'elle aura dans toute son etendue le même plan tangent. Mais on sent qu'il y a telles generations de surfaces courbes ou telles proprietes que ne sauraient avoir lieu simultanement: alors tous les points qui satisfont a la question ne sont pas lies entr'eux par la loi de continuité, mais ils ont cela de remarquable qu'ils appartiennent a l'assemblage des courbes de contact des surfaces proposées.

Nous allons verifier ces faits par des exemples.

1.° Soient les deux equations aux differences partielles $\left.\begin{aligned} p\,y - q\,x = 0 \\ p\,(x-a) + q\,(y-b) = z - c \end{aligned}\right\}$.
En eliminant p, et q entre ces deux equations et $dz = p\,dx + q\,dy$ on a pour resultat $[x(x-a) + y(y-b)]dz = (z-c)\{x\,dx + y\,dy\}$; et l'equation de condition devient $(z-c)\{ay - bx\} = 0$ qui ne saurait etre identique a moins qu'on n'ait ou $z = c$ ou a, et $b = 0$. La premiere solution appartient au plan paralle[le] a celui des x, y et il n existe pas d autre surface qui jouisse <a la fois> des propriétés exprimées par les deux equations aux differences partielles, puisque l'une appartient aux cônes qui ont leur sommet au point dont les coordonnées sont a, b, c et l'autre aux surfaces de revolution dont l'axe coïncide avec celui des z. Mais lorsque a et b sont nuls, le resultat aux differences ordinaires devient $\frac{dz}{z-c} = \frac{x\,dx + y\,dy}{x^2 + y^2}$ dont l'integrale $z - c = k\sqrt{x^2 + y^2}$ appartient au cone droit qui a son sommet dans l axe des z. M. Monge dans les Memoires de l'academie pour 1784 a traité cette equation qui appartient, lorsque a et b se sont pas nuls, a toutes les courbes formées par les contacts des surfaces de la famille des cones avec celles de revolution, et qui ne sont pas liees entr'elles par loi de continuite, tandis qu'en supposant le sommet de ces cones dans l'axe des z, toutes les courbes de contact se trouvent assujetties

<a cette loi> puisque leur assemblage constitue la surface du cône droit que nous venons de trouver.

2.° Soient encore proposées les deux equations $\left.\begin{array}{l} p\,x + q\,y = 0 \\ 1 + p^2 + q^2 = \frac{a^2}{z^2} \end{array}\right\}$ qu'on suppose appartenir a la même surface courbe. Si <on> met dans $dz = p\,dx + q\,dy$ les valeurs de p et de q tirées de ces equations on aura pour resultat $\frac{z\,dz}{\sqrt{a^2 - z^2}} = x\,dy - y\,dx$; l'equation de condition n'est pas identique, elle devient $z\sqrt{a^2 - z^2}[(x^2 + y^2) - z(x^2 + y^2)] = 0$. La question que nous traitons ne saurait appartenir a d'autre surface qu'au plan paralle[le] a celui des x, y et pour lequel on $a^2 - z^2 = 0$. En effet il s'agit de trouver la surface qui jouit a la fois des deux proprietées suivantes, 1° d etre formée de lignes droites paralleles au plan des x, y et assujetties a passer toujours par l'axe des z; 2° d'avoir toutes les normales constantes parapport a ce plan.

3.° Nous prendrons pour dernier exemple de ce genre les equations $\left.\begin{array}{l} p\,y - q\,x = 0 \\ 1 + p^2 + q^2 = a^2 \end{array}\right\}$ En operant comme precedemment on obtient $\frac{dz}{a^2 - 1} = \frac{x\,dx + y\,dy}{\sqrt{x^2 + y^2}}$, qui a pour integrale $\frac{z - c}{\sqrt{a^2 - 1}} = \sqrt{x^2 + y^2}$; équation qui appartient au cone droit dont l'axe coïncide avec celui des z et qui est seule surface comprise a la fois dans la famille des surfaces de revolution ayant pour axe celui des z, et dans celle dont l'aire d'une partie quelleconque est dans un rapport constant avec sa projection.

XIII

Lorsque l'equation resultante de l'elimination des coefficiens differentiels, ne peut appartenir a une surface courbe, ou que son integrale ne peut pas etre exprimée par une seule equation finie a trois variables, on sait qu'elle represente une infinite de courbes a doubles courbures qui ont toutes une propriete commune; si l'on se donne a volonté une relation entre deux quelconques des variables, ou même entre les trois et qu'on l'employe pour simplifier la proposée; il arrivera, ou qu'on aura une equation qui tombant sur des quantites constantes fera voir qu'il y a impossibilité de satisfaire a la question par la relation qu'on a choisie a moins que des conditions particulieres ne soient remplies; ou bien cette equation etant a deux variables aura une integrale transcendante et le plus souvent <echapera> aux methodes connues. Ce procédé d'ailleurs ne conduit qu'a une seule solution et il faut ainsi les chercher l'une après l'autre sans appercevoir d'autre liaison entrelles que l'equation differentielle proposée.

Le point de vue sous lequel nous avons envisagé les equations a trois variables mene a des solutions qui reunissent a la plus grande generalite l'avantage de renfermer dans deux equations toutes les solutions algebriques que peuvent avoir les proposées. C'est M. Monge qui le premier les a presentees dans les memoires de l'academie année 1784.

Lorsqu'on regarde ces equations comme appartenant a des courbes qui soient[?] le lieu de tous les contacts qui peuvent exister entre deux familles de surfaces courbes, cette consideration fait disparaître les differentielles et donne le moyen de satisfaire <a la question> en prenant des fonctions algebriques, sans etre assujetti a de nouvelles integrations lorsqu'on veut passer d'une solution a une autre.

La methode se presente d'elle même, il faut integrer l'une quelleconque des equations aux differences partielles qui representent la proposée et assujettir le resultat a satisfaire a l'autre. C'est ainsi qu'on aura pour le 1er exemple

$$\left. \begin{array}{l} z = \varphi(x^2 + y^2) \\ 2\varphi'\{x^2 + y^2\} \cdot \{x(x-a) + y(y-b)\} = \varphi(x^2 + y^2) - c \end{array} \right\}$$

(Jai cru devoir designer ces systemes d equations sous le nom d'ensemble des solutions de la proposée)

Pour le 2.^{e}

$$\left. \begin{array}{l} z = \varphi(\frac{x}{y}) \\ 1 + \varphi^4(\frac{x}{y})\left\{\frac{1}{y^2} + \frac{1}{y^4}\right\} = \frac{a^2}{\varphi^2(\frac{x}{y})} \end{array} \right\}$$

Pour le 3e on aurait

$$\left. \begin{array}{l} z = \varphi(x^2 + y^2) \\ 1 + 4\varphi'^2(x^2 + y^2) \cdot \{x^2 + y^2\} = a^2 \end{array} \right\}$$

d'ou on tirerait $\varphi'(x^2 + y^2) = \frac{\sqrt{a^2-1}}{2(x^2+y^2)^{\frac{1}{2}}}$ et multipliant les deux membres par $2x\,dx + 2y\,dy$ il en resulterait $\varphi(x^2 + y^2) = (\sqrt{a^2-1})\sqrt{x^2+y^2}$, resultat qui s'accorde avec celui de l'article precedent.

Ces equations sont aussi generales que les equations differentielles qu'elles representent puisqu'elles n'en sont que des transformations et qu'on reviendra au dernieres en eliminant la fonction arbitraire introduite par l integration aux differences partielles. D'ailleurs si on voulait d[et]erminer cette fonction arbitraire en se donnant[?] une relation telle que $y = \int P dx$ on retomberait encore dans la proposée. On voit encore qu'en prenant pour φ – une fonction algebrique on arrivera toujours a un resultat algebrique. <Lorsqu'on determine la forme de φ, alors on considere seulement une des surfaces de la premiere famille, et il s'en suit la [?] determination de celle de la seconde qui touche l'autre dans une des courbes a double courbure cherchee. On voit par la que les surfaces sont liees deux a deux par l'equation differentielle proposée.>

J'ai donc cru devoir presenter cette nouvelle question sur les equations a trois ou un plus grand nombre de variables qui ne satisfont pas aux equations de condition, *"trouver parmi le nombre infini de solutions dont elles sont susceptibles celles qui sont algebriques"*: et si nous nous bornons a trois variables *"trouver autant de courbes algebriques qu'on voudra qui satisfassent au probleme proposé"*. On apperçoit ici une analogie entre cette partie du calcul integral et l'analyse algebrique indeterminée, où on limite le nombre des solutions en exigeant qu'elles soient en nombres entiers.

XIV

Je vais exposer ici quelques remarques qui pourront conduire a la solution des problêmes que je viens d'indiquer dans beaucoup de cas.

Prenons l'equation $M\,dz + P\,dx + Q\,dy = 0$; si on y substitue pour dz, $p\,dx + q\,dy$ elle pourra etre representée par les deux equations suivantes qui sont aux differences partielles $\left.\begin{array}{l} Mp + P = 0 \\ Mq + Q = 0 \end{array}\right\}$ et si l'on elimine M entre les deux dernieres on aura $Pq - Qp = 0$. (M. Monge a donné ces equations dans les Mem. de l'academie pour 1784.) Si l'on integre l'une quelleconque d'entre'elles et qu'on assujetisse le resultat a satisfaire a l'une des des [sic] deux autres on aura l'ensemble des solutions de la proposée, qui sera sous une forme algebrique si l equation qui aura de integrer a pu l'etre algebriquement. Mais l'integration des equations aux differences partielles que nous venons de poser depend par le theorême de M. Delagrange de celle des equations a deux variables $\left.\begin{array}{l} M\,dz + P\,dx = 0 \\ M\,dz + Q\,dy = 0 \\ P\,dx + Q\,dy = 0 \end{array}\right\}$. Si l'une d'entr'elles a une integrale algebrique on determinera l'ensemble des solutions de la proposée sous une forme algebrique.

Au reste je ne crois pas qu'on puisse conclure de cequ'aucune des equations precedentes n'aurait une integrale algebrique, que la proposé ne saurait avoir de solutions algebriques, car le systeme d'equations que nous avons employé pour la representer n'est pas d'une forme necessaire; on peut prendre a sa place deux autres equations aux differences partielles, telles qu'etant combinées avec $dz = p\,dx + q\,dy$ elles produisent la proposée par l'elimination de p et de q.

Il peut arriver qu'en eliminant entre $\left.\begin{array}{l} Mp + P = 0 \\ Mq + Q = 0 \end{array}\right\}$ quelque fonction commune on obtienne un resultat integrable algebriquement.

Si la proposée ne renferme pas de radicaux, on peut prendre pour la representer deux equations lineaires aux differences partielles, avec des coefficiens indeterminées telles que $\left.\begin{array}{l} Kp + Gq + h = 0 \\ K'p + G'q + h' = 0 \end{array}\right\}$ desquelles eliminant p et q conjointement avec $dz = p\,dx + q\,dy$ on obtiendra un resultat qu'on comparera avec la proposée mise sous la forme suivante $dz = -\frac{P}{M}dx - \frac{Q}{M}dy$. Il viendra deux equations et on pourra essayer de determiner les coefficiens qui resteront arbitraires pour que $\left.\begin{array}{l} K\,dz + h\,dx \\ K\,dy - G\,dx \end{array}\right\}$ ou $\left.\begin{array}{l} K'dz + h'dx \\ K'dy - G'dx \end{array}\right\}$ soient integrables algebriquement puisque c'est a ces equations que se reduit l'integration de celles qu'on vient de poser aux differences partielles.

XV

Les equations elevées a trois variables qu'on peut mettre sous une forme lineaire relativement aux differentielles appartiennent aux courbes de contact des familles de surfaces courbes. En effet si on a deux equations algebriques entre

p, q, x, y et z et qu'on en tire les valeurs de p et de q pour les substituer ensuite dans $dz = p\,dx + q\,dy$ on aura un resultat qui pourra se presenter sous une forme elevee <en raison de> l evanouissement des radicaux, mais qui sera toujours susceptible d'etre remis sous une forme lineaire relativement aux differentielles. L'equation $z^2\{(y\,dx - x\,dy)^2 + (y\,dz - z\,dy)^2 + (z\,dx - x\,dz)^2\}^2 - a^2\{y\,dx - x\,dy\}^2 = 0$ est dans ce cas. En la resolvant paraport a dz on aura

$$dz = \frac{z(y\,dy + x\,dx)}{y^2 + x^2} \pm \left\{ \frac{y\,dx - x\,dy}{z(y^2 + x^2)} \right\} \sqrt{-z^4 + (a^2 - z^2)(y^2 + x^2)} \text{ et parconse-}$$

quent $-\frac{P}{M} = \frac{zx}{x^2+y^2} \pm \frac{\frac{y}{z}\sqrt{-z^4+(a^2-z^2)(y^2+x^2)}}{y^2+x^2}$, $-\frac{Q}{M} = \frac{zy}{x^2+y^2} \pm \frac{\frac{x}{z}\sqrt{-z^4+(a^2-z^2)(y^2+x^2)}}{y^2+x^2}$.

Si nous mettons au lieu de $-\frac{P}{M}, -\frac{Q}{M}$ les coefficiens differentiels p et q nous aurons pour representer la proposée des equations analogues a celles de l'article précedent. Elles se presentent sous une forme très compliquée et qui probablement echapperait aux methodes, mais si on elimine le radical on arrivera cette equation tres simple $z = px + qy$ qui appartient aux surfaces coniques dont le sommet est a l'origine. Si on assujettie l'integrale de cette derniere a satisfaire a l'une des equations en p ou en q on aura l ensemble des solutions de la proposée sous la forme suivante:

$$\left. \begin{array}{l} z = y\varphi\!\left(\frac{y}{x}\right) \\[4pt] \varphi'\!\left(\frac{y}{x}\right) = \frac{zx}{x^2+y^2} \pm \frac{\frac{y}{z}\sqrt{-z^4+(a^2-z^2)(x^2+y^2)}}{x^2+y^2} \end{array} \right\} \text{ mais on peut obtenir un resultat sans}$$

radicaux en ajoutant ensemble les valeurs de p^2 et de q^2 prises dans les equations primitives car[?] on a alors $p^2 + q^2 = \frac{a^2 - z^2}{z^2}$. L'equation proposée peut donc etre regardée comme appartenant a toutes les courbes de contact des deux familles de surfaces courbes representees par $\left. \begin{array}{l} px + qy = z \\[2pt] 1 + p^2 + q^2 = \frac{a^2}{z^2} \end{array} \right\}$; ce qui donne pour

l'ensemble de ses solutions $\left. \begin{array}{l} z = y\varphi\!\left(\frac{x}{y}\right) \\[4pt] 1 + \varphi'^2\!\left(\frac{x}{y}\right) + \left\{ \varphi\!\left(\frac{x}{y}\right) - \frac{x}{y}\varphi'^2\!\left(\frac{x}{y}\right) \right\}^2 = \frac{a^2}{y^2\varphi^2\left(\frac{x}{y}\right)} \end{array} \right\}$. En

prenant pour φ des fonctions algebriques on aura autant de solutions algebriques de la proposée qu'on le voudra. La question qui nous occuppe maintenant ne peut etre resolue par d'autre surface que par le plan des x, y; car c'est la seule qui soit commune a la famille des cônes dont le sommet est a l'origine, et a celle des surfaces courbes dont toutes les normales sont constantes paraport au plan des x, y.

Quant aux equations elevées qui ne peuvent etre ramenées a la forme lineaire, il suit de ce qu'on a vu au commencement de cet article qu'elles ne sauraient appartenir a des courbes de contact, et l'on n'est pas encore parvenu a trouver generalement les ensembles de leurs solutions. Mais M. Monge a donné dans le memoire deja cité des theorêmes, qui dans beaucoup de cas font connaitre l'ensemble des solutions de ce genre d'equation sous une forme algebrique. Il a remarqué de plus une correspondance singulière entr'elles et les equations aux differences partielles, telle que lorsqu'on a sous une forme donnée l'integrale de celles-ci on arrive a l'ensemble des solutions des autres et reciproquement.

Les moyens sont encore plus bornés pour les equations des ordres superieures et M. Monge a traité celles qui appartiennent aux courbes que nous avons remarquées a la fin de l article VI et dont la courbure est constante. Les questions que nous avons indiqués sur les courbes a double courbure tracées sur des surfaces developpables conduisent a des equations differentielles a trois variables, elevees ou des ordres superieures dont il serait peutetre interessant de connaitre l'ensemble des solutions sous une forme algebrique.

Appendix B

Lacroix's historical appraisal of his own *Traité*

Sometime around 1803 Lacroix wrote a "Compte rendu à la section de Géométrie de l'Institut national, des progrès que les mathématiques ont faits depuis 1789 jusqu'au 1.er Vendemiaire an 10" (that is, a "report to the Geometry section of the *Institut National*, on the progress made in mathematics from 1789 to *Vendemiaire* 1st, year 10 [= September 23rd, 1801]").[1] At some point Lacroix revised it and changed its title to "Essai sur l'histoire des Mathematiques, pendant les dernieres années du 18me siecle et le [sic] premieres du 19eme"[2]. It was never published under any of these titles, but most of it was incorporated in [Delambre *1810*][3]; Delambre [*1810*, 43] admitted that all that concerned "pure mathematics and transcendental analysis" had been taken from Lacroix's work.[4]

One of the interesting points in Lacroix's report is that it had to address his own *Traité*, pointing out the aspects that "should find a place in the history of science" (see page 399 below). Transcribed below (from the manuscript kept in Lacroix's *dossier biographique* at the archive of the Paris *Académie des Sciences*)

---

[1] An order of the consular government demanding such a report was read by Laplace to the Physical and Mathematical Sciences class of the *Institut* in 16 Ventose year 10 (7 February 1802) [Acad. Sc. Paris *PV*, II, 476]. It is likely that Lacroix prepared the report between 1 Germinal year 10 (22 March 1802) and 11 Pluviose year 11 (31 January 1803), that is, while he was *secrétaire* of the Mathematical Sciences section of the *Institut* [Acad. Sc. Paris *PV*, II, 479, 625]. Still, the original title suggests that it was only ready after this, as the "Geometry section" appeared only in the reorganization of 1803, presented to the class precisely in the session of 11 Pluviose year 11 [Acad. Sc. Paris *PV*, 619-625; Grattan-Guinness *1990*, I, 79].

[2] "Essay on the history of mathematics during the final years of the 18th century and the first of the 19th"

[3] That is, the eventual compliance with the 1802 demand. Delambre was *secrétaire perpétuel* since 1803.

[4] For the question of other contributions to [Delambre *1810*], and some comparisons between Lacroix's manuscript and the corresponding sections in [Delambre *1810*], see the Introduction and endnotes by Jean Dhombres to the 1989 edition.

are the four references to the *Traité*: a short one in the chapter on algebra, fl. 5v [Delambre *1810*, 90]; a long one in the chapter on differential and integral calculus, fls. 23r-25v [Delambre *1810*, 100-102]; two in the chapter on finite differences and series – a long one, fls. 31r-32v [Delambre *1810*, 109-111], and a short one, fl. 35r [Delambre *1810*, 116]. Sentences or expressions omitted in [Delambre *1810*] are within braces {} here (when a footnote mark is within braces, all of its text was omitted).

Algèbre

[5v] La considération des fonctions symétriques des racines, offrant le moyen {le plus clair et} le plus fécond pour traiter la résolution des équations {et l'élimination}, {ce n'est peut être pas sans quelque avantage pour la science, au moins pour en faciliter l'étude que, dans ces derniers tems, on a donné}[5] du théorême de Newton sur la somme des puissances semblables des racines qui sert de base à cette théorie, une démonstration indépendante des séries{, et qu'on peut regarder avec quelque fondement comme la plus simple qu'il soit possible de former}.[6]

Calculs différentiel et intégral

[23r ...] Les travaux dont je vais parler maintenant remontent à un mém.^{re} inséré dans le vol. de Berlin pour 1772, où le C.^{en} Lagrange donnait au calcul différentiel et intégral une origine purement analytique, à la fois simple, rigoureuse, reposant sur les formes du développement des fonctions en séries, et assez analogue à la manière dont Newton présenta dans le livre *des principes* sa méthode des fluxions.

Le desir de populariser des considérations aussi élégantes, de rapprocher sous un même point de vue, et de réduire pour ainsi dire à la même echelle, tous les procédés dont l'analyse transcendante s'étoit enrichi depuis la publication des traités généraux d'Euler, donna naissance à un *traité du calcul différentiel et* [23v] *du calcul intégral*,{[7]} médité pendant longtemps et dont le 1.er volume parut en l'an V. {Pour le rattacher aux élemens existans lors de sa publication,} on les fit précéder d'une introduction dans laquelle le développement des fonctions exponentielles, logarithmiques et circulaires, en series, est déduit de considérations entièrement indépendantes des notions d'infini, de limites; et par le moyen d'un calcul simple effectué sur les indices des coëfficiens à déterminer, on est parvenu à vérifier toutes les équations de condition dont on se debarassait ordinairement en assignant des

---

[5][Delambre *1810*, 90]: étoit donc important de donner

[6]Voyez le 1er vol. du *Traité du calcul differentiel et du calcul integral* de Lacroix, ou le *Complement des Elem. d'alg.* [In [Delambre *1810*, 90] an equivalent reference is included in the main text, followed by "ouvrages qui ont opéré une révolution heureuse dans l'enseignement, et ont mérité d'être adoptés pour les lycées et l'École polytechnique".]

[7]en 3 vol in 4° par S F Lacroix

valeurs particulières aux variables introduites dans le calcul.[8]

{La généralité de cette méthode rachette bien, à ce qu'il semble, un peu de longueur sans la quelle d'ailleurs on ne parvient jamais à satisfaire entièrement les esprits difficiles sur l'exactitude des démonstrations.}

[24r] Les mêmes procédés s'appliquent avec autant de succès au theorème de Taylor, qui forme la base du calcul différentiel, et qui en fait l'introduction, lorsqu'on s'appuie sur les {notions lumineuses} qu'en a données le C. Lagrange, et citées plus haut.

{Il serait déplacé dans le compte que [crossed out: nous rendons] je remis à la [crossed out: classe] section, de parler de tous les perfectionnemens de détail que doit exiger un Traité fondé sur une nouvelle manière d'envisager le calcul différentiel et intégral, et dans le quel on a rassemblé sous un même point de vue, et} assujetti à un enchainement méthodique les divers résultats ou procédés analytiques, épars dans les colletions académiques{[9]; nous citerons ici [24v] quelques points qui semblent devoir trouver place dans l'histoire de la science.}

L'auteur s'empressa d'exposer dans son ouvrage et de ramener à des formes purement analytiques l'espèce d'intégration des équations différentielles à trois variables qui ne satisfont par aux conditions d'intégrabilité, que le C.[en] Monge avait déduites de la considération des courbes à double courbure et des surfaces, et rendit évidente la liaison de ces intégrales avec la théorie g.[le] des intégrales et des solutions particulières que le C. Lagrange a fait connaitre le premier dans les mém.[res] de l'académie de Berlin pour l'année 1774, et il rapprocha cette théorie d'une classe de questions dont Euler s'occupa {dans plusieurs mémoires particuliers} et qu'il nomma *calcul intégral indéterminé*, [25r] {parcequ'il s'agit d'établir entre une fonction, et la variable indépendante, des relations qui rendent integrables algébriquement certaines expressions différentielles relatives, soit aux arcs ou aux aires des courbes, aux aires ou aux volumes des surfaces.[10]} C'est à ce genre de

---

[8]In [Delambre *1810*, 101] there is a change in the division of sentences which I think alters the intended meaning: "[...] est déduit de considérations entièrement indépendantes des notions d'infini, de limites, et par le moyen d'un calcul simple effectué sur les indices des coefficiens à déterminer. L'auteur, M. Lacroix, est parvenu à vérifier toutes les équations de condition [...]".

[9]Par exemple, après avoir montré que les accroissements eux-mêmes n'entraient pour rien dans le but et les applications du calcul différentiel, il fallait expliquer ce que signifiaient, dans le nouvel ordre des propositions, les transformations qui servent à rendre variable une différentielle qu'on regardait comme constante, et *vice versâ*. Cet objet a paru assez important au C. Lagrange pour qu'il s'en soit occupé dans la *Théorie des fonctions analytiques*; [24v] mais on observera que l'article du traité dont on parle ci dessus était composé, imprimé, et entre les mains de plusieurs personnes, entr'autres du C. Prony, avant que le C. Lagrange fit à l'école polytechnique les leçons qui ont donné naissance à la théorie des fonctions. Il en est de même des autres endroits du *Traité du calcul différentiel et du calcul intégral*, où la *théorie des fonctions analytiques* n'est pas citée. Le premier de ces ouvrages projeté et preparé 10 ans avant sa publication et reposant sur le mém.[re] contenant le germe du second, a du nécessairement mener à des développemens analogues. La lenteur de l'impression ayant permis d'enrichir la notation différentielle usitée des choses nouvelles que lillustre Géomètre, auteur de la théorie des fonctions, avait publiées dans un algorithme particulier, le C. Lacroix l'a fait, mais en citant avec le plus grand soin la source d'où il avait tiré ces précieuses additions.

[10][Sidenote, difficult to read: L'acad. de Petersbourg a publiée dans ses derniers volumes deux

questions que se rapporte le problême de la voûte quarrable, proposé par Viviani, et un théorême nouveau que le C. Bossut a communiqué en l'an IV à l'Institut, {et dont voici l'énoncé: *Si on perce une sphère perpendiculairement au plan de l'un de ses grands cercles, par deux cylindres droits, en forme de tarrière, dont les axes passent par les milieus de deux rayons qui composent un diamètre de ce grand cercle, les deux portions qu'on enlevera par là, du solide entier de la sphère, laisseront un reste égal aux 2/9 du cube du diamètre de la sphère.*}[11]

{L'introduction des fonctions arbitraires dans les intégrales des équations différentielles, ne parait pas suivre les mêmes loix que celles des constantes arbitraires dans les intégrales [25v] des équations différentielles totales. Dans le second ordre et les ordres supérieures, on ne peut introduire en général sous la forme finie et faire disparaitre successivement à chaque ordre de différentiation, une nouvelle fonction arbitraire, de même qu'on fait évanouir une constante. Ces remarques, qui n'avaient pas encore été publiées, se rattachent aisément avec la théorie g[le] des intégrales dont elles sont le complément.}

Du calcul aux differences (finies) et des series

[31r ...] La convenance qu'il y avait à séparer des premiers principes du calcul differentiel, le calcul aux differences afin de ne pas le morceler, et de n'en faire qu'un seul corps avec la doctrine des series, résulte bien nécessairement du mémoire de 1772 sur l'origine du calcul différentiel et intégral, et fut saisie par l'auteur du *Traité du Calcul différentiel et du calcul intégral*, qui rassembla en un seul volume, sous le titre de Traité des différences et des séries tout ce qui concernait ces deux branches de l'analyse et quelques méthodes pour ainsi dire *anomales*, qu'on ne pouvait rapporter que difficilement aux procédés d'intégration déduits du renversement de la différentiation.

C'est le premier ouvrage dans lequel on trouve toutes les méthodes relatives aux séries réunies en un seul [31v] corps de doctrine et liées entr'elles. L'auteur y a présenté de la manière la plus générale l'interpolation des séries, dont il a rapporté les diverses formules tant anciennement connues que récemment publiées dans les leçons que le C. Prony a données à l'Ecole polytechnique sur le calcul des différences; les divers procédés pour intégrer les équations aux différences et pour obtenir le terme général des séries recurrentes; l'usage des intégrales définies dans la sommation des séries, et pour l'intégration des équations différentielles et différentielles partielles; et {a cette occasion} l'auteur rend compte du procédé du C. Parseval, {publié par le C. Prony dans la *mécanique philosophique*}. Enfin, il a donné avec beaucoup de détails la théorie des intégrales directes et indirectes des équations aux différences. En remarquant ces dernières, et en poussant trop loin les conséquences de l'analogie quelles ont avec les [32r] solutions particulières des

---

memoires d'Euler sur ce sujet, restés inédits, et notamment sur les courbes dont les arcs peuvent être exprimés par des arcs dellipse, de parabole: il[?] n'[?] [?] réussi à en trouver que [?] des arcs d'hyperbole, M. Fuss en a indiqué de cette espèce (dans le T. XIV des nova acta)]

[11][Sidenote: M. Fuss (dans le T XIV des Nova acta Petrop), {a demontre ce théorême dont il ne connaissait que l'enonce, et} en a decouvert un grand nombre d'autres sur le même sujet]

équations différentielles, feu Charles tomba dans des paradoxes très singuliers, que le C.^{en} Biot a éclaircis dans un mém^{re} {présenté à l'Institut et imprimé} dans le {11.^{eme} cahier du} journal de l'Ecole polytechnique. Le C. [crossed out: Brisson] Poisson ayant considéré ensuite ce même sujet sous un point de vue purement analytique a donné une explication très simple et très générale de la multiplicité des intégrales dont une équation aux différences est susceptible et de leur nature.

Les CC. Laplace et Condorcet avaient imaginé de considérer des équations contenant à la fois des coëfficiens différentiels et des différences. Je les a[i] fait connaitre sous le nom d'équations aux différences mêlées, dans le *traité des séries et des différences*, et j'y ai inséré l'extrait d'un mém^{re} {présenté} par le C.^{en} Biot {à l'Institut}. Ce mém^{re} où l'on trouve quelques principes généraux sur la nature des intégrales aux différences mêlées, contient en outre la solution de plusieurs [32v] questions géométriques qu'Euler avait résolues dans un mém^{re} ayant pour titre *de insigni promotione methodi tangentium inversæ*, mais qui se rapportent plus naturellement aux équations aux différences mêlées, dont la nature est d'exprimer les propriétés des courbes qui établissent en même temps des relations entre plusieurs points infiniment voisins, et entre des points placés à des distances finies.

[…]

[35r Arbogast traite aussi] les produits de facteurs equidifferens – aux quels il donne le nom de *factorielles*.

Ce genre de fonctions, que les géomètres ont eu de fréquentes occasions de considérer et que Vandermonde a représenté par une notation très ingénieuse et très expressive, qui met en évidence leur analogie avec les puissances, a été traité presqu'en même tems sous ce point de vue, sous le nom de *facultés numériques*, dans l'analyse des réfractions astronomiques de M. Krampt [sic] et dans le Traité des *différences et des séries*, servant d'appendice au *Traité du calcul différentiel et du calcul intégral*.

Appendix C

Syllabi of Lacroix's course of analysis at the *École Polytechnique*

C.1 Lacroix's lectures on differential and integral calculus at the *École Polytechnique* in 1799-1800

The Wellcome Library for the History and Understanding of Medicine, in London, possesses a set of notebooks which once belonged to Aimé Marie Gaspard, marquis de Clermont-Tonnerre (1779-1865), a student at the *École Polytechnique* (entry of 1799 [Fourcy *1828*, 408]). These notebooks (mss. 1663-1670) contain notes from lectures at the *Polytechnique* dated from Frimaire to Thermidor, year 9 (November 1800 to August 1801) – Clermont-Tonnerre's second year there; as the library catalogue indicates, these notes (at least those on mathematics) are "very rough pencilled notes", and it is not easy at all to follow them.[1]

Happily, included in ms. 1668 is also a four-page set of summaries of first-year calculus lectures, in ink (much easier to read than the second-year notes). These summaries should then refer to Clermont-Tonnerre's first year, that is 1799-1800 – Lacroix's first year as an *instituteur* at the *École Polytechnique*.

Next to each lecture there is an indication of several numbers which clearly correspond to the relevant articles in Lacroix's large *Traité*. This is a precious source for Lacroix's pedagogical use of his large *Traité du calcul...* before the publication of the *Traité élémentaire de calcul....*

---

[1]Ms. 1666 seems to be the only one containing some lectures by Lacroix. The little I could understand from them is consistent with second-year lectures in analysis: they are mostly on integral calculus, but there is also one (21 Frimaire) on the roots of $x^m - 1 = 0$ and probably on Cotes' theorem.

This set of summaries is incomplete: the lectures on differential calculus are numbered from 1 to 19 and those on integral calculus from 1 to 10; but an extra numbering next to the first and last of each of them suggests that there had been 25 lectures before the first one on differential calculus. Those 25 missing lectures were certainly on algebraic analysis.

[Cours d'?] analyse[?]

Calcul Integral et differentiel

Calcul differentiel

26ᵉ Leçon { 1ᵉʳᵉ Definition des fonctions et leur division en explicites et implicites ; distinction entre le developpement et la serie qui donne la valeur

Definition du mot limite

Expressions algebriques susceptibles de limites

Propositions fondamentales de la theorie des Limites

Forme du developpement d'une fonction de x, lorsqu'on change x en $x + k$.

N.ᵒˢ
1, 2, 3,
4, 5, 6, 7,
11, 12, 13 de l'introduction
1, 2, 3 du Calcul differentiel excepté ce qui regarde les fonctions circulaires dans le N.ᵒ 2

——

2ᵉ Indication de la maniere dont les coefficients des puissances de l'accroissement sont liés entreux.

Le 1ᵉʳ exprime la limite du rapport des accroissᵐᵗˢ de la fonction et de la variable independante. C'est par cette consideration qu'on l'obtient lorsqu'on n'a pas l'expression analytique de la fonction proposée

Explication[?] de la notation independamment d'aucune hypothese sur l'origine du calcul

Regle[s?] pour differentier

N.ᵒˢ
3, 9 ?, 13
14, 15, 16,
17, 18.

——

3ᵉ Differentiation des fonctions transcendantes

Developpement d'une fonction en serie par des differentiations repetées.

Application aux fonctions $\sin x, \cos x$; impossibilité de developper ainsi $\log x$.

N.ᵒˢ
19, 20, 21, [22]²
93, 109[, 10]³
103 (1ᵉʳᵉ alinea)

——

4ᵉ Theoreme de Taylor

Son usage pour developper en serie

Developpement des fonctions rationelles au moyen des differentielles logarithmiques

N.ᵒˢ
109, 100, 98

——

²Crossed out (?) in pencil.
³In pencil.

5ᵉ L.   Developpement des fonctions de deux variables   N.ᵒˢ

Definitions des differences et des different.ᵉˡˡᵉˢ par-   24, 25, 26

     tielles.   28, 29, 30

Formation successive des differentielles des fonc-   31, 91.

     tions de plusieurs variables.

Proprieté des fonctions homogenes.

6ᵉ L.   Differentiation des Equations   N.ᵒˢ

De l'Elimination des constantes et des transcen-   40, 42, 43

     dantes   45, 46, 47

                                                                    48, 50, 51

                                                                      52, 53, 55.

7ᵉ L.   Passage des coefficients differentiels d'une variable   N.ᵒˢ

     à ceux de l'autre, c'est à dire dans les equations   56 (1ᵉʳᵉ alinea)

     du sécond ordre, prendre pour constante telle   57 (2ᵉ alinea)

     differentielle que l'on voudra   56 (3ᵉ alinea)

8ᵉ L.   Differentielle de l'arc au moyen du sinus du cosinus   N.ᵒˢ

     et de la tangente   23, 104,

Diverses series qui expriment l'arc   Note de la page 202

Notions generales sur la liaison des lignes et des   Introd.   page   64

     Equations à deux indeterminées.   ([?])

                                                                                       195.

9ᵉ L.   Correspondance de l'Intersection des Courbes avec   N.ᵒˢ

     l'Elimination et la Resolution des Equations.   283, 284

Recherche des lignes osculatrices des courbes   240, 241

Formules des soutangentes, tangentes, sousnor-

     males et normales

Reponse à quelques objections faites contre l'appli-

     cation de la methode des limites

10ᵉ L.   Continuation de la Methode des Tangentes   N.ᵒˢ

Recherche des asymptotes   242, 243,

                                                                  240, 246, 24[4?]

_____

[4]Crossed out.

Calcul Integral

45 et 1ere	De l'Integration des fonctions rationelles et entieres, et commencement de celle des fractions rationelles	N.ºs 358, 359, 360, 361, 362, 363, 364.
2^{e} L.	Continuation de L'Integration des fractions rationelles.	N.ºs 366, 367
3^{e} Leç.	De l'Integration des fractions irrationelles contenant le $\sqrt{(A + Bx + Cx^2)}$	N.ºs 376, 377, 378, 379 (1re, 3e et 4e alin.)
4e L.	De l'Integration des differentielles Binomes.	N.ºs 385, 387, 388, 389
5e L.	De l'Integration par les series	N.ºs 406, 407, 408, 409, 410, (3 Prem. alin.) 439 (1re et 2e alin.)
6e L.	De la determination des Constantes dans les Intégrales, de la quadrature des courbes.	N.ºs 470, 471, 476, 477, 478, 490.
7e L.	Suite de la quadrature des courbes et [de^5] leur Rectification	N.ºs 491, 492, 493, 495, 496, 498.
8e L.	Suite de la rectification des courbes, de l'Evaluation des Volumes et des aires des corps engendrés par la Revolution d'une courbe plane autour d'un axe	N.ºs 501, 513, 514, 515, 516, 517.
9e L.	De l'Integration des Equations du 1er ordre à 2 variables ; de la separation des variables dans les Equations homogenes ; des Equations immediatement integrables ; du facteur et sa determination lorsqu'il ne doit renfermer qu'une des variables.	N.ºs 543, 544, 545 546, 547 552, 553.
54e 10e et der^{niere} L.	Continuation de l'integration des Equations differentielles du 1er ordre ; principe de celle des Equations du second.	N.ºs 554, 555, 556 567, 568, 609, 610, 61[1 ?] 615

---

^{5}Crossed out.

C.2 The establishment of the first programme of analysis of the *École Polytechnique*

C.2.1 Lacroix's views on the syllabus of analysis in 1800

The following text is kept at the Archives of the *École Polytechnique* [Éc. Pol. *Arch*, III3b]. It is unsigned. The handwriting is much better than that of Lacroix – possibly professional. But there are some corrections and additions, and these are unmistakably in Lacroix's hand. Moreover, the ideas expressed here are consistent with those in [Lacroix *1805*]. The context is the discussion on the first official programme of analysis at the *École Polytechnique* [Belhoste *2003*, 248-249].

Crossed out :

Sur le cours d'analyse de l'École Polytechnique

New title

Bases proposées par le Conseil d'Instruction de l'École Polytechnique au Conseil de perfectionnement, pour servir à la formation des programmes de l'analyse a fournir aux Examinateurs pour l'examen des deux divisions

Il est constaté que dès qu'on a passé les premiers élémens, il faut s'élever très haut pour pouvoir trouver dans l'analyse des objets d'une application vraiment utile. Il suit de lá qu'un cours d'analyse fait à des élèves qui savent déjà leurs élémens, et dans la vue de les initier dans les principales théories des sciences physiques et mathématiques, doit être assés étendu.

Les efforts des Géomètres ont multiplié beaucoup les méthodes pour parvenir au même résultat ; les unes paraissent plus directes, les autres, plus rigoureuses ; mais toutes sont a peu près arrêtées par les mêmes difficultés. Si la connaissance de ces diverses méthodes importe à celui qui se propose d'enseigner ou qui veut se livrer exclusivement aux mathématiques, dans la vüe de les perfectionner, elle ferait perdre beaucoup de tems à l'élève qui doit diriger tout son travail vers la méchanique. Celui-ci préferera sans doute la connaissance d'un résultat qu'il ignore à celle d'un nouveau chemin pour arriver à l'un de ceux qu'il possède déjà.

Le cours d'analyse de l'École Polytechnique ne doit donc renfermer aucun double emploi, soit dans les objets qu'il embrasse soit dans ceux qui on été éxigés pour l'admission.

Avant qu'on se fût autant familiarisé avec le calcul différentiel et intégral qu'on l'a fait dans ces derniers tems, on s'efforçait de faire entrer dans l'algèbre le plus de choses qu'il était possible. On sacrifice souvent à cette vue la brieveté des démonstrations. Il serait convenable[?] d'en user[?] encore ainsi par rapport à des élèves à qui l'on n'enseignerait que l'algèbre ; puisqu'il n'aurait que cet instrument entre les mains, il faudrait leur apprendre à en tirer le meilleur parti. Il n'est pas de même pour les élèves de l'École Polytechnique ; le cours de leurs études embrassant le calcul différentiel et le calcul intégral, la vraie place d'une proposition ou d'un résultat est celle où ils se présentent le plus facilement et se lient avec un plus grand nombre d'autres.

Cela posé on doit exiger pour l'admission les élémens d'algèbre complets, en les reduisant néanmoins à ce qui est utile, et parconséquent en substituant la résolution générale des *équations numériques*, à la résolution particulière des *équations littérales* du troisième et du 4.ᵉ dégré qui est si compliqué que même pour les équations numériques de ce dégré on a recours à la première.

Il ne faut pas non plus exiger la démonstration du binôme pour le cas de l'exposant fractionnaire ou négatif, ni les séries des fonctions logarithmiques et circulaires, parceque le développement des fonctions se lie naturellement au calcul différentiel, comme une application spéciale du théorême de Taylor.

On ne laisserait pas néanmoins ignorer aux élèves les principales circonstances de la résolution des *équations littérales*, mais elle ne ferait pas la matière de l'examen non plus que quelques autres digressions que l'on pourrait seulement indiquer aux élèves studieux que leur goût ou d'heureuses dispositions porteraient vers les mathématiques pures.

On commencerait donc le cours d'analyse de l'École Polytechnique par les premiers élémens du calcul différentiel présentés ainsi qu'il suit :

1ère année.

1.° La théorie purement analytique du calcul différentiel des fonctions à une seule variable, et des fonctions à deux variables, autant seulement qu'il en faut pour la différentiation des équations à deux variables.

Ce calcul serait présenté par la méthode des limites [crossed out : comme il est indiqué dans le Programme que j'ai donné cette année].

On se hâterait de parvenir au théorême de Taylor qu'on peut prouver de plusieurs manières qui ne supposent la connaissance du développement des puissances du binôme que pour le cas de l'exposant entier. On montrerait ensuite que les deux premiers termes du développement de $(1+z)^{\frac{m}{n}}$ et de $(1+z)^{-m}$ sont $1 + \frac{m}{n}z$ et $1 - mz$ et delà on déduirait par le théorême de Taylor l'expression générale du développement de $(1+z)^m$.

Le même théorême conduirait au développement des fonctions circulaires et logarithmiques, lorsqu'on aurait obtenu leurs différentielles premières en prenant la limite du rapport de leurs accroissements à celui de leur variable.

On passerait à l'examen des valeurs particulières que prennent les coëfficiens différentiels dans certains cas, à la recherche des fonctions qui se présentent sous la forme de $\frac{0}{0}$ et à la théorie des maxima et de minima des fonctions d'une seule variable, soit *explicites* soit *implicites*.

Viendrait ensuite l'application du calcul différentiel à la théorie des courbes.
1.° À la recherche des osculations, et en particulier de celle de la tangente, & des cercles osculateurs.

2.° À la recherche des points singuliers, comprenant celle des inflexions, des rebroussemens et des limites des courbes, ou des maxima et des minima de leurs coordonnées.

[2.° ?] On passerait de là aux premiers élémens du calcul intégral. On a pensé qu'il ne fallait pas exposer d'abord tout ce qu'on doit dire sur le calcul différentiel, non seulement dans le dessein de donner assez de calcul intégral pour mettre les élèves en état de suivre les élémens de mécanique de la première année, mais encore pout diminuer la sécheresse de l'étude de l'analise pure, en faisant connaitre les applications dont elle est susceptible, avant de s'enfoncer dans ce qu'elle offre de plus transcendant.

Voici ce qu'on pourrait enseigner cette année :

L'intégration des fonctions rationnelles et entières.

L'intégration des fonctions rationnelles, n'indiquant pour la décomposition des fractions proposées en fractions simples que la méthode des coëfficiens indéterminées.

L'intégration des fonctions rationnelles [sic] contenant le radical $\sqrt{a + bx + cx^2}$.

Les transformations pour rendre rationnelles, quand cela est possible la différentielle binôme, $x^{m-1}dx(a + bx^n)^{\frac{p}{q}}$.

Les formules pour réduire cette différentielle à d'autres plus simples (soit relativement à l'exposant de x hors de la parenthèse, soit à celui de la parenthèse) déduites de l'intégration par parties.

L'intégration des formules comprises dans la différentielle binôme, au moyen des séries, et obtenir par ce moyen le logarithme, l'arc par son sinus et par sa tangente.

La détermination des constantes dans les intégrales.

La quadrature des courbes et leur rectification.

L'évaluation des volumes et des aires des corps engendrés par la révolution d'une courbe plane autour de son axe.

L'intégration des équations différentielles à deux variables du premier ordre, 1° lorsque ces variables sont séparées, 2° lorsque les équations sont homogènes.

Le caractère des équations différentielles du 1er ordre qui sont immédiatement intégrables.

La détermination du facteur propre à rendre une équation différentielle du 1er ordre intégrable, lorsque ce facteur ne doit renfermer que l'une des variables.

L'intégration des équations différentielles du 2.e ordre qui ne referment que le coëfficient différentiel de cet ordre et l'une des variables ou qui ne contiennent que les coëfficients différentiels du 1.er et du 2.e ordre et des quantités constantes.

[Crossed out : On a fait voir] Montrer aussi que l'équation $\frac{d^2y}{dx^2} + P\frac{dy}{dx} + Qy = R$ se ramène à l'équation $\frac{d^{[2]}y}{dx^2} + P\frac{dy}{dx} + Qy = 0$.

2.ᵉ année.

On donnerait dans cette année les développemens du calcul différentiel, savoir : la différentiation des fonctions de deux et d'un plus grand nombre de variables, l'élimination des fonctions arbitraires, les maxima et les minima des fonctions de deux et d'un plus grand nombre de variables, les équations de condition (pour le 1.ᵉʳ ordre seulement).

Les développemens du calcul intégral des fonctions d'une seule variable.

Quelques notions sur la transcendante qui donne les oscillations du pendule conique.

La théorie complète des équations différentielles du 1.ᵉʳ degré d'un ordre quelconque (équations linéaires).

L'intégration des équations simultanées.

Une idée succinte des méthodes d'approximation qu'on employe pour intégrer les équations différentielles du second ordre et notamment celles qui se rapportent aux mouvemens des corps.

La théorie des solutions particulières.

Ceci ne serait pas exigé à l'examen.

$\left\{\begin{array}{l} \text{2 leçons} \\ \\ \\ \\ \text{3 leçons} \\ \text{3 leçons} \end{array}\right.$

2 leçons Quelques notions purement analytiques sur l'intégration des équations différentielles partielles du 1.ᵉʳ ordre et sur les équations du 1.ᵉʳ degré du 2. ordre

3 leçons L'abrégé de la méthode des variations.

3 leçons L'intégration des fonctions aux différences (finies) les plus simples, et des équations du premier dégré à deux variables et a coëfficiens constants.

Approuvé par le Conseil d'Instruction pour etre soumis a l'approbation du Conseil de Perfectionnement. Le 15 Brumaire an 9

[Crossed out : Le 27 Vendemiaire an 9]

C.2.2   The approved programme of analysis for 1800-1801

The following is the programme of analysis that was approved by the *Conseil de Perfectionnement* in 1800. It is reproduced from [Éc. Pol. *Rapport*, an 9, 28-34].

PROGRAMMES D'ANALYSE.

1.re DIVISION.

ANALYSE ALGÉBRIQUE.

MONTRER qu'une équation peut se décomposer en autant de facteurs $x - a$, $x - b$, &c., qu'il y a d'unités dans le plus haut exposant de l'inconnue.

Composition des équations, ou expression des coëfficiens en fonction des racines.

Démontrer, tant par l'analyse que par la considération des courbes, que, si deux valeurs substituées à la place de x, dans le premier membre d'une équation, donnent deux résultats de signes contraires, il y a une racine comprise entre ces deux valeurs.

Conclure de là que les équations de degré impair ont toujours une racine réelle, et que celles de degré pair, dont le dernier terme est négatif, en ont toujours deux.

Méthode d'élimination réduite à la partie la plus élémentaire.

Équation qui a lieu concurremment avec la proposée, dans le cas où celle-ci a des racines égales. (On démontrera, dans le calcul différentiel, que le commun diviseur de ces deux équations contient les racines égales de la proposée, élevées chacune à une puissances moindre d'une unité.)

Faire voir quels sont les cas dans lesquels le premier membre d'une équation ne peut jamais changer de signe, quelque valeur qu'on attribue à x.

Démontrer, au contraire, que, lorsqu'il y a des racines réelles et inégales dans la proposée, on parviendra toujours à des résultats de signe contraire, en substituant à la place de x les termes successifs d'une progression arithmétique dont la raison est moindre que la plus petite différence des racines.

Faire voir comment on trouverait, par l'élimination, l'équation aux différences des racines, et quel est son degré.

Exposer ce qu'il y a de plus simple sur les limites des racines.

Reprendre en peu de mots la méthode des diviseurs commensurables pour les équations numériques, ainsi que la méthode qui sert à trouver les racines approchées.

Développement de quelques fonctions en séries, par la méthode des coëfficiens indéterminés.

Loi générale des suites récurrentes, observée dans le développement des fractions rationnelles.

Les progressions géométriques sont des suites récurrentes dont l'échelle de relation n'a qu'un terme ; les progressions arithmétiques sont des suites récurrentes dont l'échelle de relation est composée des deux termes 2 et -1.

Examen particulier des suites récurrentes, dont l'échelle de relation a deux termes. Leur décomposition en deux progressions géométriques, et de là leur terme général.

Notions générales et succinctes des suites à différences constantes.

Faire voir comment on trouve le terme général d'une suite dont les différences secondes sont constantes.

Application à diverses interpolations, et particulièrement à celle des tables de sinus et de logarithmes.

Sommer les carrés, les cubes, &c., des nombres naturels, par une méthode simple ; par exemple, par la méthode des coèfficiens indéterminés, et d'après le principe que le nombre des termes est élevé à une puissance plus grande d'une unité dans le terme sommatoire que dans le terme général.

Application à différentes piles de boulets.

Revue des formules trigonométriques les plus utiles, et de l'équation exponentielle qui a lieu entre un nombre et son logarithme.

Déduire de cette équation les propriétés générales des logarithmes ; comparer les différens systèmes entre eux, et faire voir comment on peut passer de l'un à l'autre.

Quelques notions sur les fonctions en général, et sur leur division en fonctions entières, rationnelles, &c.

CALCUL DIFFÉRENTIEL.

Établir les notions des différentielles sur la théorie des limites. Règles de la différenciation pour un nombre quelconque de variables, et pour des fonctions explicites et implicites. (On donnera la différentielle de x^m en général, d'après la formule du binome, démontrée seulement pour le cas de l'exposant entier.)

Différentielles des fonctions circulaires, logarithmiques et exponentielles, tant simples que combinées.

Différences secondes, troisièmes, &c.

Démonstration du théorème de *Taylor* par une méthode simple, telle que la méthode des coèfficiens indéterminés.

Démonstration de le formule du binome dans le cas de l'exposant fractionnaire ou négatif.

Complément de la théorie des racines égales. (*Voyez* ci-dessus le théorème à démontrer.)

Développement des séries qui donnent les logarithmes, les exponentielles, les sinus et cosinus en fonctions de l'arc, et réciproquement ; le tout pouvant être considéré comme des applications du théorème de *Taylor*.

Ce même théorème étendu à deux variables, c'est-à-dire, au développement de $F(x + i, y + k)$.

Loi du résultat. Conséquence qu'on tire par rapport à l'égalité des coèfficiens différentiels $\frac{ddz}{dxdy}$, $\frac{ddz}{dydx}$.

Condition pour que $Mdx + Ndy$ soit une différentielle complète ;

Item, pour que $Mdx + Ndy + Pdz$ en soit une.

Notion des différences partielles.

Théorie des *maxima* et des *minima* pour les fonctions d'une et de deux variables. Manière de distinguer le *maximum* du *minimum*. Application à des exemples choisis.

Formules des sous-tangentes, sous-normales, tangentes, &c., déduites de la considération des limites. Détermination des asymptotes.

D'un point donné hors d'une courbe, mener une tangente ou une normale à cette courbe.

Expression du rayon de courbure, par une méthode facile et qui mène promptement au résultat. Propriétés générales de la dévelopée ; manière d'en trouver l'équation.

Application aux sections coniques, à la cycloïde, &c ; donner la développée de la parabole, et la rectification de cette développée.

Traiter sommairement des exceptions que présente le calcul différentiel, c'est-à-dire, des cas où, pour une abscisse déterminée $x = a$, les coèfficiens différentiels $\frac{dy}{dx}$, $\frac{ddy}{dx^2}$, &c. deviennent $\frac{0}{0}$ ou infinis. Il suffira de faire $x = a + \omega$, en considérant ω comme très-petit, et de déterminer par l'analyse algébrique, la valeur de y, qui sera de la forme $y = b + c\omega^m$: alors, suivant les valeurs particulières de c et de m, on connaîtra si la courbe a un point multiple, un point d'inflexion ou de rebroussement. Deux ou trois exemples suffisent pour expliquer cette théorie, qui d'ailleurs ne doit occuper que très-peu de place dans le cours.

On démontrera de même succinctement que la fraction $\frac{P}{Q}$, dont les deux termes sont supposés s'évanouir lorsque $x = a$, est égale à $\frac{dP}{dQ}$; ce qui suffira presque toujours pour en déterminer la valeur.

CALCUL INTÉGRAL.

Notions sur l'intégration en général.

Intégration des différentielles monomes, et des fonctions entières.

Cas d'intégrabilité des différentielles binomes.

Intégration des fractions rationnelles, dans les cas les plus simples, et par la méthode des coèfficiens indéterminés. (On réservera pour la seconde année le développement des cas plus composés.)

Manière de rendre rationnelles les différentielles affectées du radical $\sqrt{(a + bx + cxx)}$.

Intégration des formules qui contiennent des sinus ou des exponentielles, dans les cas les plus simples.

Réduction des différentielles binomes, appliquée principalement aux formules qui s'intègrent par les arcs de cercle.

Montrer quelles sont ces formules principales auxquelles on rapporte les autres, et comment on en exprime l'intégrale.

Intégration par séries.

Formules pour la quadrature des courbes, leur rectification, les surfaces et les solidités des solides de révolution : insister, dans les applications, sur la détermination des constantes.

Intégration des équations différentielles du premier ordre, dans les cas les plus simples, savoir, ceux des équations séparables, des équations pour lesquelles la condition d'intégrabilité est satisfaite, des équations homogènes, et des équations linéaires.

Intégration des équations différentielles du second ordre, dans quelques-uns des cas les plus simples, qui sont nécessaires pour le cours de mécanique.

DEUXIÈME DIVISION.

ANALYSE ALGÉBRIQUE.

RÉSOLUTION des équations du troisième et du quatrième degré, par les méthodes les plus directes et les plus simples.

Manière de déterminer la somme des carrés, celle des cubes, et autres fonctions invariables des racines d'une équation donnée.

Démonstration du théorème qui fait connaître le nombre des racines positives et celui des racines négatives d'une équation dont toutes les racines sont réelles.

Démontrer que toute équation de degré pair est décomposable en facteurs réels du second degré.

Établir les formules $(\cos .x + \sqrt{(-1)} \sin .x)^m = \cos .mx + \sqrt{(-1)} \sin .mx$, $e\rho^{x\sqrt{(-1)}} = \cos .x + \sqrt{(-1)} \sin us\, x$.

Usage de la premiere, pour avoir toutes les racines des équations $x^m - 1 = 0$, $x^m + 1 = 0$, ce qui conduit au théorème de *Cotes.* (On pourra démontrer aussi ce théorème par la voie des constructions géométriques.)

Faire voir comment on peut résoudre, par la table des sinus, toute équation du troisième degré qui tombe dans le cas irréductible.

Démontrer, sur quelques exemples pris dans les fonctions algébriques circulaires et logarithmiques, que toute quantité imaginaire se réduit toujours à la forme $a + b\sqrt{-1}$, a et b étant réels.

CALCUL INTÉGRAL.

Complément de la méthode donnée dans la première partie, pour intégrer les fractions rationnlles. Manière de trouver directement les coëfficiens des fractions partielles.

Développemens sur l'intégration des formules qui contiennent des fonctions circulaires, logarithmiques ou exponentielles.

Formule générale de l'aire d'une surface courbe quelconque, avec des applications à la sphère, au cône droit, &c.

Déterminer, dans quelques cas particuliers, la solidité d'un corps terminé par une surface courbe donnée, et par des plans donnés de position ; application aux solides considérés par *Mascheroni*, dans son petit ouvrage intitulé, *Problemi per gli agrimensori, con varie soluzioni; in Pavia, 1793.*

Intégration de l'équation différentielle $y - p\,x = f{:}p$; p étant égal à $\frac{dy}{dx}$.

Intégration de l'équation différentielle linéaire du second ordre, avec le développement du cas où les coèfficiens sont constans.

Faire voir comment une équation différentielle du second ordre, où l'on a supposé une différentielle première constante, peut être changé en une autre qui ne suppose aucune différence constante.

Intégrer par approximation les équations différentielles du premier et du second ordre.

Intégration des équations linéaires simultanées du premier et du second ordre, à coèfficiens constans.

Faire voir quelle doit être la relation entre P et Q, pour que $dz - p\,dx - q\,dy = 0$ soit l'équation différentielle d'une surface continue.

Donner une idée du calcul aux différences finies, et des élémens du calcul des variations.

C.3 The syllabus of Lacroix's course of analysis at the *École Polytechnique* in 1805-1807

C.3.1 The official programme

These are the official programmes of analysis for the first year in 1805-1806 and second year in 1806-1807. Both are for the "1st division", because it was in 1806 that this expression started to apply to the second year instead of the first.

They are reproduced from [Éc. Pol. *Arch*, an 14, 39-42; year 1806, 25-26]. They are also in [Gilain *1988*, 97-99], because these happen to be the years when a certain Augustin-Louis Cauchy studied at the *École Polytechnique*.

1805-1806

PROGRAMME D'ANALYSE

—————

1.re DIVISION.

ANALYSE ALGÉBRIQUE.

DÉVELOPPEMENT de quelques fonctions en séries, par la méthode des coèfficiens indéterminés.

Loi générale des suites récurrentes observée dans le développement des fractions rationnelles.

Les progressions géométriques sont des suites récurrentes dont l'échelle de relation n'a qu'un terme ; les progressions arithmétiques sont des suites récurrentes dont l'échelle de relation est composé de deux termes 2 et -1.

Examen particulier des suites récurrentes dont l'échelle de relation a deux termes ; leur décomposition en deux progressions géométriques, et de là leur terme général.

Revue des formules trigonométriques les plus utiles, et de l'équation exponentielle qui a lieu entre un nombre et son logarithme.

Déduire de cette équation les propriétés générales des logarithmes ; donner les séries qui servent à les calculer ; comparer les différens systèmes entre eux ; et faire voir comment on peut passer de l'un à l'autre.

Quelques notions sur les fonctions en général, et sur leur division en fonctions entières, rationnelles, &c.

CALCUL DIFFÉRENTIEL.

Établir les notions des différentielles sur la théorie des limites.

Donner les différentielles des formules x^m, xy, $\frac{x}{y}$, d'après lesquelles on trouve aisément celles de toute fonction algébrique proposée d'une ou de plusieurs variables, implicite ou explicite.

Différentielles des fonctions circulaires, logarithmiques et exponentielles, tant simples que combinées.

Différentielles seconde, troisième, &c.

Démonstration du théorème de *Taylor*.

Démonstration de la formule du binome, dans le cas de l'exposant fractionnaire ou négatif.

La théorie des racines égales, par le calcul différentiel.

Application du théorème de *Taylor*, au développement des séries qui donnent les logarithmes, les exponentielles, les sinus et cosinus en fonctions de l'arc, et réciproquement.

Ce même théorème étendu à deux variables.

Notions des différentielles partielles.

Théorie des *maxima* et des *minima* pour les fonctions d'une et de deux variables. Manière de distinguer le *maximum* du *minimum*.

Application des exemples choisis.

Formules des sous-tangentes, sous-normales, tangentes, &c., déduites de la considération des limites. Détermination des asymptotes.

Expression du rayon de courbure.

Propriétés générales de la développée ; manière d'en trouver l'equation.

Application aux sections coniques, à la cycloïde, &c. ; donner la développée de la parabole, et la rectification de cette développée.

Changer une fonction ou une équation différentielle du second ordre, où une différentielle première a été supposée constante, en une autre qui ne suppose aucune différentielle constante.

On démontrera succinctement que la fraction $\frac{P}{Q}$, dont les deux termes s'évanouissent lorsque $x = a$, est égale à $\frac{dP}{dQ}$; ce qui suffira presque toujours pour en déterminer la valeur.

CALCUL INTÉGRAL.

Notions sur l'intégration en général.

Intégration des différentielles monomes, et des fonctions entières.

Cas d'intégrabilité des différentielles binomes.

Intégration des fractions rationnelles, dans les cas les plus simples et par la méthode des coèfficiens indéterminés. (On réservera pour la seconde année le développement des cas les plus composés.)

Manière de rendre rationnelles les différentielles affectées du radical $\sqrt{(a + bx + cxx)}$.

Réduction des différentielles binomes, appliquée principalement aux formules qui s'intègrent par les arcs de cercle et les logarithmes.

Montrer quelles sont les formules principales auxquelles on rapporte les autres, et comment on en exprime l'intégrale.

Intégration par séries.

Formules pour la quadrature des courbes, leur rectification, les surfaces et les solidités des solides de révolution ; insister, dans les applications, sur la détermination des constantes.

1806-1807

PROGRAMME D'ANALYSE

1.ʳᵉ DIVISION.

ANALYSE ALGÉBRIQUE.

RÉSOLUTION algébrique des équations du troisième et du quatrième degré.

Établir les formules $[\cos. x + \sqrt{(-1)}\sin. x]^m = \cos. mx + \sqrt{(-1)}\sin. mx$; et $e^{x\sqrt{(-1)}} = \cos. x + \sqrt{(-1)}\sin. x$.

Usage de la première, pour avoir toutes les racines des équations $x^m - 1 = 0$, $x^m + 1 = 0$, ce qui conduit au thérème de *Côtes*.

Faire voir comment on peut résoudre, par les tables des sinus, toute équation du troisième et du quatrième degré.

CALCUL INTÉGRAL.

COMPLÉMENT de la méthode donnée dans la première partie, pour intégrer les fractions rationnelles. Manière de trouver directement les coèfficiens des fractions partielles.

Intégration des formules qui contiennent des fonctions circulaires, logarithmiques ou exponentielles.

Formules générales du volume et de l'aire d'un corps terminé par une surface courbe quelconque, avec des applications à la sphère ; au cône droit, &c.

Condition pour que $Mdx + Ndy$ soit une différentielle complète, et pour que $Mdx + Ndy + Pdz$ en soit une aussi.

Intégration de ces différentielles lorsqu'elles satisfont à ces conditions.

Intégration de l'équation linéaire du premier ordre.

Théorème des fonctions homogènes.

Intégration des équations homogènes du premier ordre.

Intégration de l'équation différentielle $y - px = f.p$, p étant égal à $\frac{dy}{dx}$.

Intégration de l'équation différentielle linéaire d'un ordre quelconque dans les cas où les coèfficiens sont constans.

Nombre des constantes arbitraires qui doivent entrer dans l'intégrale complète d'une équation différentielle d'un ordre quelconque.

Intégrer par approximation les équations différentielles du premier et du second ordre, à coèfficiens contans.

Donner les élémens du calcul des différences finies, et les formules d'interpolation ; insister sur cette dernière partie.

Application de ces formules à la rectification des courbes, à la quadrature des surfaces, et à la cubature des solides, par approximation.

C.3.2 Summaries of lectures

From the year 1805-1806 onwards the *inspecteur des élèves* Gardeur-Lebrun kept records of the lectures given at the *École Polytechnique* (as well as of interrogations made to the students). In 1808 he used them to make a table summarizing Lacroix's first-year course of 1805-1806 for Ampère's benefit – Ampère was going to be responsible for the first-year course for the first time. This table is kept at [Ampère *AS*, cart. 5, chap. 4, chem. 100], and is reproduced below.

After that comes a table made by me, modelled on Gardeur-Lebrun's, summarizing the records of Lacroix's second-year lectures in 1806-1807 [Éc. Pol. *Arch*, X2c/6].

Cours d'analyse de l'année **14 – 1806**

Marche du Cours d'analyse fait par M.ʳ Lacroix pour la 2.ᵉ Division

Indication des matières	nombre des leçons
Analyse	
Des séries recurrentes	4.
Revue des principales formules trigonométriques	1.
Usage de ces formules	1.
Calcul différentiel	
Notions préliminaires et principes du Calcul différentiel (du N.º 1 au N.º 17 du cours de M. Lacroix)	3.
Des différentiations successives (de 17 a 23)	2.
De la différentiation des fonctions transcendantes (de 23 à 38)	6.
De la différentiation des équations quelconques à deux variables (de 38 à 47)	5.
Recherche des *Maxima* et des *Minima* des fonctions d'une seule variable (de 47 à 52)	2.
Des valeurs que prennent dans certains cas les coëfficiens différentiels, et des expressions qui deviennent 0/0 (de 52 à 60)	4.
Application du calcul différentiel à la théorie des courbes (de 60 à 77)	6.
Recherche des points singuliers des courbes (de 77 à 87)	2.
Exemple de l'analyse d'une courbe (de 87 à 94)	3.
Des courbes osculatrices (de 94 à 101)	4.
Des courbes trancendantes (de 101 à 115)	7.
Du changement de la variable indépendante (de 115 à 120)	1.
De la différentiation des fonctions de deux ou d'un plus grand nombre de variables (de 120 à 133)	4.
Recherche des *Minima* et des *Maxima* des fonctions de deux variables (de 133 à 137)	2.
Calcul intégral	
Intégration des fonctions rationnelles d'une seule variable (de 145 à 160)	4.
De l'intégration des fonctions irrationnelles (de 160 à 169)	2.
De l'intégration des différentielles binomes (de 169 à 175)	1.
De l'intégration par les séries (de 175 à 181)	1.
De la quadrature des courbes, de leur rectification ; de la quadrature des surfaces courbes, et de l'elévation[6]des volumes qu'elles comprennent (de 222 à 243)	2.
total du nombre des leçons	67.

---

[6] *Sic* ; should be "évaluation", as in [Lacroix *1802a*, xxxviii] and in the original lecture record.

1806 – 1807

Marche du Cours d'analyse fait par Lacroix pour la 1.ère Division

Indication des matières	nombre des leçons

Calcul intégral

Suite de l'intégration des fonctions rationelles d'une seule variable, depuis ou on a quitté cette matière dans le cours de la 1.e année (de 155 à 160)	2.
De l'intégration des fonctions irrationnelles (de 160 à 169)	2.
De l'intégration des quantités logarithmiques et exponentielles (de 181 à 192)	3.
De l'intégration des fonctions circulaires (de 192 à 209)	7.
Méthode générale pour obtenir la valeur approchée des intégrales (de 209 à 222, et le 384)	4.
De la cubature des corps terminés par des surfaces courbes &c. (de 243 à 253)	3.
De l'intégration des équations différentielles à 2 variables. De la séparation des variables dans les équations différentielles du 1.er ordre (de 253 à 261)	3.
Recherche du facteur propre à rendre intégrable une équation différentielle du 1.er ordre (de 261 à 268)	$3\frac{1}{2}$.
Des équations du 1.er ordre dans lesquelles les différentielles passent le 1.er degré (de 268 à 271)	$1\frac{1}{2}$.
De l'intégration des équations différentielles du 2.^{d} ordre et des ordres supérieurs (de 271 à 287)	8.
Méthode pour résoudre par approximation les équations différentielles du 1.er et du 2.^{d} ordre (de 288 à 293)	2.
Des solutions particulières des équations différentielles du 1.er ordre (de 293 à 300 inc.)	$1\frac{1}{2}$
Résolution de quelques problêmes géométriques, dependans des équations différentielles (le 304)	$\frac{1}{2}$.
De l'intégration des fonctions de deux ou d'un plus grand nombre de variables (de 305 à 311 ; $Mdx + Ndy = 0$ &c.)	1.
De la méthode des variations (de 323 à 330)	3.
Des maxima et des minima des formules intégrales indéterminées (de 331 à 340)	5.
Du calcul direct des différences (de 340 à 346)	2.
Application du calcul des différences à l'interpolation des suites (de 349 à 356)	3.
Du calcul inverse des différences, par rapport aux fonctions explicites d'une seule variable (de 356 à 361, 363 et 366)	$2\frac{1}{2}$.
Application du calcul des différences à la sommation des suites (le 367)	$\frac{1}{2}$.
Application au calcul des piles de boulets *fin du cours*	1.
De la solution générale des équations des 3e et 4e degré (Ampère)	3.
total du nombre des leçons	62.

Appendix D

Biographical data on a few obscure characters

Many mathematicians are mentioned in this book. Most of them have entries in the *Dictionary of Scientific Biography* [Gillispie *DSB*]. A few more obscure ones do not. In this appendix I give some biographical data on the most relevant ones who are in the latter case.

Jacques Charles, *le géomètre*

Charles is the most obscure of these characters, having been often confused with his contemporary, the physicist and balloonist Jacques-Alexandre-César Charles; both Charles were members of the *Académie des Sciences de Paris* (although not simultaneously) and in its archives the distinction is made by referring to *our* Charles as "Charles, *le géomètre*". Charles, *le géomètre* appears to have been born in Cluny (Burgundy), possibly in 1752. When he was 20 years old he became a teacher of mathematics in a school in Nanterre (on the outskirts of Paris), and a few years later he moved to the capital. He started submitting works to the *Académie des Sciences* in 1770 – very elementary at first, but of increasing sophistication over the years. He was finally elected a member in 1785. The following year Bossut, who held a chair of hydrodynamics at the Louvre, appointed Charles as his assistant. But just a few years later he was affected by serious health problems, including hand paralysis. He died in 1791. Most of his scientific work was on integral calculus, and especially on finite difference equations. He collaborated in the *Encyclopédie Méthodique*. [Gough *1979*; Hahn *1981*]

Jacques-Antoine-Joseph Cousin

Cousin was born in Paris on 29 January 1739. He was professor of physics at the *Collège Royal de France* from 1766 onwards; he was also a teacher at the *École Royale Militaire* from 1769. In 1772 he became a member of the *Académie des*

Sciences de Paris. After the Revolution Cousin got involved in politics: he was elected a municipal officer in 1791, member of the *Corps Législatif* in 1798, and became a senator in 1799. During the Terror, he was imprisoned for eight and a half months. Cousin published several textbooks, including one on physics written in jail. His most famous book is the *Leçons de Calcul Différentiel et de Calcul Intégral* [*1777*] – known especially for its second, enlarged edition, bearing the title *Traité de Calcul Différentiel et de Calcul Intégral* [*1796*]; according to Lacroix, this second edition was also prepared while he was in prison [Delambre *1810*, 96]. Cousin died on 29 December 1800. [Michaud *Biographie*, X, 127-128]

Jean-Guillaume Garnier

Garnier was born in Reims (Champagne) on 13 September 1766. He was above all a teacher. After studying in Reims and Paris, Garnier taught for a year in the military school at Colmar (Alsace), where he met Arbogast. This school being closed in 1789, he returned to Paris. There, he worked for six years at Prony's industrial project for construction of logarithmic and trigonometric tables [Grattan-Guinness *1990*, I, 179-183]; apparently this was his only job not related to education. From year 3 to year 8 (1794-1795 to 1799-1800) he was an examiner of candidates to the *École Polytechnique*. But he also established a private residential school for preparing those candidates. It was as examiner that he went in year 3 to Auxerre where, he later claimed, he *discovered* Fourier, later arranging for his acceptance at the *École Normale* [Quetelet *1867*, 210]. In 1798, when Fourier joined the Egyptian campaign, Garnier was employed as temporary replacement, teaching analysis at the *École Polytechnique*. But when Fourier returned and was appointed prefect at Grenoble, the minister of the interior, Laplace, appointed Poisson, rather than Garnier, to Fourier's post at the *École Polytechnique*; Poisson had been staying for some years at Garnier's school. For the next 13 years Garnier dedicated himself exclusively to his preparatory school. From 1800 to 1814 he also published several textbooks, from arithmetic to integral calculus. Following difficulties with his school in the final period of the Empire and early Restoration, he was invited in 1816 to become professor of mathematics at the University of Gand, in the newly-formed kingdom of united Netherlands; he accepted and never returned to France. In Gand he met Adolphe Quetelet, whom he helped attain the doctorate, and in 1825 they jointly launched a journal called *Correspondance mathématique et physique*. In Gand he also published new editions of some of his textbooks; but he left manuscripts of several other books unpublished. For some reason, when the Belgian universities were reformed in 1835 (following independence in 1830), Garnier was excluded from the teaching body; but at least this time he got a pension. He died in Brussels on 20 December 1840 or 1841. [Quetelet *1867*, 203-243]

Pietro Paoli

Paoli was born in Livorno (Tuscany) on 2 March 1759. He studied first in a Jesuit college in Livorno, and then in the University of Pisa, where he graduated in Law in 1778, studying mathematics and physics at the same time. He taught from 1780 to 1782 at a school in Mantua, from 1782 to 1784 in the University of Pavia, and from 1784 to 1814 at the University of Pisa. After 1814 he held administrative posts related to education in the Grand Duchy of Tuscany. He published many research memoirs, from 1780 to 1836, mostly on differential and/or finite difference equations, but also on series, definite integrals, and other topics. Nearly all of these memoirs were published by the *Società Italiana*, a scientific society of which Paoli was a founding member (1782). However, his most successful work was not a research memoir, but rather a treatise on analysis, the *Elementi d'Algebra* (1794); this is not a book on algebra only, as the title suggests – its first volume covers *lower* algebra and algebraic analysis, but the second volume is on differential and integral calculus (paying much attention to differential equations, and including the calculus of variations and even finite difference equations); this book was much praised by Lacroix and by Lagrange. Paoli died in Florence (the capital of the Grand Duchy of Tuscany) on 21 February 1839. [Nagliati *1996*, 80-82, ch. 3; *2000*, 828-830]

Bibliography

Notes on internet references

By default, internet references appear as $<url>$ (*date of access*). However, there are cases where this would not be appropriate, because: 1 – there is not a precise url for the document (which may be divided into several files, typically one image per page); 2 – the document was accessed several times, often with long intervals of time in between. Below I use the following abbreviations to indicate that a digitalized version of a document is available at an online library, and that this version was consulted during the preparation of this work (although usually also the original, paper version was consulted). Only stable libraries are included (*Google Book Search*, for instance, is not). In the case of memoirs, it is the corresponding volume of the journal that is available.

BBAW: <http://bibliothek.bbaw.de/bibliothek-digital/digitalquellen/schriften>
Gallica: <http://gallica.bnf.fr>
GDZ: <http://gdz.sub.uni-goettingen.de>

Notes on 18th-century academic collections

Dates of 18th-century academic collections are notoriously confusing. Here, "*Mém. Acad. Berlin, 1786* (1788)" refers to the volume for 1786, which was published in 1788; "*Commentarii Academiae Scientiarum Petropolitanae* 5 (*1730-1731*), 1738" refers to volume 5, which is for 1730-1731, but was published only in 1738; generally speaking, years in roman type are publication dates, while years in italic are those to which the volumes refer (and which, in a sense, are part of the title).

A few titles of academic collections have been abbreviated:

    Mém. Acad. Berlin, <year> stands for 1 – *Histoire de l'Académie Royale des Sciences et Belles Lettres, Année <year>. Avec les Mémoires pour la même Année*, Berlin ($1745 \leq year \leq 1769$); 2 – *Nouveaux Mémoires de l'Académie Royale des Sciences et Belles-Lettres, Année <year>, avec l'Histoire pour la même année*, Berlin ($1770 \leq year \leq 1786$); 3 – *Mémoires de l'Académie Royale des Sciences et Belles-Lettres*, Berlin ($1786 - 1787 \leq year \leq 1804$); often, the pages indicated are in the second pagination (the first being dedicated to the history of the academy, rather than the memoirs).

Mém. Acad. Paris, <year> stands for *Histoire de l'Académie Royale des Sci-ences. Année <year>. Avec les Mémoires de Mathématique & de Phy-sique*, Paris, 2nd pagination.

Savans Étrangers stands for *Mémoires de Mathématique et de Physique, pré-sentés a l'Académie Royale des Sciences, par divers Savans, et lus dans ses Assemblées*, Paris.

Archival sources

[Acad. R. Sc. *PV*] Registre [or Procès-verbaux] de l'Académie Royale des Scien-ces, *Archives de l'Académie des Sciences de l'Institut de France* (Paris, France) [*Gallica*, as "Procès-verbaux"].

[Ampère *AS*] Papiers de Adrien-Marie Ampère, *Archives de l'Académie des Sci-ences de l'Institut de France* (Paris, France).

[Arbogast *1789*] Louis-François-Antoine Arbogast, "Essai sur de nouveaux Princi-pes de Calcul différentiel et de Calcul intégral", *Biblioteca Medicea Lau-renziana* (Florence, Italy), Ashburnham Appp. 1840 [photocopies sup-plied by Marco Panza].

[Condorcet *Traité*] [Marie-Jean-Antoine-Nicolas de Caritat,] Marquis de Condor-cet, *Traité du Calcul Intégral*, printed proofs of the first few sections of an unpublished treatise, *Bibliothèque de l'Institut de France* (Paris, France), ms. 879.

[Éc. Pol. *Arch*] *Archives de l'École polytechnique* (Palaiseau, France).

[Garnier *1800-1802*] Jean-Guillaume Garnier, *Leçons* [or *Cours*] *d'Analyse algé-brique, différentielle et intégrale*, printed set(s) of lecture notes distri-buted to students of the *École Polytechnique*; a few copies (partial or bound in wrong order) are kept in [Éc. Pol. *Arch*]; 6 parts, each compo-sed of numbered leaves with (usually four) unnumbered pages: I – Cours d'Analyse Algébrique fait en l'an 9, II – Cours d'Analyse différentielle fait in l'an 9, III – Cours de Calcul intégral fait en l'an 9, IV – Cours d'Analyse algébrique fait en l'an 10, V – Cours d'Analyse différentielle fait in l'an dix, VI – Cours de Calcul intégral fait en l'an dix; parts I-III also exist as a set, with preface and titlepage, Paris: Baudouin, Floreal year 9 (April-May 1801) [the full set of six parts is announced as one complete work in *Journal de l'École Polytechnique*, Tome IV, 11ème cahier, (Messidor year 10 = June/July 1802), p. 358].

[Lacroix *AS*] Dossier biographique de Sylvestre-François Lacroix, *Archives de l'Académie des Sciences de l'Institut de France* (Paris, France).

[Lacroix *IF*] Papiers de Sylvestre François Lacroix, *Bibliothèque de l'Institut de France* (Paris, France), ms. 2396-2403.

[Lacroix *LH*] Dossier of Silvestre François de Lacroix as a member of the Légion d'Honneur, *Centre Historique des Archives nationales* (Paris, France), LH1429073.

[Lacroix *UF*] Dossier of Silvestre François de Lacroix as a member of the Université de France, *Centre Historique des Archives nationales* (Paris, France), F/17/21043.

Published sources

[Acad. Berlin *1786*] "Prix", *Mém. Acad. Berlin, 1786* (1788), 1st pagination, pp. 8-9 [*BBAW*].

[Acad. Sc. Paris *Guide*] Éric Brian and Christiane Demeulenaere-Douyère (eds.), *Histoire et mémoire de l'Académie des sciences: Guide de recherches*, Paris, London and New York: Lavoisier Tec et Doc, 1996.

[Acad. Sc. Paris *PV*] Institut de France – Académie des Sciences, *Procès-verbaux des scéances de l'Académie tenues depuis la fondation de l'Institut jusqu'au mois d'août 1835*, 10 vols., Hendaye: Observatoire d'Abbadia, 1910-1922; 1 vol. of index ("Tables générales alphabétiques"), Paris: Imprimerie nationale, 1979 [*Gallica*].

[Ackerberg-Hastings *2004*] Amy Ackerberg-Hastings, "From Cambridge to Cambridge: The Mathematical Significance of John Farrar's European Sojourns", *Proceedings of the Canadian Society for History and Philosophy of Mathematics / Société Canadienne d'Histoire et Philosophie des Mathématiques* 17 (2004), pp. 6-15.

[Aggarwal *2006*] Abhilasha Aggarwal, *British higher education in mathematics for and in India, 1800 - 1880*, PhD thesis, submitted in November 2006 to Middlesex University.

[Ampère *1806*] Adrien-Marie Ampère, "Recherches sur quelques points de la théorie des fonctions dérivées qui conduisent à une nouvelle démonstration de la série de *Taylor*, et à l'expression finie des termes qu'on néglige lorsqu'on arrête cette série à un terme quelconque", *Journal de l'École Polytechnique*, vol. VI, 13ème cahier (April 1806), pp. 148-181.

[Ampère *1815*] Adrien-Marie Ampère, "Considérations générales sur les intégrales des équations aux différentielles partielles", *Journal de l'École Polytechnique*, vol. X, 17ème cahier (January 1815), pp. 549-611.

[Anonymous *1818*] Ot., "Lacroix (Silvestre-François)", *Biographie des Hommes Vivants*, vol. 4, Paris: Michaud, 1818, pp. 24-25.

[Anonymous *1900*] "[Anfrage] 88. On the technical terms 'Differential Quotient, Definite Integral'", *Bibliotheca Mathematica* (3rd series) 1 (1900), p. 517.

[Arbogast *1791*] Louis-François-Antoine Arbogast, *Mémoire sur la nature des Fonctions Arbitraires qui entrent dans les intégrales des équations aux différentielles partielles*, St. Pétersbourg: Imprimerie de l'Académie Impériale des Sciences, 1791.

[Banionis *2001*] Juozas Banionis, *Matematinė mintis Lietuvoje (istorinė apžvalga iki 1832 m.)*, Vilnius: Vilniaus Pedagoginis Universitetas, 2001 [available online <http://www.vpu.lt/bibl/elpvu/28097.pdf> (27 December 2006)].

[Beckers *2000*] Danny J. Beckers, "Positive Thinking. Conceptions of Negative Quantities in the Netherlands and the Reception of Lacroix's Algebra Textbook", *Revue d'Histoire des Mathématiques* 6 (2000), pp. 95-126.

[Belhoste *1992*] Bruno Belhoste, "Sylvestre-François Lacroix et la géométrie descriptive", appendix 18 to J. Dhombres (ed.), *L'École normale de l'an III − Leçons de mathématiques*, Paris: Dunod, 1992, pp. 564-568.

[Belhoste *1995*] Bruno Belhoste (ed.), *Les sciences dans l'enseignement secondaire français - textes officiels*, Paris: Institut national de recherche pédagogique and Éditions Économica, 1995.

[Belhoste *2003*] Bruno Belhoste, *La Formation d'une Technocratie – L'École polytechnique et ses élèves de la Révolution au Second Empire*, Paris: Belin, 2003.

[Belhoste & Taton *1992*] Bruno Belhoste and René Taton, "L'invention d'une langue des figures", in J. Dhombres (ed.), *L'École normale de l'an III − Leçons de mathématiques*, Paris: Dunod, 1992, pp. 269-303.

[Jac. Bernoulli *Series*] Jacob (I) Bernoulli, "Tractatus de Seriebus Infinitis", in *Ars Conjectandi...*, Basilea: Thurnisii fratres, 1713, pp. 241-306.

[Jac. Bernoulli *1696*] Jacob (I) Bernoulli, "Problema Beaunianum universalius conceptum", *Acta Eruditorum*, Jul. 1696, pp. 332-337 = *Jacobi Bernoulli Basileensis Opera*, vol. II, Geneva: Cramer, 1744, pp. 731-739.

[Joh. Bernoulli *Integralium*] Johann (I) Bernoulli, "Lectiones Mathematicæ de Methodo Integralium", in *Johannis Bernoulli Opera Omnia*, vol. III, Lausanna & Geneva: Marcus-Michael Bousquet, 1742, pp. 385-558 [composed in 1691-1692].

[Joh. Bernoulli *1694*] Johann (I) Bernoulli, "Modus generalis construendi omnes æquationes differentiales primi gradus", *Acta Eruditorum*, Nov. 1694, pp. 435-437 = *Johannis Bernoulli Opera Omnia*, vol. I, Lausanna & Geneva: Marcus-Michael Bousquet, 1742, pp. 123-125.

[Nic. Bernoulli *1720*] Nicolaus (II) Bernoulli, "Exercitatio geometrica de Trajectoriis Orthogonalibus", 1st part *Acta Eruditorum*, May 1720, pp. 223-237, 2nd part *Actorum Eruditorum Supplementa* 7 (1721), pp. 303-326, 3rd part *Actorum Eruditorum Supplementa* 7 (1721), pp. 337-353; also in *Johannis Bernoulli Opera Omnia*, vol. II, Lausanna & Geneva: Marcus-Michael Bousquet, 1742, pp. 423-472 [page references are to this later edition].

[Bézout *1779*] Étienne Bézout, *Théorie générale des équations algébriques*, Paris: Pierres, 1779.

[Bézout *1796*] Étienne Bézout, *Cours de Mathématiques, à l'usage des Gardes du Pavillon et de la Marine*, 6 vols., Paris: Baudelot & Eberhart, year IV (1796) [first ed. is from 1764-1769].

[Bézout *1824*] Étienne Bézout, *First Principles of the Differential and Integral Calculus [...]*, Cambridge, Massachusetts: University Press, 1824.

[Binet *1809*] Paul René Binet, "Mémoire sur la fonction dérivée, ou coefficient différenciel du premier ordre", *Nouveau Bulletin des Sciences, par la Société Philomatique*, Vol. I, n. 16 (Jan 1809), pp. 275-278.

[Biot *1797*] Jean-Baptiste Biot, "Considérations sur les Intégrales des Équations aux différences finies", *Journal de l'École Polytechnique*, vol. IV, 11ème cahier (1802), pp. 182-198 [submitted to the Institut national on 6 Ventose year 5 = 24 February 1797].

[Biot *1799*] Jean-Baptiste Biot, "Mémoire sur les équations aux différences mêlées", *Mémoires présentés à l'Institut des Sciences, Lettres et Arts par divers Savans et lus dans ses Assemblées. Sciences Mathématiques et Physiques* I (1806), pp. 296-327.

[Biot *1800*] Jean-Baptiste Biot, "Considérations sur les équations aux différences mêlées", *Bulletin des Sciences, par la Société Philomatique*, vol. II, n.° 35 (Pluviose year 8 = Jan.-Feb. 1800), pp. 86-88 [*Gallica*].

[Blanc *1957*] Charles Blanc, "Preface des volumes II/6 et II/7", [Euler *Opera*, series 2, VI, vii-xxxv].

[Bos *1974*] Henk J. M. Bos, "Differentials, higher-order differentials and the derivative in the Leibnizian calculus", *Archive for History of Exact Sciences* 14 (1974), pp. 1-90.

[Bos *1980*] Henk J. M. Bos, "Newton, Leibniz and the Leibnizian Tradition", in Ivor Grattan-Guinness (ed.), *From the Calculus to Set Theory*, London: Duckworth, 1980; 2nd ed.: Princeton, New Jersey: Princeton University Press, 2000, pp. 49-93.

[Bos *1984*] Henk J. M. Bos, "Arguments on Motivation in the Rise and Decline of a Mathematical Theory; the 'Construction of Equations', 1637 - ca. 1750", *Archive for History of Exact Sciences* 30 (1984), pp. 331-380.

[Bos *1986*] Henk J. M. Bos, "The Concept of Construction and the Representation
         of Curves in Seventeenth-Century Mathematics" (invited lecture at the
         International Congress of Mathematicians at Berkeley, CA, USA, 1986),
         Lectures in the History of Mathematics, American Mathematical Society
         and London Mathematical Society, 1993, pp. 23-36.

[Bos *2001*] Henk J. M. Bos, *Redifining geometrical exactness: Descartes' transfor-
         mation of the early modern concept of construction*, New York: Springer,
         2001.

[Bos *2004*] Henk J. M. Bos, "The role of the concept of construction in the tran-
         sition from inverse tangent problems to differential equations", *Oberwol-
         fach Reports* 1 (2004), pp. 2735-2736 [extended abstract of a talk given at
         the workshop *The History of Differential Equations, 1670-1950*, Mathe-
         matisches Forschungsinstitut Oberwolfach, October 31st – November
         6th, 2004].

[Bossut *1798*] Charles Bossut, *Traités de Calcul Différentiel et de Calcul Intégral*,
         2 vols., Paris: de l'Imprimerie de la République, An VI (1798).

[Boyer *1939*] Carl B. Boyer, *The Concepts of the Calculus – A Critical and His-
         torical Discussion of the Derivative and the Integral*, New York: Colum-
         bia University Press, 1939; 2nd printing with minor corrections, 1949;
         facsimile reprint under the title *The History of the Calculus and its Con-
         ceptual Development*, New York: Dover, 1959.

[Boyer *1956*] Carl B. Boyer, *History of Analytic Geometry*, New York: Scripta
         Mathematica, 1956; facsimile reprint, Mineola, New York: Dover, 2004.

[Brezinski *1991*] Claude Brezinski, *History of Continued Fractions and Padé Ap-
         proximants*, Berlin: Springer, 1991.

[Cajori *1919*] Florian Cajori, *A History of Mathematics*, New York: MacMillan,
         1919 (2nd. ed.).

[Cajori *1928-1929*] Florian Cajori, *A History of Mathematical Notations*, 2 vols.,
         Chicago: Open Court, 1928, 1929; reprinted, 2 vols. bound as one, New
         York: Dover, 1993.

[Cantor *1879*] Moritz Cantor, "Gruson: Johann Philipp G., auch Grüson",
         Allgemeine Deutsche Biographie, X, Leipzig: Duncker & Humblot,
         1879, pp. 65-66; <http://mdz.bib.bvb.de/digbib/lexika/adb/images/
         adb010@ebt-link?target=idmatch(entyref,adb0100067)> (21 January
         2007).

[Carnot *1797*] Lazare Carnot, *Réflexions sur la Métaphysique du Calcul Infinité-
         simal*, Paris: Duprat, An V (1797); also in *Oeuvres Mathématiques du
         Citoyen Carnot*, Basle: J. Decker, 1797, pp. 125-204.

[Cauchy *1821*] Augustin-Louis Cauchy, *Cours d'Analyse de l'École Royale Polyte-chnique. 1re Partie: Analyse Algébrique*, Paris: Debure, 1821; facsimile reprint, Paris: Jacques Gabay, 1989.

[Cauchy *1823*] Augustin-Louis Cauchy, *Résumé des leçons donnés à l'École Royale Polytechnique sur le calcul infinitésimal*, Paris: Debure, 1823; facsimile reprint, Paris: ACL-éditions, 1987.

[Cauchy *1981*] Augustin-Louis Cauchy, *Équations Différentielles Ordinaires*, Paris and Saint-Laurent: Études Vivantes / New York and London: John-son Reprint, 1981 (reprint of an incomplete set of notes from lectures given to second-year students of the *École Polytechnique* in the 1820's, with an introduction by Christian Gilain).

[Chabert *1999*] Jean-Luc Chabert (ed.), *A History of Algorithms — From the Pebble to the Microchip*, Berlin: Springer, 1999.

[Charles *1785a*] Jacques Charles, "Recherches sur les intégrales des équations aux différences finies, et sur d'autres sujets", *Savans Étrangers* 10 (1785), pp. 573-588.

[Charles *1785b*] Jacques Charles, "Théorème sur les équations en différences fi-nies", *Mém. Acad. Paris, 1783* (1786), pp. 560-562 [read on 23 November 1785; *Gallica*].

[Charles *1785c*] Jacques Charles, "INTÉGRAL (Calcul intégral des équations en différences finies)", *Encyclopédie Méthodique - Mathématiques*, vol. II, Paris: Panckoucke and Liège: Plomteux, 1785, facsimile reprint, Paris: ACL-éditions, 1987, pp. 221-225 [*Gallica*].

[Charles *1785d*] Jacques Charles, "INTÉGRAL (Calcul intégral des équations en différences mêlées)", *Encyclopédie Méthodique - Mathématiques*, vol. II, Paris: Panckoucke and Liège: Plomteux, 1785, facsimile reprint, Paris: ACL-éditions, 1987, pp. 225-226 [extracted from [Charles *1785a*, 584-585]; *Gallica*].

[Charles *1788*] Jacques Charles, "Recherches sur les Principes de la différenciation, et sur les Intégrales connues jusqu'ici sous le nom d'*Intégrales particuliè-res*" and "Suite du Mémoire sur les Principes de la différenciation, &c.", *Mém. Acad. Paris, 1788* (1791), pp. 115-139.

[Clairaut *1731*] Alexis Claude Clairaut, *Recherches sur les courbes à double cour-bure*, Paris, 1731 [*Gallica*].

[Clairaut *1734*] Alexis Claude Clairaut, "Solution de plusieurs Problemes où il s'agit de trouver des Courbes dont la proprieté consiste dans une certaine relation entre leurs branches, exprimée par une Equation donnée", *Mém. Acad. Paris, 1734* (1736), pp. 196-215 [*Gallica*].

[Clairaut *1740*] Alexis Claude Clairaut, "Sur l'Intégration ou la Construction des Équations différentielles du premier ordre", *Mém. Acad. Paris, 1740* (1742), pp. 293-323 [*Gallica*].

[Condorcet *1765*] [Marie-Jean-Antoine-Nicolas de Caritat,] Marquis de Condorcet, *Du Calcul Intégral*, Paris: Didot, 1765.

[Condorcet *1768*] [Marie-Jean-Antoine-Nicolas de Caritat,] Marquis de Condorcet, *Le Marquis de Condorcet à M^r. d'Alembert, sur le systême du monde et sur le calcul intégral* [part of *Essais d'Analyse*, vol. I (and only)], Paris: Didot, 1768.

[Condorcet *1770*] [Marie-Jean-Antoine-Nicolas de Caritat,] Marquis de Condorcet, "Mémoire sur les équations aux différences partielles", *Mém. Acad. Paris, 1770* (1773), pp. 151-178 [*Gallica*].

[Condorcet *1771*] [Marie-Jean-Antoine-Nicolas de Caritat,] Marquis de Condorcet, "Sur la détermination des fonctions arbitraires qui entrent dans les intégrales des équations aux différences partielles", *Mém. Acad. Paris, 1771* (1774), pp. 49-74 [*Gallica*].

[Condorcet *1770-1773*] [Marie-Jean-Antoine-Nicolas de Caritat,] Marquis de Condorcet, "Sur les solutions particulières des équations différentielles", *Mélanges de Philosophie et de Mathématique de la Société Royale de Turin* 5 (*1770-1773*), 2nd pagination (classe mathématique), pp. 12-15.

[Coolidge *1940*] Julian Lowell Coolidge, *A history of geometrical methods*, Mineola, New York: Dover, 2003 [reprint of the 1940 OUP edition].

[Cousin *1777*] Jacques-Antoine-Joseph Cousin, *Leçons de Calcul Différentiel et de Calcul Intégral*, Paris: Jombert, 1777.

[Cousin *1796*] Jacques-Antoine-Joseph Cousin, *Traité de Calcul Différentiel et de Calcul Intégral*, 2 vols., Paris: Régent & Bernard, year 4 – 1796.

[Craik *1999*] Alex D. D. Craik, "Calculus and Analysis in Early 19th-Century Britain: The Work of William Wallace", *Historia Mathematica* 26 (1999), pp. 239-267.

[Craik *2005*] Alex D. D. Craik, "Prehistory of Faà di Bruno's Formula", *American Mathematical Monthly* 112 (2005), pp. 119-130.

[Cramer *1750*] Gabriel Cramer, *Introduction à l'analyse des lignes courbes algébriques*, Genève: fréres Cramer & Cl. Philibert, 1750.

[d'Alembert *1747*] Jean le Rond d'Alembert, "Recherches sur la courbe que forme une corde tendüe mise en vibration" and "Suite des Recherches sur...", *Mém. Acad. Berlin, 1747* (1749), pp. 214-219 and 220-249 [*BBAW*].

[d'Alembert *1750*] Jean le Rond d'Alembert, "Addition au Memoire sur la courbe que forme une corde tendüe, mise en vibration", *Mém. Acad. Berlin, 1750* (1752), pp. 355-360 [*BBAW*].

[d'Alembert *1754-1756*] Jean le Rond d'Alembert, *Recherches sur différens points importans du systême du Monde*, 3 vols., Paris: David, 1754, 1754, 1756.

[d'Alembert *1761*] Jean le Rond d'Alembert, "Recherches sur les vibrations des Cordes Sonores", *Opuscules Mathématiques* I, Paris: David, 1761, pp. 1-64 [*Gallica*].

[d'Alembert *1768*] Jean le Rond d'Alembert, "Réflexions sur les suites divergentes ou convergentes", *Opuscules Mathématiques* V, Paris: Briasson, pp. 171-183 [*Gallica*].

[d'Alembert *1780*] Jean le Rond d'Alembert, "Sur les Fonctions discontinues", *Opuscules Mathématiques* VIII, Paris: Claude-Antoine Jombert, 1780, pp. 302-308 [*Gallica*].

[Deakin *1985*] Michael A. B. Deakin, "Euler's Invention of Integral Transforms", *Archive for History of Exact Sciences* 33 (1985), pp. 307-319.

[Debnath *2004*] Lokenath Debnath, "A brief historical introduction to fractional calculus", *International Journal of Mathematical Education in Science and Technology* 35 (2004), pp. 487-501.

[Delambre *1810*] Jean Baptiste Joseph Delambre, *Rapport historique sur les progrès des sciences mathématiques depuis 1789, et sur leur état actuel*, Paris: Imprimerie Impériale; facsimile reprint, Amsterdam: B. M. Israël, 1966; modern edition with preface by Denis Woronoff and introduction and notes by Jean Dhombres, *Rapports à l'Empereur sur le progrès des sciences, des lettres et des arts depuis 1789: I. Sciences mathématiques*, Paris: Belin, 1989.

[Demidov *1982*] Serguei S. Demidov, "The Study of Partial Differential Equations of the First Order in the 18th and 19th Centuries", *Archive for History of Exact Sciences* 26 (1982), pp. 325-350.

[De Morgan *1836-1842*] Augustus De Morgan, *The Differential and Integral Calculus*, London: Baldwin, 1836-1842 [published in 25 parts].

[Descartes *Géométrie*] René Descartes, *La Géométrie*, appendix to *Discours de la méthode*, 1637, pp. 297-413; facsimile reprint, with a translation into English by David Eugene Smith and Marcia L. Latham, as *The Geometry of René Descartes*, Chicago and London: Open Court, 1925 [page references up to 241 are to the English translation, while from 297 upwards refer to the original page numbers, included in the facsimile].

[Dhombres *1985*] Jean Dhombres, "French Mathematical Textbooks from Bézout to Cauchy", *Historia Scientiarum* 28 (1985), pp. 91-137.

[Dhombres *1986*] Jean Dhombres, "Quelques aspects de l'histoire des équations fonctionnelles liés à l'évolution du concept de fonction", *Archive for History of Exact Sciences* 36 (1986), pp. 91-181.

[Dhombres *1987*] Jean Dhombres, Introduction ("L'École polytechnique et ses historiens") and "Annexes", accompanying a facsimile reprint of [Fourcy *1828*], Paris: Belin, 1987 [the pagination is independent from that of the reprint, which is placed between pages 70 and 71].

[Dhombres & Pensivy *1988*] Jean Dhombres and Michel Pensivy, "Esprit de rigueur et présentation mathématique au XVIIIème siècle: le cas d'une démonstration d'Aepinus", *Historia Mathematica* 15 (1988), pp. 9-31.

[Domingues *2004a*] João Caramalho Domingues, "Variables, limits, and infinitesimals in Portugal in the late 18th century", *Historia Mathematica* 31 (2004), pp. 15-33.

[Domingues *2004b*] João Caramalho Domingues, "A puzzling remark by Euler on constant differentials", *Oberwolfach Reports* 1 (2004), pp. 2747-2750 [extended abstract of a talk given at the workshop *The History of Differential Equations, 1670-1950*, Mathematisches Forschungsinstitut Oberwolfach, October 31st – November 6th, 2004].

[Domingues *2005*] João Caramalho Domingues, "S.F. Lacroix, *Traité du calcul différentiel et du calcul intégral*, first edition (1797-1800)", in Ivor Grattan-Guinness (ed.), *Landmark Writings in Western Mathematics, 1640-1940*, Amsterdam: Elsevier, 2005, pp. 277-291.

[Doyle *1996*] J. F. Doyle, "Some Thoughts on Fractional Calculus and the Abolition of Integration Formulae", *Teaching Mathematics and its Applications* 15 (1996), pp. 16-19.

[Dugac *1983*] Pierre Dugac, "Euler, d'Alembert et les fondements de l'analyse", in J. J. Burkhardt, E. A. Fellman and W. Habicht (eds.), *Leonhard Euler 1707-1783: Beiträge zu Leben und Werk*, Basel: Birkhäuser, 1983, pp. 171-184.

[Éc. Pol. *Concours 1802*] "Programme des connaissances exigées des Candidats qui se présenteront au concours pour l'École Polytechnique, ouvert le 1.er complémentaire de l'an 10", *Journal de l'École Polytechnique*, vol. IV, 11ème cahier (Messidor year 10 = June-July 1802), p. 381.

[Éc. Pol. *Cours 1804*] "Tableau qui indique l'ordre des Cours, leur durée, et les instituteurs qui en sont chargés", *Correspondance sur l'École Impériale Polytechnique*, vol. I, n.º 1 (April 1804), pp. 3-7.

[Éc. Pol. *Extraits Conseil*] Conseil de l'École Polytechnique, "Extraits des Registres", transcribed by E. L. Dooley, *Bulletin de la Société des Amis de la Bibliothèque de l'Ecole polytechnique* 12 (November 1994), pp. 9-63.

[Éc. Pol. *Rapport*] Conseil de Perfectionnement de l'École polytechnique, *Rapport sur la situation de l'École polytechnique* [published annually from year 9 (1800-1801) onwards; there are slight variations in the title].

[*Encyclopédie*] Denis Diderot and Jean le Rond d'Alembert (eds.), *Encyclopédie ou Dictionnaire Raisonné des Sciences, des Arts et des Métiers*, 28 vols., Paris, 1751-1780.

[Eneström *1899*] Gustaf Eneström, "Sur la découverte de l'équation générale des lignes géodésiques", *Bibliotheca Mathematica* (2nd series) 13 (1899), pp. 19-24.

[Engelsman *1980*] Steven B. Engelsman, "Lagrange's early contributions to the theory of first-order partial differential equations", *Historia Mathematica* 7 (1980), pp. 7-23.

[Engelsman *1984*] Steven B. Engelsman, *Families of Curves and the Origins of Partial Differentiation*, North-Holland Mathematics Studies 93, Amsterdam: Elsevier, 1984.

[Enros *1983*] Philip C. Enros, "The Analytical Society (1812-1813): precursor of the renewal of Cambridge mathematics", *Historia Mathematica* 10 (1983), pp. 24-47.

[Euler *Opera*] Leonhard Euler, *Opera Omnia*, 4 series, $29 + 31 + 12 + 4$ vols. [at least 6 more vols. in series 4 are planned], Leipzig and Berlin: Teubner [early volumes], Zürich: Orell Füssli [mid volumes], Basel: Birkhäuser [recent volumes], 1911-[in progress].

[Euler *Differentialis*] Leonhard Euler, *Institutiones Calculi Differentialis*, St. Petersburg: Academia Imperialis Scientiarum, 1755 [two parts in one volume; articles numbered separately in each part] = [Euler *Opera*, series 1, X]; 2nd ed., posthumous, with a supplement ("Dilucidationes" and "Adnotationes"), 2 vols., Ticinum: Petrus Galeatius, 1787.

[Euler *Integralis*] Leonhard Euler, *Institutionum Calculi Integralis...*, 3 vols., St. Petersburg: Academia Imperialis Scientiarum, 1768, 1769, 1770 = [Euler *Opera*, series 1, XI-XIII]; 2nd ed., posthumous, with a fourth volume collecting several memoirs of Euler on the subject, St. Petersburg: Academia Imperialis Scientiarum, 1792, 1792, 1793, 1794 [references are usually to article, but when page references are required, they are to the second edition].

[Euler *Introductio*] Leonhard Euler, *Introductio in Analysin Infinitorum*, 2 vols., Lausanna: Marcus-Michael Bousquet, 1748 = [Euler *Opera*, series 1, VIII-IX]; facsimile reprint, Bruxelles: Culture et Civilisation, 1967 [*Gallica*].

[Euler *1730-1731*] Leonhard Euler, "De progressionibus transcendentibus, seu quarum termini generales algebraice dari nequeunt", *Commentarii Academiae Scientiarum Petropolitanae* 5 (*1730-1731*), 1738, pp. 36-57 [also available at <http://math.dartmouth.edu/~euler/

docs/originals/E019.pdf> (26 September 2006)] = [Euler *Opera*, series 1, XIV, 1-24].

[Euler *1736*] Leonhard Euler, *Mechanica sive Motus Scientia Analytice Exposita*, 2 vols., St. Petersburg: Academia Scientiarum, 1736 = [Euler *Opera*, series 2, I-II] [*Gallica*].

[Euler *1748*] Leonhard Euler, "Sur la Vibration des Cordes", *Mém. Acad. Berlin*, *1748* (1750), pp. 69-85 [*BBAW*] = [Euler *Opera*, series 2, X, 63-77].

[Euler *1756*] Leonhard Euler, "Exposition de quelques Paradoxes dans le Calcul intégral", *Mém. Acad. Berlin*, *1756* (1758), pp. 300-321 [*BBAW*] = [Euler *Opera*, series 1, XXII, 214-236].

[Euler *1760*] Leonhard Euler, "Recherches sur la courbure des surfaces", *Mém. Acad. Berlin*, *1760* (1767), pp. 119-143 [*BBAW*] = [Euler *Opera*, series 1, XXVIII, 1-22].

[Euler *1764*] Leonhard Euler, "De motu corporis ad duo centra virium fixa attracti", *Novi comentarii academiae scientiarum Petropolitanae* 10 (*1764*), 1766, pp. 207-242 [also available at <http://math.dartmouth.edu/ ~euler/docs/originals/E301.pdf> (8 August 2005)] = [Euler *Opera*, series 2, VI, 209-246].

[Euler *1765a*] Leonhard Euler, "De usu functionum discontinuarum in analysi", *Novi comentarii academiae scientiarum Petropolitanae* 11 (*1765*), 1767, pp. 3-27 [also available at <http://math.dartmouth.edu/~euler/ docs/originals/E322.pdf> (12 February 2006)] = [Euler *Opera*, series 1, XXIII, 74-91].

[Euler *1765b*] Leonhard Euler, "Sur le mouvement d'une corde qui au commencement n'a été ébranlée que dans une partie", *Mém. Acad. Berlin*, *1765* (1767), pp. 307-334 [*BBAW*] = [Euler *Opera*, series 2, X, 426-451].

[Euler *1765c*] Leonhard Euler, "Éclaircissemens plus détaillés sur la génération et la propagation du son et sur la formation de l'écho", *Mém. Acad. Berlin*, *1765* (1767) pp. 335-363 [*BBAW*] = [Euler *Opera*, series 3, I, 540-567].

[Euler *1771*] Leonhard Euler, "Evolutio formulae integralis $\int x^{f-1}dx(lx)^{\frac{m}{n}}$, integratione a valore $x = 0$ ad $x = 1$ extensa", *Novi comentarii academiae scientiarum Petropolitanae* 16 (*1771*), 1772, pp. 91-139 = [Euler *Integralis*, IV, 78-121] = [Euler *Opera*, series 1, XVII, 316-357]; [page references are to [Euler *Integralis*, IV]].

[Euler *1774a*] Leonhard Euler, "De valore formulae integralis $\int \frac{z^{\lambda-\omega} \pm z^{\lambda+\omega}}{1 \pm z^{2\lambda}} \cdot \frac{dz}{z}(lz)^{\mu}$ casu quo post integrationem ponitur $z = 1$", *Novi comentarii academiae scientiarum Petropolitanae* 19 (*1774*), 1775, pp. 30-65 = [Euler *Integralis*, IV, 122-154] = [Euler *Opera*, series 1, XVII, 384-420]; [page references are to [Euler *Integralis*, IV]].

[Euler *1774b*] Leonhard Euler, "Nova Methodus quantitates integrales determinandi", *Novi comentarii academiae scientiarum Petropolitanae* 19 (*1774*), 1775, pp. 66-102 = [Euler *Integralis*, IV, 260-294] = [Euler *Opera*, series 1, XVII, 421-457]; [page references are to [Euler *Integralis*, IV]].

[Euler *1775*] Leonhard Euler, "Methodus inveniendi formulas integrales quae certis casibus datam inter se teneant rationem", in *Opuscula analytica*, vol. II, St. Petersburg: Academia Imperialis Scientiarum, 1785, pp. 178-216 [presented to the St. Petersburg Academy in September 1775] = [Euler *Integralis*, IV, 378-415] = [Euler *Opera*, series 1, XVIII, 209-243]; [page references are to [Euler *Integralis*, IV]].

[Euler *1776*] Leonhard Euler, "Comparatio valorum formulae integralis $\int \frac{x^{p-1}dx}{\sqrt[n]{(1-x^n)^{n-q}}}$ a termino $x = 0$ vsque ad $x = 1$ extensae", *Nova acta academiae scientiarum Petropolitanae* 5 (*1787*), 1789, pp. 86-117 [presented to the St. Petersburg Academy in October 1776] = [Euler *Integralis*, IV, 295-326] = [Euler *Opera*, series 1, XVIII, 392-423]; [page references are to [Euler *Integralis*, IV]].

[Euler & Lagrange *Correspondance*] Adolf P. Juškevič and René Taton (eds.), "Correspondance d'Euler avec J. L. Lagrange", [Euler *Opera*, series 4A, V, 359-518].

[Ferraro *1998*] Giovanni Ferraro, "Some aspects of Euler's theory of series: *inexplicable* functions and the Euler-Maclaurin summation formula", *Historia Mathematica* 25 (1998), pp. 290-317.

[Ferraro *2000*] Giovanni Ferraro, "Functions, Functional Relations, and the Laws of Continuity in Euler", *Historia Mathematica* 27 (2000), pp. 107-132.

[Fontaine *1764*] Alexis Fontaine [des Bertins], *Mémoires donnés à l'Académie Royale des Sciences, non imprimés dans leur temps*, Paris: Imprimerie Royale, 1764 = *Traité de calcul différentiel et integral*, Paris: Imprimerie Royale, 1770.

[Fourcy *1828*] A. Fourcy, *Histoire de l'École polytechnique*, Paris: "chez l'auteur, à l'École polytechnique", 1828; facsimile reprint, with an introduction and appendices by Jean Dhombres, Paris: Belin, 1987.

[Fourier *1796*] Jean-Baptiste Joseph Fourier, *Leçons d'un cours d'analyse rédigées par C. L. Donop*, transcription by Anne-Marie Lorrain of a manuscript of lectures given at the École Polytechnique in the year IV, with an introduction by Anne-Marie Lorrain and Luigi Pepe, Ferrara: Dipartamento di Matematica dell'Universitá di Ferrara, 1989.

[Frankel *1978*] Eugene Frankel, "Career-making in post-revolutionary France: the case of Jean-Baptiste Biot", *The British Journal for the History of Science* 11 (1978), pp. 36-48.

[Fraser *1985*] Craig G. Fraser, "J. L. Lagrange's Changing Approach to the Foundations of the Calculus of Variations", *Archive for History of Exact Sciences* 32 (1985), pp. 151-191. Reprinted in Craig G. Fraser, *Calculus and Analytical Mechanics in the Age of Enlightenment*, Variorum Collected Studies Series, Aldershot, England and Brookfield, Vermont: Ashgate, 1997.

[Fraser *1987*] Craig G. Fraser, "Joseph Louis Lagrange's Algebraic Vision of the Calculus", *Historia Mathematica* 14 (1987), pp. 38-53. Reprinted in Craig G. Fraser, *Calculus and Analytical Mechanics in the Age of Enlightenment*, Variorum Collected Studies Series, Aldershot, England and Brookfield, Vermont: Ashgate, 1997.

[Fraser *1989*] Craig G. Fraser, "The Calculus as Algebraic Analysis: some observations on mathematical analysis in the 18th century", *Archive for History of Exact Sciences* 39 (1988-1989), pp. 317-335. Reprinted in Craig G. Fraser, *Calculus and Analytical Mechanics in the Age of Enlightenment*, Variorum Collected Studies Series, Aldershot, England and Brookfield, Vermont: Ashgate, 1997.

[Fraser *1992*] Craig G. Fraser, "Isoperimetric Problems in the Variational Calculus of Euler and Lagrange ", *Historia Mathematica* 19 (1992), pp. 4-23. Reprinted in Craig G. Fraser, *Calculus and Analytical Mechanics in the Age of Enlightenment*, Variorum Collected Studies Series, Aldershot, England and Brookfield, Vermont: Ashgate, 1997.

[Fraser *1994*] Craig G. Fraser, "The Origins of Euler's Variational Calculus", *Archive for History of Exact Sciences* 47 (1994), pp. 103-141. Reprinted in Craig G. Fraser, *Calculus and Analytical Mechanics in the Age of Enlightenment*, Variorum Collected Studies Series, Aldershot, England and Brookfield, Vermont: Ashgate, 1997.

[Friedelmeyer *1993*] Jean-Pierre Friedelmeyer, *Le calcul des dérivations d'Arbogast dans le projet d'algébrisation de l'analyse à la fin du XVIIIe siècle*, Cahiers d'histoire et de philosophie des sciences, nouvelle série, n.º 43, Nantes: Presses de l'Université de Nantes, 1994 [publication of a 1993 doctoral thesis].

[Garnier *1800*] Jean-Guillaume Garnier, *Notes sur le Calcul différentiel et sur le Calcul intégral, faisant suite à la Méchanique de Bezout*, Paris: Courcier, year IX [because of (abandoned) plans for binding these notes to the fourth and fifth volumes of an edition of Bézout's *Cours*, the pages of differential calculus were numbered 367-501 and those of integral calculus 403-662; there is also an edition dated of the same year but with the title omitting the reference to Bézout's *Méchanique*, some differences in the text, and pages numbered consecutively 367-826; page references are to the latter edition].

[Garnier *1801*] Jean-Guillaume Garnier, "Discours préliminaire" and "Notes d'algèbre", in Alexis-Claude Clairaut, *Élémens d'Algèbre*, 6th ed, Paris: Courcier, year X = 1801, pp. vii-ix and 221-449; also printed independently as *Cours d'Analyse Algébrique*, Paris: Courcier, year X = 1801.

[Garnier *1811*] Jean-Guillaume Garnier, *Leçons de calcul différentiel*, Paris: Ve Courcier, 1811 [presented as 3rd edition, the first being [Garnier *1800*, I]; and the second being [Garnier *1800-1802*, II and V] – see above under **Archival sources**].

[Garnier *1812*] Jean-Guillaume Garnier, *Leçons de calcul intégral*, Paris: Ve Courcier, 1812 [continuation of [Garnier *1811*]; presented as 3rd edition, the first being [Garnier *1800*, II]; and the second being [Garnier *1800-1802*, III and VI] – see above under **Archival sources**].

[Gauss *1814-1815a*] Carl Friedrich Gauss, "Methodus nova integralium valores per approximationem inveniendi", *Commentationes Societatis Regiae Scientiarum Gottingensis recentiores* 3 (*1814-1815*), 1816, 4th pagination, pp. 39-76 [*GDZ*].

[Gauss *1814-1815b*] Carl Friedrich Gauss, "Demonstratio nova altera theorematis omnem functionem algebraicam rationalem integram unius variabilis in factores reales primi vel secundi gradus resolvi posse", *Commentationes Societatis Regiae Scientiarum Gottingensis recentiores* 3 (*1814-1815*), 1816, 4th pagination, pp. 107-134 [*GDZ*].

[Gauss *1814-1815c*] Carl Friedrich Gauss, "Theorematis de resolubilitate functionum algebraicarum integrarum in factores reales demonstratio tertia. Supplementum commentationis praecedentis", *Commentationes Societatis Regiae Scientiarum Gottingensis recentiores* 3 (*1814-1815*), 1816, 4th pagination, pp. 135-142 [*GDZ*].

[Gilain *1981*] Christian Gilain, "Introduction", [Cauchy *1981*, xiii-lvi].

[Gilain *1988*] Christian Gilain, "Condorcet et le Calcul intégral", in Roshdi Rashed (ed.), *Sciences a l'Époque de la Révolution Française – recherches historiques*, Paris: Albert Blanchard, 1988, pp. 87-147.

[Gilain *1989*] Christian Gilain, "Cauchy et le Cours d'analyse de l'Ecole polytechnique", *Bulletin de la Société des Amis de la Bibliothèque de l'Ecole polytechnique*, 5 (July 1989).

[Gilain *2004*] Christian Gilain, "Équations différentielles et systèmes différentiels: de d'Alembert à Cauchy", *Oberwolfach Reports* 1 (2004), pp. 2743-2745 [extended abstract of a talk given at the workshop *The History of Differential Equations, 1670-1950*, Mathematisches Forschungsinstitut Oberwolfach, October 31st – November 6th, 2004].

[Gilain *to appear*] Christian Gilain, "Mathématiques mixtes et mathématiques pures chez d'Alembert: le cas des systèmes différentiels linéaires", *Studies in History of Mathematics dedicated to A. P. Yushkevich*, Académie internationale d'histoire des sciences, to appear.

[Gillispie *DSB*] Charles Coulston Gillispie (ed.), *Dictionary of Scientific Biography*, 16 vols., New York: Charles Scribner's Sons, 1970-1980.

[Gillispie *1971*] Charles Coulston Gillispie, *Lazare Carnot Savant*, Princeton, New Jersey: Princeton University Press, 1971.

[Gillispie *1997*] Charles Coulston Gillispie (with the collaboration of Robert Fox and Ivor Grattan-Guinness), *Pierre-Simon Laplace, 1749-1827: a life in exact science*, Princeton, New Jersey: Princeton University Press, 1997; 2nd printing, Princeton, New Jersey: Princeton University Press, 2000.

[Goldstine *1977*] Herman Heine Goldstine, *A History of Numerical Analysis from the 16th through the 19th Century*, New York: Springer, 1977.

[Goudin & du Séjour *1756*] [Anonymous, in fact Mathieu-Bernard Goudin & Achille-Pierre Dionis du Séjour], *Traité des courbes algébriques*, Paris: Jombert, 1756.

[Gough *1979*] J. B. Gough, "Charles the obscure", *Isis* 70 (1979), pp. 576-579.

[Grabiner *1966*] Judith V. Grabiner, *The Calculus as Algebra: J.L. Lagrange, 1736-1813*, Harvard Dissertations in the History of Science, New York and London: Garland, 1990 [publication of a 1966 PhD thesis].

[Grabiner *1981*] Judith V. Grabiner, *The origins of Cauchy's rigorous calculus*, Cambridge, Massachussetts and London, England: The MIT Press, 1981.

[Grattan-Guinness *1970*] Ivor Grattan-Guinness, *The Development of the Foundations of Mathematical Analyisis from Euler to Riemann*, The MIT Press, 1970.

[Grattan-Guinness *1972*] Ivor Grattan-Guinness (in collaboration with J. R. Ravetz), *Joseph Fourier, 1768-1830*, Cambridge, Massachussets and London, England: MIT Press, 1972.

[Grattan-Guinness *1980*] Ivor Grattan-Guinness, "The Emergence of Mathematical Analysis and its Foundational Progress, 1780-1880", in Ivor Grattan-Guinness (ed.), *From the Calculus to Set Theory*, London: Duckworth, 1980; 2nd ed.: Princeton, New Jersey: Princeton University Press, 2000, pp. 94-148.

[Grattan-Guinness *1990*] Ivor Grattan-Guinness, *Convolutions in French Mathematics, 1800-1840*, 3 vols., Basel, Boston and Berlin: Birkhäuser, 1990.

[Grattan-Guinness *1994*] Ivor Grattan-Guinness, "Functional equations", in Ivor Grattan-Guinness (ed.), *Companion Encyclopedia of the History and*

Philosophy of the Mathematical Sciences, vol. 1, New York and London: Routledge, 1994; 2nd ed.: Baltimore and London: Johns Hopkins Press, undated, pp. 557-562.

[Grattan-Guinness *1997*] Ivor Grattan-Guinness, "Laplace's Integral Solutions to Partial Differential Equations", in [Gillispie *1997*, 259-269].

[Grattan-Guinness *2002*] Ivor Grattan-Guinness, "The End of Dominance: The Diffusion of French Mathematics Elsewhere, 1820-1870", in Karen Hunger Parshall & Adrian C. Rice (eds.), *Mathematics Unbound: The Evolution of an International Mathematical Research Community, 1800-1945*, American Mathematical Society and London Mathematical Society, 2002, pp. 17-44.

[Grattan-Guinness & Engelsman *1982*] Ivor Grattan-Guinness and Steven B. Engelsman, " The manuscripts of Paul Charpit", *Historia Mathematica* 9 (1982), pp. 65-75.

[Greenberg *1981*] John L. Greenberg, "Alexis Fontaine's 'Fluxio-differential Method' and the Origins of the Calculus of Several Variables", *Annals of Science* 38 (1981), pp. 251-290.

[Greenberg *1982*] John L. Greenberg, "Alexis Fontaine's Integration of Ordinary Differential Equations and the Origins of the Calculus of Several Variables", *Annals of Science* 39 (1982), pp. 1-36.

[Grison *1996*] Emmanuel Grison, *L'étonnant parcours du républicain J.H. Hassenfratz (1755-1827)*, Paris: Les Presses de l'École des Mines, 1996.

[Gruyter *History*] Anonymous, "The history of the five publishing houses which became Walter de Gruyter publishers" <http://www.gruyter.de/downloads/verlagsgeschichte_lang_e.pdf> (22 January 2007).

[Gua de Malves *1740*] Jean-Paul de Gua de Malves, *Usages de l'Analyse de Descartes, Pour découvrir, sans le secours du Calcul Différentiel, les Propriétés, ou Affections principales des Lignes Géométriques de tous les Ordres*, Paris: Briasson, 1740.

[Guicciardini *2003*] Niccolò Guicciardini, "Newton's Method and Leibniz's Calculus", in Hans Niels Jahnke (ed.), *A History of Analysis*, History of Mathematics, vol. 24, Providence, Rhode Island: American Mathematical Society and London Mathematical Society, 2003, pp. 73-103.

[*GV*] Peter Geils, Willi Gorzny, and Hilmar Schmuck (dir.), *Gesamtverzeichnis des deutschsprachigen Schrifttums (GV) 1700-1910*, 161 vols., München: Saur, 1979-1987.

[Gyachyauskas *1979*] É. Gyachyauskas, "Mathematics at Vilnius University before 1832", *Lithuanian Mathematical Journal* 19 (1979), pp. 165-171.

[Hahn *1964*] Roger Hahn, "L'enseignement scientifique des Gardes de la Marine au XVIIIe siècle", in René Taton (ed.), *Enseignement et diffusion des sciences en France au XVIIIe siècle*, Paris: Hermann, 1964, pp. 547-558.

[Hahn *1981*] Roger Hahn, "More light on Charles the obscure", *Isis* 72 (1981), pp. 83-86.

[Houtain *1852*] Louis Houtain, "Des solutions singulières de équations différenti-elles", *Annales des Universités de Belgique, années 1851 et 1852*, Bru-xelles: Th. Lesigne, 1854, pp. 971-1323.

[Inocêncio *DBP*] Inocêncio Francisco da Silva, Brito Aranha, Gomes de Brito, Ál-varo Gomes and Ernesto Soares, *Diccionario Bibliographico Portuguez*, 23 vols., Lisboa: Imprensa Nacional, 1858-1958; *facsimile* reprint, Lis-boa: Imprensa Nacional − Casa da Moeda, 1973.

[Itard *1973*] Jean Itard, "Lacroix, Sylvestre François", in [Gillispie *DSB*, VII, 549-551].

[Jahnke *1993*] Hans Niels Jahnke, "Algebraic Analysis in Germany, 1780-1840: Some Mathematical and Philosophical Issues", *Historia Mathematica* 20 (1993), pp. 265-284.

[Johnson *2002*] Warren P. Johnson, "The Curious History of Faà di Bruno's For-mula", *American Mathematical Monthly* 109 (2002), pp. 217-234.

[Jordan *1947*] Charles Jordan, *Calculus of finite differences*, 2nd ed, New York; Chelsea, 1965 (reprint of the 1947 edition).

[Kline *1972*] Morris Kline, "Mathematical Thought from Ancient to Modern Ti-mes", 3 vols., New York and Oxford: Oxford University Press, 1990 (2nd ed.; first ed., in one vol., is from 1972).

[Knobloch *1994*] Eberhard Knobloch, "Determinants", in Ivor Grattan-Guinness (ed.), *Companion Encyclopedia of the History and Philosophy of the Mathematical Sciences*, vol. 1, New York and London: Routledge, 1994; 2nd ed.: Baltimore and London: Johns Hopkins Press, undated, pp. 766-774.

[La Caille *1764*] Nicolas Louis de la Caille, *Leçons Élémentaires d'Astronomie*, Paris: Guerin & Delatour, 1764 (first edition is from 1746).

[La Caille & Marie *1772*] Nicolas Louis de la Caille & Joseph François Marie, *Le-çons Élémentaires de Mathématiques*, Paris: Desaint, 1772 [new edition, highly revised and enlarged by Marie, of a book first published by La Caille in 1741; many other editions].

[Lacroix *Traité*] Silvestre François Lacroix, *Traité du calcul différentiel et du calcul intégral*, 3 vols., Paris: Duprat, year V = 1797, year VI = 1798, year VIII = 1800; 2nd ed.: 3 vols., Paris: Courcier, 1810, 1814, 1819; in the 1st ed. vol. III bears the title *Traité des Différences et des Séries* [*Gallica*].

[Lacroix *1795*] Silvestre François Lacroix, *Essais de Géométrie sur les plans et les surfaces courbes*, 2nd ed. Paris: Duprat, an X - 1802 (1st ed. is from 1795) [this book also bears the subtitle *Élémens de Géométrie descriptive* and is often referred to as *Complément des Élémens de Géométrie* (namely as a part of Lacroix's *Cours élémentaire de Mathématiques pures*)].

[Lacroix *1797*] Silvestre François Lacroix, "Précis historique sur l'Astronomie-Physique (lu à la Société Philomatique.)", *La Décade Philosophique, Littéraire et Politique*, 10 Prairial year 5 = 29 May 1797, pp. 386-392.

[Lacroix *1798a*] Silvestre François Lacroix, "Supplément à la théorie des solutions particulières des équations différentielles", *Bulletin des Sciences, par la Société Philomathique* I (2nd part), n.º 11 (Pluviose year 6 = Febr. 1798), pp. 86-88 [*Gallica*].

[Lacroix *1798b*] Silvestre François Lacroix, *Traité élémentaire de trigonométrie rectiligne et sphérique, et d'application de l'algèbre à la géométrie*, Paris: Courcier, 1813 (6th ed; first ed. is from 1798) [*Gallica*: 4th ed, 1807].

[Lacroix *1799*] Silvestre François Lacroix, *Élémens d'algèbre*, Paris: Courcier, 1807 (6th ed; first ed. is from year 8 = 1799-1800) [*Gallica*: 6th ed, 1807].

[Lacroix *1799-1800*] Silvestre François Lacroix, *Lehrbegriff des Differential- und Integralcalculs*, 2 vols., Berlin: F. T. Lagarde, 1799, 1800; German translation of [Lacroix *Traité*, I] by Johann Philipp Grüson.

[Lacroix *1800*] Silvestre François Lacroix, *Complément des élémens d'algèbre*, Paris: Courcier, year XIII = 1804 (3rd ed; first ed. is from year 8 = 1799-1800).

[Lacroix *1802a*] Silvestre François Lacroix, *Traité élémentaire de calcul différentiel et de calcul intégral*, Paris: Duprat, year X = 1802; 2nd ed., Paris: Courcier, 1806; 3rd ed., Paris: Courcier, 1820; 4th ed., Paris: Bachelier, 1828; 5th ed., Paris: Bachelier, 1837.

[Lacroix *1802b*] Silvestre François Lacroix, "Note sur la résistance des fluides", *Bulletin des Sciences, par la Société Philomathique*, III, n.º 69 (Frimaire year 11 = Nov.-Dec. 1802), pp. 161-163 [Gallica].

[Lacroix *1805*] Silvestre François Lacroix, *Essais sur l'enseignement en général, et sur celui des mathématiques en particulier*, Paris: Courcier, 1805; 2nd ed., Paris: Courcier, 1816; 3rd ed., Paris: Bachelier, 1828; 4th ed., Paris: Bachelier, 1838 [Gallica: 4th ed].

[Lacroix *1812-1814*] Silvestre François Lacroix, *Tratado Elementar de Calculo Differencial, e Calculo Integral*, 2 vols., Rio de Janeiro: Impressão Regia, 1812, 1814; Portuguese translation by Francisco da Silva Torres of parts 1 and 2 (minus the method of variations) of [Lacroix *1802a*, 1st ed].

[Lacroix *1813*] Silvestre François Lacroix, "Notice historique sur la Vie et les Ouvrages de CONDORCET", *Magasin Encyclopédique*, 1813, n.º 6, pp. 54-77 [*Gallica*].

[Lacroix *1816*] Silvestre François Lacroix, *An elementary treatise on the differential and integral calculus*, Cambridge: J. Deighton and sons, 1816; English translation by Charles Babbage, George Peacock, and William Herschel of parts 1 and 2 of [Lacroix *1802a*, 2nd ed], with notes by Peacock and Herschel and an appendix on "differences and series" by Herschel.

[Lacroix *1829*] Silvestre François Lacroix, *Trattato Elementare del Calcolo Differenziale e del Calcolo Integrale*, Firenze: Francesco Cardinali, 1829; Italian translation of [Lacroix *1802a*, 4th ed].

[Lacroix *1830-1831*] Silvestre François Lacroix, *Handbuch der Differential- und Integral-Rechnung*, 3 vols., Berlin: Reimer, 1830, 1831, 1831; German translation by Dr. Fr. Baumann of [Lacroix *1802a*, 4th ed].

[Lacroix & Bézout *1826*] Silvestre François Lacroix & Étienne Bézout, *An Elementary Treatise on Plane and Spherical Trigonometry, and on the Application of Algebra to Geometry; from the Mathematics of Lacroix and Bézout*, Cambridge, Massachussets: Hilliard and Metcalf, at the University Press, 1826 (2nd ed.; first ed. is from 1820).

[Lagrange *Œuvres*] Joseph-Louis Lagrange, *Œuvres de Lagrange*, 14 vols., Paris: Gauthier-Villars, 1867-1892; facsimile reprint, Hildesheim and New York: Georg Olms, 1973 [*Gallica*].

[Lagrange *Calcul*] Joseph-Louis Lagrange, *Leçons sur le calcul des fontions*, 2nd ed. Paris: Courcier, 1806 [*Gallica*] = [Lagrange *Œuvres*, X] (first ed. is from 1801, reprinted with slight changes in 1804 in the *Journal de l'École Polytechnique*, vol. V, 12ème cahier) [page references are to the 1806 printing].

[Lagrange *Fonctions*] Joseph-Louis Lagrange, *Théorie des fonctions analytiques*, Paris: Imprimerie de la République, Prairial year 5 (May/June 1797) = *Journal de l'École Polytechnique*, vol. III, 9ème cahier [*Gallica*]; 2nd ed. Paris, 1813 = [Lagrange *Œuvres*, IX] [page references to the 2nd ed. are to the *Œuvres* printing].

[Lagrange *1759a*] Joseph-Louis Lagrange, "Recherches sur la Méthode de Maximis et Minimis", *Miscellania Philosophico-Mathematica Societatis Privatae Taurinensis* I (1759), 2nd pagination, pp. 18-32 = [Lagrange *Œuvres*, I, 3-20].

[Lagrange *1759b*] Joseph-Louis Lagrange, "Sur l'intégration d'une équation diffé-
        rentielle à différences finies, qui contient la théorie des suites recurrentes",
        Miscellanea Philosophico-Mathematica Societatis Privatae Taurinensis I
        (1759), 2nd pagination, pp. 33-42 = [Lagrange *Œuvres*, I, 23-36].

[Lagrange *1759c*] Joseph-Louis Lagrange, "Recherches sur la Nature et la Propa-
        gation du Son", *Miscellania Philosophico-Mathematica Societatis Priva-
        tae Taurinensis* I (1759), 3rd pagination, pp. i-112 = [Lagrange *Œuvres*,
        I, 39-148].

[Lagrange *1760-61a*] Joseph-Louis Lagrange, "Nouvelles Recherches sur la Nature
        et la Propagation du Son", *Miscellania Taurinensia* II (*1760-1761*) =
        [Lagrange *Œuvres*, I, 151-316].

[Lagrange *1760-61b*] Joseph-Louis Lagrange, "Note sur la Métaphysique du
        Calcul Infinitésimal", *Miscelania Taurinensia* II (*1760-1761*) =
        [Lagrange *Œuvres*, VII, 597-599].

[Lagrange *1766*] Joseph-Louis Lagrange, "Recherches sur les inégalités des sa-
        tellites de Jupiter", *Recueil des Pièces qui ont remporté les Prix de
        l'Académie Royale des Sciences* [de Paris], IX (1777) [winning entry for
        the 1766 prize] = [Lagrange *Œuvres*, VI, 67-225].

[Lagrange *1772a*] Joseph-Louis Lagrange, "Sur une nouvelle espèce de calcul re-
        latif à la différentiation et à l'intégration des quantités variables", *Mém.
        Acad. Berlin, 1772* (1774), pp. 185-221 [*BBAW*] = [Lagrange *Œuvres*,
        III, 441-476].

[Lagrange *1772b*] Joseph-Louis Lagrange, "Sur l'intégration des équations a dif-
        férences partielles du premier ordre", *Mém. Acad. Berlin, 1772* (1774),
        pp. 353-372 [*BBAW*] = [Lagrange *Œuvres*, III, 549-575].

[Lagrange *1773a*] Joseph-Louis Lagrange, "Sur l'attraction des sphéroïdes el-
        liptiques", *Mém. Acad. Berlin, 1773* (1775), pp. 121-148 [*BBAW*] =
        [Lagrange *Œuvres*, III, 619-658].

[Lagrange *1773b*] Joseph-Louis Lagrange, "Solutions analytiques de quelques pro-
        blèmes sur les pyramides triangulaires", *Mém. Acad. Berlin, 1773*
        (1775), pp. 149-176 [*BBAW*] = [Lagrange *Œuvres*, III, 661-692].

[Lagrange *1774*] Joseph-Louis Lagrange, "Sur les intégrales particulières des équa-
        tions différentielles", *Mém. Acad. Berlin, 1774* (1776), pp. 197-275
        [*BBAW*] = [Lagrange *Œuvres*, IV, 5-108].

[Lagrange *1775*] Joseph-Louis Lagrange, "Recherches sur les suites recurren-
        tes dont les termes varient de plusieurs manieres différentes, ou sur
        l'intégration des équations linéaires aux différences finies & partielles;
        & sur l'usage de ces équations dans la théorie des hasards", *Mém. Acad.
        Berlin, 1775* (1777), pp. 183-272 [*BBAW*] = [Lagrange *Œuvres*, IV, 151-
        251].

[Lagrange *1776*] Joseph-Louis Lagrange, "Sur l'usage des fractions continues dans le Calcul Intégral", *Mém. Acad. Berlin, royale des Sciences et Belles-Lettres, Année 1776* (1779), pp. 236-264 [*BBAW*] = [Lagrange *Œuvres*, IV, 301-332].

[Lagrange *1779*] Joseph-Louis Lagrange, "Sur différentes questions d'analyse relatives à la théorie des intégrales particulières", *Mém. Acad. Berlin, 1779* (1781), pp. 121-160 [*BBAW*] = [Lagrange *Œuvres*, IV, 585-634].

[Lagrange *1781*] Joseph-Louis Lagrange, "Théorie des variations séculaires des élémens des Planètes: Première Partie. Contenant les principes et les formules générales pour déterminer ces variations", *Mém. Acad. Berlin, 1781* (1783), pp. 199-276 [*BBAW*] = [Lagrange *Œuvres*, V, 125-207].

[Lagrange *1783*] Joseph-Louis Lagrange, "Sur la manière de rectifier les méthodes ordinaires d'approximation pour l'intégration des équations du mouvement des Planètes", *Mém. Acad. Berlin, 1783* (1785), pp. 224-243 [*BBAW*] = [Lagrange *Œuvres*, V, 493-514].

[Lagrange *1784-1785*] Joseph-Louis Lagrange, "Sur une nouvelle méthode de calcul intégral pour les différentielles affectées d'un radical carré sous lequel la variable ne passe pas le quatrième degré", *Mémoires de l'Académie royale des Sciences de Turin* I (*1784-1785*), 1786 = [Lagrange *Œuvres*, II, 253-312].

[Lagrange *1785*] Joseph-Louis Lagrange, "Méthode Générale pour intégrer les équations aux différences partielles du premier ordre, lorsque ces différences ne sont que linéaires", *Mém. Acad. Berlin, 1785*, (1787), pp. 174-190 [*BBAW*] = [Lagrange *Œuvres*, V, 543-562].

[Lagrange *1792-1793*] Joseph-Louis Lagrange, "Sur l'expression du terme général des séries recurrentes, lorsque l'équation génératrice a des racines égales", *Mém. Acad. Berlin, 1792-1793* (1798), pp. 247-257 [*BBAW*] = [Lagrange *Œuvres*, V, 627-641].

[Lamandé *1988*] Pierre Lamandé, "Deux manuels mathématiques rivaux: le Bézout et le Lacroix. Ancien contre nouveau régime en calcul infinitésimal", *Wissenschaftliche Zeitschrift der Wilhelm-Pieck-Universität Rostock*, G-Reihe 37 (1988), pp. 16-25.

[Lamandé *1988-1989*] Pierre Lamandé, *La mutation de l'enseignement scientifique en France (1750-1810) et le rôle des écoles centrales: l'exemple de Nantes*, Sciences et Techniques en Perspective, vol. 15, Nantes: Université de Nantes - Centre d'Histoire des Sciences et des Techniques, 1988-1989.

[Lamandé *1998*] Pierre Lamandé, "Les traités de calcul du Marquis de l'Hôpital et de Sylvestre François Lacroix. Une même mathématique?", *Contribution à une approche historique des mathématiques. Actes de la 7ᵉ université d'été interdisciplinaire sur l'histoire des mathématiques*, Nantes: IREM des pays de la Loire, 1998, pp. 207-236.

[Lamandé *2004*] Pierre Lamandé, "La conception des nombres en France autour de 1800: l'œuvre didactique de Sylvestre François Lacroix", *Revue d'histoire des mathématiques* 10 (2004), pp. 45-106.

[Langins *1981*] Janis Langins, "Une lettre inédite de Fourier sur l'enseignement destiné aux ingénieurs en 1797", *Revue d'histoire des sciences* 34 (1981), pp. 193-207.

[Langins *1987a*] Jānis Langins, *La République avait besoin de savants – Les débuts de l'École polytechnique: l'École centrale des travaux publics et les cours révolutionnaires de l'an III*, [Paris:] Belin, 1987.

[Laplace *Œuvres*] Pierre-Simon Laplace, *Œuvres complètes de Laplace*, 14 vols., Paris: Gauthier-Villars, 1878-1912 [*Gallica*].

[Laplace *1772a*] Pierre-Simon Laplace, "Mémoire Sur les Solutions particulières des Équations différentielles, et sur les inégalités séculaires des Planètes", *Mém. Acad. Paris, 1772* part 1 (1775), pp. 343-377 and 651-656 [*Gallica*] = [Laplace *Œuvres*, VIII, 325-366].

[Laplace *1772b*] Pierre-Simon Laplace, "Recherches sur le Calcul intégral et sur le Système du Monde", *Mém. Acad. Paris, 1772* part 2 (1776), pp. 267-376 and 533-554 [*Gallica*] = [Laplace *Œuvres*, VIII, 369-477].

[Laplace *1773a*] Pierre-Simon Laplace, "Recherches sur l'intégration des Équations différentielles aux différences finies, et sur leur usage dans la théorie des hasards", *Savans Étrangers* 7 (*1773*), 1776, pp. 37-162 = [Laplace *Œuvres*, VIII, 69-197].

[Laplace *1773b*] Pierre-Simon Laplace, "Mémoire sur l'inclinaison moyenne des orbites des comètes; sur la figure de la Terre, et sur les Fonctions", *Savans Étrangers* 7 (*1773*), 1776, pp. 503-540 = [Laplace *Œuvres*, VIII, 279-321].

[Laplace *1773c*] Pierre-Simon Laplace, "Recherches sur le calcul intégral aux différences partielles", *Mém. Acad. Paris, 1773* (1777), pp. 341-402 [*Gallica*] = [Laplace *Œuvres*, IX, 5-68].

[Laplace *1774*] Pierre-Simon Laplace, "Mémoire sur les suites récurro-récurrentes et sur leurs usages dans la théorie des hasards", *Savans Étrangers* 6 (1774), pp. 353-371 [*Gallica*] = [Laplace *Œuvres*, VIII, 5-24].

[Laplace *1777*] Pierre-Simon Laplace, "Mémoire sur l'intégration des équations différentielles par approximation", *Mém. Acad. Paris, 1777* (1780), pp. 373-397 [*Gallica*] = [Laplace *Œuvres*, IX, 357-379].

[Laplace *1779*] Pierre-Simon Laplace, "Mémoire sur les suites", *Mém. Acad. Paris, 1779* (1782), pp. 207-309 [*Gallica*] = [Laplace *Œuvres*, X, 1-89].

[Laplace *1782*] Pierre-Simon Laplace, "Mémoire sur les approximations des Formules qui sont fonctions de très-grands nombres", *Mém. Acad. Paris, 1782* (1785), pp. 1-88 [*Gallica*] = [Laplace *Œuvres*, X, 209-291].

[Laplace *1812*] Pierre-Simon Laplace, *Théorie analytique des probabilités*, Paris: Courcier, 1812; 2nd ed, 1814; 3r ed, 1820 = [Laplace *Œuvres*, VII] [page references are to the *Œuvres* printing of the 3rd ed].

[Legendre *1787*] Adrien-Marie Legendre, "Mémoire sur l'intégration de quelques Équations aux différences partielles", *Mém. Acad. Paris, 1787* (1789), pp. 309-351.

[Legendre *1790*] Adrien-Marie Legendre, "Mémoire sur les intégrales particulières des équations différentielles", *Mémoires de l'Académie des Sciences, année MDCCLXXXX*, Paris: Imprimerie de Du Pont, year V (1797), pp. 218-241 [according to the "Avertissement", the memoirs contained in this odd volume of *Mém. Acad. Paris* were already printed in July 1794].

[Leibniz *1694*] Gottfried Wilhelm Leibniz, "Constructio propria problematis de curva isochrona paracentrica", *Acta Eruditorum*, Aug. 1694, pp. 364-375; French translation by Marc Parmentier in G. W. Leibniz, *La naissance du calcul différentiel*, Paris: Vrin, 1989, pp. 287-305.

[Leibniz *1700*] Gottfried Wilhelm Leibniz, "Responsio ad Dn. Nic. Fatii Duillerii Imputationes, Accessit nova Artis Analytica promotio specimine indicata; dum Designatione per numeros assumtitios loco literarum, Algebra ex Combinatoria Arte lucem capit", *Acta Eruditorum*, May 1700, pp. 198-208; French translation by Marc Parmentier in G. W. Leibniz, *La naissance du calcul différentiel*, Paris: Vrin, 1989, pp. 368-382.

[l'Hôpital *1696*] [Anonymous, in fact Guillaume-François-Antoine, Marquis de l'Hôpital], *Analyse des infiniment petits*, Paris: Imprimerie Royale, 1696; facsimile reprint, Paris: ACL-éditions, 1988 [*Gallica*].

[l'Huilier *1786*] Simon l'Huilier, *Exposition Élémentaire des Principes des Calculs Supérieurs*, Berlin: Georges Jacques Decker, [1786?; *Gallica*].

[l'Huilier *1795*] Simon l'Huilier, *Principiorum Calculi Differentialis et Integralis Expositio Elementaris*, Tubinga: J. G. Cottam, 1795.

[Libri *1843*] "Discours de M. Libri, membre de l'Académie, prononcé aux funerailles de M. Lacroix, le 27 mai 1843", [Lacroix *AS*].

[Lorgna *1786-87*] Anton Mario Lorgna, "Théorie d'une nouvelle espèce de calcul fini et infinitésimal", *Mémoires de l'Académie Royale des Sciences de Turin* III (*1786-1787*), 1788, pp. 409-448.

[Lützen *1982*] Jesper Lützen, *The prehistory of the theory of distributions*, New York: Springer, 1982.

[Lützen *1983*] Jesper Lützen, "Euler's Vision of a General Partial Differential Calculus for a Generalized Kind of Function", *Mathematics Magazine* 56 (1983), pp. 299-306; reprinted in M. Anderson, V. Katz and R. Wilson (eds.), *Sherlock Holmes in Babylon*, Washington, DC: The Mathematical Association of America, 2004, pp. 354-360.

[Lützen *1990*] Jesper Lützen, *Joseph Liouville, 1809-1882: Master of Pure and Applied Mathematics*, New York: Springer, 1990.

[Lycées *1802*] "Arrêté concernant l'organisation de l'enseignement dans les lycées. Du 19 frimaire an 11 (10 décembre 1802)", *Recueil de lois et réglemens concernant l'instruction publique*, 1ère série, tome 2nd, Paris: Brunot-Labbe, 1814, pp. 304-311 [relevant excerpts also in [Belhoste *1995*, 77-78]].

[Lycées *1803*] "[Ouvrages proposés pour l'enseignement des Lycées.] Série mathématique" [dated 20 germinal an 11 − 10 April 1803], *Recueil de lois et réglemens concernant l'instruction publique*, series 1, vol. 2, Paris: Brunot-Labbe, 1814, pp. 398-401 = [Belhoste *1995*, 78-81].

[Meusnier *1785*] Jean-Baptiste Meusnier de la Place, "Mémoire sur la courbure des surfaces", *Savans Étrangers* 10 (1785), pp. 477-510.

[Michaud *Biographie*] *Biographie Universelle, Ancienne et Moderne*, 52 vols., Paris: Michaud, 1811-1828.

[Monge *Feuilles*] Gaspard Monge, *Feuilles d'Analyse appliqué à la Géométrie à l'usage de l'École Polytéchnique*, 2nd ed. Paris: Baudouin, Thermidor year 9 (Jul-Aug 1801); facsimile reprint on demand, Ann Arbor, Michigan: UMI [composed of 34 numbered leaves ("feuilles") of 4 pages each, plus one (n° 3 *bis*) of 2 pages; within each leaf pages are not numbered, but are referred here in lowercase roman]. First edition (never published as a volume) is from 1795-1796. Later editions under the title *Application de l'Analyse à la Géométrie*: 3rd ed., Paris: Bernard, 1807 (facsimile reprint, Paris: Ellipses, 1994); 4th ed., Paris : Bernard, 1809; 5th ed. (with additions by Liouville and Gauss), Paris: Bachelier, 1850; [4th and 5th ed. exclude the preliminary section on analytic geometry (leaves n°s 1 − 3*bis* in the 2nd ed.); 3rd ed. replaces it with a much larger "first part" on "application de l'algèbre a la géométrie"].

[Monge *Stéréotomie*] Gaspard Monge, "Stéréotomie", *Journal Polytechnique*, vol. I, 1er cahier (Germinal year 3 = Mar/Apr 1795), pp. 1-14.

[Monge *1770-1773*] Gaspard Monge, "Mémoire sur la détermination des fonctions arbitraires dans les intégrales de quelques équations aux différences partielles", *Mélanges de Philosophie et de Mathématique de la Société Royale de Turin* 5 (*1770-1773*), 2nd pagination (classe mathématique), pp. 16-78.

[Monge *1771*] Gaspard Monge, "Réflexions sur les équations aux différences partielles", in [Taton *1950*, 49-58] [second part of a memoir read by Monge to the *Académie Royale des Science de Paris* in 27 November 1771, but not published at the time; the first part seems to be lost].

[Monge *1773a*] Gaspard Monge, "Mémoire sur la Construction des Fonctions arbitraires qui entrent dans les intégrales des Équations aux différences partielles", *Savans Étrangers* 7 (*1773*), 1776, pp. 267-300.

[Monge *1773b*] Gaspard Monge, "Mémoire sur la Détermination des Fonctions arbitraires qui entrent dans les intégrales des Équations aux différences partielles", *Savans Étrangers* 7 (*1773*), 1776, pp. 305-327.

[Monge *1780*] Gaspard Monge, "Mémoire sur les Propriétés de plusieurs genres de Surfaces courbes, particulièrement sur celles des Surfaces développables, avec une Application à la Théorie des Ombres et des Pénombres", *Savans Étrangers* 9 (1780), pp. 382-440.

[Monge *1781*] Gaspard Monge, "Mémoire sur la théorie des déblais et des remblais", *Mém. Acad. Paris, 1781* (1784), pp. 666-704 [*Gallica*].

[Monge *1784a*] Gaspard Monge, "Mémoire sur l'expression analytique de la génération des surfaces courbes", *Mém. Acad. Paris, 1784* (1787), pp. 85-117 [*Gallica*].

[Monge *1784b*] Gaspard Monge, "Mémoire sur le calcul intégral des équations aux différences partielles", *Mém. Acad. Paris, 1784* (1787), pp. 118-192 [*Gallica*].

[Monge *1784c*] Gaspard Monge, "Supplément où l'on fait voir que les Équations aux différences ordinaires, pour lesquelles les conditions d'intégrabilité ne sont pas satisfaites, sont susceptibles d'une véritable intégration, et que c'est de cette intégration que dépend celle des équations aux différences partielles élevées", *Mém. Acad. Paris, 1784* (1787), pp. 502-576 [*Gallica*].

[Monge *1784-1785*] Gaspard Monge, "Sur l'expression analytique de la génération des surfaces courbes", *Mémoires de l'Académie Royale des Sciences* [de Turin], *1784-1785*, 1786, pp. 19-30.

[Monge *1785a*] Gaspard Monge, "Mémoire sur les développées, les rayons de courbure, et les différens genres d'inflexions des courbes à double courbure", *Savans Étrangers* 10 (1785), pp. 511-550 [submitted on 31 August 1771, but probably modified between 1776 and 1785 [Taton *1951*, 114]].

[Monge *1785b*] Gaspard Monge, "Mémoire Sur une méthode d'intégrer les Équations aux Différences ordinaires, lorsqu'elles sont élevées, & dans les cas où leurs Intégrales complètes sont algébriques", *Mém. Acad. Paris, 1783* (1786), pp. 719-724 [dates from 1785, according to [Taton *1951*, 288-289]; *Gallica*].

[Monge *1785c*] Gaspard Monge, "Mémoire Sur l'intégration des Équations aux différences finies, qui ne sont pas linéaires", *Mém. Acad. Paris, 1783* (1786), pp. 725-730 [read on 30 November 1785; *Gallica*].

[Monge *1795*] Gaspard Monge, "Leçons de Monge" [on descriptive geometry], in J. Dhombres (ed.), *L'École normale de l'an III − Leçons de mathématiques*, Paris: Dunod, 1992, pp. 305-453 [first published in 1795 in *Séances des écoles normales*].

[Monge & Hachette *1799*] Gaspard Monge and Jean Nicolas Pierre Hachette, "Des courbes à double courbure (extrait des ouvrages du C.^{en} Gaspard Monge)", *Journal de l'École Polytechnique*, vol. II, 6ème cahier, Thermidor year VII (July-August 1799), pp. 345-363.

[Montucla & Lalande *1802*] J. F. [in fact, Jean-Étienne] Montucla, *Histoire des Mathématiques*, 2nd ed., vol. 3 [of 4], completed and edited by Jerôme de Lalande, Paris: Henri Agasse, an X (May 1802).

[Nagliati *1996*] Iolanda Nagliati, *Le radici della scuola matematica pisana − La matematica nell'Università di Pisa dal 1799 al 1860*, doctoral thesis, consorzio delle Università di Pisa, Bari, Ferrara, Lecce, Parma, 1996.

[Nagliati *2000*] Iolanda Nagliati, "Aspetti della matematica", in *Storia dell'Università di Pisa 2 (1737-1861)*, Pisa: Plus, 2000, pp. 823-837.

[Nastold & Forster *1980*] Hans-Joachim Nastold and Otto Forster, "Die Mathematik an der Universität Münster", in Heinz Dollinger (ed.), *Die Universität Münster, 1780 − 1980*, Münster: Aschendorff, 1980, pp. 429-432.

[Newton *Fluxions*] Isaac Newton, "Methodus Fluxionum et Serierum infinitarum", in J. Castiglione (ed.), *Isaaci Newtoni [...] Opuscula Mathematica, Philosophica et Philologica*, I, Lausanna and Geneva: Bousquet, 1744, pp. 29-199 [this is a Latin translation by Castiglione from Colson's English translation, published in 1736 as the first edition of Newton's *Method of Fluxions*, of a Latin original].

[Newton *Principia*] Isaac Newton, *Philosophiæ Naturalis Principia Mathematica*, London: Royal Society, 1687 [*Gallica*]; 2nd ed, 1713; 3rd ed, 1726 [many reprints and several translations].

[Nieuport *Mélanges*] [Charles François Le Preud'homme-d'Hailly,] Commandeur de Nieuport, *Mélanges Mathématiques*, 2 vols. + 1 suppl. Bruxelles: Lemaire, 1794, year VII = 1799, year x = 1802.

[*NUC*] Library of [USA] Congress, *National Union Catalog, pre-1956 imprints*, 754 vols., London: Mansell, 1968-1981.

[Olivier *1843*] Théodore Olivier, *Cours de Géométrie Descriptive. Première partie. Du point, de la droite et du plan*, Paris: Carilian-Gœury & V^{or} Dalmont, 1843.

[Panteki *1987*] Maria Panteki, "William Wallace and the Introduction of Continental Calculus to Britain: A Letter to George Peacock", *Historia Mathematica* 14 (1987), pp. 119-132.

[Panteki *2003*] Maria Panteki, "French 'logique' and British 'logic': on the origins of Augustus De Morgan's early logical inquiries, 1805-1835", Historia Mathematica 30 (2003), pp. 278-340.

[Panza *1985*] Marco Panza, "Il manoscritto del 1789 di Arbogast sui principi del calcolo differenziale e integrale", *Rivista di storia della scienza* vol. 2 n.º 1 (Marzo 1985), pp. 123-157.

[Paoli *1792*] Pietro Paoli, "Riflessioni sull'integrazione di quell'equazioni, le quali non soddisfanno alle condizioni d'integrabilità", *Memorie di Matematica e Fisica della Società Italiana* VI (1792), pp. 501-533 [available at <http://www.accademiaxl.it/Biblioteca/Pubblicazioni/browser.asp> (23 October 2006)]; also in Pietro Paoli, *Memorie sul calcolo integrale e sopra alcuni problemi meccanici*, Verona: Dionigi Ramanzini, 1793, pp. 1-33.

[Pedersen *1980*] Kirsti Møller Pedersen, "Techniques of the Calculus, 1630-1660", in Ivor Grattan-Guinness (ed.), *From the Calculus to Set Theory*, London: Duckworth, 1980; 2nd ed.: Princeton, New Jersey: Princeton University Press, 2000, pp. 10-48.

[Pepe *2006*] Luigi Pepe, "Insegnamenti matematici e libri elementari nella prima metà dell'Ottocento. Modelli francesi ed esperienze italiane", <bsd.unife.it/museologia/matematica/filemat/pdf/libri   elementari.pdf> (26 August 2006); also in Livia Giacardi (ed.), *Da Casati a Gentile. Momenti di storia dell'insegnamento secondario della matematica in Italia*, Lumières Internationales, 2006, pp. 65-98.

[Pfaff *1815*] Johann Friedrich Pfaff, "Methodus generalis, aequationes differentiarum partialium, nec non aequationes differentiales vulgares, utrasque primi ordinis, inter quotcunque variabiles, complete integrandi", *Abhandlungen der Königlichen Akademie der Wissenschaften in Berlin. Aus den Jahren 1814-1815*, 1818, 3rd pagination (mathematische Klasse), pp. 76-136 [submitted in May 1815; *BBAW*].

[Phili *1996*] Christine Phili, "La reconstruction des mathématiques en Grèce: l'apport de Ioannis Carandinos (1784-1834)", in Catherine Goldstein, Jeremy Gray and Jim Ritter (eds.), *L'Europe mathématique / Mathematical Europe*, Paris: Maison des sciences de l'homme, 1996, pp. 305-319.

[Philips *2006*] Christopher Philips, "Robert Woodhouse and the evolution of Cambridge mathematics", *History of Science* 44 (2006), pp. 69-93.

[Poisson *1800*] Siméon Denis Poisson, "Mémoire Sur la pluralité des Intégrales dans le calcul des Différences", *Journal de l'École Polytechnique*, vol. IV,

11ème cahier (1802), pp. 173-181 [submitted to the Institut national on 16 Frimaire year 9 = 7 December 1800].

[Poisson *1805*] Siméon Denis Poisson, "Démonstration du théorême de Taylor", *Correspondance sur l'École Imperiale Polytechnique*, vol. I, n.º 3 (Pluviose year 13 = Jan-Feb 1805), pp. 52-55.

[Poisson *1806*] Siméon Denis Poisson, "Mémoire Sur les Solutions particulières des Équations différentielles et des Équations aux différences" and "Addition au mémoire précédent", *Journal de l'École Polytechnique*, vol. VI, 13ème cahier (April 1806), pp. 60-116 and 117-125.

[Prony *1795a*] Gaspard-François-Marie Riche de Prony, "Cours d'analyse appliqué à la mécanique", *Journal [de l'École] Polytechnique*, vol. I, part I – 1er cahier, pp. 92-119; part II – 2ème cahier, pp. 1-23; part III – 3ème cahier, pp. 209-273; part IV – 4ème cahier, pp. 459-569.

[Prony *1795b*] Gaspard-François-Marie Riche de Prony, "Notice sur un cours élémentaire d'analyse fait par Lagrange", *Journal de l'École Polytechnique*, vol. I, 2ème cahier, Paris: Imprimerie de la République, Nivose year IV (Dec 1795 - Jan 1796), pp. 206-208.

[Prony *1799*] Gaspard-François-Marie Riche de Prony, "Introduction aux Cours d'Analyse pure et d'Analyse appliquée à la Mécanique", *Journal de l'École Polytechnique*, vol. II, 6ème cahier, Thermidor year VII (July - August 1799), pp. 213-218.

[Quetelet *1867*] Adolphe Quetelet, *Sciences Mathématiques et Physiques au commencement du XIXe siècle*, Bruxelles: Muquardt, 1867.

[Ross *1977*] Bertram Ross, "The development of fractional calculus 1695-1900", *Historia Mathematica* 4 (1977), pp. 75-89.

[Rothenberg *1908*] Siegfried Rothenberg, "Geschichtliche Darstellung der Entwicklung der Theorie der singulären Lösungen totaler Differentialgleichungen von der ersten Ordnung mit zwei variablen Grössen", *Abhandlungen zur Geschichte der Mathematischen Wissenschaften...* 20 (*1905-1908*), pp. 315-404.

[Sandifer *2007*] C. Edward Sandifer, "Some Facets of Euler's Work on Series", in Robert E. Bradley and C. Edward Sandifer, *Leonhard Euler: Life, Work and Legacy*, Amsterdam: Elsevier, 2007, pp. 279-302.

[Sarton *1936*] George Sarton, "Montucla (1725-1799): his life and works", *Osiris* 1 (1936), pp. 519-567.

[Schubring *1987*] Gert Schubring, "On the Methodology of Analysing Hstorical Textbooks: Lacroix as Textbook Author", *For the Learning of Mathematics*, vol. 7, n.º 3 (November 1987), pp. 41-51.

[Schubring *2005*] Gert Schubring, *Conflicts between Generalization, Rigor and Intuition*, New York: Springer, 2005.

[C.M.S. Silva *1996*] Circe Mary Silva da Silva Dynnikov, "O conceito de derivada no ensino da Matemática no Brasil do século XIX", in proceedings of *História e Educação Matemática* (2ème Université d'Été Européenne sur Histoire et Épistémologie dans l'Education Mathématique / ICME-8 sattelite meeting of HPM, 24-30 July 1996, Braga, Portugal) vol. II, pp. 80-87.

[C.P. Silva *1992*] Clóvis Pereira da Silva, *A Matemática no Brasil – Uma História de seu Desenvolvimento*, Curitiba: Editora UFPR, 1992.

[Simson *1776*] Robert Simson, *Opera Quaedam Reliqua*, Glasgow, 1776.

[Soc. Phil. *Rapp*] Augustin-François de Silvestre et al., *Rapports Généraux des Travaux de la Société Philomathique de Paris*, 4 vols., Paris, 1792-1800 [from the second volume onwards the title is *Rapport Général des Travaux de la Société Philomatique de Paris*].

[Stirling *1730*] James Stirling, *Methodus Differentialis: sive Tractatus de Summatione et Interpolatione Serierum Infinitarum*, London: Bowyer, 1730. Annotated English translation by Ian Tweddle, *James Stirling's Methodus differentialis*, London: Springer, 2003.

[Stockler *1805*] Francisco de Borja Garção Stockler, "Nota (m)" [endnote to an eulogy of d'Alembert, on vibrating strings and continuity of arbitrary functions], in *Obras*, vol. I, Lisboa: Academia Real das Sciencias, 1805, pp. 129-188.

[Struik *1933*] Dirk J. Struik, "Outline of a history of differential geometry", part I, *Isis* 19 (1933), pp. 92-120.

[Taton *1947*] René Taton, "Une correspondance mathématique inédite de Monge", *La Revue Scientifique*, 85e année (1947), pp. 963-989.

[Taton *1948*] René Taton, "Une lettre inédite de Monge sur la situation en France, en 1791, après la fuite du roi", *Revue d'histoire des sciences*, 1 (1947-1948), pp. 358-359.

[Taton *1950*] René Taton, "Un texte inédit de Monge: Réflexions sur les équations aux différences partielles", *Osiris* 9 (1950), pp. 44-61.

[Taton *1951*] René Taton, *L'Œuvre Scientifique de Monge*, Paris: Presses Universitaires de France, 1951.

[Taton *1953a*] René Taton, "Sylvestre-François Lacroix (1765-1843), mathématicien, professeur et historien des sciences", *Actes du VIIe Congrès International d'Histoire des Sciences (4-12 Août 1953)*, Paris: Acad. Int. d'Hist. Sci. & Hermann, pp. 588-593.

[Taton *1953b*] René Taton, "Laplace et Sylvestre-François Lacroix", *Revue d'histoire des sciences*, 6 (1953), pp. 350-360.

[Taton *1959*] René Taton, "Condorcet et Sylvestre-François Lacroix", *Revue d'histoire des sciences* 12 (1959), pp. 127-158 and 243-262.

[Taton *1972*] René Taton, "Fontaine (Fontaine des Bertins), Alexis", in [Gillispie *DSB*, V, 54-55].

[Taton *1990*] René Taton, "La Société Philomathique de Paris et les sciences exactes: Premier tiers du XIXème siècle", in André Thomas (ed.), *La Société philomathique de Paris et deux siècles de la Science en France*, Paris: Presses Universitaires de France, 1990, pp. 37-54; also available at <http://philomathique.org/modules/news/article.php?storyid=20> (11 March 2007).

[Taylor *1715*] Brook Taylor, *Methodus Incrementorum Directa et Inversa*, London, 1715.

[Tinseau *1780a*] Charles Tinseau, "Solution de quelques problêmes relatifs à la Théorie des Surfaces courbes, et des Courbes à double courbure", *Savans Étrangers* 9 (1780), pp. 593-624.

[Tinseau *1780b*] Charles Tinseau d'Amondans de Gennes, "Sur quelques propriétés des Solides renfermés par des Surfaces composés de Lignes droites", *Savans Étrangers* 9 (1780), pp. 625-642.

[Tisserand *1894*] François-Félix Tisserand, *Traité de Mécanique Céleste – Tome III: Exposé de l'ensemble des théories relatives au mouvement de la Lune*, Paris: Gauthier-Villars, 1894; facsimile reprint, Paris: Jacques Gabay, 1990.

[Todhunter *1861*] Isaac Todhunter, *A history of the progress of the calculus of variations during the nineteenth century*, Cambridge and London: MacMillan, 1861.

[Torlais *1964*] Jean Torlais, "Le Collège Royal", in René Taton (ed.), *Enseignement et diffusion des sciences en France au XVIIIe siècle*, Paris: Hermann, 1964, pp. 261-286.

[Tournès *2003*] Dominique Tournès, "L'intégration graphique des équations différentielles ordinaires", *Historia Mathematica* 30 (2003), pp. 457-493.

[Tournès *2004*] Dominique Tournès, "Vincenzo Riccati's treatise on integration of differential equations by tractional motion (1752)", *Oberwolfach Reports* 1 (2004), pp. 2740-2742 [extended abstract of a talk given at the workshop *The History of Differential Equations, 1670-1950*, Mathematisches Forschungsinstitut Oberwolfach, October 31st – November 6th, 2004].

[Trembley *1790-91*] Jean Trembley, "Recherches sur les équations différentielles du premier degré", *Mémoires de l'Académie Royale des Sciences* [de Turin], *1790-91*, 1793, 2nd pagination ("mémoires présentés à l'académie"), pp. 1-52.

[Troux *1926*] Albert Troux, *L'École Centrale du Doubs à Besançon*, Bibliothèque d'Histoire Révolutionnaire (nouvelle série - IX), Paris: Librairie Félix Alcan, 1926.

[Truesdell *1960*] Clifford Ambrose Truesdell III, *The Rational Mechanics of Flexible or Elastic Bodies, 1638-1788*, [Euler *Opera*, series 2, XI, part 2] (Zürich: Orell Füssli, 1960).

[Univ. Coimbra *Estatutos 1772*] *Estatutos da Universidade de Coimbra do anno de MDCCLXXII*, 3 vols., Lisbon: Regia Officina Typografica, 1772; facsimile reprint, Coimbra, 1972.

[Venclova *1981*] Tomas Venclova, "Four Centuries of Enlightenment - a historic view of the University of Vilnius, 1579-1979", *Lituanus*, vol. 27, n.º 1 (1981); <http://www.lituanus.org/1981_2/81_2_01.htm> (26 December 2006).

[Wagener *Lacroix*] Wagener, "Lacroix (Sylvestre-François)", *Biographie Universelle (Michaud)*, 2nd ed, vol. 22, Paris: Desplaces and Leipzig: Brockhaus, undated [c. 1856, according to the catalogue of the *Bibliothèque nationale de France*], pp. 396-398.

[Wallace *1815*] William Wallace, "Fluxions", in David Brewster (ed.), *Edinburgh Encyclopædia*, vol. IX, pp. 382-467.

[Wallner *1908*] C. R. [Karl Raimund] Wallner, "Totale und partielle Differentialgleichungen. Differenzen- und Summenrechnung. Variationsrechnung", in Moritz Cantor (ed.), *Vorlesungen über Geschichte der Mathematik*, IV, Leipzig: Teubner, 1908, pp. 871-1074 [available online at <http://www.hti.umich.edu/u/umhistmath> (26 August 2006)].

[Wilson *1994*] Curtis Wilson, "The dynamics of the solar system", in Ivor Grattan-Guinness (ed.), *Companion Encyclopedia of the History and Philosophy of the Mathematical Sciences*, vol. 2, New York and London: Routledge, 1994; 2nd ed.: Baltimore and London: Johns Hopkins Press, undated, pp. 1044-1053.

[Yla *1981*] Rev. Stasys Yla, "The Clash of Nationalities at the University of Vilnius", *Lituanus*, vol. 27, n.º 1 (1981); <http://www.lituanus.org/1981_2/81_2_03.htm> (26 December 2006).

[Youschkevitch *1971*] A. P. Youschkevitch, "Lazare Carnot and the Competition of the Berlin Academy in 1786 on the Mathematical Theory of the Infinite", in [Gillispie *1971*, 149-168].

[Youschkevitch *1976*] A. P. Youschkevitch, "The Concept of Function up to the Middle of the 19th Century", *Archive for History of Exact Sciences* 16 (1976), pp. 37-85.

Index

Science Networks – Historical Studies (SNHS)

Edited by
Eberhard Knobloch, Technische Universität Berlin, Germany
Helge Kragh, University of Aarhus, Denmark
Erhard Scholz, Bergische Universität Wuppertal, Germany
In cooperation with an international editorial board

BIRKHÄUSER

The publications in this series are limited to the fields of mathematics, physics, astronomy, and their applications. The publication language is preferentially English. The series is primarily designed to publish monographs. Annotated sources and exceptional biographies might be accepted in rare cases. The series is aimed primarily at historians of science and libraries; it should also appeal to interested specialists, students, and diploma and doctoral candidates. In cooperation with their international editorial board, the editors hope to place a unique publication at the disposal of science historians throughout the world.

SNHS 35: Caramalho Domingues, J.
Lacroix and the Calculus (2008)
ISBN 978-3-7643-8637-5

SNHS 34: Høyrup, J.
Jacopo da Firenze's *Tractatus Algorismi* and
Early Italian Abbacus Culture (2007)
ISBN 978-3-7643-8390-9

The abbacus books have long been supposed to be reduced versions of Leonardo Fibonacci's *Liber abbaci*. Analysis of early abbacus books, not least of the first specimen treating of algebra — Jacopo da Firenze's *Tractatus algorismi* from 1307 — shows instead that abbacus mathematics was an exponent of a more widespread culture of commercial mathematics, already known by Fibonacci, and probably flourishing in Provence and/or Catalonia before it reached Italy. Abbacus algebra was inspired from a Romance-speaking region outside Italy, most likely located in the Provençal-Catalan area, and ultimately from a similar practitioners' level of Arabic mathematics.
The book contains, along with the English translation, an edition of Jacopo's *Tractatus* and a commentary analyzing Jacopo's mathematics and its links to Provençal, Catalan, Arabic, Indian and Latin medieval mathematics. It will provide readers with a new perspective on a period and on processes which eventually reshaped the whole mathematical enterprise in the 17th century.

SNHS 33: De Risi, V.
Geometry and Monadology. Leibniz's *Analysis Situs* and Philosophy of Space (2007)
ISBN 978-3-7643-7985-8

This book reconstructs, from both historical and theoretical points of view, Leibniz's geometrical studies, focusing in particular on the research Leibniz carried out in the last years of his life. It is indeed the first ever comprehensive historical reconstruction of Leibniz's geometry that meets the interests of both mathematicians and philosophers. The main purpose of the work is to offer a better understanding of the Leibnizean philosophy of space and mature metaphysics, through a pressing confrontation with the problems of geometric foundations. Regarding the scope of these problems, the book also deals in depth with Leibniz's theory of sensibility, thus favouring the comparison and contrast between Leibniz's philosophy and Kant's transcendentalist solution. The Appendix references to a number of previously unpublished manuscripts on geometry from the Leibniz Archiv in Hannover, which disclose new theories, points of view and technicalities of Leibniz's thought.

SNHS 32: Krömer, R.
Tool and Object. A History and Philosophy of Category Theory (2007)
ISBN 978-3-7643-7523-2

SNHS 31: Keller, A.
Expounding the Mathematical Seed. Vol. 2: The Supplements (2006). ISBN 978-3-7643-7292-7

SNHS 30: Keller, A.
Expounding the Mathematical Seed. Vol. 1: The Translation (2006). ISBN 978-3-7643-7291-0

SNHS 30/31 Set: ISBN 978-3-7643-7299-6

Science Networks – Historical Studies (SNHS)

Edited by
Eberhard Knobloch, Technische Universität Berlin, Germany
Helge Kragh, University of Aarhus, Denmark
Erhard Scholz, Bergische Universität Wuppertal, Germany
In cooperation with an international editorial board

BIRKHÄUSER

SNHS 29: Guerraggio, A. / Nastasi, P.
Italian Mathematics Between the Two World
Wars (2005). ISBN 3-7643-6555-2

SNHS 28: Hesseling, D.
Gnomes in the Fog. The Reception of Brouwer's
Intuitionism in the 1920s (2003)
ISBN 3-7643-6536-6

SNHS 27: Dauben, J.W. / Scriba, C.J.
Writing the History of Mathematics – Its
Historical Development (2002)
ISBN 3-7643-6166-2 (Hardcover)
ISBN 3-7643-6167-0 (Softcover)

SNHS 26: Israel, G. / Millán Gasca, A.
The Biology of Numbers. The Correspondence
of Vito Volterra on Mathematical Biology (2002)
ISBN 978-3-7643-6514-1

SNHS 25: Siegmund-Schultze, R.
Rockefeller and the Internationalization of
Mathematics Between the Two World Wars.
(2001). ISBN 978-3-7643-6468-7

SNHS 24: Jensen, C.
Controversy and Consensus: Nuclear Beta
Decay 1911–1934. Edited by Aaserud, F., Kragh,
H., Rüdinger, E. and Stuewer, R. (2000)
ISBN 978-3-7643-5313-1

SNHS 23: Ferreirós, J.
Labyrinth of Thought. A History of Set Theory
and its Role in Modern Mathematics (2001)
ISBN 978-3-7643-5749-8

SNHS 22: Marage, P. / Wallenborn, G. (eds.)
The Solvay Councils and the Birth of Modern
Physics (1999). ISBN 978-3-7643-5705-4

SNHS 21: Sakarovitch, J.
Épures d'architecture. De la coupe des pierres à
la géométrie descriptive XVI–XIX siècles (1998)
ISBN 978-3-7643-5701-6

SNHS 20: Grattan-Guinness, I. / Bornet, G.
George Boole – Selected Manuscripts on Logic
and its Philosophy (1997)
ISBN 978-3-7643-5456-5

SNHS 19: Ullmann, D.
Chladni und die Entwicklung der Akustik
1750–1860 (1996). ISBN 978-3-7643-5398-8

SNHS 18: Hentschel, K. (ed.)
Physics and National Socialism. An Anthology of
Primary Sources (1996)
ISBN 978-3-7643-5312-4

SNHS 17: Corry, L.
Modern Algebra and the Rise of Mathematical
Structures (1996). ISBN 978-3-7643-5311-7

SNHS 16: Yavetz, I.
From Obscurity to Enigma. The Work of Oliver
Heaviside, 1872–1889 (1995)
ISBN 978-3-7643-5180-9

**SNHS 15: Sasaki, Ch. / Sugiura, M. / Dauben,
J.W.**, The Intersection of History and
Mathematics (1994). ISBN 978-3-7643-5029-1

SNHS 14: Klein, U.
Verbindung und Affinität. Die Grundlegung der
neuzeitlichen Chemie an der Wende vom 17.
zum 18. Jahrhundert (1994)
ISBN 978-3-7643-5003-1

SNHS 13: Vizgin, V.P.
Unified Field Theories in the first third of the 20th
Century (1994). ISBN 978-3-7643-2679-1

SNHS 12: Gorelik, G.E. / Frenkel, V.Y.
Matvei Petrovich Bronstein and the Soviet
Theoretical Physics in the Thirties (1994)
ISBN 978-3-7643-2752-1

SNHS 11: Reich, K.
Die Entwicklung des Tensorkalküls vom
absoluten Differentialkalkül zur
Relativitätstheorie (1994)
ISBN 978-3-7643-2814-6